CRC
Handbook
of
Furnace Atomic
Absorption Spectroscopy

Author

Asha Varma, Ph.D., F.A.I.C., C.P.C.
Naval Air Development Center
Warminster, Pennsylvania

CRC Press
Taylor & Francis Group
Boca Raton London New York

CRC Press is an imprint of the
Taylor & Francis Group, an **informa** business

First published 1990 by CRC Press
Taylor & Francis Group
6000 Broken Sound Parkway NW, Suite
300 Boca Raton, FL 33487-2742

Reissued 2018 by CRC Press

© 1990 by Taylor & Francis
CRC Press is an imprint of Taylor & Francis Group, an Informa business

No claim to original U.S. Government works

A Library of Congress record exists under LC control number: 89015721

Publisher's Note
The publisher has gone to great lengths to ensure the quality of this reprint but points out that some imperfections in the original copies may be apparent.

Disclaimer
The publisher has made every effort to trace copyright holders and welcomes correspondence from those they have been unable to contact.

ISBN 13: 978-1-138-10507-2 (hbk)
ISBN 13: 978-1-138-55834-2 (pbk)
ISBN 13: 978-1-315-15078-9 (ebk)

Visit the Taylor & Francis Web site at http://www.taylorandfrancis.com and the CRC Press Web site at http://www.crcpress.com

PREFACE

During the last 30 years the progress and development of atomic absorption spectroscopy has generally been covered satisfactorily and enthusiastically in modern textbooks and other scientific and technical publications. However, during the last 15 to 20 years, there has been tremendous expansion of graphite furnace atomic absorption spectroscopy (FAAS) technique. The revolutionary progress in instrumentation, automation, and computerization has played a major role in acceptance of this technique as a highly sensitive method for the determination of trace and ultra-trace elements.

This handbook is intended to supplement and compliment all the FAAS - technique related publications. The material presented here is of practical importance not only for the professional analysts, but also for the clinical, environmental, and pharmaceutical chemists, who often are faced with the problem of identifying, detecting, and quantifying the chemical ingredients of the starting materials, intermediates, final products, and contamination products in a very small amount of sample.

The first section of this handbook is an introduction to the furnace atomic absorption spectroscopy. A general principle of atomic absorption spectroscopy and instruments required to perform the analysis is described. Instrumentation set-up for the technique is described in detail including sample preparation, time and temperature programming, parameters, optimization, verification of analytical results, and interferences common to this method.

The next section contains references pertaining to the developments in instrumentation for the atomic absorption spectroscopy, such as, spectrophotometers, accessories, sampling devices, light sources, etc.

In the following sections the elements which have been determined successfully by the FAAS technique are arranged in the alphabetical order. Because of fewer references, some of the metals have been grouped together: Alkali metals (lithium, sodium, potassium, rubidium, and cesium), Platinum group metals (iridium, osmium, palladium, platinum, rhodium, and ruthenium), and Rare Earth metals (dysprosium, erbium, europium, gadolinium, holmium, lanthanum, neodymium, samarium, terbium, thulium, uranium, ytterbium, and yttrium).

Due to application importance of elements in certain types of sample environments, separate sections have been provided for convenience and easy reference (i.e., Clinical Analysis, Heavy Metals, Trace Metals, and Wear Metals). Some of the miscellaneous applications of the FAAS technique, which could not be categorized under any of the sections, are placed in the section entitled Miscellaneous Analysis.

In each section, the instrumental parameter set-up for a particular element is given which will help an analyst in developing a method suitable for the sample. References are followed by author and subject indices, which will enable the location of any item desired in that particular section. There are some large sections on cadmium, copper, lead etc. as well as very short ones on boron, sulfur etc., but each section is complete in itself for easy reference purposes.

No mention has been made of elements that are not possible to determine by the FAAS technique. No attempt has been made to provide an alternate method of analysis for these elements either.

For convenience, a large number of appendices are included in this handbook. These appendices are Glossary of Terms and Definitions, Abstracts and Reviews, a list of books on atomic absorption spectroscopy, and a list of manufacturers for spectrophotometer and related accessories, chemical supplies, lamps, etc.

Like any other instrumental analysis technique, FAAS has limitations such as slow heating rates giving lower sensitivity, non-isothermal conditions, "memory effects" causing erroneous results and dependence on background correction for accuracy. Despite all these limitations, FAAS has been widely accepted as a highly sensitive analytical technique. It is highly successful in the (1) contract Lab Program of United States Environmental Protection Agency through

Superfund resources, (2) methods for regulations of food and drugs, (3) atmospheric and industrial pollution caused by metallic elements, (4) forensic sciences, (5) pharmaceuticals, and (6) most importantly, clinical chemistry.

Although, the frontiers of FAAS have extended to other analytical techniques such as Chromatography (gas and liquid), mass spectrometry, Fourier Transform Infrared Spectroscopy etc. through interfacing, I have not extended the scope of this handbook to include these techniques at this time. These are subject to wider use and considerations.

Because of its relatively wide scope, there is every reason to anticipate that this handbook on Furnace Atomic Absorption Spectroscopy will likewise find a wider circle of usability in various spheres of trace analysis applications.

I am pleased to express my appreciations to John Sotera, Steve Sourhoff, and Frank Fernandez for making suggestions. I must thank my professional acquaintances, advisory board members, and fellow chemists world wide who have encouraged and supported me for my first attempt at writing *The Handbook of Atomic Absorption Analysis* (CRC Press Inc., FL, 1984). I anxiously look forward for their continued support for this handbook on Furnace Atomic Absorption Spectroscopy.

My thanks are also due to my son, Vinay Agarwala, who spent his summer vacation typing, correcting, and making author indices for this manuscript. Special thanks are due to my daughter, Veena Agarwala, who encouraged me to complete this monumental task and consoled me during the loss of my precious and irreplaceable mother.

I would appreciate any comments and suggestions to improve upon my further adventures in analytical chemistry.

Asha
October 2nd, 1988

AUTHOR

Asha Varma, Ph.D. is affiliated with the Technology Base Office at the Naval Air Development Center, Warminster, Pennsylvania.

Dr. Varma received her B.S. (1958) and M.S. (1960) from Bareilly College, Bareilly and obtained her Ph.D. (1963) from Banaras Hindu University, Varanasi, India. She was appointed Senior Research Fellow (1963 to 1966) by the Council of Scientific and Industrial Research, Government of India at Banaras Hindu Universtiy and National Chemical Laboratory, Poona. In 1966 she became the first woman to be selected by the Madhya Pradesh Government as an Assistant Director of the Forensic Science Laboratory, Sagar. She held a postdoctoral fellowship at the Chemistry Department and Institute of Materials Science, University of Connecticut, Storrs. She has served as the Supervisor of Analytical Laboratories at the Laboratory for Research on the Structure of Matter, University of Pennsylvania, Philadelphia. She was appointed Leader of Analytical Research Group (1983) at the Naval Air Development Center, Warminster, Pennsylvania.

Dr. Varma is a Certified Professional Chemist. She is a fellow of the American Institute of Chemists; a member of the American Chemical Society, IUPAC, Coblentz Society, Electrochemical Society, Society of Spectroscopy; and an associate member of the American Museum of Natural History and the National and International Federation of Wildlife. In 1986 she was elected the President of Federally Employed Women, Buxmont Chapter, and during 1987 to 1988 she served as chairperson of the Federal Women's Program Committee at the center. Dr. Varma is listed in American Men and Women in Science, Who's Who in Technology Today, Who's Who in Frontiers of Science and Technology, The International Book of Honor, The Directory of Distinguished Americans, Two Thousand Notable Americans, etc.

Dr. Varma has published over 40 research and technical publications. In 1984, she published a best seller the *CRC Handbook of Atomic Absorption Analysis*. At present, she is involved with independent research and exploratory development projects at the Naval Air Development Center.

ACKNOWLEDGMENTS

The author is greatly indebted to Analyte Corporation, Great Pass, Oregon, Perkin-Elmer Corporation, Norwalk, Connecticut , Thermo Jarrell Ash, Franklin, Massachusetts, and CRC Press Inc., Boca Raton, Florida for providing materials for inclusion in this book.

TABLE OF CONTENTS

IN MEMORY OF MY BELOVED MOTHER

FURNACE ATOMIC ABSORPTION SPECTROSCOPY (FAAS)

Introduction

The atomic absorption technique as an analytical tool for chemical analysis was introduced by Walsh[1,2] and Alkemade and Milatz.[3,4] Walsh suggested that measurement of the absorbance at the peak of the absorption line profile would give a linear relationship with concentration over a wide range of absorbance values. By presenting this simple absorption technique combined with modern instrumentation, more reliable sources of resonance radiation, and a high temperature flame, Walsh demonstrated the advantage of this technique over the well-established emission spectroscopy method for element analysis.

During the early commercialization of atomic absorption spectroscopy (AAS), extravagant proclamations were made concerning the general lack of interferences encountered in this technique. A great deal of time and effort was spent in rediscovering interferences in flames which had been known to emission spectroscopists for at least 60 years.[5]

The traditional flame AAS also imposed few stringent requirements upon the instrumentation. Even with relatively modest spectral resolution, performance was provided that could not be distinguished from that available with a high resolution spectrophotometer. It would seem that sensitivities could be improved if the atomic vapor could be constrained to remain longer in the resonance beam and if the sample could be introduced without any pretreatment involving dilutions etc.

Many attempts have been made to produce the atomic vapor in a completely neutral or unreactive medium and to introduce the necessary amount of heat energy into the system; various electrical methods have been proposed. L'vov[6,7] introduced electrothermal atomizers into atomic absorption analysis. He used a simple graphite tube furnace, into which the sample was completely vaporized by heating with d.c. arc, and enabled the atomic vapor to be maintained in a highly reducing atmosphere thus eliminating the formation of refractory oxides. Fuller[8] and L'vov[9] have reviewed this technique extensively.

The introduction of electrothermal atomization into AAS had one of the greatest impacts on analytical chemistry in the last 3 decades. This method is ideal for ultratrace determination of 64 elements and requires only a few microliters or micrograms of a sample. It can provide detection limits up to 1000 times more sensitive than obtained by the conventional flame AAS method. The necessary investment costs for instrumentation represent only a fraction of that needed for other methods with similar analytical performance and detection power such as mass spectrometry and neutron activation analysis.

The graphite furnace is a flameless, electrothermal sampling device which can be used with almost all make atomic absorption spectrophotometers. The energy required for atomization is supplied by applying a high electrical current through a graphite tube, where the sample has been placed. The furnace is located in the sample compartment so that light from the light source passes through the graphite tube. When the furnace is fired, the generated atomic vapors absorb light from the source. The absorption signal is transient and a peak shaped signal is produced as the atom concentration within the furnace rises, and then falls as the atoms diffuse from the furnace. Either peak height may be used for quantitation, or peak area may offer some advantages. A programmable power supply provides precise control of temperature programming for drying and atomization steps of the analysis. Further automation of the technique has been achieved by using an autosampler for sample introduction to the furnace. The most useful advantage of this technique is that solid samples can be analyzed directly without prior dissolution.

Chakrabarti et. al.[10] have reported the use of the capacitive discharge technique (CDT) for graphite furnaces. The CDT employs an anisotropic pyrolytic graphite tube atomizer which is heated at very high heating rates (up to 100 K/ms) by capacitive discharge resulting in isothermal atomization at high temperatures (up to 3500 K). Since the CDT is relatively free from matrix interferences and the sensitivity is independent of matrix, analytical calibration curves are not required, or used.

Until 1980, furnace atomic absorption spectroscopy has been historically a single-element tecnique. Recently, a simultaneous, multielement atomic-absorption continuum spectro-photometer has been developed[11-16] for the measurement of at least 16 elements. The instrument incorporates the use of hollow cathode lamps as the energy source for resonant emission lines. During one firing of the furnace four elements can be measured and eight elements can be measured automatically using two firings. Some of the other features of this instrument are polarized Zeeman background correction, total automation, a flexible furnace design, and computer compatibility.

Several terms such as flameless, non-flame, electrothermal atomization (ETA), graphite furnace (GFAAS), or furnace atomic absorption spectroscopy (FAAS) have been used by various researchers for this technique. In this book the only term used will be the Furnace Atomic Absorption Spectroscopy (FAAS), so as not to confuse the technique with other flameless methods such as hydride generation, cold vapor, sampling boat, etc.

Principle of Atomic Absorption Spectroscopy
Absorption Spectra

The most important advance in the study of atomic spectra was made by Bohr, who postulated that any atom is allowed only certain discrete and characteristic energy values, and that absorption or emission of radiation by an atom occurs when it undergoes a transition between these stationary states or energy levels. These energy levels are directly proportional to the empirical spectral terms of the atom. The frequency of radiation corresponding to a transition of an atom between a higher energy level, E_m, to a lower energy level, E_n, can be determined by

$$\nu = (E_m - E_n)/h = \Delta E/h$$

In terms of wavelength,

$$\lambda = c/\nu = hc/\Delta E$$

where h is Planck's constant and c is the velocity of light.

Electronic transitions, therefore can be discussed in terms of frequency, ν, energy, E, and wavelength, λ. These parameters have unique values for a given electronic transition. An element can undergo several electronic transitions, which result in a series of sharp lines called a spectrum. This spectrum is uniquely characteristic of each element. The visible region of a spectrum extends from 770 nm (red) to 380 nm (violet), although the optical spectra can be studied for a considerable range on either side of this range.

In absorption, electrons of an atom in its lowest energy state, the ground state, can absorb a quantum of energy and undergo transition to a low-lying excited state. Emission occurs when this quantum of energy is released and the electron returns to the ground state. A transition to and from the ground state is called a resonance transition.

Energy + ground state atom $\rightarrow$ Excited state atom $\rightarrow$ Energy + ground state atom

The properties of the excited to ground state atoms at a given temperature can be determined with the aid of the Boltzmann relation,

$$N_m = N_n(G_m/G_n) \exp. \left(- \frac{E_m - E_n}{kT} \right)$$

where N is the number of atoms in first excited state, m, or the ground state, n; G_m and G_n are the statistical weights of the excited and ground states, respectively; T is the absolute temperature; and k is the Boltzmann constant. Walsh[2] calculated the ratio N_m/N_n for the most populated energy states of several elements as a function of temperature. The very low proportion of atoms in the first excited state (even at 3000 K) shows that absorption of radiation, other than that which originates from a transition involving the ground state, would be very small. Spectral lines resulting from resonance transitions are known as resonance lines. These lines are the most useful analytical lines for atomic absorption spectroscopy. Since absorption is a direct measurement of ground state atoms, it will give better sensitivity than emission spectroscopy. There are less chances of spectral interference, since the number of absorption spectra produced is very small.

Alkemade[11] has shown that for a given analysis line, the absorption sensitivity is better only if the spectral radiance of the lamp source exceeds that of a black body at the temperature of the flame. Since the sharp line sources used in atomic absorption are invariably excited by nonthermal means, their spectral radiance exceeds that of a flame atomizer by several orders of magnitude. The effective radiation temperature for resonance lines produced in gas discharge tubes is about 10,000 K.

There are two other types of spectra encountered in atomic absorption spectroscopy: the molecular or band spectra and continuum spectra. The molecular spectra are associated with the energy changes of molecules which give rise to closely packed bands of spectral lines. These spectra are complex in nature and cause spectral interferences in the atomic absorption spectroscopy. In continuum spectra, energy is distributed in an uninterrupted manner between all wavelengths within a given domain. This is often observed as the black body radiation from a hot solid or liquid[17] or as the result of an atomic or molecule energy transition in which one of the energy states is unquantized.[18]

Spectral Linewidths

The AAS technique relies on measuring the "peak absorbance" of a spectral line by using a spectral light source emitting a sharp line. Even the sharpest line produced has a finite width. The natural linewidth of an atomic spectral line is of the order of 10^{-5} nm (10^{-4} Å). Broadening of these lines occurs due to a number of factors depending on the source. The most common linewidth broadening effects are as follows:

Doppler effect — This effect arises because the atoms will have different components of velocity along the line of observation, and is given by

$$D = (1.7\lambda/c)\left(\frac{2RT}{M}\right)^{1/2}$$

where R is the universal gas constant, T is the absolute temperature, and M is the atomic weight. Doppler broadening varies with the element, the wavelength of the line, and the temperature.

Lorentz Effect — The Lorentz effect occurs as a result of the concentration of foreign atoms present in the environment of the emitting or absorbing atoms. The magnitude of the Lorentz broadening varies with the pressure of the foreign gases and their physical properties.

Stark Effect — Stark broadening results from the splitting of the electronic levels of an atom due to the presence of strong nonuniform electric fields or large densities of moving ions or electrons. This effect is negligible in flames, hollow cathodes and electrodeless discharge lamps used in atomic absorption spectroscopy.

Zeeman Effect — Zeeman broadening is a splitting effect observed in the presence of an applied strong magnetic field. This effect is deliberately applied to produce a shift in a source or absorption line to obtain better overlap of the two.

Self-Absorption or Self-Reversal Effect — This effect results in broadening of the spectral line simply because the atoms of the same kind as that emitting radiation will absorb more radiation at the center of the line than at the wings, resulting in the change of shape of the line as well as its intensity. This effect becomes serious if the vapor which is absorbing radiation is considerably cooler than that which is emitting radiation. Measurements made over wider bandwidths result in absorption values smaller than the true values.

The absorption coefficient, kv, the total absorption factor, A_T, and absorbance, A, quantities are used to relate to abnormalities observed in the practical measurement of atomic absorption, e.g., the relationship between concentration and the energy absorbed. The absolute analysis by atomic absorption spectroscopy involves the measurement of total absorption, A_T, with a continuum source, or the measurement of absorbance, A, with a sharp line source.

General Instrumentation

An experimental device is required to measure the absorption of a given element under examination. The simple components of an atomic absorption spectrophotometer are the resonance line source, atom reservoir (flame or furnace), monochromator, photomultiplier or detector, and a signal processor, a readout device, or a data system.

The principle of operating an atomic absorption spectrophotometer is that the hollow cathode lamp, a sharp line source consisting of a cathode containing the element to be determined, is used as the resonance line source. The light beam from this source consists of resonance radiation which is electronically or mechanically pulsed. When a sufficient voltage is impressed across the electrodes, the filler gas inside the lamp is ionized and the ions are accelerated towards the cathode. As these ions bombard the cathode, they cause the cathode material to "sputter" and form an atomic vapor in which atoms exist in an electronic state. In returning to the ground state, the line characteristics of the element are emitted and passed through the atom reservoir (flame, furnace, or nonflame), where they may be absorbed by the atomic vapor. Analyte atoms are produced thermally in the atom reservoir. The ground state atoms, which dominate under the experimental conditions, absorb resonance radiation from the lamp, reducing the intensity of the incident beam. A monochromator isolates the desired resonance line and allows this radiation to fall on a detector or a photomultiplier tube. An electrical current, whose magnitude depends on the light intensity, is produced. The electronics of the unit are designed to respond selectively to the pulsed radiation emanating from the radiation source, and measure the amount of light attenuation in the sample cell and convert these readings to the actual sample concentration.

There are two basic types of atomic absorption spectrophotometers.

Single-Beam Type — The light source emits a spectrum specific to the element to be determined, which is focused through the sample cell (atom reservoir) into the monochromator. The monochromator disperses the light and passes the isolated wavelength to the detector or a photomultiplier tube. Depending on the light intensity, an electric current, which is proportional to the light intensity, is produced and processed electronically.

Double-Beam Type — The light from the source lamp is divided into a sample beam, which is focused through the sample cell, and a reference beam, which is directed around the sample cell. The readout presents the ratio of the sample and the reference beams.

Instrumentation for atomic absorption has essentially two parts with different functions.

1.   The assembly, to produce the population of ground-state atoms from the sample, e.g., atomization system (flame/furnace/nonflame).
2.   The optical system which includes the resonance and the spectrophotometer.

Atomization Systems

The production of free ground-state atoms of the element of interest occurs in an atomizer, and the process is called atomization. The most common systems are the flames and the furnaces.

Flame Atomization System

A flame is simple, inexpensive, easy to use, and provides a stable environment for atomic absorption. In a typical flame system, the sample is drawn into the nebulizer by the low pressure created around the end of the capillary by the flow of the carrier gas. The liquid stream is broken into a droplet spray and is ejected with carrier gas into the spray chamber. During nebulization, some liquids break into a finer mist than others. The droplets with a diameter greater than about 5 µm fall out onto the sides of the chamber and flow to waste. The fuel gas is introduced into the chamber along with the carrier gas, and an intimate mixture of sample mist, fuel, and carrier gas leaves the spray chamber to enter the burner. Since the amount of fine mist per unit volume reaching the flame affects the signal magnitude, it is important to nebulize samples and solvents of similar solvent compositions.

Nonflame Atomization Systems

Although the use of a flame to produce an atomic vapor is convenient, stable, and economical, it has certain disadvantages due to the presence of other reactive species in the atomic vapor. Another disadvantage is the necessity to convert every solid sample to solution form for the analysis. Some nonflame atomization methods commonly used are listed below.

1. Arc, spark and plasma torch methods[18-24]
2. Cold vapor absorption method[25] or mercury analysis system
3. Flash-Lamp method[26]
4. Laser beam method[27-30]
5. Solid-propellant atomization method[3]
6. Hydride generation or vapor generation method[32]
7. Sputtering method[33-35]
8. Electrothermal atomization method

Electrothermal Atomization

This system consists of the vaporizer component, a power supply, electrodes that deliver power to the atomizer, and a casing or a frame which provides cooling and inert gas sheathing of the atomizer. The power supply usually provides three heating cycles:

1. The sample is dried
2. The charring and volatilization takes place
3. The atomization or pyrolysis of the sample (in this cycle, the device can be heated up to an atomizaton temperature of 3000°C)

The high purity gas is used to purge the device. Samples in microliter amounts are used for analysis.

Electrothermal atomizers are of two types (1) the carbon rod furnace and (2) the graphite tube furnace. The most common graphite tube furnaces in use are listed below.

Heated Carbon Rod or Filament Furnace — Samples are placed on a carbon rod or a filament which is directly heated by the use of a high current. The device is placed within a bell-jar container which has optical windows and is flushed with inert gas. The complete system is placed just below the optical axis so that the vapor is carried upwards and absorbance can be recorded. Since the atomic vapors are not placed directly in the optical path, the sensitivity obtained is low. The method is convenient as samples can be changed quickly.

Resistance Heated Graphite Tube Furnace — Massman[36] introduced a graphite tube furnace (1.55 mm thick, 6.5 mm diameter, and 55 mm length) which is supported within a water-cooled metal container flushed with argon. A temperature of up to 3000°C is produced in the tube by passing 400 A at low voltage directly through the tube. Reproducibility of this technique is excellent and absolute detection limits of $g \times 10^{-12}$ range are obtainable. Sample solutions of 5 to 200 μL volume can be easily handled to determine concentratons in ng/L of several elements.

L'vov Graphite Furnace — L'vov[6] described a simple graphite tube furnace into which the sample was completely vaporized by heating with dc arc at 2500°C and enabled atomic vapor to be maintained in a highly reducing atmosphere.

Optical System
Radiation or Light Source

It is important that the light source for atomic absorption spectrometry must emit a steady uniform level of radiation to obtain an analytical signal of low noise level for better precision and detection limits. It should produce a "clean" spectrum with minimum interference due to the filler gas or impurities in the lamp. It should have long operating and storage lives. Spectral lamps usually used in FAAS are referred to as line or continuum sources. The most commonly used lamps are vapor discharge, hollow cathode, and electroless discharge lamps.

Vapor Discharge Lamps

This type of lamp consists of a glass or silica tube containing an inert gas at a pressure of several torr, a quantity of the metal whose spectrum is to be produced, and two tungsten electrodes for discharge. These lamps can be operated with AC or DC and at current considerably below the recommended operating current to minimize self-reversal of the resonance lines. Use of these lamps is dying out because of the availability of a large number of hollow cathode lamps. Lamps for alkali metals (sodium, potassium, cesium, rubidium), cadmium, mercury, thallium, and zinc are manufactured by OSRAM and Philips.

Hollow Cathode Lamps (HCL)*

Since its introduction by Paschen,[37] the hollow cathode lamp has been used as an excellent, bright and stable line source, and is widely used to produce a fine line spectrum. Crosswhite et al.[38] produced a successful sealed-off iron lamp and Walsh[2] suggested its use for AAS.

The essential components of a HCL are a cathode and an anode sealed in a glass tube or a cylinder filled with argon or neon at about 7.5 mbar (10 Torr). The cathode is a hollowed-out cylinder constructed entirely or in part of the metal whose spectrum is to be produced. A small current of 1 to 50 mA at 150 to 500 V is passed between the electrodes resulting in ionizaton of inert gas atoms. The positively charged gas ions are accelerated to collide with the cathode, where they produce an atom cloud of the cathode material, a process known as sputtering. The atoms in the "atom cloud" are then excited by further collisions with gas ions and on falling to the ground state emit the resonance lines which are the most useful lines for AAS and FAAS. The cathode temperature is usually 300 to 400°C. When HCL is lighted, a blue-green glow is observed with argon-filled and a red glow with neon-filled lamps.

HCL provides a spectrum mainly consisting of resonance and nonresonance lines of the element of interest. Its average life expectancy is 1 to 2 years and depends on the element, amount of use, average current employed, and problems due to leakage of gas-fill. These lamps are of the following types.

1.   Demountable hollow cathode lamps
2.   High-intensity, high-brightness, or boosted-output lamps
3.   Single-element lamps
4.   Multi-element lamps

The main requirements for a HCL as a spectral source for AAS and FAAS are

---

* See Appendix for types of lamps and a list of manufacturers.

1. Intense emission of resonance (ground state derived) lines
2. Narrow line width, for maximum sensitivity and linear response
3. Minimal spectral interference from continuum emission, fill-gas lines, or from other elements or impurities present in the cathode material
4. A quick warm-up to produce stable long-term light emission
5. Noise-free operation
6. Longevity
7. A cathode geometry which matches with the optics of the instrument to be used

*Electrodeless Discharge Tubes or Lamps (EDL)**

The electrodeless discharge lamps are constructed by sealing a small amount of the pure metal salt (usually the iodide salt of the element of interest) into a quartz bulb or tube of 3 to 8 cm in length and 1 cm or less in diameter. This bulb or tube is filled with inert gas and then placed in the cavity of a radiofrequency (RF) coil powered at 27 MHz. A separate power supply and 30 min of warm-up time is required to operate these lamps.

EDL may be excited at both radiofrequencies (100 kHz to 100 MHz) and microwave frequencies (>100 MHz). The thermally uniform nature of the discharge and even a "skin effect" (discharge concentration near the tube walls) ensures that there is little or no self-reversal of the emitted resonance lines. The line profile obtained is extremely sharp. Due to these characteristics EDLs are highly recommended for elements with high vapor pressure such as arsenic, antimony, bismuth, lead, selenium, tellurium, and tin.

Continuum Sources

Gibson et al.[39] have suggested that continuum sources (incandescent lamps, etc.) can be used for atomic absorption analysis. With the use of a high resolution monochromator, sensitivity achieved is similar to that of a sharp-line source. The brightness of a continuum source is less than that of a HCL. Tungsten iodide lamps have been examined as absorption sources by Fassel and Massotti.[40] Several workers[41,42] have attempted to use the hydrogen and deuterium discharge lamps, while others succeeded with the high pressure xenon arc.[43,44]

Lasers

Lasers have been used quite widely for a number of years. The current nature and unidirectional beam of radiation obtained from a laser can provide improvements in atomic absorption sensitivities.

Monochromator

The function of a monochromator is to isolate the resonance lines emitted by the primary radiation source from the nonabsorbing lines situated close to it in the source spectrum. It should be able to isolate the measured resonance lines from molecular emissions.

Pre-Slit

A monochromator aperture of f:10 is usually suitable for atomic absorption measurements. As much light from the source as possible should pass through the furnace or else sensitivity will be low. Since the source is focused onto the monochromator slit, more energy will pass into the monochromator.

Wavelength Selection

Wavelength selection is required to measure the radiation signal to be processed by the spectrometer. Radiation is frequently accompanied by unwanted light of other wavelengths.

---

* See Appendix for list of manufacturers.

This unwanted light could be due to a complex spectrum emitted by the light source, or atom cell weaker absorption lines of the element to be determined. To select a single spectral line an optical monochromator device is used.

The ability of an atomic absorption spectrometer to spread out different wavelengths is termed its dispersion. This can be measured either as an *angular dispersion* of the separated beams of light as these emerge from the dispersive device, or as *linear dispersion* of the separated images of the entrance aperture when they are brought to a focus at the exit aperture. There is a limit to the ability of a particular device to separate closely spaced adjacent wavelengths. This is termed *resolving power*. The efficiency with which a monochromator is able to carry out the wavelength selection is termed its *transmission*.

Filters

An alternate method for wavelength selection is by means of a small optical filter transmitting a specific resonance line. The relatively poor resolution of even the best filters limits their application to simple instruments only.

Slit Width

Slit width determines the amount of spectrum that falls on the photomultiplier. The slit should be kept as narrow as possible to lower the emission reaching the photomultiplier and reject the unwanted lines from the source. The minimum slit width is determined by the amount to which the resonance line signal can be reduced before noise becomes too intolerable.

The construction of the slit mechanism is important. A single fixed slit width has limited applications. The entrance and exit slit widths should be individually optimized and this can be achieved by a single adjustment available in most of the absorption spectrophotometers.

Selective Modulation

Selective modulation requires detection of the original light beam after it has passed through the absorbing cloud. During this passage only the resonance lines of the element in the cloud will be absorbed. Results are poor unless a separate optical filter or a photomultiplier is used to restrict detection of extraneous radiation and noise.

Electronic Signal Processing

Photomultiplier tubes are used to convert the light signal into an electrical signal for direct readout. The spectral sensitivity depends upon the type of photosensitive material (antimony, bismuth and/or silver, beryllium-copper, silver-magnesium, and gallium arsenide) used to coat the cathode metals. The useful feature of this tube is the large output currents at very low light levels and the wide range at which the gain of a particular tube may be varied simply by altering the applied voltage. The tubes used are sensitive to radiation over the 1900 to 8000 Å range. The tube is a current source and the intensity of the current produced is proportional to signal strength. This current is converted to a voltage using a high resistance in the feedback loop of a high-input-impedance operational amplifier. A reference waveform is used to switch the reference and absorbed beam into the respective demodulators. These devices produce signals I_o and I proportional to pulse heights of the respective signals. The processed signal is presented to the analyzer through the readout system which can be a meter, recorder, or a digital system. Most of the modern instruments have a built-in curve corrector to straighten the nonlinear calibration graph at high levels of concentraton, and a threshold corrector to give control to adjust only the nonlinear range.

Detectors

Resonance detectors are efficient for atoms that exhibit simple spectra; a number of energy levels close to the ground state will give a complex resonance spectrum. The cloud of atoms in the resonance detector can be produced thermally by embedding a heating element in a small block of the appropriate metal; but, in other cases it is necessary to use cathodic sputtering of the type used in hollow cathode lamps.

Scale Expansion System

This system involves an electrical expansion of the presented signal by a chosen factor of 2, 5, 10, or 20. Scale expansion in a single-beam spectrophotometer is achieved by increasing the amplifier gain by the required factor. The unwanted part of the signal is backed off with a "zero" control until the reading becomes 100% transmission. This control is provided in new instruments.

Readout System

The conventional technique is a recorder readout system, where pen movement records noise and drift from all the sources. This system is used in nonflame methods to improve analytical precision.

The digital readout system avoids errors encountered in the scale readings either on meters or recorder systems. All new instruments are equipped with a direct concentration readout system with a digital display.

Auxillary computers are often connected to the spectrophotometer to facilitate the data handling.[45-47] Improved software continues to aid in methods development and in trouble-shooting with furnace systems.[48,49]

Modern Graphite Furnace Technology

The modern graphite furnace atomic absorption spectroscopy (FAAS) has come a long way in instrumentation technology. A large number of research papers are available in the literature now. The new systems have better hardware, fast digital electronics and sampling frequency to process rapid furnace signals, built in background correction ability, maximum power heating for atomization, etc.

General FAAS Instrument Setup

In any furnace AA determination it is desirable to obtain the highest possible signal-to-noise ratio, and to achieve this, instrumental parameters have to be set up and optimized. Some of the considerations are mentioned below.

Lamps

Generally, single-element hollow cathode lamps provide greater radiation output and, therefore, are used with preference. The multielement lamps can be used as back-up lamps for several single element lamps and for infrequently determined elements. These lamps have short lifespans due to different rates of sputtering of various elements. For some elements such as arsenic, selenium, thallium, etc., the use of electrodeless discharge lamps provides superior performance.

It is desirable to choose the best operating current for a particular lamp. Higher current operations result in excessive line broadening and use of lower current provides adequate stability and freedom from intensity drifts. Operation at less current settings than that recommended by the manufacturer increases the lamp life. Higher lamp currents also accelerate gas depletion and cathode sputtering, and should be avoided until the age of the lamp does require a current increase. HCL failure occurs when the fill gas is gradually captured on the inner surfaces of the lamp, and the lamp will not light.

A warmup time of a few minutes to an hour is required with a single-beam spectrophotometer. No warmup time is required with a double-beam spectrophotometer.

Wavelength

For maximum sensitivity, use the analytical wavelength provided under the parameters in each section of this handbook. If the absorbance of the atomizaton peak is much greater than 0.5 absorbance with the smallest sample volume used, an alternate wavelength with reduced sensitivity should be selected rather than changing the sample concentration.

Slit Setting

The slit setting is selected to obtain optimum analytical sensitivity with the best signal-to-noise ratio and good linearity of the working curves. To reduce the noise level, a wider slit setting can be used, but with a decrease in sensitivity and linearity.

Electronics and Integrated Absorbance

In FAAS, two parameters can be proportional to the amount of analyte present in a sample

1. Peak height, the magnitude of the highest part of the output signal observed with time during the atomization part of the cycle
2. Peak area, the integrated area under the output

When high heating rates are applied for atomization, these signals may be over in about 1 sec. If the electronic data collection speed is not fast enough, this may lead to severe signal distortion. Hard copy of the signals can be obtained with a printer/plotter. High resolution peak display is extremely important when appearance time, shifts, shapes, shoulders, or double peaks are of interest, such as needed for method development.

In modern instruments, the electronic circuits are designed to measure and display the peak absorbance (height) or integrated absorbance (area) of the signal, either in absorbance for peak height, in absorbance-seconds for peak area, or directly in concentration mode. The integrated absorbance measurements provide greater linear dynamic range, allow lower atomization temperatures to be used with no loss in sensitivity, and superior analytical performance especially with respect to analytical accuracy. Interferences can affect the height, the area, or both.

Scale Expansion and Damping

Although use of scale expansion can be avoided in FAAS by changing the sample volume, the modern instruments have the capability for expansion up to 50 times due to optimized optics and electronics.

Recorder

Recorders are usually not suitable for FAAS, but if there is no alternative, a 10 mV recorder with a full scale response of 0.5 sec can be used.

Background Correction

Background correction is required more frequently with FAAS, since it subtracts nonatomic absorption signals from the total absorbance signal generated. There are two types of background correction devices available to the analyst.

1. Continuum source background correction is designed to automatically correct for broadband nonatomic absorption. A continuum source emits light over a broad spectrum of wavelengths and the light from the primary and the continuum sources is passed alternately through the furnace. The element being determined effectively absorbs light only from the primary source, while background absorption affects both beams equally. The ratio of the two beams is measured electronically thus eliminating the effect of background absorption.

2. A deuterium background corrector is used for the far UV region. The background correction with a continuum source is limited to background absorption that is constant with wavelength over the observed spectral range. For some determinations, continuum source background correction is adequate.

Zeeman Effect Background Correction

With Zeeman effect, atomic lines are split into three or more polarized components in a strong magnetic field. In a normal Zeeman pattern, the single pi (π) component remains at the original analyte wavelength. The sigma (σ) components are symmetrically shifted away from the central analyte wavelength. In a complicated Zeeman pattern, both the pi and sigma absorbances are split into several components. In addition, these components are polarized differently. The pi components are linearly polarized in a direction parallel to the magnetic field while the sigma components are polarized in a direction perpendicular to the magnetic field.

Zeeman effect background correction makes it possible to compensate for nonspecific absorption up to about 2.0 Å. This method also permits correction for continuous background as well as for structured or line background. The only disadvantage of it may be a slightly reduced linearity of the calibration curves and some loss in sensitivity for some elements.

Graphite Tubes

There are two types of graphite tubes available in the market. One is made of spectrally pure, high density graphite and the other is made of the same material, but coated with a thin layer of pyrolytic graphite. These tubes have capacity of 100 to 150 µL of sample volumes.

Normal graphite has a relatively coarse, layered surface, which allows certain materials (elements, solvents, etc.) to penetrate into the lattice and react with the graphite at high temperatures. Uncoated tubes have limited life and application. Strong oxidizing agents used at high temperatures also cause deterioration of the tube. Some severe interferences can also be caused by graphite retaining some species and then releasing the same at high temperatures.

The pyrolytically coated graphite tubes have long life. Due to dense surface no chemical penetration or interaction takes place. The use of these tubes is recommended for elements that form refractory carbides, thus offering higher sensitivity and reduced memory effect.

These graphite tubes have limited life expectancy, which can be increased by using lower atomization temperatures. Tubes should be replaced when the integrated absorbance values drop 20 to 25% below the original value. Precision also decreases with the aging of the tubes.

L'vov Platform

A considerable reduction of interferences can be achieved by using the L'vov platform for determinations. Here, samples are atomized from a small plate of solid pyrolytic graphite inserted into the graphite tube and not from the wall of the tube. The platform is made of pyrolytic graphite having high anisotropy and therefore low thermal conductivity in the direction perpendicular to its plane, thus minimizing heat transfer by conduction in this direction. Heating occurs predominantly by radiation from the hot tube walls.

Automated Sampling

An autosampler is a microcomputer-controlled sampling device which can be programmed for sample, standard, standard addition, and matrix modification purposes. It also provides accuracy in pipetting microliters of liquid samples. Its use increases the precision of the technique.

Cooling of the Furnace

The furnace is cooled by water to allow the graphite tube temperature to drop to room temperature in between the determinations. Special cooling systems are available which can provide an optimum temperature of 20 to 40°C and a flow rate of 2.5 L/min. A small investment in the cooling system is worth the cost.

Gases

Argon is an inert gas and is generally used as purge gas for FAAS analysis. It provides external protection to the hot graphite parts. Due to cheaper cost, use of nitrogen is possible. Use adequate ventilation, since nitrogen is known to form cyanides and cyanogens with the graphite at elevated temperatures.

Standard Conditions and Optimizaton Method for the FAAS Analysis

Once an analyst has made the decision to use furnace atomic absorption techniques (Figure 1) for sample analysis, the following considerations are important steps for the quantitative determination of elements.

1.   Sample pretreatment and preparation
2.   Time/temperature programming
3.   Atomic absorption parameters
4.   Purge gas
5.   Background correction
6.   Verification of methods and analytical results

Sample Pretreatment and Preparation

Samples can be liquid or solid, in an inorganic or organic matrix, and must be in a suitable state for a sensitive analysis. The methods used most commonly and preferably depend on the sample matrix and are described below.

1. Direct Analysis

Aqueous Liquid Samples — These samples can be injected directly into a furnace atomizer using an automatic micropipette or an autosampler which dispenses the sample either as a droplet or as an aerosol mist.

Nonaqueous Samples — Organic solvents tend to spread on the graphite surface of the cuvette and "creep" out of the plastic micropipette tip. The introduction of the nonaqueous solution into the furnace as an aerosol mist eliminates problems associated with low surface tension. The aerosol dries out on contact with the graphite cuvette or tube, which is maintained at an elevated temperature thereby avoiding sample spreading.

Solids — Solid samples must be dry, clean, and finely powdered for direct analysis. Samples can be weighed in a boat and placed directly in the furnace.

Dilution

Aqueous samples can be diluted with deionized double-distilled water either to decrease analyte concentration or sample viscosity. Organic solvents such as methylisobutylketone (MIBK), xylene, and kerosene can be used as diluents, if the sample is viscous, has too high analyte concentration, or, has an unknown composition. Standards, samples, and blank solutions should be prepared at the same dilution in the same solvent.

Matrix Modification

Matrix or chemical modification is a method of manipulating the chemical form of the analyte or of the sample to increase or decrease its volatility.

Analyte Modifiers

These are used to decrease the volatility of the analyte, e.g.,

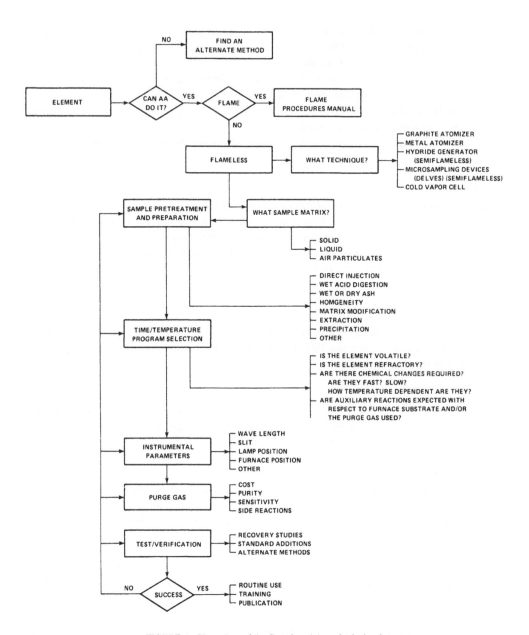

FIGURE 1. Flow chart of the flameless AA methods development.

1.   Addition of ammonium sulfide to mercury solution results in the formation of mercury sulfide which is stable at 150°C. Without the use of modifier, the mercury vapor will be lost during the drying cycle of the analysis.
2.   Addition of barium and lanthanum to silicon minimizes silicon carbide formation.
3.   Addition of nickel to arsenic and selenium results in the formation of nickel arsenide or selenide which are stable up to 900°C. This enables the use of higher pyrolysis temperatures to remove matrix interferences.
4.   Addition of phosphate ions to cadmium form cadmium phosphate which is stable up to 650°C.

Matrix Modifiers

Matrix modifiers are used to make the sample matrix more volatile. The most common use is the addition of ammonium nitrate to seawater to aid in the low temperature removal of sodium chloride by the formation of ammonium chloride and sodium nitrate.

For organic sample, matrix or chemical modification is possible, if the chemical modifier is miscible with the organic matrix of the sample, e.g., nickel octoate can be added to arsenic containing organic sample.

Preconcentration

If the analyte concentration is low, successive additions of sample solutions directly to the graphite tube (which generally has 100 μL capacity), and use of the drying cycle, will provide measurable analyte concentration.

Sample Degradation

If the analyte in the sample is tied up in particulate matter suspended in solution, then a sample degradation or solubilization is recommended. This step is usually accomplished by heating a strong oxidizing acid with the nonaqueous sample.

Dissolution in Water

In this method, deionized distilled water is used for sample dissolution. The analyte may precipitate out of solution or adhere to the wall, therefore, a small amount of mineral acid is added as a stabilizer. This technique does not remove or alter the sample matrix.

Acid Dissolution, Digestion or Wet Ashing

In this destruction technique, the oxidant is a component of the solution, typically an oxidizing acid to destroy the matrix. Sometimes the oxidant is assisted by a catalyst, such as molybdates for arsenic, or selenium, vanadates, permanganates, or a free radical generator such as hydrogen peroxide.

Wet ashing is effective to destroy or remove the sample matrix thus helping to reduce or eliminate some types of interferences.

Acid digestions with heat are rapid, occur at low temperatures (<100°C), and usually avoid retention of the analyte on the container walls. The main disadvantage of this technique is the potential introduction of contamination from the acids and catalysts into the analysis.

The most common acids are nitric, hydrochloric, sulfuric, and phosphoric acids. Perchloric acid is not suitable for furnace method. Selection should be based on the element of interest and type of sample. Various combinations and uses of acids are described below.

1.   Hydrochloric acid (HCl)

Room temperature — 120°C	Dissolves weak acid carbonates, phosphates, sulfides, and iron oxide.
250—300°C	Dissolves oxides of aluminum, beryllium, tin and silicon.

	350—400°C	Platinum metals can be dissolved when HCl is used along with perchloric acid or any other oxidizing agent.
2.	Nitric acid (HNO$_3$)	Dissolves most sulfides except mercury sulfide. Provides oxidizing attack for metals usually not dissolved by HCl. It is not appropriate for oxides except uranium oxides.
3.	Sulfuric acid (H$_2$SO$_4$)	Suitable for formation of sulfate complexes. It is generally not used for furnace analysis, because many analytes form insoluble sulfates.
4.	Hydrofluoric acid (HF)	In combination with sulfuric and perchloric acids, this acid decomposes silicates. In mixture with nitric acid, it dissolves titanium, tungsten, niobium, and zirconium by forming complex fluorides. These metals cannot be analyzed by the FAAS method. Use of HF also rapidly degrades the graphite surface of the tube.
5.	Perchloric acid	It is a strong oxidizing agent to destroy organic matters. Dissolves stainless steel and sulfides.
6.	Nitric acid/Hydrogen peroxide	1 mL HNO$_3$ + 3 mL 30% peroxide
7.	Nitric/Hydrochloric acids	Use freshly prepared 1 mL HNO$_3$ + 3 mL HCl (aqua regia) per 10 mL liquid or 500 mg of solid sample.
8.	Nitric/Sulfuric acids	5 mL HNO$_3$ + 1 mL H$_2$SO$_4$. Use cautiously for organic materials which may react vigorously and cause spattering.
9.	Nitric/Hydrochloric/Sulfuric acids	10 mL HNO$_3$ + 5 mL HClO$_3$ + 3 mL H$_2$SO$_4$ is useful for geochemical samples, fertilizers, etc. Use caution with organic matrix samples.
10.	Hydrochloric/Nitric/Water	5 mL HCl + 5 mL water + 5 mL HNO$_3$ dropwise. Dissolves stainless steel.
11.	Hydrochloric/Nitric/Hydrofluoric	A mixture of 15 mL HCl + 5 mL HNO$_3$ + 3 mL HF is used mainly for alloys, silicates, and fly ash dissolution. Use Teflon® containers for dissolving the sample.
12.	Hydrofluoric/Boric acids	6 mL HF + 50 mL 4% boric acid will dissolve alloys, silicates, fly ash etc.
13.	Nitric/Hydrofluoric/Boric acids	6 mL HNO$_3$ + 1mL HF + 50 mL 4% boric acid. Dissolves alloys and silicates.
14.	Hydrofluoric/Phosphoric acids	50 mL HF + 1 mL phosphoric acid. Used for dissolving niobium & tungsten.

Some of the recommended wet-digestion methods are described below.

Method #1

To a 25 mL sample in a 100 mL beaker add 1 mL of 30% hydrogen peroxide and 100 µL of concentrated nitric acid. Place the sample on a hot plate and digest at ≅75°C for half an hour, remove from the plate, and allow to cool at room temperature. Dilute the sample to 25 mL and transfer it to a 30 mL plastic bottle. A blank should be prepared in a similar manner. This method is useful for mercury and less-volatile elements.

This method is recommended for determination of aluminum, barium, bismuth, chromium, cobalt, copper, iron, lead, manganese, molybdenum, nickel, silver, tellurium, thallium, tin, vanadium, and zinc in waste water analysis.

Method #2

To 25 mL of sample in a 100 mL beaker, 100 μL of hydrochloric acid and 1 mL of 30% hydrogen peroxide are added. The sample is then placed on a hot plate and digested at ≅75°C for half an hour. Allow it to cool at room temperature, dilute to 25 mL and tranfer into a 30 mL plastic bottle.

This method is useful for gold, platinum, beryllium, barium, and titanium.

Method #3

To 25 mL of sample in a 100 mL beaker, add 1 mL of 30% hydrogen peroxide and 100 μL of concentrated nitric acid and place sample on a hot plate. Digest at ≅75° ± 5°C for half an hour. Remove the beaker and allow it to cool at room temperature. Add 2.5 mL of 1% ammonium phosphate and dilute to 25 mL. Transfer the solution to a 30 mL plastic bottle.

This method is useful for cadmium determination.

Method #4

To 25 mL of sample in a 100 mL beaker, add 100 μL of 5% potassium hydroxide and heat in a steam bath at ≅75° ± 5°C for 45 min. Remove the beaker from the bath and allow it to cool to room temperature. Filter the sample through Whatman #42 or equivalent filter paper. Dilute it to 25 mL and transfer to 30 mL plastic storage bottle.

Since this method prevents the formation of sodium hexafluosilicate which is extremely volatile, it is useful for insoluble aluminates and silicates. Care should be taken to prevent samples, standards, and reagents from contacting glass or silica containing materials.

Method #5

To 25 mL of sample in a 100 mL plastic beaker, add 500 μL of 48% hydrofluoric acid and place the beaker in a steam bath. Digest at 60° ± 5°C for 30 min. Cool it to room temperature and filter it. Dilute it to 25 mL and store in a 30 mL plastic bottle.

This method is useful for analytes which have stable fluorides. Make sure that samples, standards, and reagents do not come in contact with any glass- or silicon-containing materials.

Method #6

To 25 mL of sample in a 100 mL beaker, add 2.5 mL of 5% nickel nitrate and 10% nitric acid. Place it on a hot plate and digest at ≅75° ± 5°C for 45 min. Allow it to cool at room temperature and dilute to 25 mL. Store it in a 30 mL plastic bottle.

This method is useful for arsenic and selenium which form stable compounds with nickel and allows pyrolysis in the furnace without loss of the analyte.

Method #7

To 25 mL of sample in a 100 mL beaker, add 100 μL of concentrated hydrochloric acid and 2.5 mL of 5% tartaric acid. Place sample on a hot plate and digest at ≅75° ± 5°C for 45 min. Cool it at room temperature, dilute to 25 mL, and store in a 30 mL plastic bottle.

This method is recommended for tin analysis.

Dissolution Under Pressure

This method involves dissolution in a sealed, inert container under elevated pressures. The most commonly used Parr Bomb is a pressure decomposition vessel. It is a relatively expensive and tedious technique for dissolution, but is effective in destroying sample matrix. This method is recommended for volatile analytes such as mercury.

Microwave Digestion Method

Since 1975, several digestion methods for elemental analysis utilizing the microwave oven have been described in literature.[50] Matusiewicz and Sturgeon[51] have provided an updated review on the present status of microwave dissolution and decomposition method.

In the microwave oven method, only polar molecules are affected by microwave radiation. Use of acid and acid mixtures provides the necessary chemical reaction to complete the digestion. The 2450-MHz microwave field energizes the polar molecule and rapidly generates heat in samples. Therefore, the sample dissolves due to internal heating and differential polarization which mechanically agitates and ruptures the surface layers of the solid material, and in turn, exposes fresh surface to acid attack. The digestion or dissolution or decomposition completes within a few seconds to a few minutes.

There are two types of microwave digestion systems: (1) open digestion and (2) closed digestion system.

Open digestion system— In this system, the sample is placed in an open vessel (beaker, Erlenmeyer flask, crucible, etc...) along with a known amount of acid and digested in the microwave oven. This system requires a vapor scrubber or a vacuum system to evacuate acidic fumes. This method, like a conventional digestion method, also suffers from the loss of elements, incomplete digestion, and contamination.

Closed digestion system— A variety of closed vessels can be used in this type of digestion, but the best use is made of the "Teflon Bomb" Parr vessel. Sample is placed in a Teflon container of the "Bomb" with an appropriate amount of acid. The digestion parameters such as microwave power, time, size of the sample, and acid volume have to be worked out for different types of samples before using this method.

Fusion Method

This technique requires mixing of 0.1 g finely powdered sample with 0.3 g of fusion reagent (sodium peroxide, potassium ammonium persulfate, etc.) in a graphite or platinum crucible. The crucible is heated at 500 to 900°C in a muffle furnace for 2 to 4 hours. The melt is then cooled and dissolved in acidic water.

This method is not favorable for FAAS because of sample contamination and matrix problems resulting from the fusion materials.

Dry Ashing

In this method a sample is heated at 400 to 1000°C in a muffle furnace to remove organic constituents. Residue is dissolved in an acidic solution.

This method is good for dissolving tissues, grains, etc. Volatile metals are usually lost in this method.

Plasma Ashing

In this method a plasma asher is used at or near room temperature by means of radiofrequency (RF) energy. The RF energy breaks apart the oxygen molecules into free radicals (which are extremely active) which rapidly attack organic materials. The residue is then dissolved in a small amount of acid.

This method is good for volatile materials.

Solvent Extraction or Acid Extraction

In this method, the analyte is simply removed by extraction rather than destroying the matrix. This method is time consuming and error prone. Many extractions are not quantitative, but in this way analyte concentration can be increased and matrix interferences are eliminated. The solvent chosen for extraction should not be appreciably soluble in water to avoid emulsification and slow separation. Avoid using benzene (carcinogenic) or highly volatile solvents (chloroform and carbon tetrachloride-carcinogenic). The ketones and esters have been found to possess the most suitable properties for separation. The most commonly used solvents are methyl isobutyl ketone (MIBK) and ammonium pyrrolidine dithiocarbamate (APDC), because a wide range of metals can be chelated and extracted with a high degree of efficiency. Examples include (1) extraction of APDC-complexed lead from blood into water saturated MIBK; and (2) acid extraction of organic-based liquids or solids (for nickel determination in fats, the sample is shaken with hydrochloric or nitric acid and allowed to stand overnight for separation).

Precipitation or Coprecipitation Method

This is a classic technique, but is not suitable for ultratrace level analyte determinations by FAAS. Solubility product constants of analytes play a large role in this method. Precipitation with organic compounds, such as 8-hydroxy quinoline and cupferron, separates and concentrates the element of interest.

Complexation Without Extraction Method

Complexation efficiencies at trace-level analyses are questionable. This is a special version of matrix modification and can be used to stabilize the analyte, permit pyrolysis, or improve the atomization behavior. The main drawback is a concern for contamination.

Ion-Exchange Method

This method preconcentrates as well as isolates the element of interest from the complex sample matrix. Here an analyte may be adsorbed on an ion-exchange and later leached out with a suitable solvent.

Electrolytic Concentration

Preconcentration using electrolytic techniques has become a common practice for concentrating the sample analytes. Here an analyte may be plated onto a wire or foil strip. The strip can be analyzed or the analyte is removed into another solution for FAAS analysis.

Time and Temperature Program Selection

In furnace atomic absorption determinations, desolvation of the sample, dissociation from the matrix, and generation of analyte-ground state atoms occur sequentially during the three steps

1.   Dry cycle or step for drying the sample,
2.   Thermal pretreatment or char step, and
3.   Atomization or pyrolysis step.

Thus the complete analysis requires careful selection of temperatures to ensure that each process is carried out effectively. With complex samples, great care should be taken in selecting the optimum operating conditions.

Drying Time and Temperature

The purpose of this step is to evaporate low boiling liquids from the sample matrix. Drying time is largely dependent upon sample volume. Care should be taken to ensure that only rapid evaporation should occur without spattering of the sample. Boiling of the sample can be detected by an audible hissing sound emanating from the furnace. It can also be monitored visually by looking into the furnace tube with the furnace rotated out of the sample beam.*

---

* The intense light radiation generated in the furnace during atomization is a potential hazard to eyes. Do not look into the furnace if the dry step is immediately followed by the atomization step.

General tips for selecting temperature and time are given below.

1. Select a temperature slightly above the boiling point of the solvent.
2. Select a time according to the sample volume:
    10 μL — 15 seconds
    20 μL — 20 seconds
    50 μL — 40 seconds
    100 μL — 60 seconds
3. Use ramp, if required with complex samples.

Charring Time and Temperature

The thermal pretreatment or charring step is necessary to remove any components of the sample matrix which are more volatile than the elements of interest before atomization. This step decreases the possibility of interferences due to broad band absorption. There are many instances where the optimum charring temperature for elements in complex matrices are different than for the same element in an aqueous solution. It is important to consider both the matrix and the element when selecting the charring temperatures. When selecting the charring temperature and time, keep the following in mind:

1. A sufficiently long charring time and a high charring temperature must be used to volatilize the interfering sample matrix.
2. The charring time must be short enough and the temperature low enough to ensure no loss of elements during charring.
3. For more complex samples, a ramp temperature increase may be useful.
4. Use background correction to compensate for the broad band absorption occurring during atomization.

The selection of the maximum charring temperature for dilute aqueous samples is usually 30 sec. For the same amount of sample use different charring temperatures in 50 to 100°C increments. Plot a graph of charring temperature versus absorbance, where absorbance will remain constant at lower charring temperatures until the maximum charring temperature is reached.

For complex samples containing significant amounts of "smoke" producing matrix, the parameters must be selected to volatilize most of the sample without the loss of analytes of interest. Make use of the background correction.

Selection of Charring Temperature and Time for Sample

Follow the given guideline steps for setting the charring temperature and time for sample analysis.

1. Select an aliquot of sample which can give an absorption signal of 0.2 - 0.5 absorbance.
2. Select the drying parameters determined previously.
3. Go to step 13 if broad band signals are not expected from the sample.
4. Change the wavelength to that of a nonabsorbance line.
5. Char for 30 sec or longer at 200°C and atomize the sample.
6. Repeat step 5 using successively high temperatures if any atomization signals represent broad band absorbance.
7. Try ramp charring if available with the instrument. Background absorbance will decrease on atomization, if the ramp charring is effective.
8. Plot a graph of absorbance signal vs. charring temperature.
9. The minimum usable charring temperature is the lowest value (less than 0.5 absorbance) giving low background absorption signals.

10. The minimum charring temperature is the lowest value giving low broad band absorption signals.
11. Reset the wavelength to the analytical line of the element and reanalyze the sample at the minimum temperature established in steps 6 to 10. The atomization signals received will be the atomic absorbance signals of the analyte of interest.
12. Repeat step "11" using successively higher charring temperatures.
13. Plot a graph of atomic absorption atomization signal versus charring temperature. The maximum value is the highest value on the plateau of the curve.
14. Select the operating charring temperature as a value between the minimum determined in step "9" and the maximum determined in "13."

Atomization Temperature and Time

The atomization temperature may be selected by using 20 to 25 μL of an aqueous standard of the analyte of interest having a concentration to give 0.2 to 0.5 absorbance. The determination involves a series of measurements . Set up the instrument as follows:

Dry cycle	20 sec at 120°C
Char cycle	10 sec at 300°C
Atomization cycle	10 sec at 2700°C with purge

Repeat this analysis using successively lower atomization temperatures, decreasing at 100°C intervals. Determine the baseline by running the experiment without any sample in the furnace tube. Plot a graph of atomization signal versus the atomization temperatures. A plateau is observed for most of the elements when there is no increase in absorption signal with the increase in temperature. Therefore, the optimum atomization temperature would be the lowest temperature giving the maximum signal. The same procedure can be used for the complex matrix samples.

Use of high temperatures shortens the graphite-tube life, and often analytical precision is degraded. The atomization temperature is usually similar for aqueous standards and for samples. It may vary for the pyrolytically coated graphite tube from the regular graphite tube. Use the following steps for setting up the atomization temperature and time for standard and sample solutions.

Standard solutions:

1. Select a standard solution which can give 0.2 to 0.5 absorbance.
2. Dry at temperature and time established previously.
3. Char for 10 sec at 300°C.
4. Atomize for 10 sec at 2700°C. Repeat at lower successive temperatures in 100°C increments.
5. Plot a graph of atomization temperature versus absorbance.
6. Select the lowest atomization temperature giving maximum absorbance.
7. Select an atomization time sufficient to allow the atomization signal to return to the baseline. Establish the baseline by running the experiment without sample in the graphite tube.

Sample solutions:

1. Select the drying and charring parameters determined previously.
2. Repeat steps 4 through 7 and redetermine the atomization parameters for the sample.

If the atomization time is too short, some of the elements of interest may be retained by the furnace and will give erroneous results during the successive analysis. Longer atomization time results in shortening the tube life. An approximate 8 to 15 sec atomization time is sufficient for carbide-forming elements (molybdenum, titanium, and vanadium).

Summarized Operational Tips for Time and Temperature Program Selection

1. High temperature setting for drying results in loss of analyte and causes serious program error.
2. Drying too fast or at too high a temperature will cause the sample to spatter. Select a long, slow Ramp for the dry cycle. Longer drying times are required for larger samples. Doubling the sample size requires an almost double drying time.
3. An atomization cycle that is too short results in accumulation and apparent "memory" effect.
4. An atomization cycle that is too long results in a waste of time and accelerates graphite-tube aging.
5. Atomization at too high of a temperature results in a waste of time, graphite aging, and may distill unwanted contaminants.
6. Atomization at too cool a temperature results in accumulation, inaccuracy, and insufficient sensitivity.
7. A large atomization time, with respect to the residence time, endangers the validity of the analysis. Avoid step atomization for volatile elements.
8. Use the CRC *Handbook of Chemistry and Physics* or *Lange's Chemical Handbook* for the selection of an appropriate temperature for atomization, e.g., determination of silver in dilute nitric acid,

	CRC's Handbook	Lange's Handbook
Ag	m.p. 960.8°C, b.p. 1950°C	m.p. 960.15°C, b.p. 2177°C
$AgNO_3$	m.p. 1212°C, d 444°C	m.p. 208.5°C, d 444°C
Ag_2O	d 300°C	d 300°C

From this data, it can be concluded that silver salts can be atomized between 1000 to 2000°C for good results.

Atomic Absorption Parameters and Tips on Techniques

There are some variations for the instrument set-up parameters, therefore, follow the instructions given by the manufacturer in the instruction manual. Some general AA settings are recommended in each section of elements in this handbook. These can be used as a guide to develop a method for analysis. Some tips for effective use of the FAAS method are described here.

Cell Windows

Cell windows should be clean. With constant use of the furnace, there is a tendency of accumulation on cell windows which could reduce the sensitivity of the method. To make sure that windows are clean, follow the manufacturer's instructions provided in the manual. Some general instructions to minimize the accumulation on cell windows are given below.

1. Use proper purge-gas flow rate.
2. Use a gradual pyrolysis.
3. Use higher gas flow and pressurization at atomization.
4. Use micro boats to eliminate some matrix on the hot plate.
5. Use matrix modification prior to atomization.

Contamination

In any trace metal analysis, contamination is the primary source of error, especially when the actual amounts of metal to be determined are in the picogram range. Usual sources of contamination are volumetric flasks, pipettes, reagents, and sampling processes.

Glassware or Plasticware — All glasswares/plasticwares should be washed thoroughly, rinsed three times with 1:1 nitric acid, and three times with 1:1 hydrochloric acid, and finally rinsed with deionized double distilled water. The glassware should not be placed upon any type of drying rack, or table, and should be stored away from dust, cigar and cigarette smoke, etc. All pipettes should be kept in a 1:5 solution of nitric acid. These should be rinsed with deionized distilled water three times and three times with the sample solution, reagent, or standard solutions (whichever is being pipetted), and then, after use, rinsed with deionized water and returned to the 1:5 nitric acid bath. Micropipette tips should be rinsed three times with the solution to be pipetted and discarded after droplets start to build-up in the tip. Do not let solutions stand in pipet tips more than 10 to 15 sec.

Use glassware for mercury analysis, since mercury goes through the plastic walls easily. Use plasticware for most of the solutions, especially if hydrofluoric acid is used for sample preparation.

Environment — Clean areas are must for FAAS. Bench-top tents can be used for the control of dust. Smoking should be strictly prohibited and air flow in the room should be controlled.

Reagents — Use deionized double-distilled water for the preparation of any solution or dilution purposes. Be sure to check the purity of water for making standard and reagent solutions. Always run blanks to check for contamination. Use ultra-pure chemicals for sample preparation.

Micropipetting

The pipet tips should be rinsed with deionized double-distilled water and rinsed in between solutions. Conserve clean tips after washing for further use. When droplets refuse to dispense, discard the tips. For viscous solutions either modify the matrix or use microboats and weigh the sample.

Furnace Cell

The cell has some potential contamination sources such as residual contamination from the previous analysis or from various parts of the cell itself, e.g., (1) high silica in blank, which could be from the o-rings; (2) contamination could be from stainless steel parts of the cell, if iron, nickel, chromium are found; or (3) if calcium is found, it could be from electroplated rhodium coated copper electrodes to the molybdenum pins. Periodic cleaning of the cell is recommended to avoid contamination.

Purge Gas

Primarily two gases, argon and nitrogen, are used for the FAAS method. Four parameters of the purge gas are important considerations for the analysis.

1.   Flowrate: Continuous flow versus pressurization and flow rate. Pressurization can be used
     A.   To enhance both peak height and peak area linearity,
     B.   To increase peak area sensitivity,
     C.   To increase residue time,
     D.   To retard atomization rate, and
     E.   To improve precision.
2.   Physical properties: The specific heat, molecular or atomic cross-section, and molecular or atomic weight of the purge gas affect the extent of furnace cuvette cooling by the gas and also the diffusion rate of the gas and the analyte atoms after atomization.

3. Chemical properties:
    A. Argon, the most commonly used and recommended purge gas, is an inert gas and does not react with the analyte or matrix at FAA temperatures.
    B. Use of nitrogen causes depression of titanium, vanadium, and barium in graphite atomizer.
    C. Use of hydrogen enhances barium and aluminum in tantalum strip atomizers.
4. Impurities: Oxygen is a common impurity in nitrogen, which is used to purge the cell for deaeration. The presence of oxygen affects analysis in the following ways:
    A. It reacts with and damages the substrate which may be metal or graphite,
    B. It reforms oxides with analyte, acting as a vapor phase interference,
    C. As a solid phase interference on the substrate, it depresses atomization by hindering the analyte reduction reactions.

High purity nitrogen containing less than 10 ppm oxygen can be used. If a chemical or nontheoretical interaction is observed, use argon as purge gas.

Some of the stable nitrides formed at higher than 600°C are AlN, Ba_3N_2, Be_3N_2, Ca_3N_2, CrN, GaN, Ge_3N_2, HfN, Li_3N, Mg_3N_2, NbN, PuN, Si_3N_4, Sr_3N_2, TaN, ThN, TiN, UN, VN, YN, and ZrN.

Background Correction

Most sample analyses at wavelengths shorter than 350 nm require background (nonspecific) absorption correction for accurate and precise analysis. Methods of compensating for matrix effects, such as a standard addition method, do not correct for nonspecific absorption. Background correction may be accomplished in several ways.

1. The correction lamp may be an element known to be present in the sample with a suitably intense absorption line extremely close (within 2 nm at least) to the line used for analysis.
2. The source may be a hydrogen or deuterium hollow cathode lamp which produces a continuous output of varying levels from 200 to 325 or 350 nm.
3. The source may be a deuterium arc, a much more intense version of the deuterium hollow cathode lamp, which requires a separate power supply.
4. It can be achieved manually, either by,
    A. Running a sample, checking the background, and then performing the analyses and manually subtracting the background from all the subsequent data, or by,
    B. Simultaneously recording the background and analytical signals and subtracting the appropriate individual background readings.
5. Another manual procedure is to look for adjacent nonabsorbing lines in the hollow cathode lamp. An easy way to check for them is to scan the spectrum with various sized slits, then check their absorption with an aqueous standard. The spectrum depends somewhat on the manufacturer of the lamp, which changes the hollow cathode material and impurities that may be present.
6. A proper alignment of the lamp is also required for background correction. Proper function requires that the HCl and continuum beams travel the same path through the graphite cuvette. A quick way to align the two beams is to line up the cuvette by adjusting height, depth and rotational controls on the beam from one of the lamps and then move the other lamp until it is also maximum in the same physical position of the furnace cell.
7. Use the widest slit possible without harming the analysis. The principle on which background with a continuum source is based is simply that the wavelength which the analyte absorbs is very narrow, so that while the analyte will absorb very well the analytical line from the element HCl, it will absorb the continuum light very little, if at all. The background, however, will absorb both the analyte and the continuum radiation, so it may thus be subtracted.

Use of narrow slits results in over-corrections because the analyte absorbs a greater percentage of the continuum radiation, which is then subtracted as if it were background.

8. Dilute the sample or pyrolyze carefully to obtain less than one absorbance of rapidly changing background signal for precision analysis. For maximum precision, the background signal should be less than 1 absorbance in FAAS.

9. Alter the pyrolysis for thorough background removal without analyte loss. If the background is too high, add acids to decompose organic matter and reduce, if not eliminate, the background.

10. Most commercial instruments offer automatic background correction as an option. By using this mode, the instrument subtracts the background reading, point-by-point, instantaneously. This method is more accurate than the above mentioned ones.

Verification of Methods and Analytical Results

There are several ways to check a newly developed method for FAAS analysis.

1. Recovery studies
2. Standard additions
3. Artificial matrix simulation
4. Referee standards
5. An alternate analytical method

Sensitivity

Determination of an element carried out by a method is known to be sensitive if a small change of its concentration or weight causes a large change in the measurement signal. IUPAC recommended the use of "characteristic concentration," when sensitivities of different elements or different resonance lines of a single element have to be compared. As a general rule, sensitivities for the elements shown under the instrument set-up are those obtained using the most sensitive absorption wavelength.

Linearity

This is a function of absorbance and the relationship between the signal height and the amount or concentration of the element is given by the standard calibration plot. This relationship is linear only up to a certain signal height, depending on the selected resonance line, slit width, light source, rate of heating and graphite tube.

Sample Volume

For absolute sensitivities and the absorbance that a volume of sample at a particular concentration will give is calculated from the equation

$$A = \frac{4.4 \times V \times C}{S}$$

where A = sample absorbance, V = sample volume in μL, C = sample concentration in $\mu g/mL$, and S = sensitivity in picogram.

Calibration Techniques

In furnace atomic absorption the amount of an analyte can be obtained in concentration terms. The instrument can also be calibrated in terms of weight of the analyte (picograms), since the solvent is removed from the analyte during the analysis. There are two methods which can be used for calibration.

Method #1

Use 10 μL aliquots of standard solution of an element of interest in 0.05, 0.1, 0.2, and 0.5 μg/L concentrations for analysis. Plot a calibration graph of absorbance versus concentration. Same size of sample aliquots can be used for analysis and direct measurements in concentration can be made by comparison.

Method #2

Calibration can also be obtained in weight units. Use various aliquots of standard in 5, 10, 20, and 50 μL of 0.1 μg/L standard solution for analysis. In this method, the amount of analyte in the unknown sample can be measured in weight terms.

$$\mu g/mL = \frac{W_1}{V_1 \times 10^3}$$

where W_1 = weight determined in picograms (pg) and V_1 = volume pipetted into furnace (μL).

If the solution is made by dissolution of solid sample, then the concentration of an analyte in the solid can be found by

$$\mu g/g = \frac{W_1 \times V_2}{V_1 \times W_2 \times 10^3}$$

where V_2 = volume sample made up to (in mL) and W_2 = original sample weight (in g).

If the solid sample was directly analyzed in weight, then the concentration of the analyte is calculated as follows

$$\mu g/g = \frac{W_1}{W_3 \times 10^3}$$

where, W_3 is the sample weight (mg) in the furnace.

Sample Standardization

The best practice for sample standardization is to use standards that match the sample as closely as possible in matrix. The most dependable and commonly recommended method of standardization is to use samples already analyzed by a well-established technique as standards.

Standard Addition Method

This method is applied only when the calibration curve is linear over the region of interest, and background correction is made. This method compensates for physical and chemical interferences of the sample matrix by the addition of a known quantity of standard to the sample. Absorbance is measured for the sample first and then for the sample with the added aliquots of the standard. Three or four measurements are sufficient to provide greater accuracy.

The main advantage of this method is that matrix matching is achieved automatically. In this method, the first sample aliquot is diluted to volume, while the second and third are spiked with a standard solution and then diluted to a similar volume. Samples should be spiked in such a way that the second solution has a concentration of approximately twice that of the unknown, and the third, three times that of the unknown. The absorbance of each solution is plotted on a calibration curve and the concentration of the analyte is graphically derived by extrapolation to zero absorbance on the negative axis.

Dilution Method

In this method, absorbance A_1, is measured from a volume V_1, of a solution containing a known concentration C_1, of the element to be determined. A volume V_2 of the unknown sample containing a concentration C_2 of the element is added and absorbance is measured. The unknown concentration C_2 can be calculated as

$$C_2 = \frac{A_2\,(V_1 + V_2)\,(V_2 - A_1\,V_1/V_2)}{A_1}$$

Usually two measurements are enough for analysis. This method is rapid and can be used for very small samples.

Precision in FAAS

Some of the factors that can influence FAAS precision are described below.

1. Change in physical characteristics of the graphite tube or cuvette degrades the precision. The rate of degradation is dependent on the atomization temperature and the type of sample.
2. Drying, charring, and atomization temperatures and time of these steps during the analysis.
3. Purity of the purge gas.
4. Reproducible injection or placement of the sample in the furnace.
5. Use of strong acids degrades the furnace tube thus resulting in poor precision.
6. Organic solvents also cause poor precision.
7. Chemical nature of the sample. Extra care should be taken for analyzing organo-metallic samples.
8. Background absorption (1.0 absorbance) will affect precision due to incomplete compensation by the background correction system.
9. Repeated use of micropipette tips causes loss of precision.

Characterization of the Atomization Signal

Under the optimum FAAS conditions, the signal produced should be a "clear" sharp peak for any analyte determination, but sometimes peak of other characteristics are also observed. Some of these are described below.

Broad Peak — These peaks are caused by the elements with a tendency to form carbides. Other reasons could be low gas flow, low atomization temperature or background absorption.

Blank Peaks — If a peak is obtained during a blank cycle, it could be due to low atomization temperature, short time for atomization, contamination of graphite tube, etc.

Multiple Peaks — These peaks could be due to "memory" effects or due to the element being present as more than one species.

Calculations

The equations below describe the determination of quantities required for the preparation of standard solutions and the determination of the concentration of the element of interest in the original sample matrix from its concentration in the sample solution

Standard Solutions

1. From stock solution of higher concentration:

$$\text{mL of stock solution required} = \frac{(\text{conc. of dilute standard})(\text{vol. of dilute standard})}{(\text{conc. of stock solution})}$$

2.   From the metal:

$$\text{g of metal} = (C)\ (V)\ (10^{-6})$$

3.   From solid or liquid chemicals:

$$\text{g of chemical} = \frac{(C)\ (V)\ (F)\ (10^{-4})}{(A)\ (N)\ (P)}$$

4.   From liquid chemicals:

$$\text{mL of chemicals} = \frac{(C)\ (V)\ (F)\ (10^{-4})}{(A)\ (N)\ (P)\ (G)}$$

where C = concentration of the standard solution in mg/L, V = volume of the standard solution in mL, F = formula weight of the chemical used, A = atomic weight of the element of interest, N = number of atoms of the element of interest per molecule, P = purity of the chemical used in weight percent, and G = specific gravity of the liquid chemical used.

Samples:

Liquid samples	Element (mg/L) = (C) (d.f.)
Solid samples	Element (μg/g) = $\dfrac{(C)\ (V)\ (d.f.)}{(W)}$

where C= concentration of the element in the sample solution in mg/L, V = volume of the undiluted sample solution in mL, W = sample weight in grams, d.f. = dilution factor, if used, as described below, and d.f. = (volume of diluted sample solution in mL)/(volume of aliquot taken for dilution in mL).

To convert values of absorbance for percent absorption or vice versa, one can make use of Table 1.

Interferences

The literature on FAAS is replete with reports of interferences, and these analytical interferences can have severe effects on the accuracy and precision of the analyte determination. Some of these interferences have been categorized as follows:

Solid Phase Interferences — The sample drying process can cause solid phase interferences. When a sample is dispensed into the graphite cuvette, the size and shape of the crystals formed during the drying stage affect the efficiency of the atomization process. Use of an automatic pipette and autosampler will eliminate this problem.

Vapor Phase Interferences — The absence of temperature equilibrium during atomization can cause vapor phase interferences. Without temperature equilibrium, different compounds or states of a given element may require different atomization temperatures, so that analytical results are highly dependent upon the type and condition of the sample.

Use of microboat or L'vov platform will eliminate this interference. The atoms are formed in a hotter environment which tends to reduce the vapor phase interferences.

Chemical Interferences — The degree of chemical interference depends on both the sample matrix and the analyte. If the analyte has a relatively high boiling point, higher temperatures can be used during the pyrolysis stage to remove matrix interferences. Interferences are severe for

TABLE 1
Values of Absorbance for Percent Absorption

To convert percent absorption (%A) to absorbance, find the percent absorption to the nearest whole digit in the left-hand column; read across to the column located under the tenth of a percent desired, and read the value of absorbance. The value of absorbance corresponding to 26.8% absorption is thus 0.1355.

%A	0.0	0.1	0.2	0.3	0.4	0.5	0.6	0.7	0.8	0.9
0.0	0.000	0.0004	0.0009	0.0013	0.0017	0.0022	0.0026	0.0031	0.0035	0.0039
1.0	0.0044	0.0048	0.0052	0.0057	0.0061	0.0066	0.0070	0.0074	0.0079	0.0083
2.0	0.0088	0.0092	0.0097	0.0101	0.0106	0.0110	0.0114	0.0119	0.0123	0.0128
3.0	0.0132	0.0137	0.0141	0.0146	0.0150	0.0155	0.0159	0.0164	0.0168	0.0173
4.0	0.0177	0.0182	0.0186	0.0191	0.0195	0.0200	0.0205	0.0209	0.0214	0.0218
5.0	0.0223	0.0227	0.0232	0.0236	0.0241	0.0246	0.0250	0.0255	0.0259	0.0264
6.0	0.0269	0.0273	0.0278	0.0283	0.0287	0.0292	0.0297	0.0301	0.0306	0.0311
7.0	0.0315	0.320	0.0325	0.0329	0.0334	0.0339	0.0343	0.0348	0.0353	0.0357
8.0	0.0362	0.0367	0.0372	0.0376	0.0381	0.0386	0.0391	0.0395	0.0400	0.0405
9.0	0.0410	0.0414	0.0419	0.0424	0.0429	0.0434	0.0438	0.0443	0.0448	0.0453
10.0	0.0458	0.0462	0.0467	0.0472	0.0477	0.0482	0.0487	0.0491	0.0496	0.0501
11.0	0.0506	0.0511	0.0516	0.0521	0.0526	0.0531	0.0535	0.0540	0.0545	0.0550
12.0	0.0555	0.0560	0.0565	0.0570	0.0575	0.0580	0.0585	0.0590	0.0595	0.0600
13.0	0.0605	0.0610	0.0615	0.0620	0.0625	0.0630	0.0635	0.0640	0.0645	0.0650
14.0	0.0655	0.0660	0.0665	0.0670	0.0675	0.0680	0.0685	0.0691	0.0696	0.0701
15.0	0.0706	0.0711	0.0716	0.0721	0.0726	0.0731	0.0737	0.0742	0.0747	0.0752
16.0	0.0757	0.0762	0.0768	0.0773	0.0778	0.0783	0.0788	0.0794	0.0799	0.0804
17.0	0.0809	0.0814	0.0820	0.0825	0.0830	0.0835	0.0841	0.0846	0.0851	0.0857
18.0	0.0862	0.0867	0.0872	0.0878	0.0883	0.0888	0.0894	0.0899	0.0904	0.0910
19.0	0.0915	0.0921	0.0926	0.0931	0.0937	0.0942	0.0947	0.0953	0.0958	0.0964
20.0	0.0969	0.0975	0.0980	0.0985	0.0991	0.0996	0.1002	0.1007	0.1013	0.1018
21.0	0.1024	0.1029	0.1035	0.1040	0.1046	0.1051	0.1057	0.1062	0.1068	0.1073
22.0	0.1079	0.1085	0.1090	0.1096	0.1101	0.1107	0.1113	0.1118	0.1124	0.1129
23.0	0.1135	0.1141	0.1146	0.1152	0.1158	0.1163	0.1169	0.1175	0.1180	0.1186
24.0	0.1192	0.1198	0.1203	0.1209	0.1215	0.1221	0.1226	0.1232	0.1238	0.1244
25.0	0.1249	0.1255	0.1261	0.1267	0.1273	0.1278	0.1284	0.1290	0.1296	0.1302
26.0	0.1308	0.1314	0.1319	0.1325	0.1331	0.1337	0.1343	0.1349	0.1355	0.1361
27.0	0.1367	0.1373	0.1379	0.1385	0.1391	0.1397	0.1403	0.1409	0.1415	0.1421
28.0	0.1427	0.1433	0.1439	0.1445	0.1451	0.1457	0.1463	0.1469	0.1475	0.1481
29.0	0.1487	0.1494	0.1500	0.1506	0.1512	0.1518	0.1524	0.1530	0.1537	0.1543
30.0	0.1549	0.1555	0.1561	0.1568	0.1574	0.1580	0.1586	0.1593	0.1599	0.1605
31.0	0.1612	0.1618	0.1624	0.1630	0.1637	0.1643	0.1649	0.1656	0.1662	0.1669
32.0	0.1675	0.1681	0.1688	0.1694	0.1701	0.1707	0.1713	0.1720	0.1726	0.1733
33.0	0.1739	0.1746	0.1752	0.1759	0.1765	0.1772	0.1778	0.1785	0.1791	0.1798
34.0	0.1805	0.1811	0.1818	0.1824	0.1831	0.1838	0.1844	0.1851	0.1858	0.1864
35.0	0.1871	0.1878	0.1884	0.1891	0.1898	0.1904	0.1911	0.1918	0.1925	0.1931
36.0	0.1938	0.1945	0.1952	0.1959	0.1965	0.1972	0.1979	0.1986	0.1993	0.2000
37.0	0.2007	0.2013	0.2020	0.2027	0.2034	0.2041	0.2048	0.2055	0.2062	0.2069
38.0	0.2076	0.2083	0.2090	0.2097	0.2104	0.2111	0.2118	0.2125	0.2132	0.2140
39.0	0.2147	0.2154	0.2161	0.2168	0.2175	0.2182	0.2190	0.2197	0.2204	0.2211
40.0	0.2218	0.2226	0.2233	0.2240	0.2248	0.2255	0.2262	0.2269	0.2277	0.2284
41.0	0.2291	0.2299	0.2306	0.2314	0.2321	0.2328	0.2336	0.2343	0.2351	0.2358
42.0	0.2366	0.2373	0.2381	0.2388	0.2396	0.2403	0.2411	0.2418	0.2426	0.2434
43.0	0.2441	0.2449	0.2457	0.2464	0.2472	0.2480	0.2487	0.2495	0.2503	0.2510
44.0	0.2518	0.2526	0.2534	0.2541	0.2549	0.2557	0.2565	0.2573	0.2581	0.2588
45.0	0.2596	0.2604	0.2612	0.2620	0.2628	0.2636	0.2644	0.2652	0.2660	0.2668
46.0	0.2676	0.2684	0.2692	0.2700	0.2708	0.2716	0.2725	0.2733	0.2741	0.2749
47.0	0.2757	0.2765	0.2774	0.2782	0.2790	0.2798	0.2807	0.2815	0.2823	0.2832
48.0	0.2840	0.2848	0.2857	0.2865	0.2874	0.2882	0.2890	0.2899	0.2907	0.2916
49.0	0.2924	0.2933	0.2941	0.2950	0.2958	0.2967	0.2976	0.2984	0.2993	0.3002
50.0	0.3010	0.3019	0.3028	0.3036	0.3045	0.3054	0.3063	0.3072	0.3080	0.3089

TABLE 1
Values of Absorbance for Percent Absorption (continued)

To convert percent absorption (%A) to absorbance, find the percent absorption to the nearest whole digit in the left-hand column; read across to the column located under the tenth of a percent desired, and read the value of absorbance. The value of absorbance corresponding to 26.8% absorption is thus 0.1355.

%A	0.0	0.1	0.2	0.3	0.4	0.5	0.6	0.7	0.8	0.9
51.0	0.3098	0.3107	0.3116	0.3125	0.3134	0.3143	0.3152	0.3161	0.3170	0.3179
52.0	0.3188	0.3197	0.3206	0.3215	0.3224	0.3233	0.3242	0.3251	0.3261	0.3270
53.0	0.3279	0.3288	0.3298	0.3307	0.3316	0.3325	0.3335	0.3344	0.3354	0.3363
54.0	0.3372	0.3382	0.3391	0.3401	0.3410	0.3420	0.3429	0.3439	0.3449	0.3458
55.0	0.3468	0.3478	0.3487	0.3497	0.3507	0.3516	0.3526	0.3536	0.3546	0.3556
56.0	0.3565	0.3575	0.3585	0.3595	0.3605	0.3615	0.3625	0.3635	0.3645	0.3655
57.0	0.3665	0.3675	0.3686	0.3696	0.3706	0.3716	0.3726	0.3737	0.3747	0.3757
58.0	0.3768	0.3778	0.3788	0.3799	0.3809	0.3820	0.3830	0.3840	0.3851	0.3862
59.0	0.3872	0.3883	0.3893	0.3904	0.3915	0.3925	0.3936	0.3947	0.3958	0.3969
60.0	0.3979	0.3990	0.4001	0.4012	0.4023	0.4034	0.4045	0.4056	0.4067	0.4078
61.0	0.4089	0.4101	0.4112	0.4123	0.4134	0.4145	0.4157	0.4168	0.4179	0.4191
62.0	0.4202	0.4214	0.4225	0.4237	0.4248	0.4260	0.4271	0.4283	0.4295	0.4306
63.0	0.4318	0.4330	0.4342	0.4353	0.4365	0.4377	0.4389	0.4401	0.4413	0.4425
64.0	0.4437	0.4449	0.4461	0.4473	0.4485	0.4498	0.4510	0.4522	0.4535	0.4547
65.0	0.4559	0.4572	0.4584	0.4597	0.4609	0.4622	0.4634	0.4647	0.4660	0.4672
66.0	0.4685	0.4698	0.4711	0.4724	0.4737	0.4750	0.4763	0.4776	0.4789	0.4802
67.0	0.4815	0.4828	0.4841	0.4855	0.4868	0.4881	0.4895	0.4908	0.4921	0.4935
68.0	0.4948	0.4962	0.4976	0.4989	0.5003	0.5017	0.5031	0.5045	0.5058	0.5072
69.0	0.5086	0.5100	0.5114	0.5129	0.5143	0.5157	0.5171	0.5186	0.5200	0.5214
70.0	0.5229	0.5243	0.5258	0.5272	0.5287	0.5302	0.5317	0.5331	0.5346	0.5361
71.0	0.5376	0.5391	0.5406	0.5421	0.5436	0.5452	0.5467	0.5482	0.5498	0.5513
72.0	0.5528	0.5544	0.5560	0.5575	0.5591	0.5607	0.5622	0.5638	0.5654	0.5670
73.0	0.5686	0.5702	0.5719	0.5735	0.5751	0.5768	0.5784	0.5800	0.5817	0.5834
74.0	0.5850	0.5867	0.5884	0.5901	0.5918	0.5935	0.5952	0.5969	0.5986	0.6003
75.0	0.6021	0.6038	0.6055	0.6073	0.6091	0.6108	0.6126	0.6144	0.6162	0.6180
76.0	0.6198	0.6216	0.6234	0.6253	0.6271	0.6289	0.6308	0.6326	0.6345	0.6364
77.0	0.6383	0.6402	0.6421	0.6440	0.6459	0.6478	0.6498	0.6517	0.6536	0.6556
78.0	0.6576	0.6596	0.6615	0.6635	0.6655	0.6676	0.6696	0.6716	0.6737	0.6757
79.0	0.6778	0.6799	0.6819	0.6840	0.6861	0.6882	0.6904	0.6925	0.6946	0.6968
80.0	0.6990	0.7011	0.7033	0.7055	0.7077	0.7100	0.7122	0.7144	0.7167	0.7190
81.0	0.7212	0.7235	0.7258	0.7282	0.7305	0.7328	0.7352	0.7375	0.7399	0.7423
82.0	0.7447	0.7471	0.7496	0.7520	0.7545	0.7570	0.7595	0.7620	0.7645	0.7670
83.0	0.7696	0.7721	0.7747	0.7773	0.7799	0.7825	0.7852	0.7878	0.7905	0.7932
84.0	0.7959	0.7986	0.8013	0.8041	0.8069	0.8097	0.8125	0.8153	0.8182	0.8210
85.0	0.8239	0.8268	0.8297	0.8327	0.8356	0.8386	0.8416	0.8447	0.8477	0.8508
86.0	0.8539	0.8570	0.8601	0.8633	0.8665	0.8697	0.8729	0.8761	0.8794	0.8827
87.0	0.8861	0.8894	0.8928	0.8962	0.8996	0.9031	0.9066	0.9101	0.9136	0.9172
88.0	0.9208	0.9245	0.9281	0.9318	0.9355	0.9393	0.9431	0.9469	0.9508	0.9547
89.0	0.9586	0.9626	0.9666	0.9706	0.9747	0.9788	0.9830	0.9872	0.9914	0.9957

cadmium, lead, and zinc. During atomization, silicon (in high concentrations), tantalum and tungsten, form extremely thermally stable compounds which makes their determinations virtually impossible.

The general methods to reduce or eliminate interference effects have included, (1) matrix modification, (2) addition of various acids, (3) addition of organic compounds, (4) use of reactive gases, (5) coating of the furnace tubes, and (6) vaporization from a platform placed within the furnace tube.

REFERENCES

1. **Walsh, A.,** Australian Patent 23041, 1953.
2. **Walsh, A.,** *Spectrochim. Acta,* 7, 108, 1955.
3. **Alkemade, C. T. J. and Milatz, J. M. W.,** *Appl. Sci. Res. Sec. B,* 4, 289, 1955.
4. **Alkemade, C. T. J. and Milatz, J. M. W.,** *J. Opt. Soc. Am.,* 45, 583, 1955.
5. **Van Loon, J. C.,** *Analytical Atomic Absorption Spectroscopy-Selected Methods,* Academic Press, N.Y., 1980.
6. **L'vov, B. V.,** *Inzh.-Fizh. Zh.,* 2, 44, 1959.
7. **L'vov, B. V.,** *Spectrochim. Acta,* 17, 761-70, 1961.
8. **Fuller, C. W.,** *Electrothermal Atomization for Atomic Absorption Spectroscopy,* Anal. Monogr. Ser. No. 4, Chemical Society, London, 1977.
9. **L'vov, B. V.,** *Spectrochim. Acta,* 39B, 149-58, 1984.
10. **Chakrabarti, C. L., Wan, C. C., Chang, S. B., Hamed, H. A., and Bertels, P. C.,** *Recent Advances in Analytical Spectroscopy,* Fuwa, K., Ed., Pergamon Press, Oxford, U.K., 1982, 103.
11. **Alkemade, C. T. J.,** *Appl. Opt.,* 7, 1261, 1968.
12. **Harnly, J. M. et al.,** *Anal. Chem.,* 51, 2007, 1979.
13. **Harnly, J. M., Kane, J. S., and Miller-Ihli, N. J.,** *Appl. Spectrosc.,* 36, 637, 1982.
14. **Harnly, J. M. and Kane, J. S.,** *Anal. Chem.,* 56, 48, 1984.
15. **Harnly, J. M.,** *Fresnius' Z. Anal. Chem.,* 323, 759, 1986.
16. **Retzik, M. and Bass, D.,** *Am. Lab.,* 70, Sept., 1988.
17. **Wendt, R. H. and Fassel, V. A.,** *Anal. Chem.,* 38, 337, 1966.
18. **Robinson, J. W.,** *Anal. Chim. Acta.,* 27, 465, 1962.
19. **Marinkovic, M., Bojovic, B., and Oesic, D.,** *13th Colloq. Spectrosc. Int.,* Hilger & Watts, London, 1967, 1181.
20. **Belyaev, U. I., Ivantsov, L. M., Karyakin, A. V., Phi, P. H., and Shemet, V. V.,** *Zh. Anal. Khim.,* 23, 508, 1968.
21. **Kantor, T. and Erdey, L.,** *Spectrochim. Acta,* 24B, 283, 1969.
22. **Marinkovic, M. and Vickers, T. J.,** *Appl. Spectrosc.,* 25, 319, 1971.
23. **Malykh, V. D., Men'shikov, V. I., Morozov, V. N., and Shipitsyn, S. A.,** *Zh. Prikl. Spektrosk.,* 16, 12, 1972.
24. **Dorofeev, V. S., Fishman, I. I., and Fishman, I. S.,** *Pribl. Tekh. Eksp.,* No. 3, 203, 1973.
25. **Hatch, W. R. and Ott, W. L.,** *Anal. Chem.,* 40, 285, 1968.
26. **Nelson, L. S. and Kuebler, N. A.,** *Spectrochim. Acta,* 19, 781, 1963.
27. **Mossotti, V. S., Lagua, K., and Hagenah, W. D.,** *Spectrochim Acta,* 23B, 197, 1967.
28. **Slavin, W.,** *Atomic Absorption Spectroscopy,* John Wiley & Sons, N.Y., 1968, 48.
29. **Petukh, M. L., Shirokanov, A. D., and Yankovskii, A. A.,** *Zh. Prikl. Spektrosk.,* 32, 414, 1980.
30. **Venghiattis, A. A.,** *At. Absorp. Newsl.,* 6, 19, 1967.
31. **Holak, W.,** *Anal. Chem.,* 41, 1712, 1969.
32. **Nakashima, S.,** *Bunseki Kagaku,* 6, 418, 1982.
33. **Gatehouse, B. M. and Walsh, A.,** *Spectrochim. Acta,* 16, 602, 1960.
34. **Goleb, J. A.,** *Anal. Chem.,* 35, 1978, 1963.
35. **Goleb, J. A. and Yokoyama, Y.,** *Anal. Chim. Acta,* 30, 213, 1964.
36. **Massman, H.,** *Spectrochim Acta,* 23B, 215, 1968.
37. **Paschen, F.,** *Ann. Physik,* 50, 901, 1916.
38. **Crosswhite, H. M., Dieke, G. H., and Legagneur, C. S.,** *J. Opt. Soc. Am.,* 45, 270, 1955.
39. **Gibson, J. H., Grossman, W. E. L., and Cooke, W. D.,** *Anal. Chem.,* 35, 266, 1963.
40. **Fassel, V. A. and Massotti, V. G.,** *Anal. Chem.,* 35, 252, 1963.
41. **Ivanov, N. P. and Kozyreva, N. A.,** *Zh. Anal. Khim.,* 19, 1266, 1964.
42. **Ginsburg, V. L. and Satarina, G. P.,** *Zavod. Lab.,* 31, 249, 1965.
43. **Massotti, V. G. and Fassel, V. A.,** *Spectrochim. Acta,* 20, 1117, 1964.

44. DeGalan, L., McGee, W. W., and Winefordner, J. D., *Anal. Chim. Acta,* 37, 436, 1967.
45. Barnett, W. B. and Ediger, R. D., *At. Absorp. Newsl.,* 17, 125, 1978.
46. Barnett, W. B. and Cooksey, M. M., *At. Absorp. Newsl.,* 18, 61, 1979.
47. Beaty, M., Barnett, W., and Grobenski, Z., *At. Spectrosc.,* 1, 72, 1980.
48. Bayunov, P. A., Savin, A. S., and L'vov, B. V., *At. Spectrosc.,* 3, 161, 1982.
49. de Loos-Vollebgret, M. T. C., Oosterlin, R. A. M., Boudewijn, F. B., and de Galan, L., *At. Spectrosc.,* 4, 160, 1983.
50. Kingston, H. M. and Jessie, L. B., Eds., *Introduction to Microwave Sample Preparation: Theory and Practice,* American Chemical Society, Washington, D.C., 1988
51. Matusiewicz, H. and Sturgeon, R. E., Present status of microwave sample dissolution and decomposition for elemental analysis, *Progr. Analyst. Spectrosc.,* 12, 21-39, 1989.

AUTHOR INDEX

GENERAL INSTRUMENTATION

The atomic absorption spectrophotometry with a graphite furnace atomizer features a small sample volume and high sensitivity. A rapid progress in the development of instrumentation seen in the early years seems to have been stabilized. More attention has now been concentrated on the improvement of light source, atomizers, and data handling services. Furnace atomic absorption spectroscopy, being a single element analysis technique, also has a disadvantage in being low in measurement efficiency. This disadvantage has now been considerably improved by the introduction of simultaneous multi-element analysis. Multi-channel/multi-element atomic absorption analysis has been eagerly awaited, but difficulty in keeping the baseline stable for a long duration has been a major problem in the realization of this analysis. The new instruments reported by the Analyte Corporation, Oregon, Perkin-Elmer Corporation, California, and Hitachi Ltd., Japan, ensure accelerated qualitative and quantitative analyses and amazingly improved efficiency in elemental profiling.

There are some standard specifications for a furnace atomic absorption spectrophotometer. A typical unit should have a photometric system, wavelength range, wavelength correction, diffraction grating, focal length dispersion, and a slit and photo detector. A complete state-of-the-art instrument system should have the following components.

1. Optical System: Optical temperature control unit, lamp turret for using hollow cathode lamps for determining more than one element in a sample.
2. Graphite Atomizer Unit: Heating control, temperature programming, cooling system, and cuvette or tube (accessory).
3. Photometric Unit: Calibration curve, curvature correction, standard addition capabilities, etc.
4. Signal Processing/Display Unit: Background correction, display signals, calibration curves, measurement of enhancing accuracy, etc.
5. Communication Capability: RS-232C, bidirectional communication, and thermosensitive graphic printer.
6. Optional Accessories: Autosampler function with capability of pipetting volume, turn table, diluent, etc.

In this section, progress in development of a spectrophotometric unit and its accessories are referenced. The references begin with the earliest attempt in development of the instrument and technique and end with the latest information. At the end, the author and subject indices are provided for cross-referencing purposes.

GENERAL REFERENCES

1. **Gatehouse, B. M. and Walsh, A.,** Analysis of metallic samples by atomic absorption spectroscopy, *Spectrochim. Acta,* 16, 602, 1960.
2. **L'vov, B. V.,** Theory and method of atomic absorption analysis *Zavodsk. Lab.,* 28, 931, 1962.
3. **L'vov, B. V.,** Possibility of absolute analysis using atomic absorption spectra, *Tr. Gos. Inst. Prikl. Khim.,* 49, 256, 1962.
4. **L'vov, B. V.,** Determination of the absolute values of oscillator strengths by combined measurements of the total and linear absorption of vapor layers in graphite cell, *Optics and Spectroscopy,* 19, 507, 1965 (in Russian).
5. **Veghiattis, A. A. and Whitlock, L.,** Determination of metal ores by solid sampling, *At. Ab. Newsl.,* 6, 135, 1967.
6. **Venghiattis, A. A.,** A new method for the atomization of solid samples for atomic absorption spectroscopy, *Spectrochim. Acta,* 23B, 67, 1967.
7. **L'vov, B. V. and Lebedev, G. G.,** A contact method of pulsed electrode heating in atomic absorption spectroscopy with a graphite cell, *Zh. Prikl. Spektrosk.,* 7, 264, 1967.
8. **Thiel, R.,** Atomic absorption — theory and applications, *Bull. Centre Rech. PAU-SNPA,* 1, 207, 1967 (in German).
9. **Venghiattis, A.,** A technique for the direct sampling of solids without prior dissolution, *At. Ab. Newsl.,* 6, 19, 1967.
10. **Woodriff, R., Stone, R. W., and Held, A. M.,** Electrothermal atomization for atomic absorption analysis, *Appl. Spectrosc.,* 22, 408, 1968.
11. **Woodriff, R. and Stone, R. W.,** Hot tube atomic absorption spectrochemistry, *Appl. Opt.,* 7, 1337, 1968.
12. **Slavin, W.,** *Atomic Absorption Spectroscopy,* John Wiley & Sons, NY, 1968.
13. **L'vov, B. V., Katskov, D. A., and Lebedev, G. G.,** The integral absorption method with specimen atomization in a graphite cell, *Zh. Prikl. Spektrosk.,* 9, 558, 1968.
14. **Rubeska, I. and Moldan, B.,** The use of heated Fuwa tubes, *Appl. Opt.* 7, 1341, 1968.
15. **Massman, H.,** Studies of atomic absorption and atomic fluorescence in a graphite cell, *Spectrochim. Acta,* 23B, 215, 1968 (in German).
16. **L'vov, B. V., Kabanova, M. A., Katskov, D. A., Lebedev, G. G., and Sokolov, M. A.,** Spectral background correction for atomic absorption spectroscopy with a graphite cuvete, *J. Appl. Spectrosc.,* 8, 200, 1968 (in Russian).
17. **L'vov, B. V.,** A graphite cuvette for atomic absorption spectroscopy, *J. Appl. Spectrosc.,* 8, 517, 1968.
18. **L'vov, B. V. and Khartsyzov, A. D.,** Atomic absorption determination of phosphorus in a graphite cuvet, *Zh. Prikl. Spektrosk.,* 11, 9, 1969.
19. **L'vov, B. V. and Khartsyzov, A. D.,** Atomic absorption determination of iodide in a graphite cuvet, *Zh. Anal. Khim.,* 24, 799, 1969.
20. **L'vov, B. V.,** The potentialities of the graphite crucible method in atomic absorption spectroscopy, *Spectrochim. Acta,* 24B, 53, 1969.
21. **Katskov, D. A., Lebedev, G. G., and L'vov, B. V.,** Spectrophotometer for atomic absorption measurements with a graphite cuvette, *Zavod. Lab.,* 35, 1001, 1969.
22. **Borzov, V. P., L'vov, B. V., and Plyushch, G. V.,** Atomic absorption analysis of liquids and solids by vaporizing samples in the flame of a graphite oven and measuring integral absorption, *Zh. Prikl. Spektrosk.,* 10, 217, 1969.
23. **L'vov, B. V.,** Progress in atomic absorption spectroscopy employing flame and graphite cuvette techniques, *Pure Appl. Chem.,* 23, 11, 1970.
24. **Manning, D. C. and Fernandez, F.,** Atomization for atomic absorption using a heated graphite tube, *At. Ab. Newsl.,* 9, 65, 1970.
25. **L'vov, B. V. and Khartsyzov, A. D.,** Radial loss of vapors during atomic absorption measurements in a graphite cuvette, *Zh. Anal. Khim.,* 25, 1824, 1970.
26. **L'vov, B. V., Kruglicova, L. P., and Plyushch, G. V.,** Optimization of atomic absorption measurement conditions with a furnace flame atomizer, *Zh. Prikl. Spektrosk.,* 15, 975, 1971.
27. **Katskov, D. A. and L'vov, B. V.,** Optimum conditions for recording analytical signals during atomic absorption measurements using a graphite cuvette, *Zh. Prikl. Spektrosk.,* 15, 783, 1971.
28. **Woodriff, R. and Shrader, D.,** Furnace atomic absorption with reference channel, *Anal. Chem.,* 43, 1918, 1971.
29. **Segar, D. A. and Gonzalez, J. G.,** Greater flexibility with the Perkin-Elmer HGA-70 heated graphite atomizer for use in selective volatilization analysis, *At. Ab. Newsl.,* 10, 94, 1971.
30. **Mashireva, L. G. and Seregin, N. V.,** Simplified variant of a graphite cuvet for atomic absorption analyses, *Khim. Tekhnol. Masel.,* 16, 62, 1971.
31. **Kahn, H. L. and Slavin, S.,** Static method of analysis with graphite furnace (HGA-70), *At. Ab. Newsl.,* 10, 125, 1971.
32. **Kahn, H. L.,** Graphite furnace applications in atomic absorption, *Am. Lab.,* 3, 35, 1971.

33. **Hwang, J. Y., Ullucci, P. A., and Smith, S. B., Jr.,** A simple flameless atomizer, *Am. Lab.,* 3, 41, 1971.
34. **Talmi, Y. and Morrison, G. H.,** Induction furnace method in atomic spectrometry, *Anal. Chem.,* 44, 1455, 1972.
35. **Molnar, C. J., Reeves, R. D., Winefordner, J. D., Glenn, M. T., Ahlstrom, J. R., and Savory, J.,** Construction and evaluation of a versatile graphite filament atomizer for atomic absorption spectrometry, *Appl. Spectrosc.,* 26(6), 606, 1972.
36. **Slavin, S.,** An atomic absorption bibliography for Jan-Jun. 1971, *At. Ab. Newsl.,* 11, 74, 1972.
37. **L'vov, B. V.,** Use of a graphite cell in atomic absorption spectroscopy. A survey, *J. Appl. Spectrosc.,* 321, 1972.
38. **Fernandez, F. J.,** Accessory "grooved" tubes for the graphite furnace, *At. Ab. Newsl.,* 11, 123, 1972.
39. **Kuzovlev, I. A., Sverdlina, O. A., and Kovykova, N. V.,** Methods for raising the detection limit of atomic absorption analysis using a graphite cuvet, *Zh. Prikl. Spektrosk.,* 16, 5, 1972.
40. **Amos, M. D.,** Nonflame atomization in AAS-a current review, *Am. Lab.,* 4, 57, 1972.
41. **Shimazu, M., Takubo, Y., and Yoshii, M.,** Atomic absorption spectroscopy with a tunable dye laser, *Oyo Butsuri,* 42, 1234, 1973.
42. **Slavin, S.,** An atomic absorption bibliography for Jan-June 1972, *At. Ab. Newsl.,* 12, 77, 1973.
43. **Kerber, J. D., Russo, A. J., Peterson, G. E., and Ediger, R. D.,** Performance improvements with the graphite furnace, *At. Ab. Newsl.,* 12, 106, 1973.
44. **Kerber, J. D., Koch, A., and Peterson, G. E.,** The direct analysis of solid samples by atomic absorption using a graphite furnace, *At. Ab. Newsl.,* 12, 104, 1973.
45. **Slavin, S.,** An atomic absorption bibliography for July-December 1972, *At. Ab. Newsl.,* 12, 9, 1973.
46. **Reeves, R. D., Patel, B. M., Molnar, C. J., and Winefordner, J. D.,** Decay of atom populations following graphite rod atomization in atomic absorption spectroscopy, *Anal. Chem.,* 45, 246, 1973.
47. **Campbell, W. C. and Ottaway, J. M.,** Formation of atoms in carbon furnace atomizers in atomic absorption spectroscopy, *Proc. Soc. Anal. Chem.,* 11, 161, 1974.
48. **Yanagisawa, M. and Takeuchi, T.,** Flameless atomic absorption spectroscopy with glassy carbon strip atomizer, *Bunseki Kagaku,* 23, 364, 1974.
49. **Woodriff, R.,** Atomization chambers for atomic absorption spectrochemical analysis. Review., *Appl. Spectrosc.,* 28, 413, 1974.
50. **Willis, J. B.,** Atomic absorption, atomic fluorescence and flame emission spectroscopy. *Handbook of Spectroscopy,* Robinson, J. W., Ed., CRC Press, Boca Raton, FL, 1974.
51. **Walsh, A.,** Atomic absorption spectroscopy — stagnant or pregnant?, *Anal. Chem.,* 46, 698A, 1974.
52. **Slavin, S.,** An atomic absorption bibliography for Jan-June 1974, *At. Ab. Newsl.,* 13, 84, 1974.
53. **Siemer, D. D., Woodriff, R., and Watne, B.,** A simple technique for coating carbon atomic absorption atomizer components with pyrolytic carbon, *Appl. Spectrosc.,* 28, 582, 1974.
54. **Nikolaev, G. I. and Podgornaya, V. I.,** Effectiveness of the use of substance B in a graphite cuvette during atomic absorption analysis, *Zh. Prikl. Spektrosk.,* 21, 593, 1974.
55. **Montaser, A. and Crouch, S. R.,** Analytical applications of the graphite braid nonflame atomizer, *Anal. Chem,* 46, 1817, 1974.
56. **Lundgren, G., Lundmark, L., and Johansson, G.,** Temperature controlled heating of the graphite tube atomizer in flameless atomic spectrometry, *Anal. Chem.,* 46, 1028, 1974.
57. **Kantor, T., Clayborn, S. A., and Veillon, C.,** Continuous sample introduction with graphite atomization systems for atomic absorption spectroscopy, *Anal. Chem.,* 46, 2205, 1974.
58. **Kaegler, S. H. and Erdoel Kohle, E.,** Atomic absorption spectroscopy, *Brennst. Chem.,* 27, 514, 1974.
59. **Johnson, G. W. and Skogerboe, R. K.,** A simple device for renewing contact surfaces on carbon rod atomizers, *Appl. Spectrosc.,* 28, 590, 1974.
60. **Hwang, J. Y. and Thomas, G. P.,** New generation flameless atomic absorption atomizer, *Am. Lab.,* 6, 55, 1974.
61. **Grushiko, L. F., Ivanov, N. P., and Chupahkin, M. S.,** Analytical possibilities of an atomizer with a tantalum ribbon as vaporizer, *Zh. Anal. Khim.,* 29, 1842, 1974.
62. **Clayburn, S. A., Kantor, T., and Veillon, C.,** Pyrolysis treatment for graphite atomization systems, *Anal. Chem.,* 46, 2213, 1974.
63. **Slavin, S.,** An atomic absorption bibliography for July-Dec.1973, *At. Ab. Newsl.,* 13, 11, 1974.
64. **Pelieva, L. A., Muzykov, G. G., and Prushko, I. V.,** Furnace in flame atomizer studied on the Saturn atomic absorption spectrophotometer, *Zh. Prikl. Spektrosk.,* 20, 771, 1974.
65. **Findlay, W. J., Zdrojewski, A., and Quickert, N.,** Temperature measurements of a graphite furnace used in flameless atomic absorption, *Spectrosc. Lett.,* 7, 63, 1974.
66. **Ottaway, J. M. and Hough, D. C.,** Carbon furnace atomic absorption analysis of atmospheric particulates, *Proc. Anal. Div. Chem. Soc.,* 12, 319, 1975.
67. **Sturgeon, R. E., Chakrabarti, C. L., and Bartels, P. C.,** Atomization in graphite furnace atomic absorption spectroscopy. Peak height method vs integration method of measuring absorbance. Heated Graphite Atomizer 2100, *Anal. Chem.,* 47, 1250, 1975.

68. **Slavin, S. and Lawrence, D. M.**, An atomic absorption bibliography for Jan-June 1975, *At. Ab. Newsl.*, 14, 81, 1975.

69. **Issaq, H. J. and Zielinski, W. L., Jr.**, Modification of a graphite tube atomizer for flameless atomic absorption spectroscopy, *Anal. Chem.*, 47, 2281, 1975.

70. **Garnys, V. P. and Smythe, L. E.**, Fundamental studies on improvement of precision and accuracy in flameless atomic absorption spectroscopy using the graphite tube atomizer. Lead in whole blood, *Talanta*, 22, 881, 1975.

71. **Ediger, R. D.**, Atomic absorption with the graphite furnace using matrix modification, *At. Ab. Newsl.*, 14, 127, 1975.

72. **Surskii, G. A. and Avdeenko, M. A.**, Calibration of atomic absorption spectrophotometers during the measurement of the absorption of a fixed column of atomic vapor in a graphite furnace, *Zh. Prikl. Spektrosk.*, 22, 758, 1975.

73. **Thompson, K. C., Godden, R. G., and Thomerson, D. R.**, A method for the formation of pyrolytic graphite coatings and enhancement by calcium addition techniques for graphite rod flameless atomic absorption spectroscopy, *Anal. Chim. Acta*, 74, 289, 1975.

74. **Hendrikx-Jongerius, C. and De Galan, L.**, Practical approach to background correction and temperature programming in graphite furnace atomic absorption spectroscopy, *Anal. Chim. Acta*, 87, 259, 1976.

75. **Stoeppler, M. and Kampel, M.**, Long range tests with the autosampler AS-1 for automated sample injection into the HGA-74 and HGA-76 graphite furnace, Report, Kernforschungsanlage Julich GmbH, Nov. 1976.

76. **Bevan, D. G. and Kirkbright, G. F.**, The influence of operating parameters on the profile of the calcium 422.67 nm resonance line emitted by a demountable hollow cathode lamp, *Appl. Spectrosc.*, 30(2), 163, 1976.

77. **Epstein, M. S., Rains, T. C., and O'Haver, T. C.**, Wavelength modulation for background correction in graphite furnace atomic absorption spectroscopy, *Applied Spectrosc.*, 30(3), 324-329, 1976.

78. **Watne, B. and Woodriff, R.**, A very inexpensive temperature monitor for flameless AA apparatus, *Appl. Spectrosc.*, 30, 71, 1976.

79. **Cox, L. C.**, Solid-state power controller for electrothermal atomizers, *Appl. Spectrosc.*, 30, 225, 1976.

80. **Skudaev, Y. D., Shipitsin, S. A., and Morozov, V. N.**, Experience of using a graphite boat for atomic absorption analysis by the graphite furnace-flame method, *Zh. Prikl. Spektrosk.*, 25, 771, 1976.

81. **Sturgeon, R. E., Chakrabarti, C. L., and Langford, C. H.**, Studies on the mechanism of atom formation in graphite furnace and atomic absorption spectroscopy, *Anal. Chem.*, 48, 1792, 1976.

82. **Slavin, S. and Lawrence, D. M.**, An atomic absorption bibliography for Jan.-June 1976, *At. Ab. Newsl.*, 15, 77, 1976.

83. **Pelieva, L. A., Muzykov, G. G., and Sharnopol'skii, A. I.**, Optimization of measuring conditions using a graphite capsule atomizer in a flame, *Zh. Prikl. Spektrosk.*, 25, 414, 1976.

84. **Ottaway, J. M. and Shaw, F.**, Ionization interferences in carbon furnace atomic absorption and atomic emission spectroscopy, *Analyst*, 101, 582, 1976.

85. **Ottaway, J. M.**, Atom formation and interferences in flame and carbon furnace atomic spectroscopy, *Proc. Anal. Div. Chem. Soc.*, 13, 185, 1976.

86. **Hoshino, Y. and Utsunomiya, T.**, Application of an analog integrator to the measurement of signal peak areas in graphite furnace atomizer atomic absorption spectroscopy, *Nippon Kagaku Kaishi*, 11, 1781, 1976.

87. **Fuller, C. W.**, A kinetic theory of atomization for atomic absorption spectroscopy with a graphite furnace. IV. Assessment of interference effects, *Analyst*, 101, 798, 1976.

88. **Welz, B.**, Precision and accuracy in atomic absorption with graphite tube furnaces, *Z. Anal. Chem.*, 279, 103, 1976.

89. **Slavin, S. and Lawrence, D. M.**, An atomic absorption bibliography for July-Dec. 1975, *At. Ab. Newsl.*, 15, 7, 1976.

90. **Regan, J. G. and Warren, J.**, A novel approach to the elimination of matrix interferences in flameless atomic absorption spectroscopy using a graphite furnace, *Analyst*, 101, 220, 1976.

91. **Pritchard, M. W. and Reeves, R. D.**, Non-atomic absorption from matrix salts volatilized from graphite atomizers in atomic absorption spectroscopy, *Anal. Chim. Acta*, 82, 103, 1976.

92. **Manning, D. C. and Ediger, R. D.**, Pyrolysis graphite surface treatment for HGA-2100 sample tubes, *At. Ab. Newsl.*, 15, 42, 1976.

93. **Dittrich, K.**, Atomic spectroscopic trace analysis in AIIIBV semiconductor microsamples. I. Volatility and nonspecific absorption of AIIIBV compounds and their components in the graphite cuvette, *Talanta*, 24, 725, 1977.

94. **Woodriff, R., Marinkovic, M., Howald, R. A., and Eliezer, I.**, Sample-loss mechanism in a constant temperature graphite furnace, *Anal. Chem.*, 49, 2008, 1977.

95. **Van den Broek, W. M. G. T. and de Galan, L.**, Supply and removal of sample vapor in graphite thermal atomizers, *Anal. Chem.*, 49, 2176, 1977.

96. **Sturgeon, R. E., Chakrabarti, C. L., and Bartels, P. C.**, Atomization under pressure in graphite furnace atomic absorption spectroscopy, *Spectrochim. Acta*, 32B, 231, 1977.

97. **Sturgeon, R. E. and Chakrabarti, C. L.**, The temperature of atomic vapor in graphite furnace atomic absorption spectroscopy, *Spectrochim. Acta*, 32B, 257, 1977.

98. **Sturgeon, R. E. and Chakrabarti, C. L.,** Mechanism of atom loss in graphite furnace atomic absorption spectroscopy, *Anal. Chem.,* 49, 1100, 1977.

99. **Sturgeon, R. E.,** Factors affecting atomization and measurement in graphite furnace atomic absorption spectroscopy, *Anal. Chem.,* 49, 1255A, 1977.

100. **Slavin, S. and Lawrence, D. M.,** An atomic absorption bibliography for Jan.-June 1977, *At. Ab. Newsl.,* 16, 89, 1977.

101. **Shcherbakov, V. I., Zorov, N. B., and Belyaev, Y. I.,** Processes in flameless atomizers for atomic absorption analysis and their effect on the size of analytical signal. I. Thermochemical features of atomizers. Comparison of properties of graphite and metallic surfaces of atomization, *Vestn. Mosk. Univ. Ser. 2 Khim.,* 18, 246, 1977.

102. **Matousek, J. P.,** Aerosol deposition in furnace atomization, *Talanta,* 24, 315, 1977.

103. **Lindahl, P. C.,** HGA-2100 graphite furnace baseplate modification for the models 403, 503 and 603 atomic absorption spectrophotometers, *At. Ab. Newsl.,* 16, 113, 1977.

104. **Kirkbright, G. F. and Snook, R. D.,** The use of volatile organic solvents as internal standards to improve reproducibility of sample introduction in AAS using electrothermal atomization, *At. Ab. Newsl.,* 49, 1636, 1977.

105. **Hoshino, Y., Utsunomiya, T., and Fukui, K.,** Graphite furnace atomic absorption utilizing selective concentration onto tungsten wire, *Nippon Kagaku Kaishi,* 6, 808, 1977.

106. **Harnly, J. M. and O'Haver, T. C.,** Background correction for the analysis of high-solid samples by graphite furnace atomic absorption, *Anal. Chem.,* 49, 2187, 1977.

107. **Gregoire, D. C. and Chakrabarti, C. L.,** Atomization from a platform in graphite furnace atomic absorption spectroscopy, *Anal. Chem.,* 49, 2018, 1977.

108. **Grassam, E., Dawson, J. B., and Ellis, D. J.,** Application of the inverse Zeeman effect to background correction in electrothermal atomic absorption analysis, *Analyst,* 102, 804, 1977.

109. **Anand, V. D., Ducharme, D. M., Troxler, R. G., and Lancaster, M. C.,** A simple device for repetitive precise alignment of the graphite tube in the HGA-200, *At. Ab. Newsl.,* 16, 112, 1977.

110. **Alder, J. F. and Hickman, D. A.,** The influence of mineral acid and hydrogen peroxide matrices on elemental sensitivity in graphite furnace atomic absorption spectroscopy, *At. Ab. Newsl.,* 16, 110, 1977.

111. **Wegscheider, W., Knapp, G., and Spitz, H.,** Statistical investigations of interferences in graphite furnace atomic absorption spectroscopy. I. Methods and instrumentation, *Z. Anal. Chem.,* 283, 9, 1977.

112. **Sturgeon, R. E. and Chakrabarti, C. L.,** Evaluation of pyrolytic graphite coated tubes for graphite furnace atomic absorption spectroscopy, *Anal. Chem.,* 49, 90, 1977.

113. **Slavin, S. and Lawrence, D. M.,** An atomic absorption bibliography for July-Dec. 1976, *At. Ab. Newsl.,* 16, 4, 1977.

114. **Poldoski, J. E.,** Computer-assisted furnace atomic absorption spectrometric analysis, *Anal. Chem.,* 49, 891, 1977.

115. **Katskov, D. A., L'vov, B. V., Polzik, L. K., and Semenov, Y. V.,** Study of the formation of an absorbing layer of atoms in graphite furnaces during atomic absorption analysis, *Zh. Prikl. Spektrosk.,* 26, 598, 1977.

116. **Grushko, L. F., Krasil'shchik, V. Z., Lifshits, M. G., and Chupakhin, M. S.,** Some characteristics of a tube graphite atomizer for atomic absorption spectroscopy, *Zh. Anal. Khim.,* 32, 218, 1977.

117. **Epstein, M. S.,** A timing circuit to monitor baseline absorbance using the AS-1 graphite furnace auto-sampling system, *At. Ab. Newsl.,* 16, 75, 1977.

118. **Szyddlowski, F. J., Peck, E., and Bax, B.,** Optimization of pyrolytic coating procedures for graphite tubes used in AAS, *Appl. Spectrosc.,* 32, 402, 1978.

119. **Salin, E. D. and Ingle, J. D., Jr.,** Design and performance of a time multiple, multiple slit, multielement flameless atomic absorption spectrometer, *Atom. Spectrosc.,* 32(6), 579, 1978.

120. **Matousek, S. P. and Smythe, L. E.,** An experimental study of lithium furnace emission, *Appl. Spectrosc.,* 32(11), 54-56, 1978.

121. **Frigieri, P. and Trucco, R.,** Beam-shaped electrothermal graphite tube furnace for atomic absorption spectrophotometry, *Analyst,* 103, 1089, 1978.

122. **Beaty, R. D. and Cooksey, M. M.,** The influence of furnace conditions on matrix effects in graphite furnace atomic absorption, *At. Ab. Newsl.,* 17, 53, 1978.

123. **Walsh, A.,** Atomic spectroscopy. What next?, *At. Ab. Newsl.,* 17, 97, 1978.

124. **Van den Broek, W. M., De Galan, L., Matousek, J. P., and Czobik, E. J.,** The gas temperature inside graphite furnaces used for atomic absorption spectroscopy, *Anal. Chim. Acta,* 100, 121, 1978.

125. **Slavin, S. and Lawrence, D. M.,** An atomic absorption bibliography for Jan-June 1978, *At. Ab. Newsl.,* 17, 73, 1978.

126. **Rubeska, I. and Koreckova, J.,** Electrothermal atomizers in atomic absorption spectroscopy. I. Kinetics and mechanism of atomization, *Chem. Listy,* 72, 567, 1978.

127. **Montaser, A. and Mehrabzadeh, A. A.,** Atomic absorption spectroscopy with an electrothermal graphite braid atomizer, *Anal. Chem.,* 50, 1697,1978.

128. **L'vov, B. V.,** Electrothermal atomization — the way toward absolute methods of atomic absorption analysis, *Spectrochim. Acta,* 33B, 153,1978.

129. **Ivanov, B. N., Bukreev, Y. F., Alabichev, A. I., and Zolotavin, V. L.,** Use of a heat stabilizer in atomic absorption spectroscopy with flameless atomization of the sample, *Zh. Prikl. Spektrosk.,* 28, 782, 1978.

130. **Guiochon, G., Hircq, B., and Tailland, C.,** Effect of oxygen in atomic absorption spectroscopy with the use of a graphite rod and detection with a double argon flow cell, *Anal. Chim. Acta,* 99, 125, 1978.

131. **Faithful, N. T.,** Tungsten filaments as electrothermal atomizers in atomic absorption spectrophotometry. II. Choice of solvent, gases and nebulizer assembly, *Lab. Pract.,* 27, 25, 1978.

132. **Faithful, N. T.,** Tungsten filaments as electrothermal atomizers in atomic absorption spectrophotometry. III. Design of filament atomizer and power supply, *Lab. Pract.,* 27, 26, 1978.

133. **Chakrabarti, C. L.,** Electrothermal atomization in graphite furnace atomic absorption spectroscopy, *Can. J. Spectrosc.,* 23, 134, 1978.

134. **Zatka, V. J.,** Tantalum treated graphite atomizer tubes for atomic absorption spectroscopy, *Anal. Chem.,* 50, 538, 1978.

135. **Wall, C. D.,** The use of *in situ* pyrolytic coating with the HG-70 graphite furnace, *At. Ab. Newsl.,* 17, 61, 1978.

136. **Slavin, S. and Lawrence, D. M.,** An atomic absorption bibliography for July-Dec. 1977, *At. Ab. Newsl.,* 17, 7, 1978.

137. **Nakano, K. and Miura, T.,** Multifold concentration and improvement of the tube in carbon furnace atomic absorption spectrophotometry, *Bunseki Kagaku,* 27, 121, 1978.

138. **L'vov, B. V., Lelieva, L. A., and Sharnopol'skii, A. I.,** Eliminating the depressing effect of chlorides during atomic absorption analysis with a graphite furnace by excess lithium additions to the sample, *Zh. Prikl. Spektrosk.,* 28, 19, 1978.

139. **Lundberg, E.,** Application of a versatile drift-compensating digital peak reader to the direct atomization of solids in graphite furnace atomic absorption spectroscopy, *Appl. Spectrosc.,* 32, 276, 1978.

140. **Czobik, E. J. and Matousek, J. P.,** Interference effects in furnace atomic absorption spectrometry, *Anal. Chem.,* 50, 2, 1978.

141. **Torsi, G. and Desimoni, E.,** Electrostatic accumulation furnace for electrothermal atomic spectroscopy, *Anal. Lett.,* 12, 1361, 1979.

142. **Rubeska, I. and Koreskova, J.,** Electrothermal atomizers in atomic absorption spectroscopy. II. Disturbing effectts, *Chem. Listy,* 73, 1009, 1979.

143. **Manning, D. C., Slavin, W., and Myers, S.,** Sampling at constant temperature in graphite atomic absorption spectroscopy, *Anal. Chem.,* 51, 2375, 1979.

144. **Zsako, J.,** Factors controlling the shape and position of the absorption curves at the electrothermal rod atomizer. Reply to comments, *Anal. Chem.,* 51, 2040, 1979.

145. **Tsujino, R., Ikeda, M., and Musha, S.,** Effect of current and voltage-regulated supplies on temperature variation of graphite atomizer in atomic absorption spectrophotometry, *Appl. Spectrosc.,* 33, 518, 1979.

146. **Tominaga, M. and Umezaki, Y.,** Effect of organic solvent in graphite furnace atomic absorption spectroscopy, *Bunseki Kagaku,* 28, 495., 1979.

147. **Tessari, G. and Torsi, G.,** Factors controlling the shape and position of the absorption curves at the electrothermal rod atomizer, *Anal. Chem.,* 51, 2041, 1979.

148. **Tessari, G. and Torsi, G.,** Factors controlling the shape and position of the absorption curves at the electrothermal rod atomizer. Comments, *Anal. Chem.,* 51, 2039, 1979.

149. **Takada, T. and Nakano, K.,** Evaluation and application of internal standardization in atomic absorption spectroscopy with electrothermal atomization, *Anal. Chim. Acta,* 107, 129, 1979.

150. **Slovak, Z.,** Direct sampling of ion exchanger suspensions for atomic absorption spectroscopy with electro-thermal atomization, *Anal. Chim. Acta,* 110, 301, 1979.

151. **Norval, E., Human, H. G., and Butler, L. R.,** Carbide coating process for graphite tubes in eletrothermal atomic absorption spectroscopy, *Anal. Chem.,* 51, 2045, 1979.

152. **Lundberg, E. and Lundmark, L.,** Automatic gas control unit for electrothermal atomizers, *Chem. Biomed. Environ. Instrum.,* 9, 91, 1979.

153. **Lawrence, D. M.,** An atomic absorption bibliography for Jul.-Dec., 1978, *At. Ab. Newsl.,* 18, 18, 1979.

154. **Krasowski, J. A. and Copeland, T. R.,** Matrix interferences in furnace atomic absorption spectroscopy, *Anal. Chem.,* 51, 1843, 1979.

155. **Yasuda, K., Toda, S., Igarashi, C., and Tamura, S.,** Extraction system for solvent extraction-graphite furnace atomic absorption spectroscopy, *Anal. Chem.,* 51, 161, 1979.

156. **Wegscheider, W., Knapp, G., and Spitz, H.,** Sequential testing as an efficient screening method for interferences in routine analysis as applied to atomic absorption spectroscopy with flame and graphite furnace atomization, *Talanta,* 26, 25, 1979.

157. **Sychra, V., Kolihova, D., Vyskocilova, O., Hlavac, R., and Pueschel, P.,** Electrothermal atomization from metallic surfaces. I. Design and performance of a tungsten-tube atomizer, *Anal. Chim. Acta,* 105, 263, 1979.

158. **Vyskocilova, O., Sychra, V., Kolihova, D., and Pueschel, P.,** Electrothermal atomization from metallic surfaces. 2. Atom formation processes in the tungsten-tube atomizer, *Anal. Chim. Acta,* 105, 269, 1979.

159. **Katskov, D. A.,** Analysis of chemical processes on the surface of thermal atomizers in atomic absorption measurements, *Zh. Prikl. Spektrosk.,* 30, 612, 1979.

160. Garnys, V. P. and Smythe, L. E., Filament in furnace atomic absorption spectroscopy, *Anal. Chem.*, 51, 62, 1979.

161. Hoyler, W. C. and Atkinson, A., Retardation of surface adsorption of trace metals by competetive complexation, *Appl. Spectrosc.*, 33, 37, 1979.

162. Siemer, D. D., Inexpensive temperature feedback controller for the Varian 63 CRA, *Atom. Spectrosc.*, 33(6), 613, 1979.

163. De Loos-Vollerbregt, M. T. C. and De Galan, L., The shape of analytical curves in Zeeman atomic absorption spectroscopy I. Normal concentration range, *At. Spectrosc.*, 33(6), 616, 1979.

164. Falk, H., A theoretical analysis of the diffusion process in flameless atomizers, *Spectrochim. Acta*, 33B, 695, 1979.

165. Eates, G. J. and Taylor, D., An unusual cause of reduced tube life with the HGA-76B graphite furnace, *At. Ab. Newsl.*, 18, 60, 1979.

166. Barnett, W. B. and Cooksey, M. M., A study of graphite furnace peak shapes with a computer, *At. Ab. Newsl.*, 18, 61, 1979.

167. Alcock, N. W., A method for between-sample cleaning of the graphite furnace, *At. Ab. Newsl.*, 18, 37, 1979.

168. Ma, I., Chang, W., and Hsu, K., Pyrolytic graphite coated tube and its application in a graphite furnace for atomic absorption, *Fen Hsi Hua Hsueh*, 8, 462, 1980 (in Chinese).

169. Slavin, W. and Manning, D. C., The L'vov platform for furnace atomic absorption analysis, *Spectrochim. Acta*, 35B, 701, 1980.

170. Nagdaev, V. K. and Bukreev, Y. F., Formation of free atoms using a graphite rod atomizer in atomic absorption spectroscopy, *Zh. Prikl. Spektrosk.*, 33, 618, 1980.

171. Lawson, S. R. and Woodriff, R., Method for reduction of matrix interferences in a commercial electrothermal atomizer for atomic absorption spectroscopy, *Spectrochim. Acta*, 35B, 753, 1980.

172. Karwowska, R., Bulska, E., Barakat, K. A., and Hulanicki, A., Some sources of errors in atomic absorption spectroscopy with electrothermal atomization of samples in organic solvents, *Chem. Anal.*, 25, 1043, 1980.

173. Hoshino, Y., Utsunomiya, T., and Fukui, K., Graphite furnace atomic absorption spectroscopy utilizing selective concentration onto tungsten wire, *Rep. Ra. Lab. Eng. Mater. Tokyo Inst. Technol.*, 5, 109,1980.

174. Dittrich, K. and Wennrich, R., Atomic absorption spectroscopy by a combination of laser evaporation and electrothermal atomization, *Spectrochim. Acta*, 35B, 731, 1980.

175. Czobik, E. J. and Matousek, J. P., Application of electrodeposition on a tungsten wire to furnace atomic absorption spectroscopy, *Spectrochim. Acta*, 35B, 741, 1980.

176. Chamsaz, M., Sharp, B. L., and West, T. S., Comparison of sample introduction techniques with a continuously heated graphite furnace atomizer for atomic absorption spectrophotometry, *Talanta*, 27, 867, 1980.

177. Belyaev, Y. I., Sheherbakov, V. I., and Karyakin, A. V., Effect of macrocomponents in atomic absorption analysis with pulse electrothermal atomization on a graphite rod, *Zh. Anal. Khim.*, 35, 2074, 1980 (in Russian).

178. Sturgeon, R. S., Berman, S. S., and Kashyap, S., Microwave attenuation determination of electron concentrations in graphite and tantalum tube electrothermal atomizers, *Anal. Chem.*, 52, 1049, 1980.

179. Slavin, W., Myers, S. A., and Manning, D. C., Reduction of temperature variation in the atomic absorption graphite furnace, *Anal. Chim. Acta*, 117, 267, 1980.

180. Robinson, J. W. and Rhodes, L. J., Development of a two state AA thermal atomizer for metal speciation analysis, *Spectrosc. Lett.*, 13, 253, 1980.

181. Robert, R. V. D., Balaes, G., and Steele, T. W., Study of the measurement by electrothermal atomization and atomic absorption spectrophotometry of hydride-forming elements, *Natl. Inst. Metall., Repub. S. Afr.*, Rep. No. 2053, 1980.

182. L'vov, B. V., Novotny, I., and Pelieva, L. A., Determination of the heat of formation of carbon molecules from absorption and emission of the Swan band in a group furnace, *Zh. Prikl. Spektrosk.*, 32, 965, 1980.

183. L'vov, B. V., Bayunov. P. A., Patrov, I. B., and Polobeiko, T. B., Study of the atomization mechanism for IB and VIIIB subgroup elements in graphite tantalum-lined furnaces by atomic absorption spectroscopy, *Zh. Anal. Khim.*, 35, 1877, 1980.

184. Lawrence, D. M., An atomic absorption spectroscopy bibliography for Jan.-June 1980, *At. Spectrosc.*, 1, 94, 1980.

185. Khaligie, J., Ure, A. M., and West, T. S., Some observations on the mechanisms of atomization in atomic absorption spectroscopy with atom-trapping and electrothermal techniques, *Anal. Chim. Acta*, 117, 257, 1980.

186. Katskov, D. A., Study of the processes of formation of an absorbing layer of atoms and of the analytical signal in electrothermal atomic absorption analysis, *Zh. Prikl. Spektrosk.*, 33, 205, 1980.

187. Ivanov, B. N., Bukreev, Y. F., and Karyakin, A. V., Spatial distribution of atoms with different volatilities over a graphite rod atomizer in atomic absorption spectroscopy, *Zh. Anal. Khim.*, 35, 1036, 1980.

188. Haynes, R. D., Comparison of two modified Kjeldahl digestion techniques for multielement plant analysis with conventional wet and dry ashing method, *Commun. Soil Sci. Plant Anal.*, 11, 459, 1980.

189. De Galan, L. and Van Dalen, J. P. J., Atomic absorption spectroscopy, *Pharm. Weekbl.*, 115, 689, 1980.

190. **Chakrabarti, C. L., Wan, C. C., Hamed, H. A., and Bertels, P. C.,** The way to an absolute method of analysis by capacitive discharge technique in graphite furnace atomic absorption spectrophotometry, *Can. Res.,* 13, 31, 1980.

191. **Ali, S. L.,** At. absorption spectrophotometry — state of the art after 25 years, *Pharm. Ztg.,* 125, 450, 1980.

192. **Smets, B.,** Atom formation and dissipation in electrothermal atomization, *Spectrochim. Acta,* 35B, 33, 1980.

193. **Siemer, D. D. and Baldwin, J. M.,** Effects of slow instrumental responses on the accuracy of furnace atomic absorption spectrometric determinations, *Anal. Chem.,* 52, 295, 1980.

194. **Price, W. J., Dymott, T. C., and Whiteside, P. J.,** The use of graphite cups for introducing solid samples with an electrothermal atomizer, *Spectrochim. Acta,* 35B, 3, 1980.

195. **Lawrence, D. M.,** An atomic absorption bibliography for July-Dec., *At. Spectrosc.,* 1, 8, 1980.

196. **Kitagawa, K., Ide, Y., and Takeuchi, T.,** Spectroscopic determination of the degree of atomization in an electrothermal atomizer, *Anal. Chim. Acta,* 113, 21, 1980.

197. **Katskov, D. A., Grinshtein, I. L., and Kruglikova, L. P.,** Use of a carbon furnace-capsule atomizer for the atomic absorption analysis of high-purity powdered materials, *Zh. Prikl. Spektrosk.,* 32, 536, 1980.

198. **Johannson, A.,** Atomic absorption spectroscopy, *Kem. Tidskr.,* 92, 26, 1980.

199. **Ishibashi, W. and Kikuchi, R.,** Effects of various sheath gases in flameless atomic absorption using the graphite furnace atomizer, *Bunseki Kagaku,* 29, 165, 1980.

200. **Chakrabarti, C. L., Hamed, H. A., Wan, C. C., Li, W. C., Bertels, P. C., Gregoire, D. C., and Lee, S.,** Capacitive discharge heating in graphite furnace atomic absorption spectroscopy, *Anal. Chem.,* 52, 167, 1980.

201. **Beaty, M., Barnett, W., and Grobenski, Z.,** Techniques for analyzing difficult samples with the HGA graphite furnace, *At. Spectrosc.,* 1, 72, 1980.

202. **Thompson, K. C.,** Atomic absorption spectrophotometry 1979 version, *Methods Exam. Waters Assoc. Mater.,* 50, 1980.

203. **Young, E. F.,** A review of the spectrophotometer, *Opr. Spectra,* 14, 44-8, 1980.

204. **L'vov, B. V., Bayunov, P. A., and Ryabchuk, G. N.,** Macrokinetic theory of volatilization of substance in atomic absorption spectroscopy. Volatilization from the surface of electrothermal atomizers, *Zh. Anal. Khim.,* 36, 1877, 1981.

205. **L'vov, B. V. and Ryabchuk, G. N.,** Atomization mechanism of substances in electrothermal atomic absorption spectroscopy based on analysis of absolute process rates. Oxygen-containing compounds, *Zh. Anal. Khim.,* 36, 2085, 1981.

206. **Akman, S., Genc, O., and Balkis, T.,** Analysis of sample-loss mechanisms in HGA-74 graphite furnace, *Spectrochim. Acta,* 36B, 1121, 1981.

207. **Yasuda, M. and Murayama, S.,** Measurement of longitudinal atom density distributions in a graphite tube furnace using coherent forward scattering, *Spectrochim. Acta,* 36B, 641, 1981.

208. **Suzuki, M., Ohta, K., and Yamakita, T.,** Improved sensitivity using a microcomputer for electrothermal atomic absorption spectroscopy with a metal microtube, *Anal. Chim. Acta,* 133, 209, 1981.

209. **Slavin, W., Manning, D. C., and Carnrick, G. R.,** The stabilized temperature platform furnace, *At. Spectrosc.,* 2, 137, 1981.

210. **Slavin, W., Manning, D. C., and Carnrick, G.,** Effect of graphite furnace substrate materials on analyses by furnace atomic absorption spectroscopy, *Anal. Chem.,* 53, 1504, 1981.

211. **Saba, C. S., Rhine, W. E., and Eisentraut, K. J.,** Efficiencies of sample introduction systems for the transport of metallic particles in plasma emission and atomic absorption spectroscopy, *Anal. Chem.,* 53, 1099, 1981.

212. **L'vov, B. V., Bayumov, P. A., and Ryabchuk, G. N.,** Macrokinetic theory of sample vaporization in electrothermal atomic absorption spectroscopy, *Spectrochim. Acta,* 36B, 397, 1981.

213. **L'vov, B. V. and Bayunov, P. A.,** Macrokinetic theory of sample vaporization in atomic absorption spectroscopy. Vaporization in porous graphite furnaces, *Zh. Anal. Khim.,* 36, 837, 1981.

214. **Lundberg, E. and Frech, W.,** Influence of instrumental response time on interference effects in graphite furnace atomic absorption spectroscopy, *Anal. Chem.,* 53, 1437, 1981.

215. **Lawrence, D. M.,** An atomic absorption spectroscopy bibliography for Jan.-June 1981, *At. Spectrosc.,* 2, 101, 1981.

216. **Kaiser, M. L., Koirtyohann, S. R., Hinderberger, E. J., and Taylor, H. E.,** Reduction of matrix interferences in furnace atomic absorption with L'vov platform, *Spectrochim. Acta,* 36B, 773, 1981.

217. **Gancer, S. and Berndt, H.,** Basic investigations on background problems in the platinum-loop method in comparison to graphite furnace and flame atomic absorption spectrophotometry, *Talanta,* 28, 334, 1981.

218. **Erspamer, J. P. and Niemczyk, T. M.,** Convenient modifications to a Varian AA-6 atomic absorption spectrometer to allow use with a graphite furnace atomizer, *Appl. Spectrosc.,* 35, 512, 1981.

219. **Chakrabarti, C. L., Wan, C. C., Teskey, R. J., Chang, S. B., Hamed, H. A., and Bertels, P. C.,** Mechanism of atomization at constant temperature in capacitive discharge graphite furnace atomic absorption spectroscopy, *Spectrochim. Acta,* 36B, 427, 1981.

220. **Broekaert, J. A. C.,** Atomic absorption spectroscopic instrumentation: an inventory, *Spectrochim. Acta,* 36B, 931, 1981.

221. **Blinova, E. S., Guzeev, I. D., Nedler, V. V., and Khokhrin, V. M.,** Atomic absorption analysis with electrothermal atomization for the rare-metal industry, *Zavod. Lab.,* 47, 31, 1981.

222. **Welz, B.,** Atomic emission spectroscopy with ICP and atomic absorption spectroscopy with flame, graphite tube, hydride and cold vapor techniques. A comparison, *Chimia,* 35, 102, 1981.

223. **Sturgeon, R. E. and Berman, S. S.,** Analyte ionization in graphite furnace atomic absorption spectroscopy, *Anal. Chem.,* 53, 632, 1981.

224. **Maney, J. P. and Luciano, V. J.,** Time resolution of interferences in electrothermal atomic absorption spectroscopy, *Anal. Chim. Acta,* 125, 183, 1981.

225. **Lawrence, D. M.,** An atomic absorption spectroscopy bibliography for July-Dec. 80, *At. Spectrosc.,* 2, 22, 1981.

226. **Koirtyohann, S. R., Glass, E. D., and Lichte, F. E.,** Some observations in perchloric acid interferences in furnace atomic absorption, *Appl. Spectrosc.,* 35, 22, 1981.

227. **Fernandez, F. J., Beaty, M. M., and Barnett, W. B.,** Use of the L'vov platform for furnace atomic absorption applications, *At. Spectrosc.,* 2, 16, 1981.

228. **Falk, H.,** Apparatus for the electrothermic atomization of a sample for analysis, German (East) DD 152, 201 (Cl. G01N21/74), 18 Nov. 1981, Appl. 222,714, 18 Jul. 1980, 10pp.

229. **Price, W. J.,** Atomic absorption spectroscopy. Benchmark achievements of a decade, A*nnu. Rep. Anal. At. Spectrosc.,* 10, 1-30, 1981.

230. **Kim, C. H.,** Principle and application of atomic absorption spectroscopy, *Hwahak Kwa Kongop U. Chinbo,* 21, 366-81, 1981 (in Korean).

231. **Soffiantini, V.,** Atomic absorption spectrophotometry in perspective, *Analytika (Johannesburg),* Oct., 20-2, 1981. Published in Chemsa 7, 10, 1981.

232. **Jackson, K., Benedik, J., and Jackson, L.,** ASTM Spec. Tech. Publ., 760, 83-98, 1981.

233. **Sullivan, J. V.,** Lamps and sources for analytical atomic spectroscopy, *Prog. Anal. At. Spectrosc.,* 4, 311-40, 1981.

234. **Rcheulishvilia, A. N.,** Analytical possibilities of a graphite boat flame atomizer in atomic absorption spectroscopy, *Zh. Anal. Khim.,* 36, 2106-10, 1981 (in Russian).

235. **L'vov, B. V. and Ryabchuk, G. N.,** Atomization mechanism of substances in electrothermal atomic absorption spectroscopy based on analysis of absolute process rates, *Zh. Anal. Khim.,* 36, 2085-96, 1981 (in Russian).

236. **Koirtyohann, S. R., Glass, E. D., and Lichte, F. E.,** Some observations on perchloric acid interferences in furnace atomic absorption, *Appl. Spectrosc.,* 35(1), 22, 1981.

237. **Erspamer, J. E. and Niemczak, T. M.,** Vaporization of some chloride matrices in graphite furnace AAS, *Anal. Chem.,* 54, 538, 1981.

238. **Kurfuerst, U.,** Zeeman atomic absorption spectroscopy, *Nachr. Chem. Tech. Lab.,* 29, 854-8, 1981 (in German).

239. **Huber, B.,** Coated graphite tubes for atomic absorption spectroscopy, Ger. Offen. DE 3,010,717 (Cl. G01N 21/31) Appl. 20 March 1980, 8 pp, 15 Oct. 1981.

240. **Brunner, W., Heckner, H., and Sansoni, B.,** Fully automatic simultaneous operation of several atomic absorption spectrometers in the analytical service laboratory of a research center by use of a process computer, *Spektrometertagung (Vortr.),* 13, 99-119, 1980 published 1981 (in German).

241. **Yasuda, M. and Murayama, S.,** Measurement of longitudinal atom density distributions in a graphite tube furnace using coherent forward scattering, *Spectrochim. Acta,* 36B, 641-7, 1981.

242. **Richardson, R. T. and Rowston,** The influence of graphite on the reduction of some metal oxides in an argon atmosphere, *Proc. Eud. Symp. Therm. Anal.,* 2, 355-8, 1981.

243. **Katskov, D. A. and Burtseva, I. G.,** Broadening of the concentration range of elements being determined in atomic absorption spectroscopy by automatic control of graphite furnace temperature, *Zh. Anal. Khim.,* 36, 1895-902, 1981 (in Russian).

244. **Broekaert, J. A. C.,** Atomic absorption spectroscopic instrumentation: an inventory, *Spectrochim. Acta,* 36B, 931-41, 1981.

245. **Salmon, S. G.,** Effect of oxygen on analyte vaporization and gas phase reaction in electrothermally heated graphite atomizers, 260 pp, 1981. Avail. Univ. Microfilms Int., Order No. 8128682. From Diss. Abstr. Int. B, 42, 2821-2, 1981.

246. **Gano, J. T.,** Zeeman background correction with an electrothermal graphite braid atomizer, 189 pp, 1981. Avail. Univ. Microfilms. Int., Order No. 8126502. From Diss. Abstr. Int. B, 42, 2355, 1981.

247. **Akman, S., Genc, O., and Balkis, T.,** Analysis of sample-loss mechanisms in HGA-74 graphite furnace, *Spectrochim. Acta,* 36B, 1121, 1981.

248. **Skriba, M. C., Gockley, G. B., and Battaglia, J. A.,** Use of on-line atomic absorption in a power plant environment, *ASTM Spec. Tech. Publ.,* 742, 156-66, 1981.

249. **Fuwa, K. and Haraguchi, H.,** *Atomic Spectroscopy in Japan,* Pergamon Press, Oxford, UK, 1981.

250. **Slavin, W., Carnrick, G. R., and Manning, D. C.,** Magnesium nitrate as matrix modifier in the stabilized temperature platform furnace, *Anal. Chem.,* 54, 621, 1982.

251. **Salmon, S. G. and Holcombe, J. A.,** Alteration of metal release mechanisms in graphite furnace atomizers by chemisorbed oxygen, *Anal. Chem.,* 54, 630, 1982.

252. **Rayson, G. D. and Holcombe, J. A.,** Tin atom formation in a graphite furnace atomizer, *Anal. Chim. Acta,* 136, 249, 1982.

253. **Martinsen, I. and Langmyhr, F. J.,** Some observations on sulfuric acid reactions in electrothermal atomic absorption spectroscopy with graphite furnaces, *Anal. Chim. Acta,* 135, 137, 1982.

254. **Lawrence , D. M.,** An atomic absorption spectroscopy bibliography for July-Dec. 1981, *At. Spectrosc.,* 3, 13, 1982.

255. **Fazakas, J.,** Influence of purge gas flow rate on sensitivity of resonance and non-resonance lines of palladium in graphite furnace atomic absorption spectroscopy, *Spectrosc. Lett.,* 15, 211, 1982.

256. **Erspamer, J. P. and Niemizyk, T. M.,** Vaporization of some chloride matrices in graphite furnace atomic absorption spectroscopy, *Anal. Chem.,* 54, 538, 1982.

257. **Chakrabarti, C. L., Wan, C. C., Hamed, H. A., and Bertels, P. C.,** Matrix interferences in graphite furnace atomic absorption spectroscopy by capacitive discharge heating. Reply to Comments, *Anal. Chem.,* 54, 137, 1982.

258. **Bragin, G. Y. and Sadagov, Y. M.,** Spatial distribution of temperature in graphite-tubular furnaces of electrothermal atomizers, *Zh. Prikl. Spektrosk,* 36, 185, 1982.

259. **Bahreyni-Toosi, M. H., Dawson, J. B., and Ellis, D. J.,** Technique for reducing the cycle time in atomic absorption spectroscopy with electrothermal atomization, *Analyst,* 107, 124, 1982.

260. **Abbey, S.,** Matrix interferences in graphite-furnace atomic absorption spectroscopy by capacitive discharge heating. Comments, *Anal. Chem.,* 54, 136, 1982.

261. **Routh, M. W.,** Comments on instrumental response factors in graphite furnace atomic absorption spectroscopy, *Appl. Spectrosc.,* 36, 585-7, 1982.

262. **Matsumoto, H.,** Microdetermination of lead in hair samples by graphite-tube atomic absorption spectrophotometry after oxine-MIBK and APDC-MIBK extraction, *Sangyo Igaku,* 24, 298-304, 1982 (in Japanese).

263. **Koirtyohann, S. R. and Kaiser, M. L.,** Furnace atomic absorption — a method approaching maturity, *Anal. Chem.,* 54, 1515A, 1982.

264. **Scherbakov, V. I., Belyaev, Y. I., and Myasoedov, B. F.,** Radioactive isotopes in the optimization of the design of graphite rod atomizers for atomic absorption analysis, *Zh. Prikl. Spektrosk.,* 36, 893-7, 1982.

265. **Lawrence, D. M.,** An atomic absorption spectroscopy bibliography for Jan.-June 1982, *At. Spectrosc.,* 3, 93-119, 1982.

266. **Ottaway, J. M.,** A revolutionary development in graphite furnace atomic absorption, *At. Spectrosc.,* 3, 89-92, 1982.

267. **Fazakas, J.,** Influence of atomizer nature on precision in graphite furnace atomic absorption spectroscopy, *Anal. Lett.,* 15, 573, 1982.

268. **Jenke, D. R. and Woodriff, R.,** Application of the Woodriff constant temperature graphite furnace atomizer to atomic spectroscopy, *Am. Lab. (Fairfield, Ct),* 14, 14, 1982.

269. **Slavin, W. and Manning, D. C.,** Graphite furnace interferences, a guide to the literature, *Prog. Anal. At. Spectrosc.,* 5, 243-340, 1982.

270. **De Galan, L., Kornblum, G. R., and Deloos-Vollebregt, M. T. C.,** Automated atomic spectrometric analysis, *Recent Adv. Anal. Spectrosc.,* Proc. 9th Inst. Conf. At. Spectrosc., Fuwa, K., Ed., Pergamon Press, Oxford, UK, 1981, 33-50.

271. **Lawson, S. R.,** Effect of furnace design on atomic absorption and emission signals in spectrochemical analysis, 174 pp, 1981. Avail. Univ. Microfilms Int., Order No. DA8208538. From Diss. Abstr. Instr. B42, 4406, 1982.

272. **Holcombe, J. A., Rayson, G. D., and Akerlind, N., Jr.,** Time and spatial absorbance profiles within a graphite furnace atomizer, *Spectrochim. Acta,* 37B, 319-30, 1982.

273. **Schaller, K. H. and Zober, A.,** Renal excretion of toxicologically relevant metals in occupationally non-exposed individuals, *Aertz. Lab.,* 28, 209-14, 1982.

274. **Huber, B.,** Atomization device for atomic absorption spectroscopy (AAS), Ger. Offen. DE 3,047,445 (Cl. G01N21/74), 6 pp, 22 July 1982. Appl. 17 Dec. 1980.

275. **Lawson, S. R. and Woodriff, R.,** A double-walled furnace for reduction of matrix interferences in graphite furnace atomic absorption spectroscopy, Ger. Offen. DE 3,044,627 (Cl. G01N21/74), 16pp, 03 June1982. Appl. 27 Nov. 1980.

276. **Sperding, K. R.,** Determination of heavy metals in seawater and in marine organisms by flameless atomic absorption spectrophotometry. XV. Matrix effects in graphite tube atomizers and ways to overcome them, *Fresnius Z. Anal. Chem.,* 311, 656-64, 1982.

277. **Chakrabarti, C. L., Wan, C. C., Chang, S. B., Hamed, H. A., and Bertels, P. C.,** A new analytical technique-capacitive discharge graphite furnace atomic absorption spectroscopy, *Recent Adv. Anal. Spectrosc.,* Proc. 9th Int. Conf. At. Spectrosc., 1981, Fuwa, K., Ed., Pergamon Press, Oxford, UK, 1982, 103-17.

278. **Routh, M. W., Doidge, P. S., Chidzey, J., and Frary, B.,** Advances in graphite furnace atomization, *Am. Lab. (Fairfield, Ct),* 14, 80, 1982.

279. **Daidoji, H. and Tamura, S.,** Atomization using a tantalum boat in graphite furnace in atomic absorption spectroscopy, *Bunseki Kagaku,* 31, 217-18, 1982 (in Japanese).

280. **Fazakas, J.,** Influence of purge gas flow rate on sensitivity of resonance and nonresonance lines of palladium in graphite furnace atomic absorption spectroscopy, *Spectrosc. Lett,* 15, 21-38, 1982.

281. **Fazakas, J.,** Aerosol deposition as a means of preventing halogen acid interferences in graphite furnace atomic absorption spectroscopy, *Spectrosc. Lett.,* 15, 221-32, 1982.

282. **Ishibashi, W. and Kikuchi, R.,** Effect of various sheath gases on absorption vs time profiles in flameless atomic absorption, *Bunseki Kagaku,* 31, E143, 1982.

283. **Komarek, J. and Sommer, L.,** Organic complexing agents in atomic absorption spectroscopy. A review, *Talanta,* 29, 159-66, 1982.

284. **Huetsch, B.,** Electrothermal atomizer for flameless atomic absorption spectroscopy, Ger. Offen. DE 3,030, 424 (Cl. G01N21/31) 18 Mar., 1982. Appl. 12 Aug. 1980.

285. **Falk, H., Jaeckel, I., and Thiemann, H. J.,** Apparatus for the flameless atomization of a sample for analysis, Ger. (East) DD 151, 224 (Cl. G01N21/71), 10 pp, 08 Oct. 1982. Appl. 221, 825, 13 June 1980.

286. **Koizumi, H., Taiti, Y., Moriya, K., Harada, K., and Sato, K.,** Apparatus for atomizing a sample for flameless atomic absorption analysis, Ger. Offen. DE 3, 110, 783 (Cl. G01N21/01), 17 pp, 04 Feb. 1982. Appl.80/35,557, 19 March 1980.

287. **Bragin, G. Y. and Sadagov, Y. M.,** Spatial distribution of temperature in graphite tubular furnaces of electrothermal atomizers, *Zh. Prikl. Spektrosk.,* 36, 185-8, 1982 (in Russian).

288. **Kahn, H. L.,** AA or ICP? Each technique has its own advantages, *Ind. Res. Dev.,* 24, 156-60, 1982.

289. **Baranov, S. V. Baranova, I. V., and Ivanov, N. P.,** Spectral lamps for atomic absorption spectroscopy. Review, *Zh. Prikl. Spektrosk.,* 36, 357-69, 1982.

290. **Martinson, I. and Langmyhr, F. J.,** Some observations on sulfuric acid reactions in electrothermal atomic absorption spectroscopy with graphite furnaces, *Anal. Chim. Acta,* 135, 137-43, 1982.

291. **Lawrence, D. M.,** An atomic spectroscopy bibliography for July-Dec. 1981, *At. Spectrosc.,* 3, 113-35, 1982.

292. **Ravreby, M.,** Analysis of long-range bullet entrance holes by atomic absorption spectrophotometry and scanning electron microscopy, *J. Forensic Sci.,* 27, 92-112, 1982.

293. **Vollkopf, U., Grobenski, Z., and Welz, B.,** Method development and optimization in graphite tube AAS aided by high peak resolution, *Atomspektrom. Spurenanal., Vortr. Kolloq.,* Welz, B., Ed., Springer-Verlag, Weinheim, Germany, 1982, 373.

294. **Grobenski, Z., Lehmann, R., Welz, B., and Wiedeking, E.,** Standard conditions for graphite tube AAS, normals and exceptions, *Atomspektrom. Spurenanal., Vortr. Kolloq.,* Welz, B., Ed., Springer-Verlag, Weinheim, Germany, 1982, 363.

294. **Shimadzu Seisakusho Ltd.,** Atomic Absorption Spectrometer, Jpn. Kokai Tokkyo Koho JP 57, 157, 143 (82, 157, 143) (Cl. G01N21/31), 4 pp, 28 Sept. 1982. Appl. 81/43, 741, 24 March 1981.

295. **Frech, W. and Jonasson, S.,** A new furnace design for constant temperature electrothermal atomic absorption spectroscopy, *Spectrochim. Acta,* 37B, 104, 1982.

296. **Dulude, G. R. and Sotera, J. J.,** Survey reveals new applications of two-channel AA spectrometry, *Can. Res.,* 15, 21, 1982.

297. **Cresser, M. S. and Sharp, B. L., Eds.,** Annual reports on Analytical Atomic Spectroscopy, Vol. II., Reviewing 1981, 375 pp, 1982.

298. **Slavin, W. and Manning, D. C.,** The graphite probe constant temperature furnace, *Spectrochim. Acta,* 37B, 955, 1982.

299. **Zhang, Z., Li, S., Ma, Y., and Wu, Z.,** MT-1 Thermometer for the graphite furnace, *Huanjing Kexue ,* 3, 61, 1982 (in Chinese).

300. **Perkin-Elmer Ltd.,** Atomic absorption spectrometer employing Zeeman effect background correction, Jpn. Kokai Tokkyo Koho JP 57, 154, 036 (82, 154, 036) (Cl. G01N21/31), 20 pp , 22 Sept. 1982. U.S. Appl. 237, 199, 23 Feb. 1981.

301. **Routh, M. W., Doidge, P. S., Chidzey, J., and Frary, B.,** Advances in graphite furnace atomization, *Int. Lab.,* 12, 101, 1982.

302. **Giddings, R. C., Barrett, P., Barnett, W., and Kisslak, G.,** Data management system for atomic absorption spectroscopy, *At. Spectrosc.,* 3, 197, 1982.

303. **Bayunov, P. A., Savin, A. S., and L'vov, B. V.,** Automation of thermochemical studies using the atomic absorption method, *At. Spectrosc.,* 3, 161, 1982.

304. **Bennett, P.,** Intelligent automation of furnace atomic absorption analysis, *VIA, Varian Instrum. Appl.,* 16, 19, 1982.

305. **Nakahara, T.,** Analytical atomic spectroscopy coupled with hydride generation technique, *Bunseki ,* 12, 904, 1982 (in Japanese).

306. **Ebdon, L.,** *An Introduction to Atomic Absorption Spectroscopy. A Self-teaching Approach,* Heyden Press, London , 1982, 138 pp.

307. **Massman, H.,** The origin of systematic errors in background measurements in Zeeman atomic absorption spectroscopy, *Talanta*, 29, 1051, 1982.

308. **Smith, S. B., Jr. and Hieftje, G. M.,** Atomic absorption spectroscopy analysis system, Fr. Demande FR 2,501, 373 (Cl. G01 N 21/31), 20 pp, 10 Sept. 1982, U. S. Appl. 240, 542, 04 Mar. 1981.

309. **Dittrich, K.,** *Scientific Pocketbooks, Vol. 276: Chemistry Series: Atomic Absorption Spectroscopy,* Springer-Verlag, Berlin, 1982, 225 pp.

310. **Von Loewis, M., Henrion, G., and Gadow, P.,** Optimization of analyses using linear calibration curves, *Z. Chem.,* 22, 427, 1982.

311. **Cantle, J. E., Ed.,** *Techniques and Instrumentation in Analytical Chemistry, Vol. 5 Atomic Absorption Spectroscopy,* Elsevier, Amsterdam, 1982, 448 pp.

312. **Jenke, D. R. and Woodriff, R.,** Simultaneous emission/absorption analysis in constant temperature furnace atomic spectroscopy, *Appl. Spectrosc.,* 36, 686, 1982.

313. **Jenke, D. R. and Woodriff, R.,** Direct aerosol introduction in constant temperature furnace atomic absorption spectroscopy, *Appl. Spectrosc.,* 36, 657, 1982.

314. **Altman, E. L., Sveshnikov, G. B., Turkin, Y. I., and Sholupov, S. E.,** Zeeman atomic absorption spectroscopy, *Zh. Prikl. Spektrosk.,* 37, 709, 1982.

315. **Giri, S. K., Littlejohn, D., and Ottaway, J. M.,** Graphite-probe atomization for carbon furnace atomic absorption and atomic emission epectrometry, *Analyst (London),* 107, 1095, 1982.

316. Support Laboratory. A computer interface for a Perkin-Elmer 5000 atomic absorption instrument, 41 pp, Report, EPA-600/4-82-050, 1982. Avail. NTIS. From Gov. Rep. Announce. Index (U.S.), 82, 4418, 1982.

317. **Kureichik, K. P.,** Compensating for additive interferences by using noise subtraction in atomic absorption spectrophotometry, *Zh. Prikl. Spektrosk.,* 37, 476, 1982.

318. **Gregoire, D. C. and Chakrabarti, C. L.,** Atomization from a tantalum surface in graphite furnace atomic absorption spectroscopy, *Spectrochim. Acta,* 37B, 611, 1982.

319. **Human, H. G. C., Ferreira, N. P., Rademeyer, C. J., and Faure, P. K.,** Calculation of the dynamic temperature characteristics of a heated graphite tube used in electrothermal atomic absorption measurements, *Spectrochim. Acta,* 37, 593, 1982.

320. **Nikolaev, G. I. and Nemets, A. M.,** Atomic absorption spectroscopy in the study of metal vaporization, 151 pp, 1982 (in Russian).

321. **Nakahara, T., Tsujino, R., and Ikeda, M.,** Tantalum carbide coating for graphite furnace in electrothermal atomic absorption spectroscopy, *Bunko Kenkyu,* 31, 181, 1982 (in Japanese).

322. **L'vov, B. V. and Savin, A. S.,** Atomization of elements in electrothermal atomic absorption spectroscopy by solid-phase reduction of oxides with carbon, *Zh. Anal. Khim.,* 37, 2116, 1982.

323. **Aristarain, A. J., Delmas, R. J., and Briat, M.,** Snow chemistry on James Ross Island (Antarctic Peninsula), *JGR, J. Geophys. Res.,* Sect. C 87, 11, 4-12, 1982.

324. **L'vov, B. V. and Ryabchuk, G. N.,** Free oxygen content inside graphite furnace for electrothermal atomic absorption analysis, *Zh. Anal. Khim.,* 37, 2125, 1982 (in Russian).

325. **Slavin, W.,** Environmental trace analyses with the stabilized temperature platform furnace and Zeeman background correction, *Pergamon Ser. Environ. Sci.,* 7, 397, 1982.

326. **Huber, B.,** Device for atomic absorption analysis of a sample, Ger. Offen. DE 3,113,678 (Cl. G01N 21/71), 15 pp, 14 Oct. 1982. Appl. 04 April 1981.

327. **Shcherbakov, V. I., Belyaev, Y. I., Myasoedov, B. F., Marov, I. N., and Kalibichenko, N. B.,** Activation of a graphite surface by products of the carbonization of organic substances in atomic absorption spectroscopy with electrothermal atomization, *Zh. Anal. Khim.,* 37, 1717, 1982 (in Russian).

328. **Tittarelli, P., Ferrari, G., and Zerlia, T.,** Use of electrothermal atomizers for vapor-phase ultraviolet absorption spectroscopy, *At. Spectrosc.,* 3, 157, 1982.

329. **Halcombe, J. A. and Sheehan, M. T.,** Graphite furnace modification for second surface atomization, *Appl. Spectrosc.,* 36, 631, 1982.

330. **Dewalt, F. G.,** Design operation and optimization of an atomic absorption spectrometer with a constant temperature furnace, 115 pp, 1982. Avail. Univ. Microfilms Int., Order No. 8222523. From Diss. Abstr. Int. B 43, 1087, 1982.

331. **Broekaert, J. A. C.,** Atomic absorption spectroscopy instrumentation: Pye Unicam AA equipment, *Spectrochim. Acta,* 37B, 732, 1982.

332. **Frech, W., Zhou, N. G., and Lundberg, G.,** A critical study of some methods used to investigate atom formation processes in GFAAS, *Spectrochim. Acta,* 37B, 691, 1982.

333. **L'vov, B. V. and Ryabchuk, G. N.,** A new approach to the problem of atomization in electrothermal atomic absorption spectroscopy, *Spectrochim. Acta,* 37B, 673, 1982.

334. **Gil'mutdinov, A. K. and Fishman, I. S.,** Formation of the absorbing layer of atoms in semienclosed atomizers for atomic absorption spectroscopy. Small atomizers, *Zh. Prikl. Spektrosk.,* 37, 541, 1982 (in Russian).

335. **De Loos-Vollebregt, M. T. C. and De Galan, L.,** Correction for background absorption and stray radiation in a. c. modulated Zeeman atomic absorption spectroscopy, *Spectrochim. Acta,* 37B, 659, 1982.

336. **Koirtyohann, S. R. and Kaiser, M. L.,** Furnace atomic absorption—a method approaching maturity, *Anal. Chem.,* 54, 1515A, 1982.

337. **Erspamer, J. P. and Niemczyk, T. M.,** Vaporization of some chloride matrixes in graphite atomic absorption spectroscopy, *Anal. Chem.,* 54, 538-40, 1982.

338. **Chakrabarti, C. L., Wan, C. C., Hamed, H. A., and Bertels, P. C.,** Matrix interferences in graphite furnace atomic absorption spectroscopy by capacitive discharge heating. Reply to comments, *Anal. Chem.,* 54, 137, 1982.

339. **Abbey, S.,** Matrix interferences in graphite furnace atomic absorption spectroscopy by capacitive discharge heating. Comments, *Anal. Chem.,* 54, 136, 1982.

340. **Grobenski, Z., Lehmann, R., Tamm, R., and Welz, B.,** Improvements in graphite furnace atomic absorption microanalysis with solid sampling, *Mikrochim. Acta,* 1, 115-25, 1982.

341. **Salmon, S. G. and Holcombe, J. A.,** Alteration of metal release mechanisms in graphite furnace atomizers by chemisorbed oxygen, *Anal. Chem.,* 54, 630-4, 1982.

342. **Sudu, E.,** Software for analytical chemistry-calibration methods, *Bunseki,* 2, 70-6, 1982 (in Japanese).

343. **Slavin, W., Carnrick, G. R., and Manning, D. C.,** Magnesium nitrate as a matrix modifier in the stabilized temperature platform furnace, *Anal. Chem.,* 54, 621, 1982.

344. **Mingorance, M. D. and Lachica, M.,** Methods for electrothermal atomization in atomic absorption spectroscopy. Review., *An Edafol. Agrobiol.,* 41, 1533-61, 1982 (in Spanish).

345. **Puschel, P., Sychra, V., Kolihova, D., and Hlavic, R.,** Metallic atomizer for flameless atomic absorption spectrophotometry, Czech. CS 209, 959 (Cl. B01 J 19/00), 3 pp, 31 Aug. 1982. Appl. 79/5, 995, 05 Sept. 1979. Addn. to Czech. 174,728.

346. **Falk, H.,** Electrothermal atomizer for atomic spectroscopy, Ger. (East) DD 157, 280 (Cl. G01 N 21/74), 10 pp, 27 Oct. 1982.

347. Matsushita Electric Industrial Co. Ltd, Carbon atomizer for flameless atomic absorption spectroscopy, Jpn. Tokkyo Koho JP 5751, 618 (8251, 618) (Cl. G 01 N 21/74), 2 pp, 02 Nov. 1982. Appl. 75187, 741, 16 July 1975.

348. **Henn, K. H., Berg, R., and Hoerner, L.,** Using AAS in the chemical laboratory of a nuclear technology plant, *Atomspektrom. Spurenanal., Vortr. Kolloq.,* 553-9, 1981, published 1982.

349. **Widjaja, I.,** Atomic absorption spectroscopy. Problems and optimization, *Acta Pharm. Indones,* 7, 109, 1982.

350. **Adams, F.,** Techniques for bulk chemical analysis. II. Atomic spectroscopy techniques, *Comm. Eur. Communities (Rep.) EUR,* 7544, 43-89, 1983.

351. **Lomdahl. G. S., Norris, T., and Sullivan, J. V.,** Demountable cathode lamp for AA and AFS, *Am. Lab. (Fairfield, Ct),* 15, 66-71, 1983.

352. **Adams, M. J., Mitchell, M. C., and Ewen, G. J.,** A microcomputer system for processing data from a three-channel atomic absorption spectrometer, *Anal. Chim. Acta,* 149, 101, 1983.

353. **Pelieva, L. A., Bukhasntsova, V. G., and Davidyuk,** Temperature-time characteristics of graphite furnaces of electrothermal atomizers, *Zh. Prikl. Spektrosk.,* 38, 533, 1983.

354. **Lawrence, D. M.,** An atomic spectroscopy bibliography for July-Dec. 1982, *At. Spectrosc.,* 4, 10, 1983.

355. **Robinson, J. W. and Jowett, P. L. H.,** Metal speciation by atomic absorption spectroscopy. The two stage atomizer, *Spectrosc. Lett.,* 16, 159, 1983.

356. **Shabushing, J. G. and Hieftje, G. M.,** Microdrop sample application in electrothermal atomization for atomic absorption spectroscopy, *Anal. Chim. Acta,* 148, 181, 1983.

357. **Behreyni-Toosi, M. H. and Dawson, J. B.,** A new design of graphite furnace for rapid cycle electrothermal atomization atomic absorption spectroscopy, *Analyst (London),* 108, 225, 1983.

358. **De Galan, L., De Loos-Vollebregt, M. T. C., and Osterling, R. A. M.,** Glassy carbon tubes in electrothermal atomization-atomic absorption spectroscopy, *Analyst (London),* 108, 138, 1983.

359. **Shabushing, J. G.,** An investigation of methods to improve the performance of electrothermal atomization systems, 252 pp, 1982. Avail. Univ. Microfilms Inst. Order No. DA 8300868. From Diss. Abstr. Int. B, 43, 2542, 1983.

360. **Liddell, P. R.,** Sequential multielement analysis by automated AA, *Am. Lab. (Fairfield, Ct),* 15, 111, 1983.

361. **Gil'mutdinov, A. K. and Fishman, I. S.,** Formation of an absorbing layer of atoms in semienclosed atomizers for atomic absorption spectroscopy. Large atomizers, *Zh. Prikl. Spektrosk.,* 38, 208, 1983 (in Russian).

362. **Katskov, D. A.,** Current concepts on the mechanism of thermal atomization of substances in atomic absorption analysis (review), *Zh. Prikl. Spektrosk.,* 38, 181, 1983.

363. **Siemer, D. D. and Lewis, L. C.,** Characterization of two modified carbon rod atomizers for atomic absorption spectroscopy, *Anal. Chem.,* 55, 99, 1983.

364. **Kurfuerst, U., Rues, B., and Wachter, K. H.,** Studies on the analysis for heavy metals in solids by direct Zeeman atomic absorption spectroscopy. I. Automatic sampler for solids, *Fresnius' Z. Anal. Chem.,* 314, 1, 1983 (in German).

365. **Siemer, D. D.,** Furnace atomic absorption spectroscopy atomizer with independent control of volatilization and atomization conditions, *Anal. Chem.,* 55, 692, 1983.

366. **Busch, K. W. and Benton, L. D.,** Multiplex methods in atomic spectroscopy, *Anal. Chem.,* 55, 445A, 1983.

367. **Siemer, D. D.,** Furnace atomic absorption spectroscopy atomizer with independent control of volatilization and atomization conditions, *Anal. Chem.,* 55, 692, 1983.
368. **Sotera, J. J., Christian, L. C., Conley, M. K., and Kahn, M. L.,** Reduction of matrix interferences in furnace atomic absorption spectroscopy, *Anal. Chem.,* 55, 204, 1983.
369. **Fijalkowski, J.,** Worldwide development trends in analytical atomic spectroscopy in recent years, *Mater. Konwersatorium Spektrom. At. Emisyjnej, Absorp. Spektrom. Mas.,* 120, 39-49, 1983.
370. **Avram, N.,** *Atomic and Molecular Spectroscopy,* 167pp, 1983 (in Roman).
371. **Welz, B.,** *Atomic Absorption Spectroscopy* (in German), 3rd ed., Springer-Verlag Weinheim, West Germany, 1983, 527pp.
372. **Zeschmasr, B., Kurfuerst, U., Lahl, U., Gabel, B., Kokenge, M., Kozicki, R., Podbelski, A., and Stachel, B.,** Zeeman-AAS-Direct analysis of soil and liquid materials, *Heavy Met. Environ. Int. Conf. 4th,* 1, 253, 1983.
373. **Barbooti, M. M. and Jasim, F.,** A review of nonflame atomization for atomic absorption spectrochemical analysis. II. Fundamental aspects of work on graphite for. atomizers, *J. Iraqi Chem. Soc.,* 8, 19-48, 1983.
374. **Barbooti, M. M. and Jasim, F.,** A review of nonflame atomization for atomic absorption spectrochemical analysis. I. Atomizers *J. Iraqi Chem. Soc.,* 8, 1-17, 1983.
375. **Petrakiev, A.,** Critical analysis of the methods and apparatus for Zeeman atomic absorption spectral analysis (AASA), *God. Sofii, Univ. Kliment Okhridski, Fiz. Fak.,* 70-71, 62-9, 1979, published 1983.
376. **Mori, K.,** Atomic absorption spectroscopy. Pretreatment, *Kagaku to Kogyo (Osaka),* 57 (12), 486-91, 1983.
377. **Cresser, M. S. and Ebdon, L., Eds.,** Annual reports on Analytical atomic spectroscopy, Vol. 12, Reviewing 1982, Royal Society of Chemistry, London, UK, 1983, 404 pp.
378. **Shang, Z.,** Tantalum ring technique in flameless atomic absorption spectrometry, *Fenx. Huaxue,* 11(9), 701-720, 1983.
379. **Eaton, D. K. and Holcombe, J. A.,** Oxygen ashing attachment for a furnace atomizer power supply, *Anal. Chem.,* 55, 1821, 1983.
380. **Chakrabarti, C. L., Chang, S. B., and Roy, S. E.,** The role of carbon in atomization in graphite furnace atomic absorption spectroscopy, *Spectrochim. Acta,* 38B, 447, 1983.
381. **Smith, S. B., Jr. and Hieftje, G. M.,** A new background-correction method for atomic absorption spectroscopy, *Appl. Spectrosc.,* 37, 419, 1983.
382. **De Loos-Vollbregt, M. T. C., Osterling, R. A. M., Boudewjin, F. B., and De Galan, L.,** Simultaneous recording of analyte and background absorbance from the Perkin-Elmer Model 5000 spectrometer using the Model 3600 data station, *At. Spectrosc.,* 4, 160, 1983.
383. **Fazakas, J.,** Atomization under pressure as a means to control spectral line overlap in graphite furnace atomic absorption spectroscopy, *Spectrochim. Acta,* 38B, 455, 1983.
384. **Katskov, D. A.,** Current concepts on the mechanism of thermal atomization of substances in atomic absorption analysis, *Zh. Prikl. Spektrosk.,* 38, 181, 1983.
385. **Belyaev, Y. I., Khozhainov, Y. M., and Shcherbakov, V. I.,** Change in the state of a graphite furnace during atomic absorption analysis, *Zh. Anal. Khim.,* 38, 1135, 1983.
386. **Guecer, S., and Massman, H.,** Background interferences due to organic solvents in atomic absorption spectroscopy using graphite furnace, *Spectrochim. Acta,* 38B, 573-80, 1983.
387. **Sumitomo Electric Industries Ltd., Jpn.,** Carrier gas flow in flameless atomic absorption spectroscopy, Kokai Tokkyo Koho JP 58 37, 539 (83 37,539) (Cl. G01 N 21/31), 3 pp, 04 Mar. 1983. Appl. 81/135, 785, 29 Aug. 1981.
388. **Sumitomo Electric Industries Ltd., Jpn.,** Automated flameless atomic absorption spectroscopy, Kokai Tokkyo Koho JP 58 37, 541 (83 37, 541) (Cl. G01 N21/31), 3 pp, 04 Mar. 1983. Appl. 81/135, 787, 29 Aug. 1981.
389. **Sumimoto Electric Industries Ltd. , Jpn.,** Automated flameless atomic absorption spectroscopy, Kokai Tokkyo Koho JP 58 37, 542 (83 37, 542) (Cl. G01 N 21/31), 3 pp, 04 Mar. 1983. Appl. 81/135 788, 29 Aug. 1981.
390. **Lersmacher, B. and Knippenberg, W. F.,** Tubular cuvette for atomic absorption spectroscopy, Ger. Offen. DE 3, 208, 744 (Cl. G01 N 21/74), 12 pp, 22 Sept. 1983. Appl. 11 Mar. 1982.
391. **Findeisen, B.,** Device for flameless atomization, Ger. (East) DD 200, 590 (Cl. G01 N21/74), 6 pp., 18 May 1983. Appl. 232, 177, 29 Jul. 1981.
392. **Findeisen, B. and Dittrich, K.,** Graphite cuvette, Ger. (East) DD 200, 589 (Cl G01 N 21/74), 5 pp , 18 May 1983. Appl. 232, 178, 29 Jul. 1981.
393. **Hitachi Ltd. Jpn.,** Flameless atomizer for atomic absorption spectrometeer, Tokkyo Koho JP 58 14, 983 (83 14, 983) (Cl. G01 N 21/74), 4 pp, 23 Mar. 1983. Appl. 76/94, 611, 09 Aug. 1976.
394. **Hitachi Ltd., Jpn.,** Flameless atomizer for atomic absorption spectroscopy, Tokkyo Koho JP 58 41, 338, (83 41, 338) (Cl. G01 N 21/74), 4 pp, 10 Mar. 1983. Appl. 81/138, 485, 04 Sept. 1981.
395. **Berndt, H. and Messerschmidt, J.,** Electrically heated tungsten loop for sample preparation, sample intake and as a "platform" in furnace AAS, *Fresnius' Z. Anal. Chem.,* 316, 201-4, 1983 (in German).
396. **De Loos-Vollebregt, M. T. C., De Galan, L., Van Uffelen, J. W. M., Slavin, W., and Manning, D. C.,** Heating characteristics of glassy carbon tubes used in electrothermal atomization-atomic absorption spectroscopy, *Spectrochim. Acta,* 38B, 799-807, 1983.

397. Van Deijck, W., Roelofsen, A. M., Pieters, H. J., and Herber, R. F. M., Temperature-controlled electrothermal atomization atomic absorption spectroscopy using a pyrometric feedback system in conjunction with a background monitoring device, *Spectrochim. Acta*, 38B, 791-7, 1983.

398. Katskov, D. A. and Burtseva, I. G., Optimization of conditions of atomic absorption analysis based on the analytical signal and temperature, *Zh. Prikl. Spektrosk.*, 39, 473, 1983.

399. Jenke, D. R. and Woodriff, R., Continued development of direct aerosol introduction in constant temperature furnace atomic absorption spectroscopy, *Appl. Spectrosc.*, 37, 470, 1983.

400. Suzuki, M. and Ohta, K., Electrothermal atomic absorption spectroscopy with metal atomizers, *Prog. Anal. At. Spectrosc.*, 6, 49, 1983.

401. Voellkopf, U. and Schulze, H., Graphite tube furnace atomic absorption spectroscopy with Zeeman effect background corrections. I. Review of Zeeman systems, *Labor Praxis*, 7, 410, 1983.

402. Cais, M., Metalloimmunoassay: principles and practice, *Methods Enzymol.*, 92, 445-58, 1983.

403. Tominaga, M. and Umezaki, Y., Evaluation of interference suppressors in electrothermal atomic absorption spectroscopy, *Anal. Chim. Acta*, 148, 285, 1983.

404. Na, H. C. and Niemczyk, T. M., Interferences in electrothermal atomization metastable transfer emission spectrometry, *Anal. Chem.*, 55, 1240, 1983.

405. Adams, F., Techniques for bulk chemical analysis.II. Atomic spectroscopy techniques, *Comm. Eur. Communities (Rep.) EUR*, 7544, 43-89, 1983.

406. Sychra, V., Kolihova, D., Hlavac, R., Dolezal, J., Puschel, P., and Formanek, Z., New experiences with electrothermal atomization in a tungsten furnace, *Analytiktreffen: Atomspektrosk., Fortschr., Anal. Anwend., Hauptvortr*, 154-63, 1982 (published 1983).

407. Dittrich, K., Diatomic molecules in plasma of a graphite tubular cuvette-disturbance of AAS (atomic absorption spectrometry) and the possibility of determination of non-metals, *Analytiktreffen: Atomspektrosk., Fortschr. Anal. Anwend., Hauptvortr.*, 76-88, 1982 (published 1983).

408. Bulska, E. and Kaczmarczyk, K., New trends in flameless atomic absorption spectrometry, *Mater. Konwersatorium Spektrom. At. Emisyjnej, Absorpc. Spektrom. Mas.*, 120, 118-32, 1983 (in Polish).

409. Glenc, T., Methods for modification of graphite tubes in flameless AAS (atomic absorption spectrometry), *Mater. Kowersatorium Spektrom. At. Emisyjnej, Absorpc. Spektrom. Mas.*, 120, 148-50, 1983 (in Polish).

410. Frech, W., Cedergren, A., and Lundberg, E., Vapor phase interference effects in electrothermal atomic absorption spectroscopy (AAS), *Analytiktreffen: Atomspektrosk., Fortschr. Anal. Anwend., Hauptovortr.*, 102-10, 1982 (published 1983).

411. Chakrabarti, C. L., Wu, S., and Bertels, P. C., New directions of graphite furnace atomic absorption spectrometry, *Analytiktreffen: Atomspektrosk., Fortschr. Anal. Anwend., Hauptvortr.*, 15-32, 1982 (published 1983).

412. Chakrabarti, C. L., Chang, S. B., Lawson, S. R., and Wong, S. W., Computer modeling of atomization processes in graphite furnace atomic absorption spectrometry, *Spectrochim. Acta*, 38B 10, 1287, 1983.

413. Schroen, W., Lange, P., and Kramer, W., Application of the LMA-10 laser microspectral analyzer in geology, *Jena Rev.*, 28 (3), 123-8, 1983.

414. Miller-Ihli, N. J., Automation and optimization of a simultaneous multielement atomic absorption spectrometer with a continuum source, 357 pp, 1982, Avail. Univ. Microfilms Int.B, 44(6), 1816, 1983.

415. Shimadzu Corp. Jpn., Spectrophotometer, Kokai Tokkyo Koho JP 58 92, 931 (83 92, 931) (Cl. G01 N 21/25), 4 pp, 02 Jun. 1983. Appl. 81/193, 240, 30 Nov. 1981.

416. Brneckner, C., Heynisch, M., Hoffman, D., and Pawlik, H., AAS-3- A new atomic absorption spectrophotometer of JENOPTIK JENA GmbH, *Jena Rev.*, 28(3), 114, 1983.

417. Eichardt, K., and Falk, H., ETA flameless electrochemical atomizer for microliter analysis on the AAS-3 atomic absorption spectrophotometer, *Jena Rev.*, 28(3), 118, 1983.

418. Raptis, S. E., Knapp, G., and Schalk, A. P., Novel method for the decomposition of organic and biological materials in an oxygen plasma excited at high frequency for elemental analysis, *Fresnius' Z. Anal. Chem.*, 316(5), 482-7, 1983.

419. Hitachi Ltd., Jpn., Atomizer for atomic absorption spectrometry, Kokai Tokkyo Koho JP 58, 140, 625 (83 140, 625) (Cl. G01 N 21/74), 3 pp, 20 Aug. 1983. Appl. 82/22,657, 17 Feb. 1982.

420. Ma, Y., Wu, Z., Zhang, Z., and Li, S., Study of the relation between apparent temperature and optimization energy in graphite furnace atomic absorption spectroscopy, *Huaxue Tongbao*, 5, 17-20, 1983 (in Chinese).

421. L'vov, B. V. and Savin, A. S., Autocatalytic mechanism of carbothermal reduction of low volatility oxides in graphite furnaces for atomic absorption analysis, *Zh. Anal. Khim.*, 38(11), 1925-32, 1983.

422. L'vov, B. V. and Savin, A. S., Kinetics of carbothermal reduction of low volatility oxides in graphite furnaces for atomic absorption analysis, *Zh. Anal. Khim.*, 38(11), 1933-8, 1983.

423. Holcombe, J. A. and Rayson, G. D., Analyte distribution and reactions within a graphite furnace atomizer, *Prog. Anal. At. Spectrosc.*, 6(3), 225-51, 1983.

424. Falk, H., Hoffman, E., Ludke, C., Ottaway, J. M., and Giri, S. K., Furnace atomization with nonthermal excitation: experimental evaluation of detection based on a high-resolution echelle monochromator incorporating automatic background correction, *Analyst (London)*, 108(1293), 1459-65, 1983.

425. **Atsuya, I. and Itoh, K.,** The use of an inner miniature cup for direct determination of powdered biological samples by atomic absorption spectrometry, *Spectrochim. Acta, Part B,* 38B(9), 1259-64, 1983.

426. **Hitachi Ltd., Jpn.,** Burner for an atomic absorption spectrometer, Kokai Tokkyo Koho JP 58 61, 444 (83 61, 444) (Cl. G 01N 21/31), 4 pp, 12 Apr. 1983. Appl. 81/158, 782, 07 Oct. 1981.

427. **Chakrabarti, C. L., Wu, S., and Bertels, P. C.,** Isothermal atomization from a platform in graphite furnace atomic absorption spectrometry, *Spectrochim. Acta, Part B,* 38B(7), 1041-60, 1983.

428. **Dyulgerova, R., Zechev, D., Popova, S., and Momchilova, E.,** Hollow cathode spectral lamps for atomic absorption analysis, *Spectrosc. Lett.,* 16(10), 765-74, 1983.

429. **Guevromont, R. and Whitman, J.,** Automation of a graphite- furnace atomic absorption spectrometry system with a Z80- based microcomputer, *Anal. Chim. Acta,* 154, 295, 1983.

430. **Tomm, R. and Tomoff, T. M.,** Apparatus for atomizing a sample in flameless atomic absorption spectroscopy, Ger. Offen. DE 3,217,417 (Cl. G 01 N21/74), 20 pp, 10 Nov. 1983. Appl. 08 May 1982.

431. **Feinberg, M. and Schnitzer, G.,** Optimization of atomization programs in flameless atomic absorption spectrometry, *Analusis,* 11(6), 299-305, 1983.

432. **Campbell, D. E. and Comperat, M.,** Ultratrace analysis of high purity glasses for double-crucible low-attenuation optical waveguides, *Glastech. Ber.,* 56, 898-903, 1983.

433. **Castledine, C. G. and Robbins, J. C.,** Practical field-portable atomic absorption analyzer, *J. Geochem. Explor.,* 19(1-3), 689-704, 1983.

434. **Kishman, J., Barish, E., and Allen, R.,** The dielectric discharge as an efficient generator of active nitrogen for chemiluminescence and analysis, *Appl. Spectrosc.,* 37(6), 545-52, 1983.

435. **Lersmacher, B. and Knippenberg, W. F.,** Cuvette for atomic absorption spectrometry, Ger. Offen. DE 3,208, 247 (Cl. G01N 21/03), 16 pp, 22 Sept. 1983. Appl. 08 Mar. 1982.

436. **Hitachi Ltd., Jpn.,** Atomic absorption spectrometer with CRT display, Kokai Tokkyo Koho JP 58 26, 249 (83 26, 249) (Cl. G01 N 21/31), 4 pp, 16 Feb. 1983. Appl. 81/124, 070, 10 Aug. 1981.

437. **Miller-Ihli, N. J., O'Haver, T. C., and Harnly, J. M.,** Time resolved electrothermal atomic absorption spectra, *Appl. Spectrosc.,* 37, 429, 1983.

438. **Sumitomo Electric Industries Ltd. Jpn.,** Integrated absorbance measurement for flameless atomic absorption spectrometry, Kokai Tokkyo Koho JP 58 37, 538 (83 37, 538) (Cl. G 01N 21/31), 3 pp, 04 Mar. 1983. Appl. 81/135,784, 29 Aug. 1981.

439. **Herber, R. F. M.,** Some instrumental improvements in electrothermal atomization atomic absorption spectroscopy. Application in biomedical research, *Spectrochim. Acta,* 38B, 783-9, 1983.

440. **Sturgeon, R. E. and Berman, S. S.,** Determination of the efficiency of the graphite furnace for AAS, *Anal. Chem.,* 55, 190-200, 1983.

441. **Sotera, J. J., Cristiano, L. C., Konley, M. K., and Kahn, H. L.,** Reduction of matrix interferences in furnace AAS, *Anal. Chem.,* 55, 204, 1983.

442. **Matsushita Electric Industrial Co. Ltd., Jpn.,** Flameless atomic absorption spectroscopy, Kokai Tokkyo Koho JP 58 47, 655 (83 47 655 (Cl. G01 N 21/31), 4 pp, 24 Oct. 1983. Appl. 75/96, 352, 07 Aug. 1975.

443. **Frech, W., Cedergren, A., Lundberg, E., and Siemer, D. D.,** Electrothermal atomic absorption spectroscopy: present understanding and future needs, *Spectrochim. Acta,* 38B, 1435, 1983.

444. **Slavin, W.,** The usefulness of signal integration for graphite furnace AAS—a response, *Z. Anal. Chem.,* 316, 319-20, 1983.

445. **Shimadzu Corp. Jpn.,** Apparatus for atomic absorption and optical absorption spectrometry, Kokai Tokkyo Koho JP 58,161,848 (83. 161, 848) (Cl. G01 N 21/31), 5 pp, 26 Sept. 1983. Appl. 82/45, 399, 20 Mar. 1982.

446. **Men'shikov, V. I., Vorob'eva, S. E., and Tsykhanskii, V. D.,** Study of the mechanism of atomization during nonstationary evaporation in graphite electrothermal atomizers by atomic absorption spectroscopy, *Metody Spektr. Anal.,* Lontsikh, S. V., Ed., Izd. Nauka, Sib. Otd., Novosibirsk, U.S.S.R., 1983, 57-61.

447. **Matsushita Electric Industrial Co. Ltd., Jpn.,** Flameless atomic absorption analysis, Tokkyo Koho JP 58 51, 615 (83 51, 615) (Cl. G01 N 21/31), 3 pp, 17 Nov. 1983. Appl. 76/62, 117, 27 May 1976.

448. **Men'shikov, V. I., Vorob'eva, S. E., Tsykhanskii, V. D., and Lozhkin, V. I.,** Theory of atomization processes in electrothermal atomic absorption spectroscopy, *Nov. Metody Spektr. Anal.,* 53-6, 1983 (in Russian).

449. **Belyaev, Y. I., Shcherbakov, V. I., and Myasoedov, B. F.,** Radiotracer study of vaporization and atomization in electrothermal atomic absorption analysis, *Nov. Metody Spektr. Anal.,* Lontsikh, S. V., Ed., Izd. Nauka, Sib. Otd., Novosibirsk, U.S.S.R., 1983, 62-5.

450. **Bel'skii, N. K., Nebol'sina, L. A., and Shubochkin, L. K.,** Feasibility of automating the atomic absorption analysis of some solid samples, *Nov. Metody Spektr. Anal.,* Lontsikk, S. V., Ed., Izd. Nauka, Sib. Otd., Novosibirsk, U.S.S.R., 1983, 118-20.

451. **Sperling, K. R.,** Comment on the preceding paper of W. Slavin on the usefulness of signal integration for graphite furnace AAS, *Fresnius' Z. Anal. Chem.,* 316, 320, 1983.

452. **Rechenberg, W.,** Extension of the service life of graphite parts in graphite parts in graphite-tube atomic absorption spectroscopy, *Fresnius' Z. Anal. Chem.,* 316, 320, 1983 (in German).

453. *Kurfuerst, U. and Wachter, K. H.,* Device for automatic introduction of samples into a graphite-tube furnace in an atomic absorption spectrometer, Ger. Offen. DE 3, 204, 873 (Cl. G01 N 21/74), 19 pp, 01 Sept. 1983. Appl. 12 Feb. 1982.

454. **Bezlepkin, A. I., Khomyak, A. S., Aleksandrov, V. V., Voronina, T. N., and Mel'nikova, O. V.,** Mass-produced spectral lamps for atomic absorption analysis, *Zh. Prikl. Spektrosk.,* 39, 367, 1983.

455. **Suzuki, M. and Ohta, K.,** Atom formation processes in the presence of thiourea in electrothermal atomic absorption spectroscopy with a molybdenum microtube atomizer, *Anal. Chim. Acta,* 151, 401-7, 1983.

456. **Littlejohn, D., Marshall, J., Carroll, J., Cormack, W., and Ottaway, J. M.,** Automatic graphite probe sample introduction for electrothermal atomic absorption spectroscopy, *Analyst (London),* 108, 893-6, 1983.

457. **Pyatova, V. N. and Ivanov, N. P.,** Basic principles of atomic absorption spectrochemical analysis with electrothermal atomization of samples, *Atomno-absorption, Metody Analiza Mineral. Syr'ya, M.,* 5-51, 1982 (in Russian). From Ref. *Zh. Metall. Abstr.* No. 5K2, 1983.

458. **Faithful, N. T.,** A rapid response recording system for electrothermal atomic absorption spectrophotometry, *Lab Pract.,* 32, 78, 1983.

459. **Grebennkiv, M. V. and Barsukov, V. I.,** All-purpose atomic absorption spectrometer based on the C- 302 flame spectrophotometer, *Zh. Prikl. Spektrosk.,* 39, 157, 1983 (in Russian).

460. **Wiseman, A. G. and Masters, M. K.,** Tubular furnace for spectroscopic apparatus, Brit. UK Pat. Appl. GB 2, 102, 589 (Cl. G01N 21/74), 11 pp, 02 Feb. 1983. AU Appl. 81/9, 946, 28 Jul. 1981.

461. **Walters, E. A. and Niemczyk, T. M.,** Effect of underground coal gasification on ground water, Report NMERDI-2-68-3118, 1983; Order No. DE83901528, 133 pp. Avail. N. M. Energy Res. Dev. Inst., 117 Richmond Drive, NE., Albuquerque, NM 87106. From Energy Res. Abstr. and Abstr. No. 22960, 1983.

462. **Glaeser, E.,** Testing pyrolytic graphite coatings for furnace atomic absorption spectroscopy, Ger. (East) DD 158, 428 (Cl. G01 N 21/74), 6 pp, 12 Jan. 1983. Appl. 229, 254, 15 Apr. 1981.

463. **Lersmacher, B., Wassal, M. P., and Connor, P. J.,** Atomization device for atomic absorption spectroscopy, Ger. Offen. DE 3, 140, 458 (Cl. G01 N 21/74), 15 pp, 21 Apr. 1983. Appl. 12 Oct. 1981.

464. **Lawson, S. R., Dewalt, F. G., and Woodriff, R.,** Influence of furnace design on operation, sensitivity and matrix interferences in electrothermal atomic absorption spectroscopy, *Prog. Anal. At. Spectrosc.,* 6, 1-48, 1983.

465. **Voellkopf, U. and Schulze, H.,** Graphite tube furnace AAS with Zeeman effect background correction. II., *Labor Praxis,* 7, 544, 1983 (in German).

466. **Slavin, W., Carnrick, G. R., Manning, D. C., and Pruszkowska, E.,** Recent experiences with the stabilized temperature platform furnace and Zeeman background correction, *At. Spectrosc.,* 4, 69, 1983.

467. **Kurfuest, U.,** Investigations on the analysis of heavy metals in solids by direct Zeeman atomic absorption spectroscopy. II. Theory, properties and efficiency of direct Zeeman AAS, *Fresnius' Z. Anal. Chem.,* 315, 304, 1983.

468. **Katskov, D. A. and Kopeikin, V. A.,** Slot burner with a thermal evaporator for atomic absorption analysis, *Zh. Prikl. Spektrosk.,* 38, 863, 1983.

469. **Kuellmer, G. and Morton, S. F. N.,** Graphite tube (furnace) AAS determination of mercury, *CLB Chem. Lab. Betr.,* 34, 243, 1983.

470. **Falk, H., Thiemann, H. J., Jaeckel, I., Schmidt, K. P., Hoffman, E., Tilch, J., and Luedke, G.,** Device for atomization of a sample for electrothermal atomic absorption measurements, Ger. (East) DD 200, 050 (Cl. G01 N 21/.74), 11 pp, 09 Mar. 1983. Appl. 237, 611, 09 July 1981.

471. **Woodriff, R. A.,** A method and apparatus for reduction of a matrix interference in an electrothermal atomizer for atomic absorption spectroscopy, U.S. US 4, 407, 582 (Cl. 356-312; Goi N21/74), 8 pp,04 Oct. 1983. Appl. 224, 627, 12 Jan. 1981.

472. **C-Huan, Chung,** Atomization mechanism with Arrhenius plots taking the dissipation function into account in graphite furnace AAS, *Anal. Chem.,* 56, 2714-20, 1984.

473. **Paveri- Fontane, S. L. and Tessari, G.,** Models in electrothermal optimization: the platform atomizer, *Prog. Anal. At. Spectrosc.,* 7, 243, 1984.

474. **L'vov, B. V., Ryabchuk, G. N., and Fernandez, G. H. A.,** Mechanism of atomization of oxygen-containing alkali metal compounds in graphite furnace atomic absorption analysis, *Zh. Anal. Khim.,* 39, 1206, 1984.

475. **Carnrick, G. R. and Lumas, B. K.,** Evaluation of new L'vov platform design, *At. Spectrosc.,* 5, 135, 1984.

476. **Barrett, P., Barnett, W., and Fernandez, F.,** Advances in atomic absorption graphite furnace analysis, *J. Test Eval.,* 12, 207, 1984.

477. **Chakrabarti, C. L., Wu, S., Karwowska, R., Chang, S. B., and Bertels, P. C.,** Studies in atomic absorption spectroscopy with a laboratory made graphite probe furnace, *At. Spectrosc.,* 5, 69-77, 1984.

478. **Hitachi Ltd., Jpn.,** Atomic spectral analyzer, Tokkyo Koho JP 59 13, 696 (84, 13, 696) (Cl. G 01 N21/31), 6 pp, 31 Mar. 1984. Appl. 77/5,851, 24 Jan. 1977.

479. **Stefanec, J.,** Atomic and molecular spectroscopy, *SVST: Bratislava Czech.* 248 pp, 1984 (in Slovak).

480. **Muzgin, V. N. and Atnashev, Y. B.,** Problem of atomization in atomic absorption and atomic fluorescence analysis, *Izv. Akad. Nauk SSSR, Ser. Fiz.,* 48, 1322, 1984.

481. **Suzuki, M.,** Atomic absorption analysis methods, *Genshi Kyuko Bunsekiho,* 184 pp, 1984.
482. **Volynskii, A. B. and Sedykh, E. M.,** Factors influencing the oxygen concentration in the protective gas of an electrothermal atomizer, *Zh. Anal. Khim.,* 39, 1197, 1984.
483. **Lersmacher, B.,** Tube cuvettes for atomic absorption spectroscopy, Ger. Offen. DE 3, 239, 253 (Cl. G 01 N 21/74), 11 pp, 26 April 1984. Appl. 23 Oct. 1982. addition to Ger. Offen. 3, 208, 744.
484. **Price, W. J.,** Atomic absorption spectroscopy — some recent perspectives, *Kem. Ind.,* 33, 239, 1984.
485. **Lawrence, D. M.,** An atomic absorption spectroscopy bibliography for Jan.-June, *At. Spectrosc.,* 5, 1984.
486. **Zhou, N. G., Frech, W., and De Galan, L.,** On the relationship between heating rate and peak height in electrothermal atomic absorption spectroscopy, *Spectrochim. Acta,* 39B, 225, 1984.
487. **Slavin, W. and Carnrick, G. R.,** The possibility of standardless furnace atomic absorption spectroscopy, *Spectrochim. Acta,* 39B, 271, 1984.
488. **Sturgeon, R. E., Siu, L. W. M., and Berman, S. S.,** Oxygen in the high temperature graphite furnace, *Spectrochim. Acta,* 39 B, 213, 1984.
489. **Falk, H., Hoffman, E., and Luedke, C.,** A comparison of furnace atomic non-thermal excitation spectrometry (FANES) with other atomic absorption spectroscopic techniques, *Spectrochim. Acta,* 39B, 283, 1984.
490. **Men'shikov, V. I., Vorob'eva, S, E., Tsykhanski, V. D., and Lozhkin, V. I.,** Theory of atomization processes in electrothermal atomic absorption spectroscopy, *Nov. Metody Spektr. Anal., Novosibirsk,* 53-6 (1983). From *Ref. Zh. Khim. Abstr.,* No. 5G87, 1984.
491. **Wendl., W. and Mueller, Vogt. G.,** Chemical reactions in graphite furnace atomic absorption spectroscopy and their application to the practice of trace analysis, *GIT Fachz. Lab.,* 28(4), 271, 1984.
492. **Matousek, J. P.,** Removal of sample vapor in electrothermal atomization studied at constant-temperature conditions, *Spectrochim. Acta,* 39B, 205, 1984.
493. **Rettberg, T. M. and Holcombe, J. A.,** A temperature controlled, tantalum second surface for graphite furnace atomization, *Spectrochim. Acta,* 39B, 249, 1984.
494. **Siemer, D. D. and Frech, W.,** Improving the performance of the CRA atomizer by reducing the rate of diffusional atom loss and delaying analyte volatilization, *Spectrochim. Acta,* 39B, 261, 1984.
495. **Pilipenko, E. P., Atnashev, Y. B., and Muzgyn, V. N.,** Direct atomic-absorption analysis of solid current-conducting samples with a tungsten atomizer, *Nov. Metody Spektr. Anal., Novosibiosk,* 68-70 (1983). From *Ref. Zh. Khim. Abstr.,* No. 5G45 (1984).
496. **Baeckstroem, K., Danielsson, L. G., and Nord, L.,** Sample work-up for graphite furnace atomic absorption spectroscopy using continuous flow extraction, *Analyst (London)* 109, 323-5, 1984.
497. **Siemer, D. D.,** Consequences of light beam misalignment in background corrected atomic absorption spectrophotometers, *Anal. Chem.,* 56, 1517, 1984.
498. **L'vov, B. V.,** The investigation of atomic absorption spectra by complete vaporization of the sample in a graphite cuvette, *Spectrochim. Acta,* 39B, 159, 1984.
499. **Herber, R. F. M., Pieters, H. J., Roelofsen, A. M., and Van Deijk, W.,** A pyrometric feedback system covering the entire temperature program for electrothermal atomization- atomic absorption spectroscopy, *Spectrochim. Acta,* 39B, 397, 1984.
500. **Lawrence, D. M.,** An atomic absorption spectroscopy bibliography for July - Dec. 1983, *At. Spectrosc.,* 5(1), 10-31, 1984.
501. **Nagourney, S. J., Negro, V. C., Bogen, D. C., Latner, N., Cassidy, M. E.,** Electrothermal atomic absorption spectroscopic analysis using a computer, *Comput. Appl. Lab.,* 2 (1), 49, 1984.
502. **Reece, P. A., McCall, J. T., Powis, G., and Richardson, R. L.,** Sensitive high-performance liquid chromatographic assay for platinum ultrafiltrate, *J. Chromatogr.,* 306, 417-23, 1984.
503. **L'vov, B. V. and Fernandez, G. A.,** Principles of thermal dissociation of oxides in graphite furnace atomic absorption analysis, *Zh. Anal. Khim.,* 39(2), 221-31, 1984.
504. **Bertram, H. P., Robbers, J., and Schmidt, R.,** Multielement analysis with ET-AAS within the scope of the Muenster human environmental specimen bank, *Fresnius' Z. Anal. Chem.,* 317, 462-7, 1984.
505. **Sperling, K. R.,** The tube-in-tube technique in electrothermal atomic absorption spectrometry. II. The three-bar tube, *Fresnius' Z Anal. Chem.,* 317(3-4), 261-3, 1984.
506. **Schmidt, P., Busche, C., and Keuscher, G.,** Sample carrier, especially graphite tube or boat for flameless atomic absorption spectroscopy, Ger. Offen. DE 3,234, 770 (Cl. G01 N 1/00), 8 pp, 22 Mar. 1984. Appl. 20 Sept. 1982.
507. **Katskov, D. A.,** Study of atomization processes as a way to use atomic absorption spectroscopy in physicochemical analysis, *Nov. Metody Spektr. Anal.,* Lontsikh, S. V., Ed., Izd. Nauka, Sib. ord. i Novosibirsk, U.S.S.R., 1984, 48-52 (in Russian).
508. **Judelevich, I. G., Beisel, N. F., Papina, T. S., and Dittrich, K.,** Layer-by-layer and film analysis of semiconductor materials by atomic absorption with electrothermal atomization, *Spectrochim. Acta,* 39B, 467, 1984.
509. **De Loos-Vollebregt, M. T. C. and De Galan, L.,** Analytical results for glassy tubes in electrothermal atomization atomic absorption spectroscopy at high temperatures, *Spectrochim. Acta,* 39B, 449, 1984.

510. **Pelieva, L. A. and Lakiza, Z. V.,** Effect of graphite powder introduced into the furnace on the evaporation of a series of elements, *Zh. Prikl. Khim.,* 57 (3), 485, 1984.

511. **Harnly, J. M., Miller-Ihli, N. J., and O'Haver T. C.,** Simultaneous multielement atomic absorption spectroscopy with graphite furnace atomization, *Spectrochim. Acta,* 39B, 305, 1984.

512. **Smith, N. J.,** Automated atomic absorption spectroscopy in occupational health monitoring, *Anal. Proc. (London),* 21, 111, 1984.

513. **Marinescus, D. M.,** External sampling tube atomizer as a means of lowering interferences in electrothermal atomic absorption spectroscopy, *Spectrosc. Lett.,* 17, 63, 1984.

514. **Ducheyne, P., Willems, G., Martens, M., and Helsen, J.,** *In vivo* metal-ion release from porous titanium-fiber material, *J. Biomed. Mater. Res.,* 18, 293, 1984.

515. **Bayunov, P. A., Savin, A. S., and L'vov, B. V.,** Use of a computer in thermochemical studies by the method of atomic absorption spectroscopy, *Zh. Prikl. Khim.,* 57, 357, 1984.

516. **Littlejohn, D., Duncan, I., Marshall, J., and Ottaway, J. M.,** Analytical evaluation of totally pyrolytic graphite cuvettes for electrothermal atomic absorption spectroscopy, *Anal. Chim. Acta,* 157, 291, 1984.

517. **Wennrich, R., Dittrich, K., and Bonitz, U.,** Matrix interference in laser atomic absorption spectroscopy, *Spectrochim. Acta,* 39B, 657, 1984.

518. **Men'shikov, V. I., Vorob'eva, S. E., and Tsykhanskii, V. D.,** Theory of atom source in electrothermal atomic absorption spectroscopy, *Zh. Anal. Khim.,* 39, 591, 1984 (in Russian).

519. **Gregoire, D. C. and Hall, G. E. M.,** Evaluation and characterization of a commercial Zeeman-modulated tungsten-strip electrothermal atomic absorption spectrometer, *Anal. Chim. Acta,* 158, 257, 1984.

520. **Nagdaev, V. K. and Pupyshev, A. A.,** Study of the formation of free atoms using an open-type graphite atomizer, *Nov. Metody Specktr. Anal., Novosibirsk,* 74, 1983 (in Russian). From Ref. *Zh. Khim.,* Abstr. No. 6G101, 1984.

521. **Suzuki, M. and Ohta, K.,** Furnace atomic absorption spectroscopy, *Bunseki Kagaku,* 6, 412, 1984.

522. **Magyar, B., Wampfler, B., and Zihlmann, J.,** Comparison of sensitivity and precision between the tube-in-tube technique and graphite tube technique of atomic absorption spectroscopy, *GIT Fachz. Lab.,* 28, 301, 1984 (in German).

523. **Littlejohn, D., Cook, S., Duries, D., and Ottaway, J. M.,** Investigation of working conditions for graphite probe atomization in electrothermal atomic absorption spectroscopy, *Spectrochim. Acta,* 39B, 295, 1984.

524. **Slavin, W.,** Modern graphite furnace AAS technology and standardless analyses, *Anal. Proc. (London),* 21, 59, 1984.

525. **Maglaty, J. L.,** Extending the linear dynamic range in flameless atomic absorption using a photodiode array detector, 271 pp, 1984. Avail. Univ. Microfilms Int. Order No. DA 8327977. From Diss. Abstr. Int. B, 44, 2415, 1984.

526. **Ottaway, J. M.,** Recent innovations in electrochemical atomization, *Anal. Proc. (London),* 21, 55, 1984.

527. **Volynskii, A. B., Spivakov, B. Y., and Zolotov, Y. A.,** Solvent extraction—electrothermal atomic absorption analysis, *Talanta,* 31, 449, 1984.

528. **Siemer, D. D., Lundberg, E., and Frech, W.,** Gas-phase temperatures in Massman furnaces equipped with L'vov platforms, *Appl. Spectrosc.* 38, 389, 1984.

529. **Almeida, M. C.,** The study and characterization of a new procedure for applying carbide coatings to graphite furnace used in atomic absorption spectroscopy, 167 pp, 1983. Avail. Univ. Microfilms Int. Order No. DA 8403933. From Diss Abstr. Int. B, 44, 3392, 1984.

530. **Sneddon, J.,** An inexpensive probe for sample introduction in electrothermal atomization atomic absorption spectroscopy, *Anal. Lett.,* 17, 665, 1984.

531. **Shimadzu Corp. Jpn.,** Atomization furnace heating system for flameless atomic absorption analysis, Tokkyo Koho JP 59 23, 376 (84 23, 376) (Cl.G01 N 21/31), 4 pp, 01 Jun. 1984. Appl. 79/126,204, 28 Sept. 1979.

532. **Briganti, L. A.,** HGA graphics software modified to generate data as concentration, *At. Spectrosc.,* 5, 131, 1984.

533. **Cai, D.,** Atomic absorption spectroscopic analysis. IV., *Huaxue Shijie,* 25, 234, 1984.

534. **Dittrich, K. and Wennrich, R.,** Laser vaporization in atomic absorption spectroscopy, *Prog. Anal. At. Spectrosc.,* 7, 139, 1984.

535. **Chung, C. H.,** Atomization mechanism with Arrhenius plots taking the dissipation function into account in graphite furnace atomic absorption spectroscopy, *Anal. Chem.,* 56, 2714, 1984.

536. **Barnett, W. B. and Carnrick, G. R.,** A graphite software package for Zeeman furnace atomic absorption spectrophotometry, *At. Spectrosc.,* 5, 210, 1984.

537. **Steele, A. W. and Hieftje, G. M.,** Microdroplet titration apparatus for analysing small sample volumes, *Anal. Chem.,* 56, 2884, 1984.

538. **Nihon Jareru Attsushu K. K., Jpn.,** Hydride generation apparatus, Kokai Tokkyo Koho JP 59, 119, 244 (84, 119, 244) (Cl. G01 N 21/31), 3 pp, 10 Jul. 1984. Appl. 82/232, 190, 25 Dec. 1982.

539. **Siemer, D. D.,** Graphite furnace atomizer, U. S. Pat. Appl. US 495, 888, 23 pp, 14 Sept. 1984. Appl. 18 May 1983.

540. **Sedykh, E. M. and Belyaev, Y. I.,** A study of sample volatilization in a graphite furnace by means of atomic and molecular absorption spectra, *Prog. Anal. At. Spectrosc.,* 7(4), 373-85, 1984.

541. **Ortner, H. M., Krabichler, H., and Wegscheider, W.,** Coating and impregnation of graphite tubes, *Fortschr. Atomspektrom. Spurenanal.,* 1, 73-114, 1984.

542. **Schlemmer, G. and Welz, B.,** High-resolution signals, a valuable aid for method development in graphite furnace AAS, *Fortschr. Atomspektrom. Spurenanal.,* 1, 157-62, 1984.

543. **Dabeka, R. W.,** Basic computer subroutine for routine application of graphite-furnace atomic absorption spectroscopy, *Can. J. Spectrosc.,* 29(5), 109-12, 1984.

544. Matsushita Electric Industrial Co. Ltd., Jpn., Flameless atomic absorption spectrometry, Tokkyo Koho JP 59 50,928 (84 50, 928) (Cl. G01 N 21/00), 3 pp, 11 Dec. 1984. Appl. 78/17, 924, 17 Feb. 1978.

545. **Li, X., Ta0, C., Jin, Z., and Sun, Z.,** A microcomputer-controlled system for an atomic absorption spectrometer, *Fenxi Huaxue,* 12(12), 1106, 1984 (in Chinese).

546. **Allain, P. and Mauras, Y.,** A study of background signals in graphite furnace atomic absorption spectrometry, *Anal. Chim. Acta,* 165, 141-7, 1984.

547. **Wibetoe, G. and Langmyhr, F. J.,** Spectral interferences and background over-compensation in Zeeman-corrected atomic absorption spectrometry. I. The effect of iron on 30 elements and 49 element lines, *Anal. Chim. Acta,* 165, 87-96, 1984.

548. Toshiba Corp. Jpn., Semiconductor crystal analyzer, Kokai Tokkyo Koho JP 59, 189, 642 (84, 189, 642) (Cl. H01 L 21/66), 5 pp, 27 Oct. 1984. Appl. 83/65, 033, 13 Apr. 1983.

549. **Dusan, B. and Skaerdin, I. L.,** A comparative study between two methods: neutron activation and flameless atomic absorption, *Inst. Vatten-Luftvaardsforsk.* (Publ.) B IVL B758, 19 pp, 1984.

550. **Harnly, J. M. and Kane, J. S.,** Optimization of electrothermal atomization parameters for simultaneous multielement atomic absorption spectrometry, *Anal. Chem.,* 56(1), 48-54, 1984.

551. **Slavin, W., Carnrick, G. R., and Manning, D. C.,** Chloride interferences in graphite furnace atomic absorption spectrometry, *Anal. Chem.,* 56(2), 163-8, 1984.

552. **Grobecker, K. H. and Kurfuerst, U.,** Prefered usage possibilities of solids analysis using Zeeman atomic absorption spectroscopy, *Fortschr. Atomspektrom. Spurenanal.,* 1, 365-74, 1984.

553. **Dungs, K. and Neidhart, B.,** Graphite furnace chemistry - ways to metal species analysis, *Fortschr. Atomspektrom. Spurenanal.,* 1, 411-19, 1984.

554. **Schimmel, U. and Lapornik, E.,** Optimization of the fully automatic monitoring of sample streams, *Fortschr. Atomspektrom. Spurenanal.,* 1, 647, 1984.

555. **Halls, D. J.,** Speeding up determinations by electrothermal atomic absorption spectrometry, *Analyst,* 109(8), 1081-4, 1984.

556. **Batz, L., Ganz, S., Hermann, G., Scharmann, A., and Wirz, P.,** The measurement of stable isotope distribution using Zeeman atomic absorption spectroscopy, *Spectrochim. Acta, Part B,* 39B(8), 993-1003, 1984.

557. **Belyaev, Y. I., Khozhainov, Y. M., Shcherbakov, V. I., and Myasoedov, B. F.,** Radiographic study of the memory effect of a graphite furnace during atomic absorption analysis, *Zh. Prikl. Spektrosk.,* 41(5), 831-2, 1984.

558. **Veber, M. and Gomiscek, S.,** On the electrothermal atomization of metals deposited electrolytically on graphite electrodes, *Vestn. Slov. Kem. Drus.,* 31(3), 313-23, 1984.

559. **Falk, H.,** Electrothermal atomization of a sample for analysis, Ger. (East) DD 213, 063 (Cl. G01 N21/74), 7 pp, 29 Aug. 1984. Appl. 247, 176, 12 Jan. 1983.

560. **Sperling, K. R.,** Device and method for studying samples by flameless atomic absorption analysis, Ger. Offen. DE 3, 307, 251 (Cl. G01 N 21/74), 8 pp, 06 Sept. 1984. Appl. 02 Mar. 1983.

561. **Jing, S., Wang, R., Yu, C., Zhang, D., Yan, Y., and Ma, Y.,** Design of model ZM-1 Zeeman effect atomic absorption spectrometer, *Huanjing Kexue,* 5(3), 41-4, 1984.

562. **Zheng, Y.,** Atomic loss mechanism in constant-temperature furnace atomic absorption spectroscopic analysis, *Fenxi Huaxue,* 12(7), 597-600, 1984 (in Chinese).

563. Fuji Electric Res. Lab., Jpn., Sample atomization device for flameless atomic absorption spectrometry, Kokai Tokkyo Koho JP 59, 126, 232 (84, 126, 232) (Cl. G01 N 21/74) , 3 pp, 20 July 1984. Appl. 93/813, 07, Jan. 1983.

564. **Sakai, T., Hanamura, S., Smith, B. W., and Winefordner, J. D.,** Use of Zeeman atomic absorption flame spectrometry for measurements in acetylene/air and acetylene/nitrous oxide flames, *Spectrosc. Lett.,* 17(12), 819-26, 1984.

565. **Ma, Y., Li, S., Zhang, Z., Wu, Z., Feng, X., Su, W., and Sun, D.,** Effect of heating rate on sensitivity in graphite-furnace atomic absorption spectroscopy, *Huaxue Tongbao,* 8, 18-21, 1984.

566. **Bahreyni-Toosi, M. H., Dawson, J. B., Ellis, D. J., and Duffield, R. J.,** An examination of instrumental systems for reducing the cycle time in atomic absorption spectroscopy with electrothermal atomization, *Analyst (London),* 109(12), 1607-12, 1984.

567. **Deng, C. and Wang, G.,** A new type of atomization system in hydride generation electrothermal atomic absorption spectrometry, *Fenxi Huaxue,* 12(11), 1034, 1984 (in Chinese).

568. **Chakrabarti, C. L., Chang, S. B., Lawson, S. R., and Bertels, P. C.,** Studies on the capacitive discharge technique in GFAAS, *Spectrochim. Acta,* Part B, 39B (9-11), 1195-208, 1984.

569. **Shekiro, J. M., Jr.,** Characterization of an electrothermal atomizer system by multiple wavelength spectrometry, 181 pp, 1984. Avail. Univ. Microfilms Int. , Order No. DA 84 17113. From Diss. Abstr. Int. B, 45(6), 1769, 1984.

570. **Nagdaev, V. K., Pupyshev, A. A., and Bukraeev, Y. F.,** Study of the mechanism of chemical interference during electrothermal atomization on a graphite rod, *Metody Spektr. Anal. Miner. Syr'ya* (Mater. Vses. Konf. Nov. Metodam Spektr. Anal. Ikh Primen.), 106-9 pp, 2nd 1981, published 1984.

571. **Szakass, O., Horvath, Z., and Lasztity, A.,** AES with a.c. interrupted arc and AAS with graphite furnace and cold vapor technique in environmental protection studies, *Ann. Ist. Super. Sanita,* 19(4), 531-5, 1983, published 1984.

572. **Pinel, R., Benabdallah, M. Z., Astruc, A., Potin-Gautier, M., and Astruc, M.,** Automated specific detection for liquid chromatography by electrothermal atomic absorption . Application to organotin compounds, *Analusis,* 12(7), 344-9, 1984.

573. **Narres, H. D., Mohl, C., and Stoeppler, M.,** Metal analysis in difficult materials with platform furnace-Zeeman atomic absorption spectrometry, *Int. J. Environ. Anal. Chem.,* 18(4), 267-79, 1984.

574. **Schulze, H.,** Atomic absorption spectrometry is now routine even for complex samples, *Labor Praxis,* 8(10), 1008, 1984.

575. **Yao, J., Jiang, B., and Huang, W.,** Technique and application of tantalum-foil lining in graphite furnace atomic absorption spectroscopy, *Guangpuxue Yu Guangpu Fenxi,* 4(4), 40-3, 1984 (in Chinese).

576. **Robinson, J. W. and Ekman, T. A.,** An improved double stage furnace for speciation and quantitative atomic absorption spectrometry, *Spectrosc. Lett.,* 17(10), 615-32, 1984.

577. **Ma, Y., Li, S., Zhang, Z., Wu, Z., Feng, X., Su, W., and Sun, D.,** Influence of heating rate on the matrix effect in graphite furnace atomic absorption spectroscopy, *Guangpuxue Yu Guangpu Fenxi,* 4(4), 31-9, 1984 (in Chinese).

578. **Voinovitch, I. A., Druon, M., and Louvrier, J.,** Effect of the absence of an injection orifice and the length of graphite furnaces on the sensitivity of determinations by atomic absorption spectrometry with electrothermal atomization, *Analusis,* 12(5), 256-9, 1984.

579. **Sturgeon, R. E. and Berman, S. S.,** Absorbance pulse shifting in graphite furnace AAS, *Anal. Chem.,* 57, 1268-75, 1985.

580. **Harnly, J. M. and Holcombe, J. A.,** Background correction errors originating from non-simultaneous sampling for graphite furnace AAS, *Anal. Chem.,* 57, 1983-85, 1985.

581. **Dungs, K. W., Hopp, D., and Neidhart, B.,** Modifications and new developments for HGA graphics software for PE 3600 data station with Zeeman/5000 spectrophotometer, *At. Spectrosc.,* 6(6), 161, 1985.

582. **Slavin, W. and Carnrick, G. R.,** A survey of applications of the stabilized temperature platform furnace and Zeeman correction, *At. Spectrosc.,* 6(6), 157, 1985.

583. **Kawashima, F. and Sasaki, K.,** Fully automated single beam atomic absorption spectrometer with blank correction, Jpn. Kokai Tokkyo Koho JP 60, 252, 243 (85, 252, 243) (Cl. 001 N 21/31), 3 pp, 12 Dec. 1985. Appl. 29 May 1984.

584. **Sasaki, K.,** A flameless atomic absorption spectrometer with better reproducibility of analysis, Jpn. Tokkyo Koho JP 60, 253, 851 (85, 253, 851) (Cl. G01 N 21/31), 3 pp, 14 Dec. 1985. Appl. 84/112, 128, 30 May 1984.

585. **Sasaki, K.,** A monitor for graphite tube consumption in a flameless atomic absorption spectrometer, Jpn. Kokai Tokkyo Koho JP 60, 253, 850 (85, 253, 850) (Cl. G01 N 21/31), 3 pp, 14 Dec. 1985. Appl. 84/112, 127, 30 May 1984.

586. **Alt, F. and Brendt, H.,** Atomic absorption spectrometry, *GIT. Fachz. Lab.,* 29(1), 17, 1985.

587. **Shimadzu Corp., Jpn.,** Flameless atomizer, Kokai Tokkyo Koho JP 60 04, 845 (85 04, 845) (Cl. G01 N 21/74), 11 pp, 11 Jan. 1985. Appl. 83/22, 709, 24 Aug. 1983.

588. **Scheer, J. and Kurfuerst, U.,** Electrothermal atomization device, Ger. Offen. DE 3, 327, 698 (Cl. G01 N 21/74), 8 pp, 21 Feb. 1985, Appl. 1 Aug. 1983.

589. **Sperling, K. R. and Bahr, B.,** The tube-in-tube technique in electrothermal atomic absorption spectrometry. I. Atomization behavior and matrix effects of cadmium using the L'vov platform and the tube-in-tube technique, *Fortschr. Atomspektrom. Spurenanal.,* 1, 145, 1984. CA 12, 102: 197154, 1985.

590. **L'vov, B. V.,** Role of free carbon in the gas phase during carbothermic reduction of oxides in graphite furnaces for atomic absorption analysis, *Zh. Anal. Khim.,* 39(11), 1953-60, 1985.

591. **Zinger, M.,** A compact Zeeman corrected electrothermal AA spectrophotometer, *Am. Lab.,* 17(3), 96-102, 1985.

592. **Brown, A. A., Whiteside, P. J., and Kuellmer, G.,** Design and application of an alignment jig for graphite furnace AAS (atomic absorption spectroscopy), *GIT Fachz. Lab.,* 29(3), 198-201, 1985.

593. **Kuellmer, G. and Whiteside, P. J.,** Pyrographite tube (TPC) for AAS (atomic absorption spectroscopy), *Labor Praxis,* 9(3), 141-148, 1985.

594. **Nakamura, S.,** Application and limitation of a mathematical distortion model of an absorbance signal in electrothermal atomic absorption spectroscopy, *Anal. Chim. Acta,* 167, 365-70, 1985.

595. **Criaud, A. and Fouillac, C.,** Use of the L'vov platform and molybdenum coating for the determination of volatile elements in thermomineral waters by atomic absorption spectroscopy, *Anal. Chim. Acta,* 167, 257-67, 1985.

596. Hitachi Ltd. Jpn., Corrosion monitoring, Kokai Tokkyo Koho JP 60 00, 351 (85 00, 351) (Cl. G01 N 27/86), 7 pp, 05 Jan. 1985. Appl. 83/ 108, 361, 16 Jun. 1983.

597. **Sturgeon, R. E. and Berman, S. S.,** Absorption pulse shifting in graphite furnace atomic absorption spectroscopy, *Anal. Chem.,* 57(7), 1268-75, 1985.

598. **Bortolaccini, M. A., D'Innocenzio, F., and Gucci, P. M. B.,** Preparation of an atmospheric aerosol pattern for metal determination, *Inquinamento,* 27(1), 57-9, 1985.

599. **Yansheng, Z. and Feng, Z.,** Atomization process in the graphite furnace atomizer. I. Effect of graphite furnace substrate materials on dissipation of aluminum, *Guangpuxue Yu Guangpu Fenxi,* 5, 52, 1985 (in Chinese).

600. **Pelieva, L. A., Lakiza, Z. V., and Bukhantsova, V. G.,** Optimization of temperature conditions for the graphite tube furnace-platform system, *Zh. Anal. Khim.,* 40, 1790, 1985.

601. **Men'shikov, V. I., Vorob'eva, S. E., and Tsykhanskii, V. D.,** Atomization of substances from the graphite furnace surface in electrothermal atomic absorption analysis, *Zh. Anal. Khim.,* 40, 1930, 1985.

602. **Littlejohn, D., Duncan, I. S., Hendry, B. M., Marshall, J., and Ottaway, J. M.,** Comparison of uncoated, pyro-coated and totally pyrolytic graphite tubes for the HGA-500 electrothermal atomizer, *Spectrochim. Acta,* 40B, 1677, 1985.

603. **Frech, W., Lundberg, E., and Cedergren, A.,** Investigation of interference effects in the presence of sodium and copper chlorides using platform graphite furnace AAS, *Prog. Anal. At. Spectrosc.,* 8, 257, 1985.

604. **Chakrabarti, C. L., Wu, S., and Karwowska, R.,** The gas temperature in and the gas expulsion from a graphite furnace used for atomic absorption spectrometry, *Spectrochim. Acta,* 40B, 1663, 1985.

605. **Findeisen, B.,** Graphite tube cuvettes for FAAS, Ger. (East) DD 217, 026 (Cl. G01 N 21/74), 4 pp, 02 Jan. 1985. Appl. 253, 610, 02 Aug. 1983.

606. **Matsunaga, H., Hirata, N., and Okada, A.,** Sample dissolution device for semiconductor thin films, Eur. Pat. Appl. EP 137, 409 (Cl. G01 N1/28), 14 pp, 17 Apr. 1985. JP Appl. 83/176, 503, 26 Sept. 1983.

607. **Burakov, V. S., Misakov, P. Y., Naumenkov, P. A., and Raikov, S. N.,** Intraresonator laser spectroscopy with resonance detection of absorption lines, *Primen. Spektr. Anal. Nar. Khoz. Nauchen. Issled., Mater. Resp. Semin. Spektr.,* Petukh, M. L. and Putrenko, O. I., Eds., Anal. 21-8, 1982, published 1983, CA 17, 103: 47383n, 1985.

608. **Kinnunen, H.,** Zeeman background correction in AAS, *Kem. Kemi,* 12(5), 431-6, 1985.

609. **Harnly, J. M. and Holcombe, J. A.,** Background correction errors originating from nonsimultaneous sampling for GFAAS, *Anal. Chem.,* 57(9), 1983-6, 1985.

610. **L'vov, B. V. and Yatsenko, L. F.,** Refinement of the value of partial pressure of C2 radicals in graphite furnace for AA analysis, *Zh. Anal. Khim.,* 40(4), 626-9, 1985.

611. **L'vov, B. V. and Bayunov, P. A.,** Influence of the penetration of sample solution into the wall of a graphite furnace and polydispersity of dry sample residue on the analytical signal shape in AAS, *Zh. Anal. Khim.,* 40(4), 614-25, 1985.

612. **Sato, Y. and Tateuchi, S.,** Comparison of the Zeeman optical polarization method and deuterium lamp background correction method in AAS, *Hiken Kaiho,* 37(3), 22-31, 1984. CA 17, 103: 36642 v, 1985.

613. IUPAC Commission on Electroanalytical Chemistry, Recommended methods for the purification of solvents and tests for impurities: methanol and ethanol, *Pure Appl. Chem.,* 57(6), 855-64, 1985.

614. **Hiraki, K., Nakaguchi, Y., Morita, M., and Hajiri, N.,** Determination of metal ions with polarized Zeeman effect and optical temperature sensor and control system flameless AA spectrophotometry, *Kinki Daigaku Rikogakuba Kenku Hokoku,* 20, 77-84, 1984. CA 20, 103:97971, 1985.

615. **Showa Denko, K. K.,** Manufacture of graphite tube for AAS analysis, Jpn. Kokai Tokkyo Koho JP 60 49, 247 (85 49, 247) (Cl. G01 N 21/73), 3 pp, 26 Mar. 1985. Appl. 83/161, 011, 31 Aug. 1983.

616. **Kitagawa, K. and Noguchi, T.,** A study on the effect of the source line profile in atomic magneto-optical rotation spectroscopy (AMORS), using a Zeeman- shifted hollow cathode lamp, *Fresnius' Z. Anal. Chem.,* 321(5), 436-41, 1985.

617. **L'vov, B. V., Fernandez, G. J. A., and Ryabchuk, G. N.,** Effect of hydrogen on atomization of oxides in graphite furnaces for AA analysis, *Zh. Anal. Khim.,* 40(5), 792-8, 1985.

618. **Chang, S. B. and Chakrabarti, C. L.,** Factors affecting atomization in graphite furnace AAS, *Prog. Anal. At. Spectrosc.,* 8(2), 83-91, 1985.

619. **Yao, J. , Zheng, Y., and Wu, T.,** Semiautomatic hydride generator used in flame AAS, *Fenxi Huaxue,* 13(6), 468-71, 1985.

620. **Holcombe, J. A. and Rettberg, T.,** Method and apparatus for concentrating a selected element for atomic spectroscopy, U.S. 4,529, 307 (Cl. 356-312: G01 N 21/74), 5 pp, 16 Jul. 1985. Appl. 482, 066, 05 Apr. 1983.

621. Sumitamo Aluminum Smelting Co. Ltd., Jpn., Graphite tube for flameless atomic absorption spectrometry, Tokkyo Koho JP 6026, 9712 (85 26, 971) (Cl. G 01 N 21/00), 5 pp, 26 Jan. 1985. Appl. 77/93, 553, 03 Aug. 1977.

622. Gao, T. and Stephens, R., Use of air-cooled solenoids for Zeeman background correction, *Anal. Chem.*, 57(2), 424-7, 1985.

623. De Galan, L., New instrumentation in atomic absorption and atomic emission spectrometry, *Analytiktreffen: Atomspektrosk. Fortschr. Anal. Anwend., Hauptvortr.*, 59-69, 1982, published 1983. CA 5, 102:71809f, 1985.

624. Belyaev, Yu. I., Sedykh, E. M., and Sherbakov, V. I. S., Atomization problems in electrothermal astomic absorption spectrometry, *Analytitreffen: Atomspektrosk., Fortschr. Anal. Anwend., Huaptovortr.*, 41-50, 1982, published 1983. CA 5, 102: 71808e, 1985.

625. Falk, H., Elementary processes in the FANES (furnace atomic nonthermal excitation spectrometry) technique, *Analytiktreffen. Atomspektrosk., Fortschr. Anal. Anwend., Hauptvortr.*, 306-12, 1982, published 1983. CA 5, 102: 71767r, 1985.

626. Barnett, W. B., Bohler, W., Carnrick, G. R., and Slavin. W., Signal processing and detection limits for graphite furnace atomic absorption with Zeeman background correction, *Spectrochim. Acta*, 40B, 1689, 1985.

627. Baez, M. E., Gonzalez, C., and Lachica, M., Lifetime of pyrolytic graphite-tube atomizers in molybdenum determination, *Analusis*, 13, 474, 1985.

628. Lakatos, I., Direct FAES and FAAS analysis of macromolecular solutions, *ATOMKI Kozol.*, 27(3), 343-9, 1985.

629. Carroll, J., Miller-Ihli, N. J., Harnly, J. M., Littlejohn, D., Ottaway, J. M., and O'Haver, T. C., Simultaneous multielement analysis by continuum source AAS with graphite probe electrothermal atomization, *Analyst (London)*, 110(9), 1153-9, 1985.

630. Volynskii, A. B., Sedykh, E. M., Spivakova, B. Y., and Khavezov, I., Factors influencing the free oxygen content in an electrothermal atomizer, *Anal. Chim. Acta*, 174, 173-82, 1985.

631. Men'shikov, V. I., Vorob'eva, S. E., and Tsykhanskii, V. D., Atomization of substances from the graphite furnace surface in electrothermal AA analysis, *Zh. Anal. Khim.*, 40(11), 1930-7, 1985.

632. Slavin, W. and Carnrick, G. R., A survey of application of the stabilized temperature platform furnace and Zeeman correction, *At. Spectrosc.*, 6(6), 157-60, 1985.

633. Matousek, J. P., Analytical furnace having preheating and constant temperature sections, PCT Int. Appl. WO 85, 04, 717 (Cl. G01 N21/74), 11 pp, 24 Oct 1985. AU Appl. 84/4498, 10 Apr. 1984.

634. Hadeishi, T. and McLaughlin, R., Direct ZAS analysis of solid samples: early development, *Z. Anal. Chem.*, 322(7), 657-9, 1985.

635. Hernandez, C. A., Nguyen, K. L., and Sneddon, J., Investigation of transparent efficiency of pneumatic nebulization for dissolved solids in flame and furnace AAS, *Spectrosc. Lett.*, 18(10), 815-25, 1985.

636. Findseisen, B., Platform for flameless AAS, Ger. (East) DD 225, 218 (Cl. G01 N 21/ 74), 7 pp, 24 Jul. 1985. Appl. 264, 868, 03 July 1984.

637. Chang, S. B., Chakrabarti, C. L., Huston, T. J., and Byrne, J. P., Estimation of a partial pressure of oxygen inside the graphite furnace used for AAS, *Fresnius' Z. Anal. Chem.*, 322(6), 567-73, 1985.

638. Takada, K., Enhancement of sensitivity in atomic absorption spectrometry by addition of graphite lid to cup furnaces, *Talanta*, 32, 921, 1985.

639. Voellkopf, U., Gorbenski, Z., Tamm, R., and Welz, B., Solid sampling in graphite furnace atomic absorption spectrometry using cup-in-tube technique, *Analyst*, 110, 573, 1985.

640. Wennrich, R., Bonitz, U., Brauer, H., Niebergall, K., and Dittrich, K., Graphite furnace AAS with ultrasonic nebulization, *Talanta*, 32, 1035, 1985.

641. He, Y., Computer applications for data acquisition and processing in AAS, *Jisuang Yu Yingyong Huaxue*, 2(3), 186-9, 1985 (in Chinese).

642. L'vov, B. V., Pyabchuk, G. N., and Fernandez, G., Effect of hydrogen on the analytical characterization of the method of atomic absorption with electrothermal atomization, *Rev. Cubana Quim.*, 1(1), 61-7, 1985.

643. Fernandez, G. J. A., Effect of gaseous phase composition in evaporation process of samples in graphite furnace for atomic absorption, *Bunsaki*, 1(3), 766-72, 1985 (in Japanese).

644. Carroll, J., Marshall, J., Littlejohn, D., and Ottaway, J. M., Automation end-entry sample atomization in electrothermal AAS, *Fresnius' Z. Anal. Chem.*, 322(2), 145-50, 1985.

645. Dougherty, J. P., Michel, R. G., and Slavin, W., Microwave excited electrodeless discharge lamps as intense sources for Zeeman AAS, *Spectrosc. Lett.*, 18(8), 627-41, 1985.

646. Takada, K., Enhancement of sensitivity in AAS by addition of a graphite lid to a cup furnace, *Talanta*, 32(9), 921-5, 1985.

647. Pelieva, L. A. and Uzunbadzhakov, A. S., Temperature stabilization method for the graphite tube furnace of the electrothermal atomizer, *Zh. Prikl. Spektrosk.*, 43(4), 538-44, 1985.

648. Niemczyk, T. M. and Yin, I. H., A method for the evaluation of background corrector performance in graphite FAAS, *Appl. Spectrosc.*, 39(5), 882-3, 1985.

649. O'Grady, C. E., Marr, I. L., and Cresser, M. S., Patterns and causes of deposition losses in a simple spray chamber, *Analyst (London)*, 110(6), 729-31, 1985.

650. **Voelkopf, U., Grobenski, Z., Tamm, R., and Welz, B.,** Solid sampling in GFAAS using the cup-in-tube technique, *Analyst (London),* 110(6), 573-7, 1985.

651. **L'vov, B. V., Savin, A. S., and Yatsenko, L. F.,** Mechanism and kinetics of carbothermal reduction of metal oxides in graphite furnaces for AA analysis, *Zh. Prikl. Spektrosk.,* 43(6), 887-92, 1985.

652. **Brown, A. A.,** Use of furnace alignment jig to decrease errors associated with background correction in a graphite furnace AAS, *Anal. Chim. Acta,* 175, 319-23, 1985.

653. **Lovett, R. J.,** The influence of temperature on absorbance in graphite furnace AAS. I. General considerations, *Appl. Spectrosc.,* 39(5), 778-86, 1985.

654. **Ortner, H. M., Schlemmer, G., Welz, B., and Wegscheider, W.,** Scanning electron microscopy studies on surfaces from electrothermal AAS. I. Polycrystalline electrographite tubes with and without pyrographite coating, *Spectrochim. Acta, Part B,* 40B(7), 959-77, 1985.

655. **Wennrich, R., Bonitz, U., Brauer, H., and Niebergall, K.,** Graphite-furnace AAS with ultrasonic nebulization, *Talanta,* 32(11), 1035-9, 1985.

656. **Pelieva, L. A., Lazika, Z. V., and Bukhantsova, V. G.,** Optimization of temperature conditions for the graphite tube furnace-platform system, *Zh. Anal. Khim.,* 40(10), 1790-6, 1985.

657. **Churchill, J. E., Flack, M. D., and Widmer, D. S.,** Electrothermal atomizer, EUR. PAT. Appl. EP 164, 804 (Cl. G01 N 21/74), 16 pp, 18 Dec. 1985. Appl. 84/14, 815, 11 Jun. 1984.

658. **Findeisen, B.,** Moldings from glassy carbon for flameless atomic absorption spectroscopy, Ger. (East) DD 227, 523 (Cl. G01 N 21/74), 7 pp, 18 Sept. 1985. Appl. 268, 362, 15 Oct. 1984.

659. **Zerlia, T. and Tittarelli, P.,** Some applications of vapor-phase molecular and atomic spectrometry, *Riv. Combust.,* 39(7), 193-9, 1985.

660. **Chakrabarti, C. L., Wu, S., Karwowska, R., Rogers, J. T., and Dick, R.,** The gas temperature in and the gas expulsion from a graphite furnace used for atomic absorption spectrometry, *Spectrochim. Acta, Part B,* 40B(10-12), 1663-76, 1985.

661. **Marshall, J., Baxter, D. C., Carroll, J., Cook, S., Corr, S. P., Giri, S. K., Durie, D., Littlejohn, D., Ottaway, J. M. et.al.,** The probe furnace in atomic spectrometry, *Anal. Proc. (London),* 22(12), 371-3, 1985

662. **Dittrich, K., Findeisen, B., and Mandry, R.,** Atomizer tube for flameless hydride atomic absorption spectroscopy, Ger. (East) DD 225, 217 (Cl. G01N 21/74), 6 pp, 24 Jul. 1985. Appl. 264. 204, 15 Jun. 1984.

663. **Littlejohn, D., Duncan, I. S., Hendry, J. B. M., Marshall, J., and Ottaway, J. M.,** Comparison of uncoated, pyro-coated and totally pyrolytic graphite tubes for the HGA-500 electrothermal atomizer, *Spectrochim. Acta, Part B,* 40B(10-12), 1677-87, 1985.

664. **Kurfuerst, U.,** Solid sample insertion systems and L'vov platform effect, *Fresnius' Z. Anal. Chem.,* 322(7), 660-5, 1985.

665. **Bazhov, A. S., Zekki, L. N., and Klimenkova, N. B.,** Use of flameless metallic atomizers for AA analysis, *Issled. v Obl. Khim. i Fiz. MetodovAnal. Mineral. Syr' ya, Alma-Ata,* 53-64, 1985. From *Zh. Anal. Khim.,* Abstr. No. 22683, 1985.

666. **Scholl, W.,** Experience with a PE Zeeman AA spectrometer 5000 in the analysis of solids, *Fresnius' Z. Anal. Khim.,* 322(7), 681-4, 1985 (in German).

667. **Falk, H., Hoffman, E., and Luedke, C.,** Multielement trace analysis of µl samples in the ppb range by FANES (furnace atomization non-thermal excitation spectrometry), *Proc. 6th High-Purity Mater. Sci. Technol. Int. Symp.,* 2, 77-78, 1985.

668. **Ohls, K. and Koch, K. H.,** Optimization of flameless AAS with Smith/Hieftje correction, *Anal. Proc. (London),* 22(12), 380-2, 1985.

669. **Hausen, M.,** Organization of the working place and weighing requirements for the analysis of solids by AAS, *Fresnius' Z. Anal. Chem.,* 322(7), 666-8, 1985.

670. **Costantini, S., Giordano, R., Rizzico, M., and Benedetti, F.,** Applicability of anodic-stripping voltammetry and graphite furnace AA spectrometry to the determination of antimony in biological matrixes: a comparative study, *Analyst (London),* 110(11), 1355-9, 1985.

671. **Falk, H., Tilch, J., Schmidt, K. P., and Thiemann, H. J.,** Electrothermal atomizer, Ger. (East) DD 221, 019 (Cl. G01 N 21/74), 8 pp, 10 Apr. 1985. Appl. 259, 069, 30 Dec. 1983.

672. Atomizer tube for flameless AAS, Ger. (East) DD 221, 279 (Cl. G01 N 21/74), 8 pp, 17 Apr. 1985. Appl. 259, 068, 30 Dec. 1983.

673. **Atnashev, Y. B., Korepanov, V. E., and Muzgin, V. N.,** Analytical signal distortion in electrothermal atomic absorption spectroscopy, *Zh. Prikl. Spektrosk.,* 45, 737, 1986.

674. **Dittrich, K. and Mandry, R.,** Investigations into the improvement of the analytical applications of the hydride technique in atomic absorption spectroscopy by matrix modification and graphite furnace atomization. I. Analytical results, *Analyst (London),* 111, 269, 1986 (in German).

675. **Dittrich, K., Hanisch, B., and Staerk, H. J.,** Molecule formation in electrothermal atomizers: interferences and analytical possibilities by absorption, emission and fluorescence processes, *Z. Anal. Chem.,* 324, 497, 1986.

676. **Schlemmer, G. and Welz, B.,** Palladium and magnesium nitrates, a more universal modifier for graphite furnace atomic absorption spectroscopy, *Spectrochim. Acta,* 41B, 1157, 1986.

677. **Nakamura, S. and Kubota, M.,** Effects of alkali and alkaline earth salts on signal absorbance in electrothermal AAS, *Bunseki Kagaku,* 35, 961, 1986 (in Japanese).

678. **Nakamura, S. and Kubota, M.,** Effect of anions and acids on signal absorbance in electrothermal atomic absorption spectroscopy, *Bunseki Kagaku,* 35, 844, 1986.

679. **Atnashev, Y. B., Korepanov, V. E., and Muzgin, V. N.,** Analytical signal distortion in electrothermal atomic absorption spectroscopy, *Zh. Prikl. Spektrosk.,* 45, 737, 1986.

680. **Calmet Fontane, J.,** Importance of the Zeeman effect correction in atomic absorption, *Tec. Lab.,* 10, 272, 1986.

681. **Macdonald, L. R., O'Haver, T. C., Ottaway, B. J., and Ottaway, J. M.,** Background atomic absorption in graphite furnace atomic absorption spectroscopy, *J. Anal. At. Spectrom.,* 1, 485, 1986.

682. **Carnrick, G. R., Lumas, B. K., and Barnett, W. B.,** Analyses of solid samples by graphite furnace atomic absorption correction, *J. Anal. At. Spectrom.,* 1, 443-7, 1986.

683. **Barzev, A., Dobreva, D., Futekov, L., Rusev, V., Bekyarov, G., and Toneva, G.,** Determination of detection limits in graphite furnace atomic absorption spectroscopy by using ensamble summation of signals, *Fresnius' Z. Anal. Chem.,* 325, 255-7, 1986.

684. **Nakamura, S., Kobayashi, Y., and Kubota, M.,** Characteristics of absorbance signal in electrothermnal atomic absorption spectroscopy. Temperature, time and absorbance at the peak, *Bunseki Kagaku,* 35, 824-6, 1986.

685. **Voth-Beach, L. M. and Shrader, D. E.,** Graphite furnace atomic absorption spectroscopy: new approach to matrix modification, *Spectroscopy (Springfield, OR),* 1, 49-50, 1986.

686. **Chang, S. B.,** Factors affecting atomization in graphite furnace atomic absorption spectroscopy, 1985, Avail. NLC From Diss. Abstr. Int. B, 47(3), 1026, 1986.

687. **L'vov, B. V., Nikolaev, V. G., Norman, E. A., Polzik, L. K., and Mojica, M.,** Theoretical calculation of the characteristic mass in graphite furnace atomic absorption spectroscopy, *Spectrochim. Acta,* 41B, 1043, 1986.

688. **Welz, B.,** Abuse of the analyte addition technique in atomic absorption spectroscopy, *Fresnius' Z. Anal. Chem.,* 325, 95-101, 1986.

689. **Jin, F. and Liu, F.,** Study on the interferences by matrix and perchloric acid and their elimination in graphite furnace atomic absorption spectroscopy, *Guangpuxue Yu Guangpu Fenxi,* 6, 45-8, 1986 (in Chinese).

690. **Sturgeon, R. E. and Arlow, J. S.,** Atomization in graphite FAAS: atmospheric pressure vis-a-vis vacuum vaporization, *J. Anal. At. Spectrom.,* 1(5), 359-63, 1986.

691. **De Loos-Vollebregt, M. T. and De Galan, L.,** Extended range Zeeman AA spectrometry based on a 3-field a.c. magnet, *Spectrochim. Acta Part B,* 41B(8), 825-35, 1986.

692. **Yang, X.,** Determination of the polarized radiation of Zeeman effect AA spectrometer, *Fenxi Ceshi Tongbao,* 4(3), 34-6, 1985. CA 23, 105: 163945n, 1986.

693. **Fuwa, K.,** Analysis of environmental materials using spectroscopic methods, *Fresnius' Z. Anal. Chem.,* 324(6), 531-6, 1986.

694. **Sturgeon, R.,** Graphite furnace AAS: fact and fiction, *Fresnius' Z. Anal. Chem.,* 324(8), 807-18, 1986.

695. **Liu, J.,** Probe atomization technique in graphite furnace AAS, *Fenxi Ceshi Tongbao,* 4(2), 1-6, 1985 (in Chinese). CA 23, 105: 163916d, 1986.

696. **He, H.,** Application of Zeeman AAS, *Fenxi Ceshi Tongvao,* 4(1), 23-7, 1985. CA 23, 105: 163914b, 1986.

697. **Slavin, W. and Carnrick, G. R.,** Interferences in graphite furnace AAS continuum background correction. A survey, *At. Spectrosc.,* 7(1), 91-3, 1986.

698. **Rohl, R.,** Objective function aiding in the selection of time windows for the integration of signal peaks near the detection limit by the base line offset correction method, *Anal. Chem.,* 58, 2891, 1986.

699. **Ma, Z. and He, J.,** Application of self-absorption effect background correction technique in graphite furnace AAS, *Fenxi Huaxue,* 14(4), 265-9, 1986.

700. **Pluhar, Z.,** Metal electrothermal atomizers for atomic spectroscopy, *Sci. Komun.,* 11, 37-43, 1986.

701. **Murphy, L. C., Almeida, M. C., Dulude, G. R., and Sotera, J. J.,** Minimizing matrix interferences in furnace atomic absorption spectrometry, *Spectroscopy (Springfield, OR),* 1(3), 39, 1986.

702. **Voth-Beach, L. M. and Shrader, D. E.,** Investigations of a reduced palladium chemical modifier for graphite furnace atomic absorption spectroscopy, *J. Anal. At. Spectrosc.,* 2, 45, 1986.

703. **Tatro, M. E.,** Methods development logic for furnace AAS analyses of clinical matrixes, *Spectroscopy (Springfield, OR),* 1, 22, 24, 1986.

704. **Schlemmer, G. and Welz, B.,** Use of alternative gases in graphite-tube furnace AAS (atomic absorption spectroscopy), *Fortschr. Atomspektrom. Spurenanal.,* 2, 65, 1986 (in German).

705. **Roesick, U., Parlow, A., and Fuerstenau, G.,** Automatic signal analysis in graphite furnace AAS, *Fortschr. Atomspektrom. Spurenanal.,* 2, 225,1986.

706. **De Galan, L.,** New directions in optical atomic spectrometryy, *Anal. Chem.,* 58(6), 697A-698A, 1986.

707. **Grobenski, Z., Lehmann, R., Radziuk, B., and Voellkopf, U.,** Graphite furnace AAS detection limits in real samples, *At. Spectrosc.,* 7(2), 61-3, 1986.

708. **Rettberg, T. M. and Holcombe, J. A.,** Interference minimization using second surface atomizer for furnace atomic absorption , *Spectrochim. Acta Part B,* 41B(4), 377-89, 1986.

709. **Sadagov, Y. M., Eristavi, V. D., and Kutsiava, N. A.,** Optimization of the analytical characteristics of a Zeeman AA spectrometer, *Soobshch. Akad. Nauk. Gruz. SSR,* 121(2), 321-4, 1986 (in Russian).

710. **Ottaway, J. M., Carroll, J., Cook, S., Corr, S. P., Littlejohn, D., and Marshall, J.,** Developments in probe design for electrothermal AAS, *Fresnius' Z. Anal. Chem.,* 323(7), 742-7, 1986

711. **Feely, R. A., Massoth, G. J., Baker, E. J., Gendron, J. F., Paulson, A. J., and Crecelius, E. A.,** Seasonal and vertical variations in the elemental composition of suspended and settling particulate matter in Puget Sound, Washington, *Estuarine Coastal Shelf Sci.,* 22(2), 215-39, 1986.

712. **Pelieva, L. A., Lazika, Z. V., and Nagorskii, V. P.,** Optimization of conditions of AA measurements in a tube furnace-tantalum boat system, *Zavods. Lab.,* 52(4), 32-5, 1986.

713. **Schlemmer, G. and Welz, B.,** Influence of the tube surface on atomization behavior in a graphite tube furnace, *Fresnius' Z. Anal. Chem.,* 323(7), 703-9, 1986.

714. **Dittrich, L. and Mandry, R.,** Investigations into the improvement of the analytical application of the hydride technique in atomic absorption spectrometry by matrix modification and graphite furnace atomization. I. Analytical results, *Analyst (London),* 111(3), 269-75, 1986.

715. **Frech, W., Baxter, D. C., and Huetsch, B.,** Spatially isothermal graphite furnace for atomic absorption spectrometry using side-heated cuvettes with integrated contacts, *Anal. Chem.,* 58(9), 1973-7, 1986.

716. **Hashimoto, M. S., Yamada, H., Ohishi, K., and Yasuda, K.,** Dynamic range enhancement for flameless Zeeman atomic absorption spectroscopy, *Anal. Sci.,* 2(2), 109-12, 1986.

717. **Dittrich, K. and Mandry, R.,** Investigations into the improvement of the analytical application of the hydride technique in atomic absorption spectroscopy by matrix modification and graphite furnace atomization. II. Matrix interferences in the gaseous phase of hydride atomic absorption spectroscopy, *Analyst (London),* 111(3), 277-80, 1986.

718. **Holcombe, J. A. and Harnly, J. M.,** Minimization of background correction errors using nonlinear estimates of the changing background in carbon furnace AAS, *Anal. Chem.,* 58, 2606, 1986.

719. **Sperling, K. R.,** The tube-in-tube technique in electrothermal AAS IV. The influence of vapor movements on the development of the analytical signal, *Fortschr. Atomspektrom. Spurenanal.,* 2, 37, 1986.

720. **Zhou, N., Frech, W., and Lundberg, E.,** Some methods for studying atom formation mechanism in graphite furnace AAs, *Guangpuxue Yu Guangpu Fenxi,* 5(5), 13-21, 1985 (in Chinese), CA 5, 104: 61117f, 1986.

721. **Tatro, M. E.,** Methods development logic for furnace AAS, *Spectroscopy (Springfield, OR),* 1(2), 22, 1986.

722. **Eristavi, V. D., Sadagov, Y. M., Ivanov, V. K., and Sharashidze, P. A.,** AA spectrometer with electrothermal atomizer in a magnetic field, *Zh. Prikl. Spektrosk.,* 44(2), 337-40, 1986 (in Russian), CA 10, 104:159202e, 1986.

723. **Parks, E. J., Manders, W. F., Johanneson, R. B., and Brinckman, F. E.,** Characterization of organometallic polymers by size exclusion chromatography on preconditioned columns, *J. Chromatogr.,* 351(3), 475-87, 1986.

724. **Murphy, L. C., Almeida, M. C., Dulude, G. R., and Sotera, J. J.,** Minimizing matrix interferences in furnace AAS, *Spectroscopy (Springfield, OR),* 1(3), 39-43, 1986.

725. **Harada, K.,** Flameless atomizer for AAS, Jpn. Tokkyo Koho JP 60 35, 022 (85 35, 022) (Cl. G01 N 21/31), 4 pp, 12 Aug. 1986. Appl. 79/80, 163, 27 Jun. 1979.

726. **Aleksenko, A. N., Guletskii, N. N., Korennoi, E. P., Turkin, Y. I., and Utgof, A. A.,** AA method for analysis of solid rock samples using electrothermal atomization, *Otsenka Prognoz. Resursov i Metody Izuch Rud. Mestorozhd., M.* 48-53, 1984. From *Zh. Geol. Abstr,* No. 9D 52, 1985. CA 11, 104: 1792313, 1986.

727. **Ortner, H. M., Birzer, W., Welz, B., Schlemmer, G., Curtius, J. A., Wegscheider, W., and Sychra, V.,** Surfaces and materials of electrothermal atomic absorption spectrometry — more than a merely morphological study, *Fresnius' Z. Anal. Chem.,* 323(7), 681-8, 1986.

728. **Huettner, W. and Busche, C.,** Structure and reactivity of carbon materials used in atomization furnaces, *Z. Anal. Chem.,* 323(7), 674-80, 1986.

729. **Kuellmer, G. and Whiteside, P. J.,** Pyrographite tubes (TPC) for AAS (atomic absorption spectroscopy), *LP Spec. Chromatogr. Spektrosk.,* 156-163, 1986 (in German).

730. **Welz, B., Schlemmer, G., and Ortner, H. M.,** Scanning electron microscopy studies on surfaces from electrothermal atomic absorption spectrometry. II. Total pyrolytic graphite platforms in pyrographite coated polycrystalline electrographite tubes, *Spectrochim. Acta Part B,* 41B, 567-89, 1986.

731. **Byrne, J. P., Chakrabarti, C. L., Chang, S. B., Tan, C. K., and Delgado, A. H.,** A thermodynamic equilibrium model for atomization in graphite furnace atomic absorption spectrometry, *Fresnius' Z. Anal. Chem.,* 324(5), 448-55, 1986.

732. **De Loos-Vollebregt, M. T. C., and De Galan, L.,** Stray light in Zeeman and pulsed hollow cathode lamp atomic absorption spectrometry, *Spectrochim. Acta Part B.,* 41B(6), 597-610, 1986.

733. **Caroli, S.,** Hollow cathode lamps as excitation sources for analytical atomic spectroscopy, *Z. Anal. Chem.,* 324(5), 442-7, 1986.

734. **Zheng, Y.,** Role of oxygen in atomization in the graphite furnace atomizer, *Guangpuxue Yu Guangpu Fenxi,* 6(3), 3308, 1986 (in Chinese). CA 21, 105: 126021n, 1986.

735. **Almeida, M. C. and Seitz, W. R.,** Carbide-treated graphite cuvettes for electrothermal atomization prepared by impregnation with metal chlorides, *Appl. Spectrosc.,* 40(1), 4-8, 1986.

736. **Besse, A., Rosopulo, A., Busche, C., and Kuellmer, G.,** Solid AAS (AA spectrophotometry) with deuterium background compensation, *Labor Praxis,* 10(1), 64-6, 71-2, 1986.

737. **Slavin, W.,** Flames, furnaces, plasmas: how do we choose?, *Anal. Chem.,* 58(4), 589A, 1986.

738. **Pelieva, L. A. and Uzunbadzhakov, A. S.,** Rapid heating program for the graphite tube furnace for an electrothermal atomizer, *Zh. Prikl. Spektrosk.,* 44(2), 293-5, 1986.

739. **Nakamura, S. and Kubota, M.,** Characteristics of fast response system for electrothermal AAS, *Bunseki Kagaku,* 35(1), 61-3, 1986.

740. **Sturgeon, R. E., Siu, K. W. M., Gardner, G. J., and Berman, S. S.,** Carbon-oxygen reaction in graphite furnace AAS, *Anal. Chem.,* 58(1), 42-50, 1986.

741. **Tsai, P. H. and Chang, T. Y.,** Chemical analysis of ESR (electroslag remelting) slags: AAS and flame emission spectroscopy, *Kung Yeh Chi Shu,* 137, 75-81, 1985. CA 7, 104: 101531u, 1986.

742. **Niemczyk, T. M. and Yin, I. H.,** A method for the evaluation of background-correction performances in graphite furnace atomic absorption spectroscopy, *Appl. Spectrosc.,* 40, 882, 1986.

743. **Lovett, R. J.,** The influence of temperature on absorbance in graphite furnace atomic absorption spectrometry, *Appl. Spectrosc.,* 40, 778, 1986.

744. **Brueggemeyer, T. W. and Fricke, F. L.,** Comparison of furnace atomization behavior of aluminum from standard and thorium treated L'vov platform, *Anal. Chem.,* 58, 1143, 1986.

745. **Baxter, D. C., Duncan, I. S., Littlejohn, D., Marshall, J., Ottaway, J. M., Fell, G. S., and Ataman, O. Y.,** Low-resolution monochromator system for electrothermal atomic spectrometry with probe atomization, *J. Anal. At. Spectrom.,* 1, 29, 1986.

746. **Katskov, D. A. and Kopeikin, V. A.,** Development of the atomic absorption method for thermochemical investigations, *Zh. Prikl. Spektrosk.,* 44(6), 922-9, 1986 (in Russian).

747. **Koshy, V. J. and Garg, V. N.,** Atomic absorption spectroscopy and trace element analysis, *J. Sci. Ind. Res.,* 45(6), 294-314, 1986.

748. **Nakamura, S. and Kubota, M.,** Effects of anions and acids on signal absorbance in electrothermal atomic absorption spectroscopy, *Bunseki Kagaku,* 35, 844, 1986 (in Japanese).

749. **Egila, J. N., Littlejohn, D., Ottaway, J. M., and Xiao, Q. S.,** Clinical applications of electrothermal atomic absorption spectroscopy with Zeeman-effect background correction, *Anal. Proc. (London),* 23, 426, 1986.

750. **Tatro, M. E.,** Methods development logic for furnace AAS analyses of clinical matrixes , *Spectroscopy (Springfield, OR),* 1, 22, 1986.

751. **Hiratsuka, H., Miyashita, M., and Ukaji, M.,** Apparatus for dissolving films for analysis, Jpn. Kokai Tokkyo Koho JP 61, 144, 545 (86, 144, 545) (Cl. G01 N 1/28), 3 pp, 02 July 1986. Appl. 84/267, 185. 18 Dec. 1984.

752. **Grobecker, K. H., Kurfuerst, U., and Stoeppler, M.,** Significance of the homogeneity of solid samples in direct analysis and in analysis after digestion, *Fortschr. Atomspektrom. Spurenanal.,* 2, 81-9, 1986.

753. **Wedl, W., Kolb, A., and Mueller-Vogt, G.,** Reactions of oxide-forming elements in a graphite-tube furnace in AAS (atomic absorption spectroscopy), *Fortschr. Atomspektrom. Spurenanal.,* 2, 15-24, 1986(in German).

754. **Sperling, K. R.,** The tube-in-tube technique in electrothermal AAS (atomic absorption spectroscopy). IV. The influence of vapor movements on the development of the analytical signal, *Fortschr. Atomspektrom. Spurenanal.,* 2, 37, 1986.

755. **Chakrabarti, C. L., Delgado, A. H., Chang, S. B., Falk, H., Huton, T. J., Runde, G., Sychra, V., and Dolezal, J.,** Temperature distribution in a tungsten electrothermal atomizer, *Spectrochim. Acta,* 41B, 1075, 1986.

756. **Bily, J. and Cermakova, L.,** Study of volatilization of some sodium salts in flameless atomic absorption spectroscopy, *Anal. Lett.,* 19, 755, 1986 (in French).

757. **Almeida, M. C. and Seitz, W. R.,** Carbide-treated graphite cuvettes for electrothermal atomization prepared by impregnation with metal chlorides, *Appl. Spectrosc.,* 40, 4, 1986.

758. **Hickson, C. J. and Juras, S. J.,** Sample contamination by grinding, *Can. Mineral.,* 24, 585-9, 1986.

759. **Carnrick, G. R., Barnett, W., and Slavin, W.,** Spectral interferences using the Zeeman effect for furnace atomic absorption spectroscopy, *Spectrochim. Acta,* 41B, 991, 1986.

760. **Akkari, K. H., Frans, R. F., and Lavy, T. L.,** Factors affecting degradation of MSMA in soil, *Weed Res..* 34, 781, 1986.

761. **Dittrich, K., Mandry, R., and Wennrich, R.,** Electrothermal evaporation in graphite tubes for halogen determination by molecule formation for microanalysis by combination with laser evaporation and for atomization for the hydride technique, *Fortschr. Atomspektrom. Spurenanal.,* 2, 91-121, 1986.

762. **Wang, P. and Lin, T.,** Concept of "appearance temperature" and its application in the study of the atomization mechanism, *Guangpuxue Yu Guangpu Fenxi,* 6, 56, 1986.

763. **Matousek, J. P. and Powell, H. K. J.,** Halogen assisted volatilization in electrothermal atomic absorption spectroscopy reduction of memory effects from refractory carbides, *Spectrochim. Acta Part B,* 41B, 1347, 1986.

764. **Agness,D.,** Recent advances in graphite furnace analysis, *Spher. Publ. R. Soc. Chem.,* 61, 223, 1986.
765. **Kawakubo, T., Koyama, T., and Odajinsa, H.,** Carbon furnace for atomic absorption analysis, Jpn. Kokai tokkyo koho JP 61, 239, 145 (86, 239, 145) (Cl. G01N 21/31), 5 pp, 24 Oct. 1986. Appl. 85/79,280, 16 Apr. 1985.
766. **Rademeyer, C. J., Human, H. G. C., and Faure, P. K.,** The dynamic wall and gas temperature distributions in a graphite furnace atomizer *Spectrochim. Acta Part B,* 41B(5), 439, 1986.
767. **Matousek, J. P., Orr, B. J., and Selby, M.,** Interferences due to easily ionized elements in a microwave-induced plasma system with graphite-furnace sample introduction, *Spectrochim. Acta Part B,* 41B (5), 415-29, 1986.
768. **Rademeyer, C. J. and Human, H. G. C.,** Wall and gas temperature distributions in a graphite furnace atomizer, *Prog. Anal. Spectrosc.,* 9(2), 167-235, 1986.
769. **Schlemmer, G. and Welz, B.,** Basic atomic spectrometric analysis, *Laboratoriumsmedizin,* 10(5), 160-5, 1986 (in German).
770. **Itho, K. and Atsuya, I.,** Evaluation of graphite miniature cup for electrothermal atomic absorption spectroscopy. I. Applications to liquid samples, *Bunseki Kagaku,* 35, 530, 1986 (in Japanese).
771. **Fengming, J. and Fengzhi, L.,** Study of the interferences by matrix and perchloric acid and their elimination in graphite furnace atomic absorption spectroscopy, *Guangpuxue Yu Guangpu Fenxi,* 6, 45, 1986.
772. **Harnly, J. M.,** Aerosol deposition-carbon furnace atomization for simultaneous multielement atomic absorption spectroscopy, *J. Anal. At. Spectrom.,* 1, 287, 1986.
773. **Hashimoto, H. S., Yamada, H., Ohishi, K., and Yasuda, K.,** Dynamic range enhancement for flameless Zeeman atomic absorption spectroscopy, *Anal. Sci.,* 2, 109, 1986.
774. **Galli, P. and Magistrell, C.,** Recent instrumental developments in the background correction systems for flame and graphite atomic absorption, *Chem. Eval.,* 2, 195, 1986.
775. **Akkari, K. H., Frans, R. E., and Lavy, T. L.,** Factors affecting degradation of MSMA in soil, *Weed Sci.,* 34, 781, 1986.
776. **Roesick, U., Parlow, A., and Fuerstanan, G.,** Automatic signal analysis in graphite furnace AAS (atomic absorption spectroscopy), *Fortschr. Atomspektrom. Spurenanal.,* 2, 225, 1986.
777. **Tamm, R.,** Contamination-free, accurate and rapid dilution in trace analysis, *Fortschr. Atomspektrom. Spurenanal.,* 2, 275, 1986.
778. **L'vov, B. V., Nikolaev, V. G., Norman, E. A., Polzik, L. K., and Mojica, M.,** Theoretical calculation of the characteristic mass in graphite furnace atomic absorption spectroscopy, *Spectrochim. Acta,* 41B, 1043,1986.
779. **Pelieva, L. A. and Lakiza, Z. V.,** Optimization of conditions of atomic absorption measurements in a tube furnace-tantalum boat system, *Zavod. Lab.,* 52(4), 32, 1986.
780. **Pelieva, L. A. and Uzunbadzhakov, A. S.,** Rapid heating program for the graphite tube furnace of an electrothermal atomizer, *Zh. Prikl. Spektrosk.,* 44, 293, 1986 (in Russian).
781. **Rademeyer, C. J. and Human, H. G. C.,** Wall and gas temperature distributions in a graphite furnace atomizer, *Human Prog. Anal. Spectrosc.,* 9, 167, 1986.
782. **Randall, L., Donard, O. F. X., and Weber, J. H.,** Speciation of N-butyltin compounds by atomic absorption spectroscopy with an electrothermal quartz furnace after hydride generation, *Anal. Chim. Acta,* 184, 197, 1986.
783. **Zhighong, M. and Jianling, H.,** Application of self-absorption effect background correction technique in graphite furnace atomic absorption spectroscopy, *Fenxi Huaxue,* 14, 265, 1986.
784. **Nakamura, S., Kobayashi, Y., and Kubuta, M.,** Characteristics of absorbance signal in electrothermal atomic absorption spectroscopy. Temperature, time and absorbance at the peak, *Bunseki Kagaku,* 35, 824, 1986.
785. **Nakamura, S., Kobayashi, Y., and Kuboto, M.,** Temperature of a tungsten ribbon furnace in electrothermal atomic absorption spectroscopy, *Spectrochim. Acta,* 41B, 817, 1986.
786. **Nakamura, S. and Kubota, M.,** Comparison of furnace materials for electrothermal atomic absorption spectroscopy on the basis of chemical thermodynamic calculation, *Bunseki Kagaku,* 35, 548, 1986.
787. **Nakamura, S. and Kubota, M.,** Characteristics of fast response system for electrothermal atomic absorption spectroscopy, *Bunseki Kagaku,* 35, 61, 1986.
788. **Sturgeon, R. E.,** Graphite furnace atomic absorption spectroscopy: fact and fiction, *Z. Anal. Chem.,* 324, 807, 1986.
789. **Sturgeon, R. E. and Arlow, J. S.,** Atomization in graphite furnace atomic absorption spectroscopy: atmospheric pressure vis-a-vis vacuum vaporization, *J. Anal. At. Spectrom.,* 1, 359, 1986.
790. **Torsi, G. and Palmisano, A.,** Particle collection mechanism and efficiency in electrostatic accumulation furnace for electrothermal atomic spectroscopy, *Spectrochim. Acta,* 41B, 257, 1986.
791. **Matsusaki, K.,** Removal-effect of carbon powder on chloride interference in the determination of metals by atomic absorption spectroscopy with electrothermal atomization, *Technol. Rep. Yamaguchi Univ.,* 3, 385, 1986.
792. **Wendl, W.,** Investigations on chemical reactions in graphite furnace AAS (atomic absorption spectrometry), *Fresnius' Z. Anal. Chem.,* 323(7), 726-9, 1986.

793. **Itoh, K. and Atsuya, I.,** Evaluation of graphite miniature cup for electrothermal AA spectrometry. I. Application to liquid samples, *Bunseki Kagaku,* 35(6), 530, 1986.

794. **Littlejohn, D., Carroll, J., Quinn, A. M., Ottaway, J. M., and Falk, H.,** Comments on the characterization of an atomizer for furnace atomic-non-thermal excitation spectrometry (FANES), *Fresnius' Z. Anal. Chem.,* 323(7), 762-6, 1986.

795. **Stockton, R. A.,** Graphite furnace atomic absorption and IC argon plasma emission spectrometers as element specific detectors for the determination of trace elements and trace element compounds, 186 pp, 1985. Avail. Univ. Microfilms Int. Order No. DA 8528385. CA select 18, 105:57294q, 1986.

796. **Guenther, H. and Findeisen, B.,** Requirements for carbon materials used in flameless AA spectrometry, *Freiberg, Forschungsh.,* A730, 130-47, 1986 (in German).

797. **Shaole, W., Chakrabarti, C. L., and Rogers, J. J.,** A theoretical and experimental study of platform furnace and probe furnace in graphite furnace atomic absorption spectroscopy, *Anal. Spectrosc.,* 10, 111, 1987.

798. **Barzev, A., Dobreva, D., Futekov, L., Rusev, V., Bekjarev, G., and Toneva, G.,** Determination of detection limits in graphite furnace atomic absorption spectroscopy by using ensemble summation of signals, *J. Anal. Chem.,* 325, 255, 1987.

799. **Yu, F.,** Solid sampling technique for graphite furnace atomic absorption spectroscopy, *Fenxi Ceshi Tongbao,* 6, 2035, 1987 (in Chinese).

800. **Allen, E. and Jackson, K. W.,** Fast-response system for signal acquisition and processing in electrothermal atomic absorption spectroscopy, *Anal. Chim. Acta,* 192, 355, 1987.

801. **Fietkau, R.,** Development of rapid slurry methods for flame and direct current plasma emission and graphite furnace atomic absorption analysis of solid animal tissue, 237 pp, 1986. Avail. Univ. Microfilms Int. Order No. DA 8705835. From Diss. Abstr. No. Int. B 47, 4486, 1987.

802. **Shuttler, I. L. and Delves, H. T.,** Between-batch variability of thermal characteristics of commercially available L'vov platform graphite tube atomizers and analytical accuracy in electrothermal atomization AAS, *J. Anal. At. Spectrom.,* 2, 171, 1987.

803. **Muzgin, V. N., Atnashev, Y. B., Korepanov, V. E., and Pupyshev, A. A.,** Electrothermal atomic absorption and atomic fluorescence spectroscopy with a tungsten-coil atomizer, *Talanta,* 34, 197, 1987.

804. **Chakrabarti, C. L., Chang, S. B., Thong, P. W., Huston, T. J., and Wu, S.,** Studies of atomization from a graphite platform in graphite furnace atomic absorption spectroscopy, *Talanta,* 34, 259, 1987.

805. **Torsi, G. and Palmisano, F.,** Spray deposition versus single-drop deposition for calibration of an electrostatic accumulation furnace for electrothermal atomization atomic absorption spectroscopy, *J. Anal. At. Spectrom.,* 2, 51, 1987.

806. **McNally, J. and Holcombe, J. A.,** Existence of microdroplets and dispersed atoms on the graphite surface in electrothermal atomizers, *Anal. Chem.,* 59, 1105, 1987.

807. **L'vov, B. V.,** Recent advances in the theory of atomization in graphite furnace atomic absorption spectroscopy: the oxygen-carbon alternatives. Plenary lecture, *Analyst (London),* 112, 355, 1987.

808. **Routh, M. W.,** A comparison of atomic spectroscopic technique: Atomic absorption , inductively coupled plasma and direct current plasma, *Spectroscopy (Springfield, OR),* 2(2), 45, 1987.

809. **Ohta, K. and Su, S. Y.,** Electrothermal atomic absorption spectroscopy with improved tungsten tube atomizer, *Anal. Chem.,* 59, 539, 1987.

810. **McGeorge, S. W.,** Imaging systems: detectors of the past, present and future, *Spectroscopy (Springfield, OR),* 2, 26, 1987.

811. **Coates, J. P.,** Computers in spectroscopy. III, *Spectroscopy (Springfield, OR),* 2, 14, 1987.

AUTHOR INDEX

SUBJECT INDEX

ALKALI METALS
(CESIUM, LITHIUM, POTASSIUM, RUBIDIUM, SODIUM)

CESIUM, Cs   (ATOMIC WEIGHT 132.905)

Instrumental Parameters:

Wavelength	852.1 nm
Slit width	320 μm
Bandpass	1 nm
Light source	Electrodeless Discharge Lamp
Purge gas	Argon
Sample size	25 μL
Furnace	Cylindrical cuvette/Pyrolytic graphite

Standard Operating Conditions:

Optimum char temperature	900°C
Optimum atomization temperature	1900°C
Sensitivity	5 pg/1% absorption
Sensitivity check	0.01 μg/mL
Working range	0.05-0.1 μg/mL
Background correction	Required only to correct light scattering or non-specific absorption from the sample containing high dissolved solids.

General Note:
1. Use commercial standard or a previously analyzed sample as a working standard.
2. Use ultra clean glass and plastic ware, soaked in 1:5 nitric acid solution and rinsed thoroughly with deionized distilled water.
3. All analytical solutions should be at least 0.2-0.5% (v/v) in nitric acid.
4. Use 0.2% (v/v) sulfuric acid as matrix modifier.
4. Check for blank values on all reagents including water.
5. Dilution is recommended when sample exhibits greater than 0.5 absorbance units.

LITHIUM, Li   (ATOMIC WEIGHT 22.9898)

Instrumental Parameters:

Wavelength	670.8 nm
Slit width	160 μm
Bandpass	0.5 nm
Light source	Hollow Cathode Lamp
Purge gas	Argon or Nitrogen
Sample size	25 μL
Furnace	Cylindrical cuvette/Pyrolytic graphite

Standard Operating Conditions:

Optimum char temperature	1000°C
Optimum atomization temperature	2600°C
Sensitivity	4 pg/1% absorption
Sensitivity check	0.01 μg/mL
Working range	0.05-0.5 μg/mL
Background correction	Required only to correct light scattering or non-specific absorption from the sample containing high dissolved solids.

General Note:

1.   Use commercial standard or a previously analyzed sample as a working standard.
2.   Use ultra clean glass and plastic ware, soaked in 1:5 nitric acid solution and rinsed thoroughly with deionized distilled water.
3.   Prepare all analytical solutions in at least 0.2 (v/v) nitric acid or 0.5% hydrochloric acid.
4.   Check for blank values on all reagents including water.
5.   Dilution is recommended when sample exhibits greater than 0.5 absorbance units.

POTASSIUM, K   (ATOMIC WEIGHT 39.102)

Instrumental Parameters:

Wavelength	766.5 nm
Slit width	320 μm
Bandpass	1 nm
Light source	Hollow Cathode Lamp
Purge gas	Argon or Nitrogen
Sample size	25 μL
Furnace	Cylindrical cuvette/Pyrolytic graphite

Standard Operating Conditions:

Optimum char temperature	1000°C
Optimum atomization temperature	2200°C
Sensitivity	4 pg/1% absorption
Sensitivity check	0.005-0.01 μg/mL
Working range	0.004-0.1 μg/mL
Background correction	Required only to correct light scattering or non-specific absorption from the sample containing high dissolved solids.

General Note:
1. Use commercial standard or a previously analyzed sample as a working standard.
2. Use ultra clean glass and plastic ware, soaked in 1:5 nitric acid solution and rinsed thoroughly with deionized distilled water.
3. All analytical solutions should be prepared in at least 0.2% (v/v) hydrochloric acid.
4. Check for blank values on all reagents including water.
5. Dilution is recommended when sample exhibits greater than 0.5 absorbance units.

RUBIDIUM, Rb   (ATOMIC WEIGHT 85.47)

Instrumental Parameters:

Wavelength	780.0 nm
Slit width	320 μm
Bandpass	1 nm
Light source	Hollow Cathode or Electrodeless Discharge Lamp
Purge gas	Argon
Sample size	25 μL
Furnace	Cylindrical cuvette/Pyrolytic graphite

Standard Operating Conditions:

Optimum char temperature	800°C
Optimum atomization temperature	1900°C
Sensitivity	3 pg/1% absorption
Sensitivity check	0.005-0.05 μg/mL
Working range	0.05-0.1μg/mL
Background correction	Required only to correct light scattering or non-specific absorption from the sample containing high dissolved solids.

General Note:
1.   Use commercial standard or a previously analyzed sample as a working standard.
2.   Use ultra clean glass and plastic ware, soaked in 1:5 nitric acid solution and rinsed thoroughly with deionized distilled water.
3.   All analytical solutions should be at least 0.2-0.5% (v/v) in nitric acid.
4.   Check for blank values on all reagents including water.
5.   Dilution is recommended when sample exhibits greater than 0.5 absorbance units.

SODIUM, Na (ATOMIC WEIGHT 132.905)

Instrumental Parameters:

Wavelength	589.5 nm
Slit width	160 μm
Bandpass	0.5 nm
Light source	Hollow Cathode Lamp
Purge gas	Argon or Nitrogen
Sample size	25 μL
Furnace	Cylindrical cuvette/Pyrolytic graphite

Standard Operating Conditions:

Optimum char temperature	1500°C
Optimum atomization temperature	2700°C
Sensitivity	4 pg/1% absorption
Sensitivity check	0.005-0.01μg/mL
Working range	0.004-0.05 μg/mL
Background correction	Required only to correct light scattering or non-specific absorption from the sample containing high dissolved solids.

General Note:
1. Use commercial standard or a previously analyzed sample as a working standard.
2. Use ultra clean glass and plastic ware, soaked in 1:5 nitric acid solution and rinsed thoroughly with deionized distilled water.
3. All analytical solutions should be prepared in at least 0.5% (v/v) in hydrochloric acid.
4. Check for blank values on all reagents including water.
5. Dilution is recommended when sample exhibits greater than 0.5 absorbance units.

REFERENCES

1. **Yasuda, S. and Kakiyama, H.,** Study of absorption spectra for alkali and alkaline earth metal salts in flameless atomic absorption spectroscopy using a carbon tube atomizer, *Bunseki Kagaku,* 24, 377, 1975.
2. **Katz, A. and Taitel, N.,** Matrix problems in the determination of lithium by flameless (HGA) atomic absorption spectroscopy and their solution, *Talanta,* 24, 132, 1977.
3. **Patel, B. M., Gupta, N., Purohit, P., and Joshi, B. D.,** Electrothermal atomic absorption spectrometric determination of lithium , sodium, potassium and copper in uranium without preliminary chemical separation, *Anal. Chim. Acta,* 118, 163, 1980.
4. **Frigieri, P., Trucco, R., Ciaccolini, I., and Pampurini, G.,** Determination of cesium in river and seawater by electrothermal atomic absorption spectroscopy. Interference of cobalt and iron, *Analyst,* 105, 651, 1980.
5. **Katskov, D. A. and Grinshtein, I. L.,** Atomic absorption study of graphite furnace, *Zh. Prikl. Spektrosk.,* 34, 773, 1981.
6. **Kantor, T., Bezur, L., and Pungor, E.,** Furnace-in-flame atomizer developed from the Varian-Techtron CRA system: determination of sodium in alumina, *Mikrochim. Acta,* 1, 289, 1981.
7. **Zhang, C. and Liang, W.,** Determination of potassium and sodium in soil and plants by atomic absorption spectrophotometry, *Fenxi Huaxue,* 10, 382, 1982 (in Chinese).
8. **Sanui, H. and Rubin, H.,** *Atomic Absorption Measurement of Cations In Cultured Cells, In Ions, Cell Proliferation, Cancer,* Boyuton, A. L., McKeechan, and Whitfield, J. F., Eds., Academic Press, New York, N.Y., 1982.
9. **Li, S. and Chen, B.,** Interferences in the determination of potassium and sodium in rocks by atomic absorption spectrophotometry, *Fenxi Huaxue,* 10, 358, 1982 (in Chinese).
10. **Hansen, R. W., Schweinsberg, D. P., and Noakes, R. J.,** Routine batch processing for multielement analysis by atomic absorption spectrophotometry, *Lab. Pract.,* 31, 1097, 1982.
11. **El Ghandour, M. F. M., Abdel Salam, M. S., Hindy, K. T., and Kamel, M. M.,** Studies on air pollution from construction plants in Helwan industrial area. II. Alkali, alkaline earth, and heavy metal constituents of dust-fall, *Environ. Pollut. Ser. B,* 4, 303, 1982.
12. **Dragon, S., Fazakas, J., Grapinine, L., Niculae, G., and Popovici, M.,** Determination of calcium, sodium and potassium 10*N* concentration in wastewater by atomic spectrometry. Effect of the chemical-biological treatment, *Rev. Chim.,* 33, 1140, 1982 (in Rumanian).
13. **Deutsch, C., Slater, L., and Goldstein, P.,** Volume regulation of human peripheral blood lymphocytes and stimulated proliferation of volume-adapted cells, *Biochim. Biophys. Acta,* 721, 262, 1982.
14. **Christensen, T. H.,** Comparison of methods for preparation of municipal compost for analysis of metals by atomic absorption spectrophotometry, *Int. J. Environ. Anal. Chem.,* 12, 211, 1982.
15. **Boutron, C.,** Atmospheric trace metals in snow layers deposited at South Pole from 1928 to 1977, *Atmos. Environ.,* 16, 2451, 1982.
16. **Aristarain, A. J., Delmas, R. J., and Briat, M.,** Snow chemistry on James Ron Island (Antarctic Peninsula), *J. Geophys. Res.,* 87C, 11004, 1982.
17. **Slavin, W., Carnrick, G. R., Manning, D. C., and Pruszkowska, E.,** Recent experiences with the stabilized temperature platform furnace and Zeeman background extraction, *At. Spectrosc.,* 4, 69, 1983.
18. **Bettinelli, M.,** Fusion procedure for the trace metal analysis of coal by atomic absorption, *At. Spectrosc.,* 4, 5, 1983.
19. **Chang, I., Bodoe, P. D., and Mohammed, R.,** Chemical analysis of seven nutrient elements in some sugar cane products and by-products, *Trop. Agric.,* 60, 41, 1983.
20. **Melucci, R. C.,** Use of an air-natural gas flame in atomic absorption, *J. Chem. Educ.,* 60, 238, 1983.
21. **Novozamsky, I., Houba, V. J. G., Van Eck, R., and Van Vark, W.,** Novel digestion technique for multielement plant analysis, Commun. Sci. Plant Anal., 14, 239, 1983.
22. **Zheng, Y., Woodriff, R., and Nichols, J. A.,** Mechanism of atomic vapor loss for aluminum and potassium in a constant-temperature carbon-tube furnace, *Anal. Chem.,* 56, 1388, 1984.
23. **Pfluger, C. E. and Nessel, T.,** Wire loop microfurnace-power supply controller system for miniature helium direct current discharge tube emission method, *Analyst,* 109, 593, 1984.
24. **Jing, S., Wang, R., Yu, C., Yan, Y., and Ma, Y.,** Design of model ZM-1 zeeman effect atomic absorption spectrometer, *Huanjing Kexue,* 5, 41, 1984 (in Chinese).
25. **Zheng, Y.,** Matrix interferences for cadmium and potassium in a constant-temperature graphite furnace, *Gaodeng Xuexiao Huaxue Xuebao,* 5, 119, 1984.
26. **Qi, W. and Wei, F.,** Direct determination of ultratrace lithium in serum by controlled-temperature graphite-furnace atomic absorption spectrophotometry, *Zhongguo Kexue Jishu Daxue Xuebao,* 68-73, 1984 (in Chinese).
27. **Hallis, K. F., Boon, N. A., Perkins, C. M., Aronson, J. K., and Grahame-Smith, D. G.,** Sensitive high-temperature electrothermal atomic absorption analysis for rubidium in erythrocytes and plasma of normal and hypertensive patients, *Clin. Chem.* (Winston-Salem, N.C.), 31, 274, 1985.
28. **Solinas, M., Angerosa, F., and Cichelli, A.,** Use of a graphite furnace in the determination of alkaline metals

in olive oils by atomic absorption spectrophotometry, *Riv. Soc. Ital. Sci. Aliment.*, 14, 271, 1985.

29. **Knecht, U.,** Routine biological monitoring of lithium in serum by electrothermal atomic absorption spectroscopy, *Aertz. Lab.*, 37, 215, 1985.

30. **Trapp, G. A.,** Matrix modifiers in GFAA analysis of trace lithium in biological fluids, *Anal. Biochem.*, 148(1), 127-32, 1985.

31. Nippon Telegraph and Telephone Public Corp., Method for determining alkali metals in metals, *Jpn. Kokai Tokkyo Koho JP 60*, 86, 447 (85 86,447) (Cl. Go1N21/31), 16 May 1985.

32. **Ekman, T. A.,** Development of a new carbon furnace for AAS. Studies on speciation of lithium in blood, 262 pp, 1984. Avail. Univ., Microfilms Int. Ord. No. DA 8511744. From *Diss. Abstr. Int. B*, 46(4), 1145, 1985.

33. **Bourret, E., Moynier, J., Bardet, L., and Fussellier, M.,** Determination of traces of lithium in blood serum by electrothermal atomic absorption spectrometry. Optimization of experimental parameters, *Anal. Chim. Acta*, 172, 157, 1985 (in French).

34. **Jeffrey, A. J. and Lyons, D. J.,** Determination of marker rubidium in heliothis moths using two analytical techniques, *Analyst*, 110, 951, 1985.

35. **Trapp, G. A.,** Matrix modifiers in graphite furnace atomic absorption analysis of trace lithium in biological fluids, *Anal. Biochem.*, 148, 127, 1985.

36. **Kantor, T.,** On the mechanism of releasing effect of alkaline earth and lanthanum chlorides in flame spectrometry, *ATOMKI Kozl.*, 27(3), 340-2, 1985.

37. **Nakamura, S. and Kubota, M.,** Effects of alkali and alkaline earth salts on signal absorbance in electrothermal AAS, *Bunseki Kagaku*, 35, 961, 1986 (in Japanese).

38. **Chapman, J. F., Dale, L. S., and Tophan, S. A.,** Improved cesium sensitivity in electrothermal atomic absorption spectroscopy, *Anal. Chim. Acta*, 187, 307, 1986.

39. **Rios, C., Valero, H., and Sanchez, T.,** Lithium determination in plasma and erythrocytes using furnace atomic absorption spectrophotometry, *At. Spectrosc.*, 8, 67, 1987.

40. **Wingo, C. S., Bixler, G. B., Park, R. H., and Straub, S. G.,** Picomole analysis of alkali metals by flameless atomic absorption spectrophotometry, *Kidney Int.*, 31, 1225, 1987.

AUTHOR INDEX

SUBJECT INDEX

ALUMINUM

ALUMINUM, Al   (ATOMIC WEIGHT 26.9815)

Instrumental Parameters:

Wavelength	309.2 nm
Slit width	160 µm
Bandpass	0.5 nm
Light source	Hollow Cathode Lamp (single element)
Lamp current	8 mA
Purge gas	Argon (high purity)
Sample size	25 µL
Furnace	Cylindrical cuvette/pyrolytic graphite

Standard Operating Conditions:

Optimum char temperature	1500°C
Optimum atomization temperature	2700°C
Sensitivity	4 pg/1% absorption
Sensitivity check	0.04-0.1 µg/mL
Working range	0.04-0.4 µg/mL
Background correction	For high dissolved solids

General Note:

1. Use clean glass and plastic ware soaked in 1:5 nitric acid solution and rinsed thoroughly with deionized distilled water prior to preparing solutions.
2. Standards can be commercially purchased or a previously analyzed sample can be used.
3. Prepare all analytical solutions and subsequent dilutions in 0.2-0.5% (v/v) nitric acid. Acidity will prevent hydrolysis of the solutions.
4. Do not use any halogen acids.
5. When sample exhibits greater than 0.5 absorbance unit, it is recommnded to make dilutions.
6. Always check for blank values on all reagents used, including water.
7. Use solutions with concentration less than 10 µg/L within 2 hr of preparation.

REFERENCES

1. **Atsuya, I. and Sugiura, N.,** Determination of microamounts of aluminum by atomic absorption spectroscopy using a heated graphite atomizer. Application to analysis of steel, *Bunseki Kagaku,* 23, 1170, 1974.
2. **Shaw, F. and Ottaway, J. M.,** Determination of trace amounts of aluminum and other elements in iron and steel by atomic absorption spectroscopy with carbon furnace atomization, *Analyst,* 100, 217, 1975.
3. **Krishnan, S. S., Quittkat, S., and Crapper, D. R.,** Atomic absorption analysis for traces of aluminum and vanadium in biological tissue. A critical evaluation of the graphite furnace atomizer, *Can. J. Spectrosc.,* 21, 25, 1976.
4. **Julshamn, K., Andersen, K. J., Willassen, Y., and Braekkan, O. R.,** A routine method for the determination of aluminum in human tissue samples using standard addition and graphite furnace atomic absorption spectrophotometry, *Anal. Biochem.,* 88, 552, 1978.
5. **Gorsky, J. G., and Dietz, A. A.,** Determination of aluminum in biological samples by atomic absorption spectrophotometry with a graphite furnace, *Clin. Chem.,* 24, 1485, 1978.
6. **Garmestani, K., Blotcky, A. J., and Rack, E. P.,** Comparison between neutron activation analysis and graphite furnace atomic absorption spectroscopy for trace aluminum determination in biological materials, *Anal. Chem.,* 50, 144, 1978.
7. **Matsusaki, K., Yoshino, T., and Yamamoto, Y.,** A method for the removal of chloride interference in determination of aluminum by atomic absorption spectroscopy with a graphite furnace, *Talanta,* 26, 377, 1979.
8. **Toda, W., Lux, J., and Van Loon, J. C.,** Determination of aluminum in solids from gel filtration chromatography of human serum by electrothermal atomic absorption spectroscopy, *Anal. Lett.,* 13, 1105, 1980.
9. **Persson, J. A., Frech, W., Polil, G., and Lundgren, K.,** Determination of aluminum in wood pulp liquors using graphite furnace atomic absorption spectroscopy, *Analyst,* 105, 1163, 1980.
10. **Katskov, D. A. and Grinshtein, J. L.,** Study of the evaporation of beryllium, magnesium, calcium, strontium, barium and aluminum from a graphite surface by an atomic absorption method, *Zh. Prikl. Spektrosk.,* 33, 1004, 1980.
11. **Tsunoda, K., Haraguchi, H., and Fuwa, K.,** Studies on the occurrence of atoms and molecules of aluminum, gallium and indium and their monohalides in an electrothermal carbon furnace, *Spectrochim. Acta,* 35B, 715, 1980.
12. **L'vov, B. V. and Ryabchuk, G. N.,** Determination of aluminum in a graphite furnace, *Zh. Prikl. Spektrosk.,* 33, 1013, 1980.
13. **Oster, O.,** Aluminum content of human serum determined by atomic absorption spectroscopy with a graphite furnace, *Clin. Chim. Acta,* 114, 53, 1981.
14. **King, S. W., Wills, M. R., and Savory, J.,** Electrothermal atomic absorption spectrometric determination of aluminum in blood serum, *Anal. Chim. Acta,* 128, 221, 1981.
15. **Gardiner, P. E., Ottaway, J. M., Fell, G. S., and Halls, D. J.,** Determination of aluminum in blood plasma or serum by electrothermal atomic absorption spectroscopy, *Anal. Chim. Acta,* 128, 57, 1981.
16. **Arafat, N. M. and Glooschenko, W. A.,** Method for the simultaneous determination of arsenic, aluminum, iron, zinc, chromium and copper in plant tissue without the use of perchloric acid, *Analyst* (London), 106, 1174-8, 1981.
17. **Bertram, H. P.,** Aluminum determination in body fluids, *Nieren-Hochdruckkr.,* 10, 188, 1981.
18. **Clavel, J. P., Joudon, M. C., and Galli, A.,** Determination of serum aluminum: new estimation of normal valves, *Ann. Biol. Clin.* (Paris), 40, 51-2, 1982.
19. **Li, C. and Yan, W.,** Graphite furnace atomic absorption determination of aluminum in serum and urine samples, *Beijing Yixueyuan Xuebao,* 14, 38-40, 1982 (in Chinese).
20. **Slavin, W., Carnrick, G. R., and Manning, D. C.,** Graphite-tube effects on perchloric acid interferences on aluminum and thallium in the stabilized-temperature platform furnace, *Anal. Chim. Acta,* 138, 103-10, 1982.
21. **Wawschinek, O., Petek, W., Lang, J., Pogglitsch, H., and Holzer, H.,** The determination of aluminum in human plasma, *Mikrochim. Acta,* 1, 335-9, 1982.
22. **Kelty, K. C., Miller, R. G., Ulmer, N. S., and Facciolo, A.,** Analysis of aluminum in water by flameless atomic absorption spectrophotometry, *Proc.-AWWA Water Qual. Technol. Conf.,* 63-72, 1982.
23. **Nakajima, H.,** Determination of aluminum in soil extracts by atomic absorption spectroscopy using the heated graphite atomizer furnace, *Pedorojisuto,* 26, 27, 1982 (in Japanese).
24. **Taddia, M.,** Determination of aluminum in silicon by electrothermal atomic absorption spectroscopy, *Anal. Chim. Acta,* 142, 333, 1982.
25. **Parkinson, I. S., Ward, M. K., and Kerr, D. N. S.,** A method for the routine determination of aluminum in serum and water by flameless atomic absorption spectroscopy, *Clin. Chim. Acta,* 125, 125-33, 1982.
26. **Frech, W., Cedergren, A., Cederberg, C., and Vessman, J.,** Evaluation of some critical factors affecting determination of aluminum in blood, plasma or serum by electrothermal atomic absorption spectroscopy, *Clin. Chem.,* 28, 2259, 1982.
27. **Isozaki, A., Kawakami, T., and Utsumi, S. M.,** Electrothermal atomic absorption spectrometry for aluminum by direct heating of aluminum-adsorbed chelating resin, *Bunseki Kagaku,* 31, 6311, 1982.

28. **Kostyniak, P. J.,** An electrothermal atomic absorption method for aluminum analysis in plasma: identification of sources of contamination in blood sampling procedures, *J. Anal. Toxicol.,* 7, 20, 1983.

29. **Drescher, V. and Schmidt, L. H.,** Studies on the standardization of aluminum-determination by AAS and electrothermal atomization in graphite tubular cuvettes, *Mengen-Spurenelem., Arbeitstag,* Anke, M., Ed., Karl-Marx University, Leipzig, GDR, 1983, 318.

30. **Yokel, R. A. and Melograma, J. M.,** Safe method to acid digest small samples of biological tissues for graphite furnace atomic absorption analysis of aluminum, *Biol. Trace Elem. Res.,* 5, 225, 1983.

31. **Alfrey, A. C.,** Aluminum determinants in biological samples, *Alum. Anal. Biol. Mater. (Proc. Conf.),* Wills, M. R., and Savory, J., Eds., University of Virginia, Charlottesville, VA, 1983, 76-84.

32. **Petiot, J., Postaire, E., Prognon, P., and Hamon, M.,** Method for the determination of trace amounts of aluminum hemodialysis and hemofiltration solutions by atomic absorption spectrophotometry, *Ann. Pharm. Fr.,* 41(3), 229-37, 1983 (in French).

33. **Leung, F. Y. and Henderson, A. R.,** Quality-control serums for routine determination of aluminum by electrothermal atomic absorption spectroscopy, *Clin. Chem.,* 29(11), 1966-8, 1983.

34. **Yu, J. and Cui, Z.,** Determination of trace amount of aluminum in pure iron and low-alloy steel by graphite-furnace atomic absorption spectrometry, *Fenxi Huaxue,* 11(8), 612-15, 1983.

35. **Yokel, R. A.,** Persistent aluminum accumulation after prolonged systemic aluminum exposure, *Biol. Trace Elem. Res.,* 5(6), 467-74, 1983.

36. **Yokel, R. A. and Melograns, J. M.,** A safe method to acid digest small samples of biological tissues for graphite furnace atomic absorption analysis of aluminum, *Biol. Trace Elem. Res.,* 5, 225-37, 1983.

37. **Berthoff, R. L., Brown, S., Renoe, B. W., Wills, M. R., and Savory, J.,** Improved determination of aluminum in serum by electrothermal atomic absorption spectrophotometry, *Clin. Chem. (Winston-Salem, N.C.),* 29, 1087, 1983.

38. **Hudnik, V., Kozak, E., and Marolt-Gomiscek, M.,** On accuracy of aluminum determination in human serum, *Vestn. Slov. Kem. Drus.,* 30, 411, 1983.

39. **Postel, W., Meier, B., and Markert, R.,** Determination of lead, cadmium, aluminum and tin in beer using atomic absorption spectrophotometry, *Monatsschr. Brauwiss,* 36, 300, 1983 (in German).

40. **Stevens, B. J.,** Determination of aluminum, copper and zinc in human hair, *At. Spectrosc.,* 4(5), 176-8, 1983.

41. **Mazzucotelli, A., Minoia, C., and Frache, R.,** Electrothermal atomic absorption spectrophotometry of aluminum, lead and zinc in CSF samples, *Trace Elem. Anal. Chem. Med. Biol., Proc. 2nd Int. Workshop,* 1982, 975-80, published 1983.

42. **Postel, W., Meier, B., and Markert, R.,** Lead, cadmium, aluminum, tin, zinc, iron and copper in bottled and canned beer, *Moatsschr. Brauwiss,* 36, 360-7, 1983 (in German).

43. **Taddia, M.,** Minimization of matrix interferences in determination of aluminum in silicon by electrothermal atomic absorption spectrometry with L'vov platform, *Anal. Chim. Acta,* 158, 131, 1984.

44. **Alemasova, A., Doroshenko, A. I., and Shevchuk, I. A.,** Determination of aluminum in bronzes by atomic absorpton in furnace with graphite segment, *Zavod. Lab.,* 50, 37, 1984 (in Russian).

45. **Brown, S., Berthoff, R. L., Wills, M. R., and Savory, J.,** Electrothermal atomic absorption spectrometric determination of aluminum in serum with new technique for protein precipitation, *Clin. Chem.,* 30, 1216, 1984.

46. **Buratti, M., Caravelli, G., Cabzaferri, G., and Colombi, A.,** Determination of aluminum in body fluids by solvent extraction and atomic absorption spectroscopy with electrothermal atomization, *Clin. Chim. Acta,* 141, 253, 1984.

47. **Hodsman, A. B., Anderson, C., and Leung, F. Y.,** Accelerated accumulation of aluminum by osteoid matrix in vitamin A deficiency. Animal model of aluminum toxicity, *Miner. Electrolyte Metab.,* 10, 309, 1984.

48. **Lindahl, P. C., Voight, K. C., Bishop, A. M., Lafon, G. M., and Huang, W. L.,** Determination of aluminum in hydrothermal reaction fluids by graphite furnace atomic absorption spectrophotometry, *At. Spectrosc.,* 5, 137, 1984.

49. **Slavin, W. and Carnrick, G. R.,** Possibility of standardless furnace atomic absorption spectroscopy, *Spectrochim. Acta,* 39B, 271, 1984.

50. **Zheng, Y., Woodriff, R., and Nichols, J. A.,** Mechanisms of atomic vapor loss for aluminum and potassium in a constant-temperature carbon-tube furnace, *Anal. Chem.,* 56, 1388, 1984.

51. **Phelan, V. J. and Powell, R. J. W.,** Combined reagent purification and sample dissolution (corpad) applied to trace analysis of silicon, silica and quartz, *Analyst,* 109, 1269, 1984.

52. **Brodie, K. G. and Routh, M. W.,** Trace analysis of lead in blood, aluminum and manganese in serum and chromium in urine by graphite furnace atomic absorption spectrometry, *Clin. Biochem.,* 17, 19, 1984.

53. **Chung, C.,** Atomization mechanism with arrhenius plots taking the dissipation function into account in graphite furnace atomic absorption spectrometry, *Anal. Chem.,* 56, 2714, 1984.

54. **Yokel, R. A.,** Persistent aluminum accumulation after prolonged systemic aluminum exposure, *Biol. Trace Elem. Res.,* 5, 467, 1984.

55. **Brown, S., Bertholf, R. L., Wills, M. R., and Savory, J.,** Electrothermal atomic absorption spectrometric determination of aluminum in serum with a new technique for protein precipitation, *Clin. Chem.,* 30, 1216, 1984.

56. **Riley, K. W.,** Significant reactions of aluminum, magnesium and fluoride during the graphite furnace atomic absorption spectrophotometric determination of arsenic in coal, *Analyst* (London), 109, 181, 1984.

57. **Fagioli, F., Scanavini, L., and Locatelli, C.,** Determination of aluminum in dialysis fluids by graphite tube furnace atomic absorption spectroscopy using the L'vov platform, *Anal. Lett.,* 17, 1473, 1984.

58. **El-Yazizi, A., Al-Saleh, J., and Al-Mefty, O.,** Concentrations of Ag, Al, Bi, Cd, Cu, Pb, Sb and Se in cerebrospinal fluid of patients with cerebral neoplasma, *Clin. Chem.,* 30, 1358, 1984.

59. **Wennrich, R., Dittrich, K., and Bonitz, U.,** Matrix interference in laser atomic absorption spectrometry, *Spectrochim. Acta,* 39B, 657, 1984.

60. **Parkinson, I. S., Chaunon, S. M., Ward, M. K., and Kerr, D. N. S.,** Determination of aluminum in hemodialysis fluid using flameless atomic absorption spectrometry, *Trace Elem. Med.,* 1, 139, 1984.

61. **Stevens, B. J.,** Electrothermal atomic absorption determination of aluminum in tissues dissolved in tetramethyl amm. hydoxide, *Clin. Chem.,* 30, 745, 1984.

62. **Allain, P. and Mauras, Y.,** Determination of aluminum in hemodialysis concentrates by electrothermal atomic absorption spectroscopy, *Anal. Chem.,* 56, 1196, 1984.

63. **Slanina, P., Falkeborn, Y., Frech, W., and Cedergren, A.,** Aluminum concentrations in the brain and bone of rats fed citric acid, aluminum citrate or aluminum hydroxide, *Food Chem. Toxicol.,* 22, 391, 1984.

64. **Thorburn-Burns, D., Dadgar, D., Harriott, M., McBride, K., and Swindall, W. J.,** Investigations on the determination of aluminum in aluminum alkoxides and carboxylates using direct carbon furnace atomization, *Analyst (London),* 109(12), 1613-14, 1984.

65. **Guillard, O., Tiphancau, K., Reiss, D., and Piriou, A.,** Improved determination of aluminum in serum by electrothermal atomic absorption spectrometry and Zeeman background correction, *Anal. Lett.,* 17(B14), 1593-605, 1984.

66. **Giordano, R., Costantini, S., Vernillo, I., Casetta, B., and Aldrightti, F.,** Comparative study for aluminum determination in bone by atomic absorption techniques and inductively coupled plasma atomic emission spectroscopy, *Microchem. J.,* 30(3), 435-47, 1984.

67. **Hudnik, V., Kozak, E., and Marolt-Gomiscek, M.,** On the accuracy of aluminum determination in human serum, *Vestn. Slov. Kem. Drus.,* 30(4), 411-19, 1983. CA 7, 100:82266m, 1984.

68. **Salvadeo, A., Minoia, C., Piazza, V., Poggio, F., and Galli, F.,** Determination of aluminum in raw water, dialyzate, ultrafiltrate and plasma in dialytic centers in Lombardia (Italy), *Med. Biol. Environ.,* 12(1), 7-17, 1984.

69. **Costantini, S., Giordano, R., and Vernillo, I.,** Application of zirconium carbide-coated tubes to determination of aluminum in serum by graphite furnace atomic absorption spectroscopy, *Microchem. J.,* 30, 425, 1984.

70. **Costantini, S., Giordano, R., Bondanini, M., and Rizzica, M.,** Applicability of spectroscopic techniques for determination of aluminum in biological samples, *Ann. Ist. Super. Sanita,* 19(4), 661-4, 1983, published 1984.

71. **Wronski, R., Esters, W., Leimenstoll, H., and Niedermayer, W.,** Significance of aluminum in internal medicine, *Fortschr. Atomspektrom. Spurenanal.,* 1, 285-95, 1984.

72. **L'vov, B. V., Norman, E. A., and Masal'tseva, L. V.,** Atomic absorption determination of aluminum in metallurgical samples using sample calibration graph, *Zh. Anal. Khim.,* 40, 275, 1985 (in Russian).

73. **Halls, D. J. and Fell, G. S.,** Determination of aluminum in dialysate fluids by atomic absorption spectrometry with electrothermal atomization, *Analyst,* 110, 243, 1985.

74. **Slanina, P., Frech, W., Bernhardson, A., and Cedergren, A., Mattsson, P.,** Influence of dietary factors on aluminum absorption and retention in the brain and bone of rats, *Acta Pharmacol. Toxicol.,* 56(4), 331-6, 1985.

76. **Van der Voet, G. B. and De Wolff, F. A.,** Intestinal absorption of aluminum in rats, *Arch. Toxicol. Suppl.,* 8, 316-18, 1985.

77. **Burnatowska-Hiedin, M. A., Mayor, G. H., and Lau, K.,** Renal handling of aluminum in rat: clearance and micropuncture studies, *Am. J. Physiol.,* 249(2, Pt.2), F192-F197, 1985.

78. **Yoshimura, C. and Huzino, T.,** Direct determination of aluminum fluoride and aluminum phosphate by flameless AAS in the presence of carbon black, *Nippon Kagaku Kaishi,* 7, 1392-7, 1985.

79. **Smeyers-Verbeke, J. and Verbeelen, D.,** Determination of aluminum in bone by atomic absorption spectroscopy, *Clin. Chem.,* 31, 1172, 1985.

80. **Van deer Voet, G. B., De Haas, E. J. M., and De Wolff, F. A.,** Monitoring of aluminum in whole blood plasma, serum and water by single procedure using flameless atomic absorption spectrophotometry, *J. Anal. Toxicol.,* 9, 97, 1985.

81. **Fleischer, M. and Schaller, K. H.,** Analytically reliable determination of aluminum in serum and plasma using the graphite furnace, *Fortschr. Atomspektrom. Spurenanal.,* 1, 297, 1984 (in German). CA 16, 103:17877j, 1985.

82. **Leung, F. Y., Hodsman, A. B., Muirhead, N., and Henderson, A. R.,** Ultrafiltration studies in vitro of serum aluminum in dialysis patients after deferoxamine chelation therapy, *Clin. Chem.,* 31(1), 20-3, 1985.

83. **Gardiner, P. E., Stoeppler, M., and Nuerenberg, H. W.,** Optimization of analytical conditions for determination of aluminum in human blood plasma or serum by graphite furnace atomic absorption, *Analyst,* 110, 611, 1985.

85

84. **Brumbaugh, W. G. and Kane, D. A.,** Variability of aluminum concentrations in organs and whole bodies of smallmouth Bass (*Micropterus dolomicui*), *Environ. Sci. Technol.,* 19, 828, 1985.
85. **Fitzgerald, E. A., Bornstein, A. A., and Davidowski, L. J.,** Determination of trace elements in negative photoresist by inductively coupled plasma atomic emission spectroscopy and atomic absorption spectroscopy, *At. Spectrosc.,* 6, 1, 1985.
86. **Gitelman, H. J.,** Electrothermal atomic absorption aluminum, *Alum. Anal. Biol. Mater. (Proc. Conf.),* Wills, M. R. and Savory, J., Eds., University of Virginia, Charlottsville, Va, 1985, 15.
87. **Lewis, S. A., O'Haver, T. C., and Harnly, J. M.,** Determination of metals at microgram-per-liter level in blood serum by simultaneous multielement atomic absorption spectrometry with graphite furnace atomization, *Anal. Chem.,* 57, 2, 1985.
88. **Visser, W. J., Van de Vyver, F. L., Verbueken, A. H., D'Haese, P., Bekaert, A. B., Van Grieken, R. E., Duursma, S. A., and De Broe, M. E.,** Evaluation of different techniques used to determine aluminum in patients with chronic failure ,*Comm. Eur. Communiti,* (REP)EUR9250, 433-41, 1985.
89. **Bradley, C., Leung, F. Y., Slavin, W., and Henderson, A. R.,** Direct-calibration method for determination of aluminum in serum is comparable with protein-precipitation technique, *Clin. Chem.,* 31, 1882, 1985.
90. **Schindler, E.,** Determination of silver, aluminum, arsenic, cadmium, chromium, copper, iron, manganese, lead, selenium and zinc in drinking water. Procedure for study of drinking water by graphite tube AAS. I., *Dtsch. Lebensm.-Rundsch.,* 81, 1, 1985 (in German).
91. **D'Hase, P. C., Van de Vyver, F. L., de Wolff, F. A., and De Broe, M. E.,** Measurement of aluminum in serum, blood, urine and tissues of chronic hemodialyzed patients by use of electrothermal atomic absorption spectrometry, *Clin. Chem.,* 31, 24, 1985.
92. **Van der Voet, G. B., De Haas, E. J. M., and de Wolff, F. A.,** Monitoring of aluminum in whole blood, plasma, serum and water by a single procedure using flameless AAS, *J. Anal. Toxicol.,* 9(3), 97-100, 1985.
93. **Bettinelli, M., Baroni, U., Fontana, F., and Poisetti, P.,** Evaluation of L'vov platform and matrix modification for determination of aluminum in serum, *Analyst,* 110, 19, 1985.
94. **Sun, H. J.,** Atomic absorption determination of aluminum in serum by a modified cuvette and low temperature ashing method, *Hua Hsueh,* 44, 90-5, 1986.
95. **Yoshimura, C. and Shinya, N.,** Direct determination of aluminum silicate by flameless atomic absorption spectroscopy in the presence of carbon black, *Nippon Kagaku Kaishi,* 10, 1363-5, 1986 (in Japanese).
96. **Ross, D. S., Bartlett, R. J., and Magdoff, F. R.,** Graphite furnace determination of aluminum in soil leachates using uncoated graphite tubes, *At. Spectrosc.,* 7, 158-60, 1986.
97. **Manning, D. C. and Slavin, W.,** The choice of an analytical Zeeman AAS wavelength for aluminum, *At. Spectrosc.,* 7(4), 123-6, 1986.
98. **Duval, G., Grubb, B. R., and Bentley, P. J.,** Aluminum accumulation in the crystalline lens of human and domestic animals, *Trace Elem. Med.,* 3(3), 100-4, 1986.
99. **Winnefeld, K., Schroeter, H., Tennigkeit, E., and Kulick, I.,** Quantitative determination of aluminum in the serum of dialysis patients with the AAS-3/EA-3 instrument system, *Jena Rev.,* 30(4), 189-90, 1985. CA 23, 105:147599x, 1986.
100. **Gitelman, H. J. and Alderman, F. R.,** Electrothermal determination of aluminum in biological samples by AAS, *Kidney Int. Suppl.,* 29(18), S28-S31, 1986.
101. **Brueggmeyer, T. W. and Fricke, F. L.,** Comparison of furnace atomization behavior of aluminum from standard and thorium-treated L'vov platform, *Anal. Chem.,* 58(6), 1143-8, 1986.
102. **Giordano, R., Costantini, S., Vernillo, I., and Scopigno, M.,** Problems in the determination of aluminum in biological tissues and fluids, *Rapp. ISTISAN,* 86/3, 1986. CA 13, 104:201661f, 1986.
103. **Andersen, J. R. and Holboe, P.,** Aluminum in antihemophilia preparations as determined by electrothermal atomic absorption spectrophotometry, *J. Pharm. Biomed. Anal.,* 4(1), 111-14, 1986.
104. **Pierson, K. B. and Evenson, M. A.,** Measurement of aluminum in neutonal tissues using electrothermal atomization atomic absorption spectrophotometry, *Anal. Chem.,* 58(8), 1744-8, 1986.
105. **Zheng, Y. and Zhu, F.,** Atomization process in the graphite furnace atomizer. I. Effect of graphite furnace substance materials on dissipation of aluminum, *Guangpuxue Yu Guangpu Fenx.,* 5(5), 52-6, 1985. CA 6, 104:81121f, 1986.
106. **Zheng, Y. and Zhu, F.,** Atomization process in the graphite furnace atomizer. II. Effects of matrixes on atomization of aluminum, *Gaodeng Xuexiao Huaxue Xuebao,* 7(1), 19-22, 1986 (in Chinese). CA 11, 104:179232t, 1986.
107. **Zyukova, N. D. and Panarina, N. A.,** Determination of metallic aluminum in steels by atomic absorption with electrothermal atomization, *Zavod. Lab.,* 52(2), 36-7, 1986.
108. **Vorberg, B., Peters, H. J., Achenbach, H., Koehler, H., and Wuerberger, G.,** Determination of aluminum in hemodialysis patients, *Z. Med. Laboratoriumsdiagn.,* 26(7), 370-3, 1986.
109. **Craney, C. L., Swartout, K., Smith, F. W., III, and West, C. D.,** Improvement of trace aluminum determination by electrothermal atomic absorption spectroscopy using phosphoric acid, *Anal. Chem.,* 58(3), 656-8, 1986.

110. **Bouman, A. A., Platenkamp, A. J., and Posma, F. D.,** Determination of aluminum in human tissues by flameless atomic absorption spectroscopy and comparison of reference values, *Ann. Clin. Biochem.,* 23(1), 97-101, 1986.

111. **Minoia, C., Tempini, G., Salvadeo, A., Vitali, M. T., and Micoli, G.,** Aluminum determination in serum: extraanalytical factors, *G. Ital. Med. Lav.,* 6(5-6), 239-49, 1984 (in Italian), CA 9, 104:84707n, 1986.

112. **McGraw, M., Bishop, N., Jameson, R., Robinson, M. J., O'Hara, M., Hewitt, C. D., and Day, J. P.,** Aluminum content of milk formulas and intravenous fluids used in infants, *Lancet,* 1(8473), 157, 1986.

113. **Allain, P., Mauras, Y., Beaudeau, G., and Hingouet, P.,** Indirect micro-scale method for the determination of desferrioximanie and its aluminum and iron chelated forms in biological samples by atomic absorption spectrometry with electrothermal atomization, *Analyst* (London), 11(5), 531-3, 1986.

114. **Alder, J. F., Batoreu, M. C. C., Pearse, A. D., and Marks, R.,** Depth concentration profiles obtained by carbon furnace spectroscopy for nickel and aluminum in human skin, *J. Anal. At. Spectrom.,* 1, 365, 1986.

115. **Carnrick, G. R. and Slavin, W.,** Use of Th-treated platforms for the determination of aluminum and palladium, *At. Spectrosc.,* 7, 175, 1986

116. **Slavin, W.,** An overview of recent developments in the determination of aluminum in serum by furnace atomic absorption spectroscopy, *J. Anal. At. Spectrom.,* 1, 281, 1986.

117. **Ross, D. S., Bartlett, R. J., and Magdoff, F. R.,** Graphite furnace determination of aluminum in soil leachates using uncoated graphite tubes, *At. Spectrosc.,* 7(5), 158, 1986.

118. **Yansheng, Z. and Feng, Z.,** Atomization process in the graphite furnace atomizer. II. Effects of matrixes on atomization of aluminum, *Gaodeng Xuexiao Huaxue Xuebao,* 7, 19, 1986.

119. **Stoeppler, M., Mohl, C., Novak, L., and Gardiner, P. E.,** Application of STPF concept to the determination of aluminum, cadmium and lead in biological and environmental materials, *Firtschr. Atomspektrom. Spurenanal.,* 2, 419, 1986.

120. **Gomez Coedo, A. and Dorado Lopez, M. T.,** Determination of tin and aluminum in zinc. Electrothermal atomic absorption spectroscopy, *Rev. Metal.* (Madrid), 22, 90, 1986 (in Spanish).

121. **Lugowski, S., Smith, D. C., and Van Loon, J. C.,** The determination of aluminum, chromium, cobalt, iron and nickel in whole blood by electrothermal atomic absorption spectrophotometry, *J. Biomed. Mater. Res.,* 21, 657, 1987.

122. **Gardiner, P. E. and Stoeppler, M.,** Optimization of the analytical conditions for the determination of aluminum in human blood plasma and serum by graphite furnace atomic absorption spectroscopy. II. Assessment of the analytical method, *J. Anal. At. Spectrom.,* 2, 401-4, 1987.

123. **McHalsky, M. L., Rabinow, B. E., Ericson, S. P., Weltzer, J. A., and Ayd, S. W.,** Reduction of aluminum levels in dialysis fluids through the development and use of accurate and sensitive analytical methodology, *J. Parenter. Sci. Technol.,* 41, 67, 1987.

124. **Andersen, J. R.,** Graphite furnace atomic absorption spectrometric screening method for determination of aluminum in hemodialysis concentrates, *J. Anal. At. Spectrom.,* 2, 257, 1987.

125. **Kratochvil, B., Motkosky, N., Duke, M., John, M., and Ng, D.,** Determination of trace aluminum concentrations and homogeneity in biological reference material TORT-1 by instrumental neutron activation analysis and graphite-furnace atomic absorption spectrometry, *Can. J. Chem.,* 65, 1047, 1987.

126. **Cedergren, A. and Frech, W.,** Critical evaluation of analytical methods for determination of trace elements in various matrixes. Determination of aluminum in biological materials by graphite furnace atomic absorption spectroscopy (GFAAS), *Pure Appl. Chem.,* 59, 221-8, 1987.

AUTHOR INDEX

SUBJECT INDEX

ANTIMONY

ANTIMONY, Sb   (ATOMIC WEIGHT 121.75)

Instrumental Parameters:

Wavelength	217.6 nm
Slit width	160 μm
Bandpass	0.5 nm
Light source	Hollow Cathode Lamp
Lamp current	10 mA
Purge gas	Argon or Nitrogen
Sample size	25 μL
Furnace	Cylindrical cuvette/Pyrolytic graphite

Standard Operating Conditions:

Optimum char temperature	1000°C
Optimum atomization temperature	2500°C
Sensitivity	20 pg/1% absorption
Sensitivity check	0.15 μg/mL
Working range	0.05-1.0 μg/mL
Background correction	Required only to correct light scattering or nonspecific absorption from the sample matrix.

General Note:

1. Use commercially purchased standard or a previously analyzed sample or established by using the standard addition method.
2. Use ultra clean glass and plastic ware.
3. Use 0.3% sulfuric acid as diluent.
4. Check for blank values on all reagents including water.
5. Dilution is recommended when sample exhibits greater than 0.5 absorbance units.
6. All analytical solutions should be used within a day of preparation.
7. Use of nitrogen as purge gas will reduce sensitivity.
8. Use alternate resonance line such as 206.8 nm or 231.2 nm when spectral interferences are observed.
9. The analytical precision of this method is 3% RSD.

REFERENCES

1. **Yasuda, S., and Kakiyama, H.,** Determination of arsenic and antimony by flameless atomic absorption spectroscopy using a carbon tube atomizer, *Bunseki Kagaku*, 23, 620, 1974.
2. **Kamada, T. and Yamamoto, Y.,** Selective determination of antimony (III) and antimony (V) with ammonium pyrrolidine dithiocarbamate, sodium diethyldithiocarbamate and dithizone by atomic absorption spectroscopy with a carbon tube atomizer, *Talanta*, 24, 330, 1977.
3. **Aruscavage, P.,** Determination of As, Sb, Se in coal by atomic absorption spectroscopy with graphite tube atomizer, *J. Res. U. S. Geol. Surv.*, 5, 405, 1977.
4. **Hocquellet, P.,** Application of electrothermal atomization to the determination of As, Sb, Se and Hg by atomic absorption spectroscopy, *Analusis*, 6, 426, 1978.
5. **Yamada, H., Uchino, K., Koizumi, H., Noda, T., and Yasuda, K.,** Spectral interference in antimony analysis with high temperature furnace atomic absorption, *Anal. Lett.*, AII, 855, 1978.
6. **Haynes, B. W.,** As, Sb, Se and Te determinations in high-purity copper by electrothermal atomization, *At. Ab. Newsl.*, 18, 46, 1979.
7. **Ward, R. J., Black, C. D., and Watson, G. J.,** Determination of antimony in biological materials by electrothermal atomic absorption spectroscopy, *Clin. Chim. Acta*, 99, 143, 1979.
8. **Katskov, D. A., Ginshtein, I. L., and Kruglikova, L. P.,** Study of the evaporation of the metals In, Ga, Tl, Ge, Sn, Pb, Sb, Bi, Se and Te from a graphite surface by the atomic absorption method, *Zh. Prikl. Spektrosk.*, 33, 804, 1980.
9. **Del Monte Tamba, M. G. and Luperi, N.,** Determination of traces of As, Sb, Bi and V in steel and cast iron by graphite furnace atomic absorption spectroscopy, *Metall Ital.*, 72, 253, 1980 (in Italian and English).
10. **Morita, K., Shimizu, M., Inoue, B., and Ishida, T.,** Graphite furnace atomic absorption spectroscopy of antimony and selenium in whole blood, *Okayama-ken Kankyo Hoken Senta Nempo*, 4, 94, 1980.
11. **Miyakawa, H.,** Determination of antimony by thorium hydroxide coprecipitation-flameless atomic absorption, *Nenpo-Fukui-ken Kogyo Shikenjo*, 120-8, 1979, published 1980.
12. **Subramanian, K. S. and Meranger, J. C.,** Determination of As (III), As (V), Sb (III), Sb (V), Se (IV) and Se (VI) by extraction with ammonium pyrrolidine dithiocarbamate-methyl isobutyl ketone and electrothermal atomic absorption spectroscopy, *Anal. Chim. Acta*, 124, 131, 1981.
13. **Saeed, K. and Thomassen, Y.,** Spectral interferences from phosphate matrixes in the determination of As, Sb, Se and Te by electrothermal atomic absorption spectroscopy, *Anal. Chim. Acta*, 130, 281, 1981.
14. **Shan, V. and Ni, Z.,** Matrix modification for the determination of volatile elements of arsenic, selenium, tellurium, silver, antimony and bismuth by graphite furnace atomic absorption spectroscopy, *Huaxue Xuebao*, 39, 575-8, 1981.
15. **Smith, B. M. and Griffiths, M. B.,** Determination of lead and antimony in urine by atomic absorption spectroscopy with electrothermal atomization, *Analyst*, 107, 253, 1982.
16. **Kantin, R.,** Chemical speciation of antimony in marine algae, *Limnol. Oceanogr.*, 28, 165, 1983.
17. **Nomura, N.,** Determination of trace antimony by atomic absorption spectroscopy. II. *Kochi Kogyo Koto Semmon Gakko Gakujutsu Kiyo*, 19, 103-7, 1983 (in Japanese).
18. **Brovko, I. A., Tursunov, A., Fidirko, E. V., and Rish, M. A.,** Atomic absorption determination of antimony and bismuth in natural waters, *Deposited Doc. SPSTL*, 1239 Khp-D82, 20 pp, 1982.
19. **Headridge, J. B. and Nicholson, R. A.,** Determination of arsenic, antimony, selenium and tellurium in nickel-base alloys by atomic absorption spectrometry with introduction of solid samples into furnaces, *Analyst (London)*, 107, 1200, 1982.
20. **Pyatkova, V. N., Kroshkina, A. B., Konstantinova, M. G., and Sidoruk, E. I.,** Atomic absorption determination of antimony using a graphite tubular atomizer, *Atomno-absorbtsion, Metody Analiza Mineral. Syr'ya M.*, 57-60, 1982 (in Russian). From *Ref. Zh. Metall.*, Abstr. No. 5K13, 1983.
21. **Liu, Y., Lei, X., Zhang, X., and Luo, C.,** Hydride generation electrothermal atomic absorption method for determination of tin and antimony in geological samples, *Fenxi Huaxue*, 12, 218, 1984 (in Chinese).
22. **Vanloo, B., Dams, R., and Hoste, J.,** Determination of antimony in steel by hydride generation and electrothermal zeeman atomic absorption spectrometry, *Anal. Chim. Acta*, 160, 329, 1984.
23. **Inui, T., Terada, S., Tamura, H., and Ichinose, N.,** Determination of antimony in solder alloy by hydride generation followed by graphite furnace atomic absorption spectrometry, *Z. Anal. Chem.*, 318, 502, 1984.
24. **Kobayashi, T., Kujirai, O., Hirose, F., and Okochi, H.,** Determination of traces of antimony in nickel-base heat-resisting alloys by graphite-furnace atomic-absorption spectroscopy, *Nippon Kinzoku Gakkaishi*, 48, 542-8, 1984 (in Japanese).
25. **Chung, C.,** Atomization mechanism with arrhenius plots taking the dissipation function into account in graphite furnace atomic absorption spectrometry, *Anal. Chem.*, 56, 2714, 1984.
26. **Chung, C. H., Iwamoto, E., Yamamoto, M., and Yamamoto, Y.,** Selective determination of arsenic (III, V), antimony (III, V), selenium (IV, VI) and tellurium (IV, VI) by extraction and graphite furnace atomic absorption spectrometry, *Spectrochim. Acta*, 39B, 459, 1984.

27. **El-Yazigi, A., Al-Saleh, I., and Al-Mefty, O.,** Concentrations of Ag, Al, Au, Bi, Cd, Pb, Sb and Se in cerebrospinal fluid of patients with cerebral neoplasma, *Clin. Chem.,* 30, 1358, 1984.
28. **Liu, Y., Lei, X., Zhang, X., and Luo, C.,** Hydride generation electrothermal atomic absorption method for the determination of tin and antimony in geological samples, *Fenxi Huaxue,* 12, 218, 1984 (in Chinese).
29. **Bombach, H., Luft, B., Weinhold, E., and Mohr, F.,** Determination of arsenic, antimony, bismuth, tin, selenium and tellurium in steels using hydride AAS, *Neue Huette,* 29, 233, 1984 (in German).
30. **Blackmore, D. J. and Stanier, P.,** Methods for measurement of trace elements in equine blood by electrothermal atomic absorption spectrophotometry, *At. Spectrosc.,* 5, 215, 1984.
31. **Pyatova, V. N., Kroshkina, A. B., Sidoruk, E. I., Ivanov, N. P., Konstantinova, M. G., and Voskre-senskaya, V. S.,** Characteristics of the atomization mechanism for antimony and determination of antimony by atomic absorption in a graphite tube furnace, *Zh. Anal. Khim.,* 39, 831, 1984 (in Russian).
32. **Sugawara, H. and Tayama, K.,** Determination of tin and antimony in cinder of zinc ore by flameless atomic absorption spectrometry, *Ryusan to Kogyo,* 37(10), 173-6, 1984 (in Japanese).
33. **Brovko, J. A., Tursunov, A., Rish, M. A., and Davirov, A. D.,** Atomic absorption determination of antimony and bismuth in natural samples by hydride formation and metal preconcentration in graphite tube, *Zh. Anal. Khim.,* 39, 1768, 1984 (in Russian).
34. **Kasimova, O. G., Shcherbina, N. I., Sedykh, E. M., Bal'shakova, L. I., and Myasoedova, G. V.,** Preconcentration of antimony and arsenic on chelating sorbent polyorgs IX for atomic absorption determina-tion , *Zh. Anal. Khim.,* 39, 1823, 1984 (in Russian).
35. **Hoenig, M., Scokart, P. D., and Van Hoeyweghen, P.,** Efficiency of L'vov platform and ascorbic acid modifier for reduction of interferences in analysis of plant samples for lead, thallium, antimony, cadmium, nickel and chromium by ETAAS, *Anal. Lett.,* 17A, 1947, 1984.
36. **Hojyo, T. and Nagumo, S.,** Determination of antimony in electrolytic chromium metal by graphite furnace Zeeman atomic absorption spectrometry, *Bunseki Kagaku,* 33(3), T22-T25, 1984 (in Japanese).
37. **Hijyo, T., Nagumo, S., and Fujiki, T.,** Determination of antimony in chromium metal by flameless atomic absorption spectroscopy, *Toyo Soda Kenkyu Hokoku,* 28, 45, 1984.
38. **Lontsikh, B. S.,** Direct atomic absorption determination of antimony using the furnace-flame electrothermal atomizer, *Metody Spektr. Anal. Miner. Syr'ya,* (Mater. Vses. Konf. Nov. Metodam Spektr. Anal. Ikh Primen.), 110-113 pp, 2nd, 1981, published 1984.
39. **Niskavaara, H., Virtasalo, J., and Lajunen, L. H. J.,** Determination of antimony in geochemical samples by graphite furnace atomic absorption spectroscopy using different matrix modifiers, *Spectrochim. Acta,* 40B, 1219, 1985.
40. **Xianlian, L. and Zhenfan, Y.,** Determination of trace antimony in geological samples by hydride-graphite furnace atomic absorption spectrometry, *Nanjing Daxue Xuebao Ziran Kexue,* 21, 563, 1985.
41. **Sturgeon, R. E., Willie, S. N., and Berman, S. S.,** Hydride generation-atomic absorption determination of antimony in seawater with *in situ* concentration in a graphite furnace, *Anal. Chem.,* 57(12), 2311-14, 1985.
42. **Niskavaara, H., Virtasalo, J., and Lajunen, L. H. J.,** Determination of antimony in geochemical samples by graphite furnace AAS using different matrix modifiers, *Spectrochim. Acta Part B,* 40B(9), 1219-25, 1985.
43. **Constantini, S., Giordano, R., Rizzica, M., and Benedetti, F.,** Applicability of anodic stripping voltammetry and graphite furnace atomic absorption spectrometry to determination of antimony in biological matrices: comparative study, *Analyst,* 110, 1355, 1985.
44. **Welz, B., Akman, S., and Schlemmer, G.,** Investigations of interferences in graphite furnace atomic absorption spectrometry using dual cavity platform. I. Influence of nickel chloride on determination of antimony, *Analyst,* 110, 459, 1985.
45. **Sturgeon, R. E., Willie, S. N., and Berman, S. S.,** Preconcentration of selenium and antimony from seawater for determination of GFAAS, *Anal. Chem.,* 57(1), 6-9, 1985.
46. **Criand, A. and Fouillac, C.,** Use of L'vov platform and molybdenum coating for determination of volatile elements in thermomineral waters by atomic absorption spectrometry, *Anal. Chim. Acta,* 167, 257, 1985
47. **Kudryasheva, E. G., Erkovich, G. E., and Malykh, V. D.,** Study of factors affecting the graphite furnace AA determination of Sb and As in technological solutions, *Zh. Anal. Khim.,* 40(12), 2173-8, 1985.
48. **Zakhariya, A. N. and Diallo, I. K.,** AA determination of Sb and Bi impurities in nonferrous metals and alloys by using furnace-flame atomizers, *Zh. Prikl. Spektrosk.,* 43(6), 1002-4, 1985.
49. **Suzuki, T., Ozaki, H., and Sawada, K.,** Combined solvent extraction-graphite furnace atomic absorption spectrometry for the determination of antimony in plant materials, *Anal. Sci.,* 2(1), 25-9, 1986.
50. **Iwamoto, E., Inoike, Y., Yamamoto, Y., and Hayashi, Y.,** Interferences of antimony (V) in the differentia-tion of antimony (III) from antimony (V) by extraction with amm.tetra methylene dithiocarbamate using graphite furnace atomic absorption spectrometry, *Analyst (London),* 111(3), 295-8, 1986.
51. **Kharlamov, I. P., Lebedev, V. I., Persits, V. Y., and Eremina, G. V.,** Influence of cations and anions on the analytical signals of zinc, cadmium, lead, tin, bismuth and antimony introduced with solutions of steels and alloys into electrothermal atomizers, *Zh. Anal. Khim.,* 41, 1004, 1986.

52. **Terashima, S.,** Determination of arsenic and antimony in eighty five geochemical reference samples by automated hydride generation and electrothermal atomic absorption spectroscopy, *Geostand. Newsl.,* 10, 127, 1986.

53. **Kharlamov, I. P., Lebedev, V. I., and Persits, V. Y.,** Complex application of atomic and molecular spectra to the study of matrix effects on the atomization of zinc, cadmium, lead, tin, bismuth and antimony, *Zh. Anal. Khim.,* 41, 1965, 1986.

54. **Castillo, J. R., Lopez-Molinero, A., and Sucunza, T.,** Determination of arsenic, antimony and bismuth in high purity copper by electrothermal atomic absorption spectroscopy, *Mikrochim. Acta,* 4, 105, 1986.

55. **Criand, A. and Fouillac, C.,** A study of carbon dioxide-rich thermal-mineral waters of French Massif Central. II. Behavior of some trace elements, arsenic, antimony and germanium, *Geochim. Cosmochim. Acta,* 50(8), 1573, 1986 (in French).

56. **Clark, J. R.,** Electrothermal atomization atomic absorption conditions and matrix modifications for determining Sb, As, Bi, Cd, Ga, Au, In, Pb, Mo, Pd, Pt, Se, Ag, Te, Tl and Sn following back-extraction of organic aminohalide extracts, *J. Anal. At. Spectrom.,* 1, 301, 1986.

AUTHOR INDEX

SUBJECT INDEX

ARSENIC

ARSENIC, As   (ATOMIC WEIGHT 74.9216)

Instrumental Parameters:

Wavelength	193.17 nm
Slit width	640 μm
Bandpass	2 nm
Light source	Hollow Cathode /Electrodeless Discharge Lamp
Lamp current	10 mA
Purge gas	Argon
Sample size	25 μL
Furnace	Cylindrical cuvette/Pyrolytic graphite

Standard Operating Conditions:

Optimum char temperature	900°C/300°C
Optimum atomization temperature	2700°C/1900°C
Sensitivity	12 pg/1% absorption
Sensitivity check	0.1 μg/mL
Working range	0.05-0.50 μg/mL
Background correction	Required only to correct light scattering or non-specific absorption from the sample containing high dissolved solids.

General Note:
1. Use commercial standard or a previously analyzed sample as a working standard.
2. Use ultra clean glass and plastic ware, soaked in 1:5 nitric acid solution and rinsed thoroughly with deionized distilled water.
3. All analytical solutions should be at least 1% (v/v) in nitric acid.
4. Check for blank values on all reagents, including water.
5. Dilution is recommended when sample exhibits greater than 0.5 absorbance units.
6. Addition of nickel (10 μg Ni/mL as nickel nitrate) to all solutions allows the use of higher temperature for pyrolysis.
7. Sulfuric acid (3%) can be used as diluent.
8. The analytical precision of this method is 6% RSD.

REFERENCES

1. **Baird, R. B. and Gabrielian, S. M.,** A tantalum foil lined graphite tube for the analysis of arsenic and selenium by atomic absorption spectroscopy, *Appl. Spectrosc.,* 28, 213, 1974.
2. **Yasuda, S. and Kakiyama, H.,** Determination of arsenic and antimony by flameless atomic absorption spectroscopy using a carbon tube atomizer, *Bunseki Kagaku,* 23, 620, 1974.
3. **Freeman, H. C., Uthe, J. F., and Flemming, B.,** A rapid and precise method for the determination of inorganic and organic arsenic with and without wet ashing using a graphite furnace, *At. Absorp. Newsl.,* 15, 49, 1976.
4. **Yamamoto, Y. and Kamada, T.,** Fractional determination of ppb levels of As (III) and As (V) in water using graphite furnace atomic absorption spectrophotometry combined with amm. pyrrolidine dithiocarbamate-nitrobenzene extraction, *Bunseki Kagaku,* 25, 567, 1976.
5. **Aruscavage, P.,** Determination of arsenic, antimony and selenium in coal by atomic absorption spectroscopy with graphite tube atomizer, *J. Res. U.S. Geol. Surv.,* 5, 405, 1977.
6. **Shaikh, A. U. and Tallman, D. E.,** Determination of sub-microgram per liter quantities of arsenic in water by arsine generation followed by graphite furnace atomic absorption spectroscopy, *Anal. Chem.,* 49, 1093, 1977.
7. **Thompson, A. J. and Thompson, P. A.,** Determination of arsenic in soil and plant materials by atomic absorption spectrophotometry with electrothermal atomization, *Analyst,* 102, 9, 1977.
8. **Hocquellet, P.,** Application of electrothermal atomization to the determination of arsenic, antimony, selenium and mercury by atomic absorption spectroscopy, *Analusis,* 6, 426, 1978.
9. **Shaikh, A. U. and Tallman, D. E.,** Species specific and for nanogram quantities of arsenic in natural waters by arsine generation followed by graphite furnace atomic absorption spectroscopy, *Anal. Chim. Acta,* 98, 251, 1978.
10. **Fleming, D. E. and Taylor, G. A.,** Improvement in the determination of total arsenic by arsine generation and atomic absorption spectrophotometry using a flame-heated silica furnace, *Analyst,* 103, 101, 1978.
11. **Haynes, B. W.,** Arsenic, antimony, selenium and tellurium determinations in high-purity copper by electrothermal atomization, *At. Ab. Newsl.,* 18, 46, 1979.
12. **Stockton, R. A. and Irgolic, K. J.,** The Hitachi graphite furnace-Zeeman atomic absorption spectrometer as an automated, element-specific detector for high pressure liquid chromatography. The separation of arsenobetaine, arsenocholine and arsenite/arsenate, *Int. J. Environ. Anal. Chem.,* 6, 313, 1979.
13. **Odanaka, Y., Matano, O., and Gato, S.,** Determination of inorganic and methylated arsenicals in environmental materials by graphite furnace atomic absorption enhancing and depressing effects of various coexisting reagents, *Bunseki Kagaku,* 28, 517, 1979.
14. **Chakraborti, D., de Jonghe, W., and Adams, F.,** Determination of arsenic by electrothermal atomic absorption spectroscopy with a graphite furnace. I. Difficulties in the direct determination, *Anal. Chim. Acta,* 119, 331, 1980.
15. **Chakraborti, D., de Jonghe, W., and Adams, F.,** Determination of arsenic by electrothermal atomic absorption spectroscopy with a graphite furnace. II. Determination of arsenic (III) and arsenic (V) after extraction, *Anal. Chim. Acta,* 120, 121, 1980.
16. **Woolson, E. A. and Aharonson, N.,** Separation and detection of arsenical pesticide residues and some of their metabolites by high pressure liquid chromatography graphite furnace atomic absorption spectroscopy, *J. Assoc. Off. Anal. Chem.,* 63, 523, 1980.
17. **Iadevaia, R., Aharonson, N., and Woolson, E. A.,** Extraction and clean-up of soil arsenical residues for analysis by high-pressure liquid chromatographic-graphite-furnace atomic absorption, *J. Assoc. Off. Anal. Chem.,* 63, 117, 1979 (in Spanish).
18. **Del Monte Tamba, M. G. and Luperi, N.,** Determination of traces of arsenic, antimony, bismuth and vanadium in steel and cast iron by graphite furnace atomic absorption spectroscopy, *Metall. Ital.,* 72, 253, 1980 (in Italian and English).
19. **Costantini, S., Giordano, R., and Ravagnan, P.,** Arsenic determination in groundwater and cistern water, *Ann. Ist. Super. Sanita,* 16, 287-94, 1980 (in Italian).
20. **Saeed, K. and Thomassen, Y.,** Spectral interferences from phosphate matrixes in the determination of arsenic, antimony, selenium and tellurium by electrothermal atomic absorption spectroscopy, *Anal. Chim. Acta,* 130, 281, 1981.
21. **Subramanian, K. S.,** Rapid electrothermal atomic absorption method for arsenic and selenium in geological materials via hydride evolution, *Z. Anal. Chem.,* 305, 382, 1981.
22. **Pacey, G. E. and Ford, J. A.,** Arsenic speciation by ion-exchange separation and graphite furnace atomic absorption spectrophotometry, *Talanta,* 28, 935, 1981.
23. **Shkinev, V. M., Khavezov, I., Spivakov, B. Y., Mareva, S., Ruseva, E., Zolotov, Y. A., and Iordanov, N.,** Solvent extraction of As(V) by dialkyl tin di nitrates. Extraction-atomic absorption determination of arsenic with a flame and graphite furnace, *Zh. Anal. Khim.,* 36, 896, 1981.
24. **Koreckova, J., Frech, W., Lundberg, E., Persson, J. A., and Cedergren, A.,** Investigations of reactions involved in electrothermal atomic absorption procedures. X. Factors influencing the determination of arsenic, *Anal. Chim. Acta,* 130, 267, 1981.

25. **Inui, T., Terada, S., and Tamura, H.,** Determination of arsenic by arsine generation with reducing tube followed by graphite furnace atomic absorption spectroscopy, *Z. Anal. Chem.,* 305, 189, 1981.

26. **Hirayama, T., Sakagami, Y., Nohara, M., and Fukui, S.,** Wet digestion with nickel ion and graphite furnace atomic absorption spectrophotometry for the determination of total arsenic in food samples, *Bunseki Kagaku,* 30, 278, 1981.

27. **Brooks, R. R., Ryan, D. E., and Zhang, H. F.,** Use of a tantalum-coated graphite furnace tube for the determination of arsenic by flameless atomic absorption spectroscopy, *At. Spectrosc.,* 2, 161, 1981.

28. **Pacey, G. E. and Ford, J. A.,** Arsenic speciation by ion-exchange separation and graphite furnace atomic absorption spectrophotometry, *Talanta,* 28, 935-8, 1981.

29. **Shan, X. and Ni, Z.,** Matrix modification for the determination of volatile elements of arsenic, selenium, tellurium, silver, antimony and bismuth by graphite furnace atomic absorption spectroscopy, *Huaxue Xuebao,* 39, 575-8, 1981.

30. **Arafat, N. M. and Glooschenko, W. A.,** The simultaneous determination of arsenic, aluminum, iron, zinc, chromium and copper in plant tissue without the use of perchloric acid, *Analyst (London),* 106, 1174-8, 1981.

31. **Cheam, V. and Asmila, K. I.,** Interlaboratory quality control study No. 26 arsenic and selenium in water, *Rep. Ser.-Inland Waters Dir. (Can.),* 68, 8, 1981.

32. **Brodie, K. G. and Rowland, J. J.,** Trace analysis of arsenic, lead and tin, *Eur. Spectrosc. News,* 36, 41-44, 1981.

33. **Brooks, R. R., Ryan, D. E., and Zhang, H. F.,** Use of a tantalum-coated graphite furnace tube for the determination of arsenic by flameless atomic absorption spectroscopy, *At. Spectrosc.,* 2, 161, 1981.

34. **Koreckova, J., Frech, W., Lundberg, E., Persson, J. A., and Cedergren, A.,** Investigations of reactions involved in electrothermal atomic absorption procedures. X. Factors influencing the determination of arsenic, *Anal. Chim. Acta,* 130, 267, 1981.

35. **Irgolic, K. J., Banks, C. H., Bottino, N. R., Chakraborti, D., Gennity, J. M., Hillman, D. C., O'Brien, D. H., Pules, R. A., Stockton, R. A., et. al.,** Analytical and biochemical aspects of the transformation of arsenic and selenium compounds into biomolecules, *NBS Spec. Publ. (U.S.),* 618, 244-63, 1981.

36. **Subramanian, K. S. and Meranger, J. C.,** Determination of As(III), As(V), Sb(III), Sb(V), Se(IV) and Se(VI) by extraction with amm. pyrrolidine dithiocarbamate-methyl isobutyl ketone and electrothermal atomic absorption spectroscopy, *Anal. Chim. Acta,* 124, 131, 1981.

37. **Hoenig, M. and Van Hoeyweghen, P.,** Application of electrothermal atomic absorption spectrophotometry to the analysis of complex matrixes: arsenic in plants, *Spectrochim. Acta,* 37B, 817, 1982.

38. **Welz, B. and Melcher, M.,** Determination of arsenic in wastewater using the hydride AAS technique, *Vom Wasser,* 59, 407, 1982.

39. **Malinina, R. D. and Artemova, T. N.,** Determination of arsenic in ferromolybdenum by flameless atomic absorption spectrophotometry, *Automatizir. Metody Ispytanii. Met., M.,* 11, 1982 (in Russian). From *Ref. Zh. Khim.,* 16G114, 1982.

40. **Headridge, J. B. and Nicholson, R. A.,** Determination of arsenic, antimony, selenium and tellurium in nickel-base alloys by atomic-absorption spectrometry with introduction of solid samples into furnaces, *Analyst (London),* 107, 1200, 1982.

41. **Kellerman, S. P.,** The use of masking agents in the determination by hydride generation and atomic absorption spectrophotometry of arsenic, antimony, selenium, tellurium and bismuth in the presence of noble metals, *Rep.-MINTEK,* M39, 14, 1982.

42. **Tam, G. K. H. and Lacroix, G.,** Dry ashing, hydride generation atomic absorption spectrometric determination of arsenic and selenium in foods, *J. Assoc. Off. Anal. Chem.,* 65, 647, 1982.

43. **Kida, A.,** Matrix effects of metal salts for the determination of arsenic by graphite furnace atomic absorption spectrophotometry and their application to water analysis, *Bunseki Kagaku,* 31, 1-6, 1982 (in Japanese).

44. **Benard, H. and Pinta, M.,** Determination of arsenic in atmospheric aerosols by atomic absorption with electrothermal atomization, *At. Spectrosc.,* 3, 8-12, 1982.

45. **Puttemans, F. and Massart, D. L.,** Solvent extraction procedures for the differential determination of arsenic (V) and arsenic(III) species by electrothermal atomic absorption spectroscopy, *Anal. Chim. Acta,* 141, 225, 1982.

46. **Takamatsu, T., Nakata, R., and Yoshida, T.,** Determination of dimethylarsinate, monomethylarsonate and inorganic arsenic in the extract of Pond sediment certifies reference material (SRM) issued by the National Institute for Environmental studies (Japan), *Bunseki Kagaku,* 31, 540, 1982 (in Japanese).

47. **Ebdon, L. and Pearce, W. C.,** Direct determination of arsenic in coal by atomic absorption spectroscopy using solid sampling and electrothermal atomization, *Analyst (London),* 107, 942, 1982.

48. **Artemova, T. N. and Malinina, R. D.,** Flameless atomic absorption spectroscopy determination of arsenic in steels, *Nov. Metody Ispyt. Met.,* Moskva No. 7K30.

49. **Schelpakowa, I. R., Schtscherbakowa, O. I., Judelewwitsch, I. G., Beisel, N. F., Dittrich, K., and Mothes, W.,** Examination of atomic absorption spectroscopic trace analysis in A(III) B(V)-semiconducting micro-samples. VI. Examination of layer separation, profile and trace analysis in indium antimonide materials, *Talanta,* 29, 577-81, 1982 (in German).

50. **Fabec, J. L.,** Direct determination of arsenic in shale oil and its products by furnace atomic absorption spectroscopy with a tetrahydrofuran solvent system, *Anal. Chem.,* 54, 2170, 1982.

51. **Riley, K. W.,** Spectral interference by aluminum in the determination of arsenic using the graphite furnace: choice of resonance lines, *At. Spectrosc.,* 3, 120, 1982.

52. **Tarui, T. and Takairin, H.,** Determination of trace metal elements in petroleum. IV. Determination of arsenic in petrtoleum by combustion in an oxygen bomb followed by graphite furnace atomic absorption spectroscopy, *Bunseki Kagaku,* 31, T45-T48, 1982.

53. **Subramanian, K. S., Leung, P. C., and Meranger, J. C.,** Determination of arsenic (III, V, total) in polluted waters by graphite furnace atomic absorption spectroscopy and anodic stripping voltammetry, *Int. J. Environ. Anal. Chem.,* 11, 121, 1982.

54. **Subramanian, K. S. and Meranger, J. C.,** Rapid hydride evolution-electrothermal atomization atomic-absorption spectrophotometric method for determining arsenic and selenium in human kidney and liver, *Analyst (London),* 107, 157-62, 1982.

55. **Yu, M. and Liu, G.,** Hydride-generation atomic absorption spectrophotometric determination of trace arsenic (III) and arsenic (V) in water by concentration and separation with sulfhydryl cotton fibers, *Fenxi Huaxue,* 10, 747, 1982.

56. **Akman, S., Genc, O., and Balkis, T.,** Atom formation mechanisms of arsenic with different techniques in atomic absorption spectroscopy, *Spectrochim. Acta,* 37B, 1982.

57. **Luo, D., Xia, Z., and Chen, M.,** Determination of arsenic, antimony and bismuth in chemical prospecting samples by hydride generation atomic absorption spectroscopy, *Fenxi Huaxue,* 10, 294, 1982 (in Chinese).

58. **Ueta, T., Nishida, S., and Yoshihara, T.,** Hygienic chemical studies on dental treatment materials. II. Determination of arsenic, lead and cadmium in dental cements by flameless atomic absorption spectrophotometry, *Tokyo-toritsu Eisei Kenkyusho Kenkyu Nempo,* 33, 110, 1982 (in Japanese).

59. **Crock, J. G. and Lichte, F. E.,** An improved method for the determination of trace levels of arsenic and antimony in geological materials by automated hydride generation-atomic absorption spectroscopy, *Anal. Chim. Acta,* 144, 223, 1982.

60. **Woolson, E. A., Aharonson, N., and Iadevaia, R.,** Application of the high-performance liquid chromatography-flameless atomic absorption method to the study of alkyl arsenical herbicide metabolism in soil, *J. Agric. Food Chem.,* 30, 580, 1982.

61. **Persson, J. A. and Irgum, K.,** Determination of dimethyl-arsinic acid in seawater in the sub-ppb range by electrothermal atomic absorption spectroscopy after preconcentration on an ion-exchange column, *Anal. Chim. Acta,* 138, 111-19, 1982.

62. **Puttemans, F., Van den Winkel, P., and Massart, D. L.,** The determination of arsenic by electrothermal atomic absorption spectroscopy after liquid-liquid extraction, *Anal. Chim. Acta,* 149, 123, 1983.

63. **Lovell, M. A. and Farmer, J. G.,** The determination of arsenic in soil and sediment digests by graphite furnace atomic absorption spectroscopy, *Int. J. Environ. Anal. Chem.,* 14, 181, 1983.

64. **Welz, B. and Melcher, M.,** Investigations on atomization mechanisms of volatile hydride-forming elements in a heated quartz cell. I. Gas-phase and surface effects, decomposition and atomization of arsine, *Analyst (London),* 108, 213, 1983.

65. **Krapf, N. E.,** Commercial scale removal of arsenite, arsenate and methane arsonate from ground and surface water, *Arsenic: Ind.,* Lederer, W. H. and Fensterheim, R. J., Eds., Van Nostrand Reinhold, New York, 1983, 269-79.

66. **Ullman, A. H.,** Determination of arsenic in glycerin by hydride generation atomic absorption spectroscopy, *J. Am. Oil Chem. Soc.,* 60, 614, 1983.

67. **Voronkova, M. A., Nemodruk, A. A., and Antonova, E. A.,** Atomic absorption determination of arsenic in mineral raw materials using a graphite furnace, *Khim. Metody Anal. Prom. Mater. Mater. Semin.,* 63-8, 1982, published 1983 (in Russian).

68. **Bolibrzuch, E.,** Determination of arsenic in plants, soils and rainwater by AAS (atomic absorption spectroscopy) using a hydride formation technique, *Mater. Konwersatorium Specktrom. At. Emisyjnej, Absorp. Spektrom. Mas,* 120th, 157, 1983.

69. **Maher, W. A.,** Determination of methylated arsenic species by use of a zinc column arsine generator, *Spectrosc. Lett.,* 16(11), 865-70, 1983.

70. **Zelentsova, L. V. and Yudelevich, I. G.,** Atomic absorption analysis of high-purity arsenic, *Zh. Anal. Khim.,* 38(8), 1404, 1983 (in Russian).

71. **Ficklin, W. H.,** Separation of arsenic (III) and arsenic (V) in ground waters by ion-exchange, *Talanta,* 30, 371, 1983.

72. **Jin, K., Ogawa, H., and Taga, M.,** Wet digestion method for the determination of total arsenic in marine organisms. II. Digestion method of marine organisms for determination of total arsenic by atomic absorption spectroscopy, *Bunseki Kagaku,* 32, E259-E264, 1983.

73. **Fish, R. H. and Brinckman, F. E.,** Organometallic geochemistry. Isolation and identification of organoarsenic and inorganic arsenic compounds from Green River formation oil shales, *Prepn. Pap-Am. Chem. Soc. Div. Fuel Chem.,* 28(3), 177-80, 1983.

74. **Kumamaru, T., Matsuo, H., and Ikeda, M.,** Effect of continuous prereduction by heating with potassium iodide and hydrochloric acid for determining arsenic (III, V) by continuous hydride generation-atomic absorption spectroscopy using sodium tetrahydroborate reduction, *Bunseki Kagaku,* 32, 357-61, 1983.
75. **Shan, X., Ni, Z., and Zhang, L.,** Determination of arsenic in soil, coal fly ash and biological samples by electrothermal atomic absorption spectroscopy with matrix modification, *Anal. Chim. Acta,* 151, 179, 1983.
76. **Yu, M. Q., Liu, G. Q., and Jin, Q.,** Determination of trace arsenic, antimony, selenium and tellurium in various oxidation states in water by hydride generation and atomic absorption spectrophotometry after enrichment and separation with thiol cotton, *Talanta,* 30, 265, 1983.
77. **Inamasu, T.,** Arsenic metabolites in urine and feces of hamsters, *Toxicol. Appl. Pharmacol.,* 71, 142, 1983.
78. **Ruseva, E., Khavezov, I., Spivakov, B. Y., and Shkinev, V. M.,** Electrothermal atomic absorption determination of traces of arsenic and phosphorus in copper-nickel alloys, *Z. Anal. Chem.,* 315, 499, 1983.
79. **Carlin, L. M., Colovos, G., Garland, D., Jamin, M. E., and Klenck, M.,** Analytical methods evaluation and validations: arsenic, nickel, tungsten, vanadium, talc, wood dust, Report NIOSH-210-79-0060; Order No. PB83-155325, 223 pp, 1981. Avail. NTIS from Govt. Rep. Announce. Index (U.S.), 83, 2355, 1983.
80. **Edgar, D. G. and Lum, K. R.,** Zeeman effect electrothermal atomic absorption spectrophotometry with matrix modification for determination of arsenic in urine, *Int. J. Environ. Anal. Chem.,* 16, 219, 1983.
81. **Solomons, E. T. and Walls, H. C.,** Analysis of arsenic in forensic cases utilizing a rapid, nonashing technique and furnace atomic absorption, *J. Anal. Toxicol.,* 7, 220, 1983.
82. **Woolson, E. A.,** Application of HPLC/GFAA for arsenic in environmental samples, *Pestic. Chem.: Hum. Welfare Environ., Proc. 5th Int. Congr. Pestic. Chem.,* 79-82, 1982, published 1983.
83. **Takamatsu, T., Aoki, H., and Yoshida, T.,** Arsenic speciation in pot soil cropped with rice plant. Fluctuations of arsenate, arsenite, monomethylarsenate and dimethylarsinate contents, *Kokuritsu Kogai Kenkyusho Kenkyu Hokoku,* 47, 153-63, 1983.
84. **Richards, W. A.,** Determination of arsenic in natural water and a study of matrix effects in electrothermal atomic absorption spectroscopy, 164 pp, 1983. Avail. Univ. Microfilms Int., Order No. DA8319280. From *Diss. Abstr. Int.B,* 44(4), 1106-7, 1983.
85. **Inamasu, T.,** Arsenic metabolites in urine and feces of hamsters pretreated with PCB, *Toxicol. Appl. Pharmacol.,* 71(4), 142-52, 1983.
86. **Solomons, E. T. and Walls, H. C.,** Analysis of arsenic in forensic cases utilizing a rapid, nonashing technique and furnace atomic absorption, *J. Anal. Toxicol.,* 7(5), 220, 1983.
87. **Ho, C. L., Tweedy, S., and Mahan, C.,** Determination of arsenic in geologic, biologic and water samples by flameless atomic absorption spectrophotometry and inductively coupled plasma atomic emission spectrometry following distillation, *J. Test. Eval.,* 12, 107, 1984.
88. **Jing, S., Wang, R., Yu, C., Yan, Y., Zhang, D., and Ma, Y.,** Design of model ZM-1 zeeman effect atomic absorption spectrometer, *Huanjing Kuxue,* 5, 41, 1984 (in Chinese).
89. **Grobenski, Z., Lehmann, R., Radziuk, B., and Voellkopf, U.,** Determination of trace metals in seawater using zeeman graphite furnace AAS, *At. Spectrosc.,* 5, 87, 1984.
90. **Banslaugh, J., Radziuk, B., Saeed, K., and Thomassen, Y.,** Reduction of effects of structural nonspecific absorption in determination of arsenic and selenium by electrothermal atomic absorption spectrometry, *Anal. Chim. Acta,* 165, 149, 1984.
91. **Kasimova, O. G., Shcherbina, N. J., Sedykh, E. M., Bal'shakova, L. I., and Myasoedova, G. V.,** Preconcentration of antimony and arsenic on chelating sorbent polyorgs IX for atomic absorption determination, *Zh. Anal. Khim.,* 39, 1823, 1984 (in Russian).
92. **Pruszkowska, E. and Barrett, P.,** Determination of arsenic, selenium, chromium, cobalt and nickel in geochemical samples using the stabilized temperature platform furnace and Zeeman background correction, *Spectrochim. Acta,* 39B, 485, 1984.
93. **Hudnik, V. and Gomiscek, S.,** The atomic absorption spectrometric determination of arsenic and selenium in mineral waters by electrothermal atomization, *Anal. Chim. Acta,* 157, 135-42, 1984.
94. **Chung, C. H., Iwamoto, E., Yamamoto, M., and Yamamoto, Y.,** Selective determination of arsenic (III, V), antimony (III, V), selenium (IV, VI) and tellurium (IV, VI) by extraction and graphite furnace atomic absorption spectroscopy, *Spectrochim. Acta,* 39B, 459, 1984.
95. **Shan, X., Ni, Z. and Zhang, L.,** Use of arsenic resonance line of 197.2 nm and matrix modification for determination of arsenic in environmental samples by graphite furnace atomic absorption spectrometry using palladium as a matrix modifier, *At. Spectrosc.,* 5(1), 1-4, 1984.
96. **Chakraborti, D., Irgolic, K. J., and Adams, F.,** Atomic absorption spectrometric determination of arsenite in water samples by graphite furnace after extraction with amm. sec.-butyldithiophosphate, *J. Assoc. Off. Anal. Chem.,* 67, 277, 1984.
97. **Meranger, J. C., Subramanian, K. S., and McCurdy, R. F.,** Arsenic in Nova Scotian groundwater, *Sci. Total Environ.,* 39(1-2), 49-55, 1984.
98. **Subramanian, K. S., Meranger, K. S., and McCurdy, R. F.,** Determination of As(III) in some Nova Scotian groundwater samples, *At. Spectrosc.,* 5(4), 192-4, 1984.

99. **Aneva, Z., Marinova, E., and Pankova, M.,** Determination of arsenic in hydrogen by combustion in a wickbold apparatus and spectrophotometric measurement, *God. Vissh. Khim.-Tekhnol. Inst. Sofia,* 29, 254-9, 1983, published 1984.

100. **Torsi, G., Palmisano, F., Nardelli, A., and Zambonin, P. G.,** Calibration problems in the determination of arsenic in the atmosphere by the "electrostatic accumulation furnace for electrothermal atomic spectrometry" method, *Ann. Chim. (Rome),* 74(11-12), 811, 1984.

101. **Belyaev, Y. I., Khozhainov, Y. M., Shcherbakov, V. I., and Myasoedov, B. F.,** Radiographic study of memory effect of graphite furnace during atomic absorption analysis, *Zh. Prikl. Spektrosk.,* 41, 831, 1984 (in Russian).

102. **Chi, X., Xu, J., and Li, X.,** Determination of trace arsenic (III) and arsenic (V) in water by atomic absorption spectroscopy using hydride generation and electrothermal atomization, *Fenxi Huaxue,* 12, 278, 1984.

103. **Chakraborti, D., Irgolic, K. J., and Adams, F.,** Matrix interferences in arsenic determinations by graphite furnace atomic absorption spectroscopy: recommendations for the determination of arsenic in water samples, *Int. J. Environ. Anal. Chem.,* 17, 241, 1984.

104. **Ikeda, M., Nakata, F., Matsuo, H., and Kumamaru, T.,** Suction flow hydride generation-heated quartz cell atomic absorption spectroscopy of arsenic (III, V) by utilizing sensitivity enhancement effect of air introduction, *Bunseki Kagaku,* 33, 416, 1984 (in Japanese).

105. **Bozsai. G., Czegeny, I., and Karpati, Z.,** Determination of arsenic in drinking water and human hair by hydride generation and atomic absorption spectroscopy, *Magy. Kem. Lapja,* 39, 121, 1984 (in Hungarian).

106. **Welz, B. and Melcher, M.,** Mechanisms of transition metal interferences in hydride generation atomic absorption spectroscopy. III. Releasing effect of iron (III) on nickel interference on arsenic and selenium, *Analyst (London),* 109, 577, 1984.

107. **Voellkopf, U. and Grobenski, Z.,** Interference in analysis of biological samples using stabilized temperature platform furnace and zeeman background correction, *At. Spectrosc.,* 5, 115, 1984.

108. **Riley, K. W.,** Significant reactions of aluminum, magnesium and fluoride during graphite furnace atomic absorption spectrophotometric determination of arsenic in coal, *Spectrochim. Acta,* 39B, 249, 1984.

109. **Hocquellet, P.,** Use of atomic absorption spectrometry with electrothermal atomization for direct determination of trace elements in oils: cadmium, lead, arsenic and tin, *Rev. Fr. Corps Gras,* 31, 117, 1984 (in French).

110. **Narasaki, H. and Ikeda, M.,** Automated determination of arsenic and selenium by atomic absorption spectroscopy with hydride generation, *Anal. Chem.,* 56, 2059, 1984.

111. **Bombach, H., Luft, B., Weinhold, E., and Mohr, F.,** Determination of arsenic, antimony, bismuth, tin, selenium and tellurium in steels using hydride-AAS, *Neue Huette,* 29, 233, 1984 (in German).

112. **Slavin, W. and Carnrick, G. R.,** Possibility of standardless furnace atomic absorption spectroscopy, *Spectrochim. Acta,* 39B, 271, 1984.

113. **Harswell, S. J., O'Neill, P., and Bancroft, K. C. C.,** Arsenic speciation in soipore waters from mineralized and unmineralized areas of South-West England, *Talanta,* 32, 69, 1985.

114. **Dabeka, R. W. and Gladys, G. M. A.,** Graphite furnace atomic absorption spectrometric determination of arsenic in foods after dry-ashing and coprecipitation with amm. pyrrolidine dithiocarbamate, *Can. J. Spectrosc.,* 30, 154, 1985.

115. **Takamatsu, T., Nakata, R., Yoshida, T., and Kawashima, M.,** Depth profiles of dimethylarsinate, monomethylarsonate and inorganic arsenic in sediment from Lake Biwa, *Kokuritsu Kogai Kenkyusho Kenkyu Hokoku,* 75, 39-45, 1985.

116. **Van Hoeyweghen, P. and Hoenig, M.,** Comparative study of different mineralization methods of animal tissues for arsenic determination by electrothermal AAS, *Analysis,* 13(6), 275-8, 1985.

117. **Buchet, J. P. and Lauwerys, R.,** Study of inorganic arsenic methylation by rat liver *in vitro*: relevance for the interpretation of observations in man, *Arch. Toxicol.,* 57(2), 125-9, 1985.

118. **Pegon, Y.,** Direct determination of arsenic in blood serum by electrothermal atomic absorption spectrometry, *Anal. Chim. Acta,* 172, 147, 1985.

119. **Branch, C. H. and Hutchinson, D.,** Simultaneous determination of arsenic and selenium in geochemical samples by hydride evolution and atomic absorption spectrometry: success and failure, *Analyst,* 110, 163, 1985.

120. **Okubo, N., Kawabata, N., Koshida, K., and Miyazaki, M.,** Determination of total arsenic in food samples by AAS with a graphite furnace after nickel ion-added wet digestion, *Eisei Kagaku,* 31(4), 274-7, 1985.

121. **Eaton, D. K. and McCutcheon, J. R.,** Matrix modification for furnace atomic absorption spectrometric determination of arsenic in whole human blood, *J. Anal. Toxicol.,* 9(5), 213-16, 1985.

122. **Chakraborti, D. and Irgolic, K. J.,** Separation and determination of arsenic and selenium compounds by high pressure liquid chromatography with a graphite furnace atomic absorption spectroscopy as the element-specific detector, *Heavy Met. Environ., 5th Int. Conf.,* Lekkas, T. D., Ed., CEP Consult., Edinburgh, U.K., 1985, 484.

123. **Criand, A. and Foullac, C.,** Use of L'vov platform and molybdenum coating for determination of volatile elements in thermomineral waters by atomic absorption spectrometry, *Anal. Chim. Acta,* 167, 257, 1985.

124. **Kudryasheva, E. G., Erkovich, G. E., and Malykh, V. D.,** Study of factors affecting the graphite furnace atomic absorption determination of antimony and arsenic in technological solutions, *Zh. Anal. Khim.,* 40(12), 2173-8, 1985.

125. **Schindler, E.,** Determination of silver, aluminum, arsenic, cadmium, copper, iron, manganese, lead, selenium and zinc in drinking water. Procedure for study of drinking water by graphite tube AAS. I., *Dtsch. Lebensm. Rundsch.,* 81, 1, 1985 (in German).

126. **Clark, J. R.,** Electrothermal atomization atomic absorption conditions and matrix modification for determining Sb, As, Bi, Cd, Ga, Au, In, Pb, Mo, Pd, Pt, Se, Ag, Te, Tl and Sn following back-extraction of organic aminohalide extracts, *J. Anal. At. Spectrom.,* 1, 301, 1986.

127. **El-Yazizi, A., Al-Saleh, I., and Al-Mefty, O.,** Concentrations of zinc, iron, molybdenum, arsenic and lithium in cerebrospinal fluid of patients with brain tumors, *Clin. Chem.,* 32, 2187, 1986.

128. **Harako, A.,** Arsenic in environment around the Osorezan Volcano region. III. Arsenic in soil, *Hirosaki Igaku,* 38, 232-43, 1986.

129. **George, G. M. and Frahm, L. J.,** Graphite furnace atomic absorption spectrophotometric determination of 4-hydroxy-3-nitrobenzenearsonic acid in finished animal feed: collaborative study, *J. Assoc. Off. Anal. Chem.,* 69, 838-43, 1986.

130. **Manneh, V. A., McGowan, J. P., and Shirachi, D. Y.,** The determination of arsenic by HPLC-electrothermal atomic absorption spectrophotometry, *Proc. West. Pharmacol. Soc.,* 29, 137-9, 1986.

131. **Tang, Y. and He, J.,** Determination of trace arsenic in edible vegetable oil by flameless Zeeman atomic absorption spectroscopic analysis—comparison between direct measurement and acid digestion methods, *Wuhan Daxue Xuebao Ziran Kexueban,* 86, 87-90, 1985 (in Chinese), CA 8, 104:108023q, 1986.

132. **Iwamoto, E., Chung, C. H., Yamamoto, M., and Yamamoto, Y.,** Arsenic determination with graphite-cloth ribbon in graphite-furnace atomic absorption spectrometry, *Talanta,* 33(7), 577-82, 1986.

133. **Sarx, B. and Baechmann, K.,** Determination of arsenic (III) and arsenic (V) in soil and airborne dust samples, *Fortschr. Atomspektrom. Spurenanal.,* 2, 619, 1986.

134. **Manneh, V. A., McGowen, J. P., and Shirachi, D. Y.,** The determination of arsenic by HPLC-electrothermal atomic absorption spectrophotometry, *Proc. West. Pharmacol. Soc.,* 29, 137, 1986.

135. **Hagen, J. A. and Lovett, R. J.,** Determination of arsenic by graphite furnace atomic absorption spectrometry: an iodine-based trapping solution for arsine, *At. Spectrosc.,* 7(3), 69-71, 1986.

136. **Paschal, D. C., Kimberly, M. M., and Bailey, G. C.,** Determination of urinary arsenic by electrothermal atomic absorption spectrometry with the L'vov platform and matrix modifications, *Anal. Chim. Acta,* 181, 179-86, 1986.

137. **Brzezinska-Paudyn, A., Van Loon, J. C., and Hancock, R.,** Comparison of the determination of arsenic in environmental samples by different analytical techniques, *At. Spectrosc.,* 7(3), 72, 1986.

138. **Hoenig, M. and Van Hoeyweghen, P.,** Determination of selenium and arsenic in animal tissues with platform furnace atomic absorption spectroscopy and deuterium background correction, *Int. J. Environ. Anal. Chem.,* 24(3), 193-202, 1986.

139. **Castillo, J. R., Lopez-Molinero, A., and Sucunza, T.,** Determination of arsenic, antimony and bismuth in high-purity copper by electrothermal atomic absorption spectroscopy, *Mikrochim. Acta,* 4, 105, 1986.

140. **Terashima, S.,** Determination of arsenic and antimony in eighty-five geochemical reference samples by automated hydride generation and electrothermal atomic absorption spectroscopy, *Geostand. Newsl.,* 10, 127-30, 1986.

141. **Welz, B. and Schlemmer, G.,** Determination of As, Se and Cd in marine biological tissue samples using a stabilized temperature platform furnace and comparing deuterium arc with Zeeman-effect background correction AAS, *J. Anal. At. Spectrom.,* 1(2), 119-24, 1986.

142. **El-Yazizi, A., Al-Saleh, I., and Al-Mefty, O.,** Concentrations of zinc, iron, molybdenum, arsenic and lithium in cerebrospinal fluid of patients with brain tumors, *Clin. Chem.,* 32, 2187, 1986.

143. **Welz, B. and Schlemmer, G.,** Determination of As, Se and Cd in marine biological tissue samples using a stabilized temperature platform furnace and comparing deuterium arc with Zeeman effect background correction atomic absorption spoectroscopy, *Spectrochim. Acta,* 41B, 567, 1986.

144. **Sturgeon, R. E., Willie, S. N., and Berman, S. S.,** Hydride generation-graphite furnace atomic absorption spectrometry. New prospects, *Fresnius Z. Anal. Chem.,* 323(7), 788-92, 1986.

145. **Dams, R., Alluyn, F., Vanloo, B., Wauters, G., and Vandecasteele, C.,** Gray cast-iron reference material certified for antimony, arsenic, bismuth and lead, *Fresnius' Z. Anal. Chem.,* 325, 163-7, 1986.

146. **Maher, W. A.,** Determination of arsenic by graphite furnace atomic absorption spectroscopy after zinc-column or sodiumtetrahydroborate (III) arsine generation and trapping in potassium iodide-iodine solution, *At. Spectrosc.,* 8, 88, 1987.

147. **Atsuya, I., Itoh, K., Akatsuka, K., and Jin, K.,** Direct determination of trace amounts of arsenic in powdered biological samples by atomic absorption spectroscopy using an inner miniature cup for solid sampling techniques, *Fresnius' Z. Anal. Chem.,* 326, 53, 1987.

148. **Tsalev, D., Mandzhukov, P., and Stratis, J. A.,** ETA determination of inorganic and methylated arsenic after preconcentration by hydride generation and trapping the hydrides in a cerium(IV)-iodide absorbing solution, *J. Anal. At. Spectrom.,* 2, 135, 1987.
149. **Ebdon, L. and Parry, H. G. M.,** Direct atomic spectrometric analysis by slurry atomization. II. Elimination of interferences in the determination of arsenic in whole coal by electrothermal atomization atomic absorption spectroscopy, *J. Anal. At. Spectrom.,* 2, 131, 1987.
150. **Krynitsky, A. J.,** Preparation of biological tissue for determination of arsenic and selenium by graphite furnace atomic absorption spectroscopy, *Anal. Chem.,* 59, 1889, 1987.

AUTHOR INDEX

SUBJECT INDEX

BARIUM

BARIUM, Ba   (ATOMIC WEIGHT 137.34)

Instrumental Parameters:

Wavelength	553.5 nm
Slit width	320 μm
Bandpass	1 nm
Light source	Hollow Cathode Lamp
Lamp current	10 mA
Purge gas	Argon
Sample size	25 μL
Furnace	Cylindrical cuvette/Pyrolytic graphite

Standard Operating Conditions:

Optimum char temperature	1500°C
Optimum atomization temperature	2500°C
Sensitivity	4pg/1% absorption
Sensitivity check	0.5 μg/mL
Working range	0.05-0.10 μg/mL
Background correction	Required only to correct light scattering or non-specific absorption from the sample containing high dissolved solids.

General Note:

1. Use commercial standard or a previously analyzed sample as a working standard.
2. Use ultra clean glass and plastic ware, soaked in 1:5 nitric acid solution and rinsed thoroughly with deionized distilled water.
3. All analytical solutions should be at least 0.2-0.5% (v/v) in nitric acid.
4. Check for blank values on all reagents including water.
5. Dilution is recommended when sample exhibits greater than 0.5 absorbance units.
6. Use of nitrogen as purge gas and high concentrations of >3% nitric acid (v/v) will suppress the sensitivity.
7. Sensitivity may be enhanced by the use of hydrochloric acid.
8. The analytical precision of this method is 3% RSD.

REFERENCES

1. **Renshaw, G. D.,** The determination of barium by flameless atomic absorption spectrophotometry using a modified graphite tube atomizer, *At. Ab. Newsl.,* 12, 158, 1973.
2. **Sherfinski, J. H.,** A graphite tube degradation study of barium atomic gunshot residue levels, *At. Ab. Newsl.,* 14, 26, 1975.
3. **Lagas, P.,** Determination of beryllium, barium and vanadium and some other elements in water by atomic absorption spectroscopy with electrothermal atomization, *Anal. Chim. Acta,* 100, 139, 1978.
4. **Epstein, M. S. and Zander, A. T.,** Direct determination of barium in sea and estuarine water by graphite furnace atomic spectrometry, *Anal. Chem.,* 51, 915, 1979.
5. **Berggren, P. O.,** Determination of barium, lanthanum and magnesium in pancreatic islets by electrothermal atomic absorption spectroscopy, *Anal. Chim. Acta,* 119, 161, 1980.
6. **Katskov, D. A. and Grinshtein, J. L.,** Study of the evaporation of beryllium, magnesium, calcium, strontium, barium and aluminum from a graphite surface by an atomic absorption method, *Zh. Prikl. Spektrosk.,* 33, 1004, 1980.
7. **Kuga, K.,** Ionization of barium in graphite furnace atomic absorption spectroscopy, *Bunko Kenkyu,* 29, 191, 1980 (in Japanese).
8. **Jasim, F. and Barbooti, M. M.,** Electrothermal atomic absorption determination of barium in brines and caustic soda solutions, *Talanta,* 28, 353, 1981.
9. **Berggren. P. O., Andersson, B., and Hellman, B.,** Stimulation of the insulin secretory mechanism following barium accumulation in pancreatic ß-cells, *Biochim. Biophys. Acta,* 720, 320-8, 1982.
10. **Zhukova, M. P.,** Atomic absorption of barium in bearing steel using electrothermal atomization, *Zh. Anal. Khim.,* 37, 1629, 1982.
11. **Rollemberg, M. C. E. and Curtis, A. J.,** Flameless atomic absorption determination of barium in natural waters using the technique of standard additions, *Mikrochim. Acta,* 2, 441, 1982.
12. **Zhukova, M. P.,** Development and certification of a method for flameless atomic absorption determination of barium in bearing steel after ion-exchange separation of the base, *Attestsiya Metodik Vypolneniya Izmerenii n a Baze Primeneniya Standart.* Obraztsov., M., 43-6, 1982 (in Russian). From *Ref. Zh. Khim. Abstr.,* No. 9G210, 1983.
13. **Rossi, G., Omenetto, N., Pigozzi, G., Vivian, R., Mattinez, V., Mousty, F., and Crabi, G.,** Analysis of radioactive waste solutions by atomic absorption spectrometry with electrothermal atomization, *At. Spectrosc.,* 4, 113, 1983.
14. **Berggren, P. O., Andersson, T., and Hellman, B.,** The interaction betwen barium and calcium in ß-cell-rich pancreatic islets, *Bromed. Res.,* 4, 129, 1983.
15. **Roe, K. K. and Froelich, P. N.,** Determination of barium in seawater by direct injection graphite furnace atomic absorption spectrometry, *Anal. Chem.,* 56, 2724, 1984.
16. **Styris, D. L.,** Atomization mechanisms for barium in furnace atomic absorption spectrometry, *Anal. Chem.,* 56, 1070, 1984.
17. **Sugiyama, M., Fujino, O., and Matsui, M.,** Determination of barium in sea water by graphite furnace atomic absorption spectroscopy after preconcentration and separation by solvent extraction, *Bunseki Kagaku,* 33, E123, 1984.
18. **Kozusnikova, J.,** Determination of barium in steel by the AAS method with atomizer WETA 80, *Chem. Listy,* 78(11), 1209-16, 1984.
19. **Zhukova, M. P., Kostina, L. V., and Zueva, N. A.,** Electrothermal atomic-absorption spectroscopic determination of barium in iron-nickel alloys, *Metody i Svetstva Kontrolya v Cher. Metallurgii, M.* 16-20, 1984. From. *Ref. Zh. Metall. Abstr.,* No. 12K15, 1984.
20. **Wang, S.,** Analytical characteristics of graphite tube for determination of barium by graphite furnace atomic absorption spectroscopy, *Fenxi Huaxue,* 14(5), 362, 1986.
21. **Sun, S.,** Determination of barium in water by tantalum-lined graphite furnace atomic absorption spectrophotometry, *Fenxi Huaxue,* 14, 494-7, 501, 1986 (in Chinese).
22. **Suwen, W.,** Analytical characteristics of graphite tube for determination of barium by graphite furnace atomic absorption spectroscopy, *Fenxi Huaxue,* 14, 362, 1986.
23. **Shirong, S.,** Determination of barium in water by tantalum-lined graphite furnace atomic absorption spectrophotometry, *Fenxi Huaxue,* 14, 494, 1986.

AUTHOR INDEX

SUBJECT INDEX

BERYLLIUM

BERYLLIUM, Be   (ATOMIC WEIGHT 9.0122))

Instrumental Parameters:

Wavelength	234.9 nm
Slit width	320 μm
Bandpass	1 nm
Light source	Hollow Cathode Lamp
Lamp current	8 mA
Purge gas	Argon
Sample size	25 μL
Furnace	Cylindrical cuvette/Pyrolytic graphite

Standard Operating Conditions:

Optimum char temperature	1000°C
Optimum atomization temperature	2600°C
Sensitivity	1pg/1% absorption
Sensitivity check	0.05 μg/mL
Working range	0.001-0.01 μg/mL
Background correction	Required only to correct light scattering or non-specific absorption from the sample containing high dissolved solids.

General Note:

1. Use commercial standard or a previously analyzed sample as a working standard.
2. Use ultra clean glass and plastic ware, soaked in 1:5 nitric acid solution and rinsed thoroughly with deionized distilled water.
3. All analytical solutions should be at least 0.2-0.5% (v/v) in nitric acid.
4. Check the blank values on all reagents used, including water.
5. Dilution is recommended when sample exhibits greater than 0.5 absorbance units.

REFERENCES

1. **Siemer, D., Lech., J. F., and Woodriff, R.,** Direct filtration thru porous graphite for AA analysis of beryllium particulates in air, *Spectrochim. Acta,* 28B, 469, 1973.
2. **Bettger, R. G., Ficklin, A. C., and Rees, T. F.,** Determination of beryllium in air samples using the graphite furnace, *At. Ab. Newsl.,* 14, 124, 1975.
3. **Shimomura, S., Morita, H., and Kubo, M.,** Determination of beryllium by atomic absorption spectroscopy with a graphite tube atomizer, *Bunseki Kagaku,* 25, 539, 1976.
4. **Lagas, P.,** Determination of beryllium, barium, vanadium and other elements in water by atomic absorption spectroscopy with electrothermal atomization, *Anal. Chim. Acta,* 100, 139, 1978.
5. **Geladi, P. and Adams, F.,** The determination of beryllium and manganese in aerosols by atomic absorption spectrometry with electrothermal atomization, *Anal. Chim. Acta,* 105, 219, 1979.
6. **Katskov, D. A. and Grinshtein, J. L.,** Study of the evaporation of Be, Mg, Ca, Sr, Ba and Al from a graphite surface by an atomic absorption method, *Zh. Prikl. Spektrosk.,* 33, 1004, 1980.
7. **Matsusaki, K., Kameshima, N., Murakami, S., and Yoshino, T.,** Interference effect of chloride on the determination of beryllium by atomic absorption spectroscopy with graphite furnace, *Nippon Kagaku Kaishi,* 11, 1715, 1981.
8. **Matsusaki, K., Kameshima, N., Murakami, S., and Yoshino, T.,** Interference effect of chloride on the determination of beryllium by atomic absorption spectroscopy with a graphite furnace, *Nippon Kagaku Kaishi,* 11, 1715, 1981.
9. **Pilipenko, A. T. and Samchuk, A. I.,** Extraction-atomic absorption determination of beryllium in natural waters using a graphite furnace, *Zh. Anal. Khim.,* 37, 614-19, 1982 (in Russian).
10. **Toshimura, C. and Huzino, T.,** Effect of powder reductants in flameless absorption of beryllium, *Nippon Kagaku Kaishi,* 5, 659, 1983.
11. **Fazakas, J.,** Sensitivity enhancement of beryllium determination in graphite furnace-atomic absorption spectroscopy by use of atomization under pressure, *Mikrochim. Acta,* 2, 217, 1983.
12. **Yoshimura, C. and Huzino, T.,** Effect of powder reductants in flameless atomic absorption of beryllium, *Nippon Kagaku Kaishi,* 5, 659, 1983 (in Japanese).
13. **Matsusaki, K. and Yoshino, T.,** Electrothermal atomic absorption spectrometric determination of traces of chromium, nickel, iron and beryllium in aluminum and its alloys without preliminary separation, *Anal. Chim. Acta ,* 157, 193, 1984.
14. **Yoshimura, C. and Huzino, T.,** Direct determination of beryllium oxide by atomic absorption spectroscopy in the presence of carbon black, *Nippon Kagaku Kaishi,* 7, 1132, 1984.
15. **Gorlova, M. N. and Veller, N. D.,** Atomic absorption determination of beryllium in aluminum-and magnesium-based light alloys, *Zavod. Lab.,* 50(9), 3506, 1984.
16. **Jing, S., Wang, R., Yu, C., Yan, Y., and Ma, Y.,** Design of model ZM-1 zeeman effect atomic absorption spectrometer, *Huanjing Hexue,* 5, 41, 1984 (in Chinese).
17. **Matsusaki, K. and Yoshino, T.,** Electrothermal atomic absorption spectrometric determination of traces of chromium, nickel, iron and beryllium in aluminum and its alloys without preliminary separation, *Anal. Chim. Acta,* 157(1), 193-7, 1984.
18. **Samchuk, A. I.,** Behavior of extracts of thallium, indium, beryllium and molybdenum chelates in a graphite furnace during atomic absorption analysis, *Ukr. Khim. Zh.,* 51(3), 287-91, 1985.
19. **Gorlova, M. N. and Veller, N. D.,** Atomic-absorption determination of beryllium in aluminum-based alloys, *Fiz. Metody Kontrolya Khim. Sostava Mater.* 113, 1983. CA 12, 102: 197101n, 1985.
20. **Bratjer, K. and Klejny, K.,** Determination of beryllium in Pt-Be alloy by atomic absorption spectrometry after ion exchange separation, *Talanta,* 32, 521, 1985.
21. **Zorn, H., Stiefel, T., and Porcher, H.,** Clinical and analytical follow-up of 25 persons exposed accidentally to beryllium, *Toxicol. Environ. Chem.,* 12, 163, 1986.
22. **Paschal, D. C. and Bailey, G. G.,** Determination of beryllium in urine with electrothermal atomic absorption using the L'vov platform and matrix modifications, *At. Spectrosc.,* 7(1), 1-3, 1986.
23. **Voncken, J. H. L., Vriend, S. P., Kocken, J. W. M., and Jansen, J. B. H.,** Determination of beryllium and its distribution in rocks of the tin-tungsten granite of Regoufe, Northern Portugal, *Chem. Geol.,* 56(1-2), 93-103, 1986.
24. **Yao, J. and Huang, W.,** Use of different graphite tubes in the determination of beryllium in rocks by graphite furnace atomic absorption spectrometry, *Fenxi Huaxue,* 14(4), 273-6, 1986.
25. **Yingu, Y. and Weixiang, H.,** Use of different graphite tubes in the determination of beryllium in rocks by graphite furnace atomic absorption spectroscopy, *Fenxi Huaxue,* 14, 273, 1986.

AUTHOR INDEX

SUBJECT INDEX

BISMUTH

BISMUTH, Bi   (ATOMIC WEIGHT 208.980)

Instrumental Parameters:

Wavelength	223.1 nm
Slit width	80 μm
Bandpass	0.3 nm
Light source	Hollow Cathode Lamp
Lamp current	6 mA
Purge gas	Argon or Nitrogen
Sample size	25 μL
Furnace	Cylindrical cuvette/Pyrolytic graphite

Standard Operating Conditions:

Optimum char temperature	450°C
Optimum atomization temperature	2100°C/1100°C
Sensitivity	4pg/1% absorption
Sensitivity check	0.1μg/mL
Working range	0.02-3.0 μg/mL
Background correction	Required only to correct light scattering or non-specific absorption from the sample containing high dissolved solids.

General Note:
1. Use commercial standard or a previously analyzed sample as a working standard.
2. Use ultra clean glass and plastic ware, soaked in 1:5 nitric acid solution and rinsed thoroughly with deionized distilled water.
3. All analytical solutions should be at least 0.2-0.5% (v/v) in nitric acid.
4. Check for blank values on all reagents including water.
5. Dilution is recommended when sample exhibits greater than 0.5 absorbance units.
6. Use of nitrogen as purge gas will suppress the sensitivity.
7. A wider slit width can be used with reduced sensitivity.

REFERENCES

1. **Kujirai, O., Kobayashi, T., and Sudo, E.**, Determination of trace quantities of lead and bismuth in heat-resisting alloys by atomic absorption spectroscopy with heated graphite atomizer, *Trans. Jpn. Inst. Met.,* 18, 775, 1977.

2. **Djudzman, R., Van den Eeckhout, E., and De Moerloose, P.**, Determination of bismuth by atomic absorption spectrophotometry with electrothermal atomization after low-temperature ashing, *Analyst,* 102, 688, 1977.

3. **Headridge, J. B. and Thompson, R.**, Determination of bismuth in nickel-base alloys by atomic absorption spectroscopy with introduction of solid samples into an induction furnace, *Anal. Chim. Acta,* 102, 33, 1978.

4. **Ohta, K. and Suzuki, M.**, Atomic absorption spectroscopy of bismuth with electrothermal atomization from metal atomizers, *Anal. Chim. Acta,* 96, 77, 1978.

5. **Baeckman, S. and Karlssson, R. W.**, Determination of lead, boron, zirconium, silver and antimony in steel and nickel base alloys by atomic absorption spectrophotometry using direct atomization of solid sample in a graphite furnace, *Analyst,* 104, 120, 1979.

6. **Inui, T., Fudagawa, N., and Kawase, A.**, Extraction and atomic absorption spectrometric determination of bismuth with electrothermal atomization, *Z. Anal. Chem.,* 299, 190, 1979.

7. **Katskov, D. A., Grinshtein, I. L., and Kruglikova, L. P.**, Study of the evaporation of the metals In, Ga, Tl, Ge, Sn, Pb, Sb, Bi, Se and Te from a graphite surface by the atomic absorption method, *Zh. Prikl. Spektrosk.,* 33, 804, 1980.

8. **Del Monte Tamba, M. G. and Luperi, N.**, Determination of traces of As, Sb, Bi and V in steel and cast iron by graphite furnace atomic absorption spectroscopy, *Metall Ital.,* 72, 253, 1980 (in Italian and English).

9. **Jin, L. Z. and Ni, Z. M.**, Matrix modification for the determination of trace amounts of bismuth in wastewater, seawater and urine by graphite furnace atomic absorption spectroscopy, *Can. J. Spectrosc.,* 26, 219-22, 1981.

10. **Baker, A. A. and Headridge, J. B.**, Determination of Bi, Pb and Te in copper by atomic absorption spectroscopy with introduction of solid samples into an induction furnace, *Anal. Chim. Acta,* 125, 93, 1981.

11. **Chaleil, D., Lefevre, F., Allain, P., and Martin, G. J.**, Enhanced bismuth digestive absorption in rats by some sulfhydryl compounds: NMR study of complexes formed, *J. Inorg. Biochem.,* 15, 213-21, 1981.

12. **Shimizu, T., Kawamata, Y., Kimura, Y., Shijo, Y., and Sakai, K.**, Determination of bismuth in seawater by graphite furnace atomic absorption spectroscopy after solvent extraction and preconcentration of extract by a rotary evaporator, *Bunseki Kagaku,* 31, 299-303, 1982.

13. **Bertholf, R. L. and Renoe, B. W.**, The determination of bismuth in serum and urine by electrothermal atomic absorption spectroscopy, *Anal. Chim. Acta,* 139, 287-95, 1982.

14. **Lee, D. S.**, Determination of bismuth in environmental samples of flameless atomic absorption spectroscopy with hydride generation, *Anal. Chem.,* 54, 1682-6, 1982.

15. **Ohnashi, T., Watanabe, H., and Ohwa, T.**, Determination of bismuth in white metal by flameless atomic absorption spectroscopy, *Kure Kogyo Shikenjo Hokoku,* 25, 33, 1982 (in Japanese).

16. **Nakahara, T.**, Flameless atomic absorption spectrometric determination of bismuth in sulfides ores and metallurgical materials by electrothermal atomization, *Bull. Univ. Osaka Prefect, Ser. A.,* 31, 57, 1982.

17. **Dong, G.**, A horizontal Y-shaped capillary nebulizer used in hydride generation-flame atomic absorption analysis and determination of trace amounts of bismuth in geochemical (water sediment) samples, *Yankuang Ceshi,* 1, 68-71, 1982 (in Chinese).

18. **Brovko, I. A., Tursunov, A., Fidirko, E. V., and Rish, M. A.**, Atomic absorption determination of antimony and bismuth in natural waters, *Deposited Doc., SPSTL* 1239 Khp-D82, 20 pp, 1982.

19. **Takada, K. and Hirokawa, K.**, Origin of double-peak signals for trace lead, bismuth, silver and zinc in a microamount of steel in atomic absorption spectroscopy with direct electrothermal atomization of a solid sample in a graphite-cup cuvette, *Talanta,* 29, 849, 1982.

20. **Vanloo, B., Dams, R., and Hoste, J.**, Determination of lead and bismuth in steel and cast iron by Zeeman atomic absorption spectroscopy and hydride generation, *Fonderie Belge,* 53, 7-16, 1983.

21. **Terashima, S.**, Determination of bismuth in geological materials by automated hydride generation and electrothermal atomic absorption spectrometry, *Anal. Chim. Acta,* 156, 301, 1984.

22. **L'vov, B. V. and Yatsenko, L. F.**, Carbothermal reduction of zinc, cadmium, lead and bismuth oxides in graphite furnaces for atomic absorption analysis in the presence of organic substances, *J. Phys. C.,* 17(35), 6415-34, 1984.

23. **Headridge, J. B. and Riddington, I. M.**, Determination of silver, lead and bismuth in glasses by atomic absorption spectroscopy with introduction of solid samples into furnaces, *Analyst (London),* 109, 113, 1984.

24. **Brovko, I. A., Tursunov, A., Rish, M. A., and Davirov, A. D.**, Atomic absorption determination of antimony and bismuth in natural samples by hydride formation and metal preconcentration in graphite tube, *Zh. Anal. Khim.,* 39, 1768, 1984 (in Russian).

25. **L'vov, B. V. and Yatsenko, L. F.**, Carbothermal reduction of zinc, cadmium, lead and bismuth oxides in graphite furnaces for atomic absorption analysis in presence of organic substances, *Zh. Anal. Khim.,* 39, 1773, 1984 (in Russian).

26. **El-Yazigi, A., Al-Saleh, I., and Al-Mefty, O.,** Concentrations of Ag, Al, Au, Bi, Cd, Cu, Pb, Sb and Se in cerebrospinal fluid of patients with cerebral neoplasma, *Clin. Chem.,* 30, 1358, 1984.
27. **De Doncker, K., Dumarey, R., Dams, R., and Hoste, J.,** Determination of bismuth in atmospheric particulate matter by hydride generation and atomic absorption spectroscopy, *Anal. Chim. Acta,* 161, 365, 1984.
28. **Terashima, S.,** Determination of bismuth in geological materials by automated hydride generation and electrothermal atomic absorption spectrometry, *Anal. Chim. Acta,* 156, 301-5, 1984.
29. **Terashima, S.,** Determination of bismuth in eighty-three geochemical reference samples by atomic absorption spectrometry, *Geostand. Newsl.,* 8, 155, 1984.
30. **Matsusaki, K., Yoshino, T., and Yamamoto, Y.,** Interference effect of chloride on determination of bismuth by electrothermal atomic absorption spectrometry, *Anal. Chim. Acta,* 167, 299, 1985.
31. **Matsusaki, K. and Yoshino, T.,** Electrothermal atomic absorption spectrometric determination of trace bismuth in aluminum and its alloys without preliminary separation, *Japan Technol. Rep.* Yamaguchi Univ., 3, 289, 1985.
32. **Zakhariya, A. N. and Diallo, I. K.,** AA determination of Sb and Bi impurities in nonferrous metals and alloys by using furnace-flame atomizers, *Zh. Prikl. Spektrosk.,* 43(6), 1002-4, 1985.
33. **Shan, X. and Wang, D.,** X-ray photoelectron spectroscopic study of mechanism of palladium matrix modification in electrothermal atomic absorption spectrometric determination of lead and bismuth, *Anal. Chim. Acta,* 173, 315, 1985.
34. **Ni, Z., Le, X., and Han, H.,** Determination of bismuth in river sediment by electrothermal atomic absorption spectroscopy with low-temperatures atomization in argon/hydrogen, *Anal. Chim. Acta,* 186, 147, 1986.
35. **Bosnak, C. P., Carnrick, G. R., and Slavin, W.,** The determination of bismuth in nickel alloys, *At. Spectrosc.,* 7, 148-50, 1986.
36. **Meintjies, E., Strelow, F. W. E., and Victor, A. H.,** Separation of bismuth from gram amounts of thallium and silver by cation-exchange chromatography in nitric acid, *Talanta,* 34(4), 401-5, 1987.
37. **Dams, R., Alluyn, F., Vanloo, B., Wauters, G., and Vandercasteele, C.,** Gray cast iron reference material certified for antimony, arsenic, bismuth and lead, *Fresnius' Z. Anal. Chem.,* 325, 16307, 1986.
38. **Clark, J. R.,** Electrothermal atomization atomic absorption conditions and matrix modifications for determining Sb, As, Bi, Cd, Ga, Au, Sn, Pb, Mo, Pd, Pt, Se, Ag, Te, Tl and Sn following back-extraction of organic aminohalide extracts, *J. Anal. At. Spectrom.,* 1, 301-8, 1986.
39. **Frech, W., Lindberg, A. O., Lundberg, E., and Cedergren, A.,** Atomization mechanisms and gas phase reactions in graphite furnace atomic absorption spectrometry, *Fresnius' Z. Anal. Chem.,* 323(7), 716-25, 1986.
40. **Castillo, J. R., Lopez-Molinero, A., and Sucunza, T.,** Determination of arsenic, antimony and bismuth in high purity copper by electrothermal atomic absorption spectroscopy, *Mikrochim. Acta,* 4, 105, 1986.
41. **Kharlamov, I. P., Lebedev, V. I., Persits, V. Y., and Eremina, G. V.,** Influence of cations and anions on the analytical signals of zinc, cadmium, lead, tin, bismuth and antimony introduced with solutions of steels and alloys into electrothermal atomizers, *Zh. Anal. Khim.,* 41, 1004, 1986.
42. **Dittrich, K., Mandry, R., Udelnow, C., and Udelnow, A.,** Hydride atomization in graphite furnace atomizers, *Fresnius' Z. Anal. Chem.,* 323(7), 793-9, 1986.

AUTHOR INDEX

SUBJECT INDEX

BORON

BORON , B   (ATOMIC WEIGHT 10.811)

Instrumental Parameters:

Wavelength	249.7 nm
Slit width	320 μm
Bandpass	1 nm
Light source	Hollow Cathode Lamp
Lamp current	10 mA
Purge gas	Argon
Sample size	25 μL
Furnace	Cylindrical cuvette/Pyrolytic graphite

Standard Operating Conditions:

Optimum char temperature	1000°C
Optimum atomization temperature	2700°C
Sensitivity	1000pg/1% absorption
Sensitivity check	1.0-2.5 μg/mL
Working range	1.0-3.0 μg/mL
Background correction	Required only to correct light scattering or non-specific absorption from the sample containing high dissolved solids.

General Note:

1. Use commercial standard or a previously analyzed sample as a working standard.
2. Use ultra clean glass and plastic ware, soaked in 1:5 nitric acid solution and rinsed thoroughly with deionized distilled water.
3. All analytical solutions should be at least 0.2 (v/v) in nitric acid.
4. Calcium in 500 μg/mL concentration added to all analytical solutions may enhance the sensitivity.
5. Check for blank values on all reagents including water.
6. Dilution is recommended when sample exhibits greater than 0.5 absorbance units.
7. Use 0.2% (v/v) nitric acid and 500 μg/mL calcium solution as diluent.

REFERENCES

1. **Szydlowski, F. J.,** Boron in natural waters by atomic absorption spectroscopy with electrothermal atomization, *Anal. Chim. Acta,* 106, 121, 1979.
2. **Van der Geuten, R. P.,** Determination of boron in river water with flameless atomic absorption spectroscopy (graphite furnace technique), *Z. Anal. Chem.,* 306, 13, 1981.
3. **Liu, C. Y., Chen, P. Y., Lin, H. M., and Yang, M. H.,** Determination of boron in high-purity silicon and trichlorosilane indirectly by measurement of cadmium in tris (1,10-phenanthroline) cadmium tetra-fluoroborate, *Fresnius' Z. Anal. Chem.,* 320(1), 22-8, 1985.

AUTHOR INDEX

SUBJECT INDEX

CADMIUM

CADMIUM, Cd   (ATOMIC WEIGHT 112.40)

Instrumental Parameters:

Wavelength	228.8 nm
Slit width	320 µm
Bandpass	1 nm
Light source	Hollow Cathode Lamp
Lamp current	5 mA
Purge gas	Argon or Nitrogen
Sample size	25 µL
Furnace	Cylindrical cuvette/Pyrolytic graphite

Standard Operating Conditions:

Optimum char temperature	250°C
Optimum atomization temperature	1000°C
Sensitivity	0.2 pg/1% absorption
Sensitivity check	0.004 µg/mL
Working range	0.005-0.05 µg/mL
Background correction	Required only to correct light scattering or non-specific absorption from the sample containing high dissolved solids.

General Note:

1. Use commercial standard or a previously analyzed sample as a working standard.
2. Use ultra clean glass and plastic ware, soaked in 1:5 nitric acid solution and rinsed thoroughly with deionized distilled water.
3. All analytical solutions should be at least 0.2-0.5% (v/v) in nitric acid.
4. Use of 1% (w/v) ammonium phosphate $[(NH_4)_2HPO_4]$, 1% (w/v) ammonium sulfate $[(NH_4)_2SO_4]$, or 0.1% (w/v) ammonium molybdate $[(NH_4)MoO_4 \cdot 4H_2O]$ in analytical solutions will enable one to achieve higher pyrolysis temperatures (800°C). Use of 1% (v/v) sulfuric or phosphoric acid can be substituted for ammonium salt solutions.
5. Check for blank values on all reagents including water.
6. Dilution is recommended when sample exhibits greater than 0.5 absorbance units.
7. Use of nitrogen as a purge gas will suppress the sensitivity.
8. The analytical precision of this method is approximately 5% RSD.

REFERENCES

1. **Katskov, D. A. and L'vov, B. V.,** The atomic absorption determination of ultramicroamounts of cadmium using a graphite cell, *Zh. Prikl. Spektrosk.,* 10, 867, 1969.
2. **Headridge, J. B. and Smith, D. R.,** Induction furnace for the determination of cadmium in solutions and zinc base metals by atomic absorption spectroscopy, *Talanta,* 18, 247, 1971.
3. **Robinson, J. W., Wolcott, D. K., Slevin, P. J., and Hindman, G. D.,** Determination of cadmium by atomic absorption in air, water, seawater and urine with a radio frequency carbon rod atomizer, *Anal. Chim. Acta,* 66, 13, 1973.
4. **Shigematsu, T., Matsui, M., and Fujino, O.,** Determination of cadmium by atomic absorption with a heated carbon tube atomizer. Application to sea water, *Bunseki Kagaku,* 22, 1162, 1973.
5. **Kubasik, N. P. and Volosin, M. T.,** Simplified determination of urinary cadmium, lead and thallium, with use of carbon rod atomization and atomic absorption spectrophotometry, *Clin. Chem.,* 19, 954, 1973.
6. **Yasuda, S. and Kakiyama, H.,** Determination of trace amounts of cadmium by flameless atomic absorption spectroscopy using a carbon tube atomizer, *Bunseki Kagaku,* 23, 406, 1974.
7. **Schumacher, E. and Umland, F.,** Improved fast destruction method for the determination of cadmium in body fluids using a graphite tube atomizer, *Z. Anal. Chem.,* 270, 285, 1974.
8. **Ross, R. T. and Gonzalez, J. G.,** The direct determination of cadmium in biological samples by selective volatilization and graphite tube reservoir atomic absorption spectroscopy, *Anal. Chim. Acta,* 70, 443, 1974.
9. **Jensen, F. O., Dolezal, J., and Langmyhr, F. J.,** Atomic absorption spectrometric determination of cadmium, lead and zinc in salts or salt solutions by heating mercury drop electrodeposition and atomization in a graphite furnace, *Anal. Chim. Acta,* 72, 245, 1974.
10. **Janssens, M. and Dams, R.,** Determination of cadmium in air particulates by flameless atomic absorption spectroscopy with a graphie tube, *Anal. Chim. Acta,* 70, 25, 1974.
11. **Vondenhoff, T.,** Determination of lead, cadmium, copper and zinc in plant and animal material by atomic absorption in a flame and in a graphite tube after sample decomposition by the Schoniger technique, *MittBl. GDCh. Fachgr. Lebensmittelchem. Gerichtl. Chem.,* 24, 341, 1975 (in German).
12. **Evenson, M. A. and Anderson, C. T., Jr.,** Ultramicro analysis for copper, cadmium and zinc in human liver tissue by use of atomic absorption spectrophotometry and the heated graphite tube atomizer, *Clin. Chem.,* 21, 537, 1975.
13. **Wright, F. C. and Riner, J. C.,** Determination of cadmium in blood and urine with the graphite furnace, *At. Ab. Newsl.,* 14, 103, 1975.
14. **Kovats, A. and Bohm, B.,** Urinary cadmium determination by flameless atomic absorption spectrophotometry with atomization in a graphite furnace, *Stud. Cercet. Biochim.,* 19, 125, 1976.
15. **Lundren, G.,** Direct determination of cadmium in blood with a temperature-controlled heated graphite tube atomizer, *Talanta,* 23, 309, 1976.
16. **Brodie, K. G. and Stevens, B. J.,** Measurement of whole blood lead and cadmium at low levels using an automatic sample dispenser and furnace atomic absorption, *J. Anal. Toxicol.,* 1, 282, 1977.
17. **Bea-Barredo, F., Polo-Polo, C., and Polo-Diez, C.,** The simultaneous determination of gold, silver and cadmium at ppb levels in silicate rocks by atomic absorption with electrothermal atomization, *Anal. Chim. Acta,* 94, 283, 1977.
18. **Thompson, K. C., Wagstaff, K., and Wheatstone, K. C.,** Method for the minimization of matrix interferences in the determination of lead and cadmium in non-saline waters by atomic absorption spectrophotometry using electrothermal atomization, *Analyst,* 102, 310, 1977.
19. **Campbell, W. C. and Ottaway, J. M.,** Direct determination of cadmium and zinc in sea water by carbon furnace atomic absorption spectrometry, *Analyst,* 102, 495, 1977.
20. **Del-Castiho, P. and Herber, R. F.,** The rapid determination of cadmium, lead, copper and zinc in whole blood by atomic absorption spectroscopy with electrothermal atomization. Improvements in precision with a peak-shape monitoring device, *Anal. Chim. Acta,* 94, 269, 1977.
21. **Wegscheider, W., Knapp, G., and Spitzy, H.,** Statistical investigations of interferences in graphite furnace atomic absorption spectroscopy. II. Cadmium, *Z. Anal. Chem.,* 283, 97, 1977.
22. **Tanaka, T., Hayashi, Y., and Ishizawa, M.,** Simultaneous determination of cadmium and copper in water by a graphite furnace dual channel atomic absorption spectrophotometry, *Bunseki Kagaku,* 27, 494, 1978.
23. **Robinson, J. W. and Weiss, S.,** Direct determination of cadmium in whole blood using an RF-heated carbon-bed atomizer for atomic absorption spectroscopy, *Z. Anal. Chem.,* 292, 365, 1978.
24. **Kitagawa, K., Shigeyasu, T., and Takeuchi, T.,** Application of the Faraday effect to the trace determination of cadmium by atomic spectroscopy with an electrothermal atomizer, *Analyst,* 103, 1021, 1978.
25. **Hirata, S., Marushita, K., and Takimura, O.,** Determination of cadmium in sediments by atomic absorption spectrometry with a carbon tube atomizer, *Bunseki Kagaku,* 27, 543, 1978.
26. **Vesterberg, O. and Wrangskogh, K.,** Determination of cadmium in urine by graphite furnace atomic absorption spectroscopy, *Clin. Chem.,* 24, 681, 1978.
27. **Meranger, J. C. and Subramanian, K. S.,** Direct determination of cadmium in drinking water supplies by graphite furnace atomic absorption spectroscopy, *Can. J. Spectrosc.,* 24, 132, 1979.

28. **Tominaga, M. and Umezaki, Y.,** Suppression of interferences in the determination of lead and cadmium by graphite furnace atomic absorption spectroscopy, *Bunseki Kagaku,* 28, 347, 1979.
29. **Page, A. G., Godbole, S. V., Kulkarni, M. J., Shelar, S. S., and Joshi, B. D.,** Direct AAS determination of cobalt, chromium, copper, manganese and nickel in uranium oxide (U_3O_8) by electrothermal atomization, *Z. Anal. Chem.,* 296, 40, 1979.
30. **Dabeka, R. W.,** Graphite furnace atomic-absorption spectrometric determination of lead and cadmium in foods after solvent extraction and stripping, *Anal. Chem.,* 51, 902, 1979.
31. **Patel, B. M., Bhatt, P. M., Gupta, N., Pawar, M. M., and Joshi, B. D.,** Electrothermal atomic absorption spectrometric determination of cadmium, chromium and cobalt in uranium without preliminary separation, *Anal. Chim. Acta,* 104, 113, 1979.
32. **Kitagawa, K., Koyama, T., and Takeuchi, T.,** Correction system spectroscopic determination of trace amounts of cadmium using the atomic Faraday effect with electrothermal atomization, *Analyst,* 104, 822, 1979.
33. **Carmack, G. D. and Evenson, M. A.,** Determination of cadmium in urine by electrothermal atomic absorption spectroscopy, *Anal. Chem.,* 51, 907, 1979.
34. **Schmidt, W. and Dietl, F.,** Determination of cadmium in digested soils and sediments and extract using flameless atomic absorption with zirconium coated graphite tubes, *Z. Anal. Chem.,* 295, 110, 1979.
35. **Bengtsson, M., Danielsson, L. G., and Magnusson, B.,** Determination of cadmium and lead in sea water after extraction using electrothermal atomization. Minimization of interferences from coextracted sea salts, *Anal. Lett.,* 12, 1367, 1979.
36. **Hoenig, M., Vanderstappen, R., and Van Hoeyweghen, P.,** Electrothermal atomization of cadmium in the presence of complex matrixes, *Analusis,* 7, 17, 1979.
37. **Allain, P. and Mauras, Y.,** Micro determination of lead and cadmium in blood and urine by graphite furnace atomic absorption, *Clin. Chim. Acta,* 91, 41, 1979.
38. **Chakrabarti, C. L., Wan, C. C., and Li, W. C.,** Atomic absorption spectrometric determination of cadmium, lead, zinc, copper, cobalt and iron in oyster tissue by direct atomization from the solid state using the graphite furnace platform technique, *Spectrochim. Acta,* 35B, 547, 1980.
39. **Tsushida, T. and Takeo, T.,** Determination of copper, lead and cadmium in tea by graphite furnace atomic absorption spectrophotometry, *Nippon Shokuhin Kogyo Gakkaishi,* 27, 585, 1980 (in Japanese).
40. **Kruse, R.,** Multiple determination of lead and cadmium in fish by electrothermal AAS after wet-ashing in commonly available teflon beakers, *Z. Lebensm.-Unters Forsch.,* 171, 261, 1980 (in German).
41. **Pederson, B., Willems, M., and Storgaard-Joergensen, S.,** determination of copper, lead, cadmium, nickel and cobalt in EDTA extracts of soil by solvent extraction and graphite furnace atomic absorption spectrophotometry, *Analyst,* 105, 119, 1980.
42. **Legotte, P. A., Rosa, W. C., and Sutton, D. C.,** Determination of cadmium and lead in urine and other biological samples by graphite furnace atomic absorption spectroscopy, *Talanta,* 27, 39, 1980.
43. **Rantala, R. T. T. and Loring, D. H.,** Direct determination of cadmium in silicates from a fluoboric-boric acid matrix by graphite furnace atomic absorption spectroscopy, *At. Spectrosc.,* 1, 163, 1980.
44. **Guevremont, R., Sturgeon, R. E., and Berman, S. S.,** Application of EDTA to direct graphite furnace atomic absorption analysis for cadmium in seawater, *Anal. Chim. Acta,* 115, 163, 1980.
45. **Zolotovitskaya, E. S. and Fidel'man, B. M.,** Use of electrothermal atomic absorption spectroscopy for determining iron, nickel, cadmium and chromium in raw materials and single crystals of corundum, *Zh. Anal. Khim.,* 36, 1564, 1981.
46. **Oradovskii, S. G. and Kuz'michev, A. N.,** Flameless extraction-atomic absorption method for determination of ultramicro amounts of toxic metals (lead, cadmium, copper, cobalt and nickel) in seawaters and ices in labile form, *Tr. Gos Okeanogr. Inst.,* 162, 22, 1981.
47. **Carelli, G., Cecchetti, G., La Bua, R., Bergamaschi, A., and Iannaccone, A.,** Determination of selenium, cadmium, zinc, cobalt, copper and silver using atomic absorption spectophotometry in aerosols from the work place in the color television electronic industry, *Ann. Ist. Super Sanita,* 17, 505, 1981.
48. **Farmer, J. G. and Gibson, M. J.,** Direct determination of cadmium, chromium, copper and lead in siliceous standard reference materials from a fluoboric acid matrix by graphite furnace atomic absorption spectroscopy, *At. Spectrosc.,* 2, 176-8, 1981.
49. **Bouzanne, M.,** Determination of traces of inorganic pollutants (copper, lead, cadmium, zinc) in waters, by differential pulse anodic stripping voltammetry at a rotating mercury film electrode, *Analysis,* 9, 461-7, 1981 (in French).
50. **Subramanian, K. S. and Meranger, J. C.,** A rapid electrothermal atomic absorption spectrophotometric method for cadmium and lead in human whole blood, *Clin. Chem. (Winston-Salem, N.C.),* 27, 1866-71, 1981.
51. **Butterworth, F. E. and Alloway, B. J.,** Investigations into the speciation of cadmium in polluted soils using liquid chromatography, *3rd Int. Conf. Heavy Met. Environ.,* CEP Consult Ltd., Edinburgh, U.K., 1981, 4pp.
52. **Adamska-Dyniewska, H. and Trojanowska, B.,** Relationship between cadmium concentrations in liver biopsy material and in various blood, *Acta Med. Pol.,* 22, 319, 1981.
53. **Van Hattum, B. and De Voogt, P.,** An analytical procedure for the determination of cadmium in human placentas, *Int. J. Environ. Anal. Chem.,* 10, 121, 1981.

54. **Vackova, M., Kuchar, E., and Zemberyova, M.,** Cadmium in the environment and its determination, *Acta Fac. Rerum Nat. Univ. Comenianae Form Prot. Nat.,* 7, 321-31, 1981 (in Slovak).

55. **Delves, H. T. and Woodward, J.,** The determination of low levels of cadmium in blood by electrothermal atomization and atomic absorption spectrophotometry, *3rd Int. Conf. Heavy Met. Environ.,* CEP Consult. Ltd., Edinburgh, U.K., 1981, 622-7.

56. **Delves, H. T. and Woodward, J.,** Determination of low levels of cadmium in blood by electrothermal atomization and atomic absorption spectrophotometry, *At. Spectrosc.,* 2, 65, 1981.

57. **Bruhn, F. C. and Navarrete, A. G.,** Matrix modification for the direct determination of cadmium in urine by electrothermal atomic absorption spectroscopy, *Anal. Chim. Acta,* 130, 209, 1981.

58. **Shimizu, T., Shijo, Y., and Sakai, K.,** Determination of cadmium in human urine by graphite furnace atomic absorption spectroscopy, *Bunseki Kagaku,* 30, 770, 1981.

59. **Ranchet, J., Menissier, F., Lamathe, J., and Voinovitch, I.,** Interlaboratory comparison of the determinations of cadmium, chromium, copper and lead by flameless atomic absorption spectrometry, *Bull. Liasion Lab. Ponts Chaussees,* 114, 81-6, 1981 (in French).

60. **Severin, G., Schumacher, E., and Umland, F.,** Determination of cadmium by flameless atomic absorption spectroscopy. I. Surface treatment of graphite furnace tubes by carbide forming metals, *Z. Anal. Chem.,* 311, 201, 1982 (in German).

61. **Severin, G., Schumacher, E., and Umland, F.,** Determination of cadmium by flameless atomic absorption spectroscopy. II. Modification of graphite furnace tubes by aluminum oxide, *Z. Anal. Chem.,* 311, 205, 1982 (in German).

62. **Poluektov, N. S., Timinskii, Y. A., Zelyukova, Y. V., and Ul'yanova, T. M.,** Flameless atomic absorption determination of cadmium in natural waters with preliminary electrodeposition, *Khim. Tekhnol. Vody,* 4, 324, 1982 (in Russian).

63. **Sakata, M. and Shimoda, O.,** Sample and rapid method for determination of lead and cadmium in sediment by graphite furnace atomic absorption spectroscopy, *Water Res.,* 16, 231, 1982.

64. **Ranchet, J., Manissier, F., Lamathe, J., and Voinovitch, I.,** Interlaboratory comparison: the determinations of cadmium, chromium, copper and lead in standard solutions by flameless atomic absorption spectroscopy, *Analysis,* 10, 71-7, 1982.

65. **Li, R.,** Determination of trace silver and cadmium in geochemical reference samples by flameless atomic absorption spectroscopy, *Yankuang Ceshi,* 1, 60, 1982 (in Chinese).

66. **Prokof'ev, A. K., Oraderskii, S. G., and Georgievskii, V. V.,** Flameless atomic absorption method for the determination of copper, lead and cadmium in marine bottom sediments, *Tr. Gos. Okeanogr. In-ta,* 162, 51-5, 1981 (in Russian). From Ref. *Zh. Khim.,* Abstr. No. 11G258, 1982.

67. **Viala, A., Gouezo, F., Mallet, B., Fondarai, J., Grimaldi, F., and Cano, J. P.,** Air pollution by cadmium in Marseille (France), *Toxicol. Eur. Res.,* 4, 25-9, 1982.

68. **Sharma, R. P., McKenzie, J. M., and Kjellstrom, T.,** Analysis of submicrogram levels of cadmium in whole blood, urine and hair by graphite furnace atomic absorption spectroscopy, *J. Anal. Toxicol.,* 6, 135, 1982.

69. **Suzuki, M. and Ohta, K.,** Reduction of interferences with thiourea in the determination of cadmium by electrothermal atomic absorption spectroscopy, *Anal. Chem.,* 54, 1686-9, 1982.

70. **Petrov, I. and Tsalev, D.,** Determination of cadmium in soils by electrothermal atomic absorption spectroscopy, *Dokl. Bolg. Akad. Nauk,* 35, 467-70, 1982.

71. **Hoenig, M.,** Difficulties of standardization in cadmium determinations by electrothermal atomic absorption spectroscopy, *Spectrochim. Acta,* 37B, 929, 1982 (in French).

72. **Kamata, E., Nakashima, R., Goto, K., Furakawa, M., and Shibata, S.,** Electrothermal atomic absorption spectrometric determination of ng/g^{-1} levels of cadmium in bone after dithiozone extractions, *Anal. Chim. Acta,* 144, 197, 1982.

73. **Panholzer, M., Raptis, S. E., and Mueller, K.,** Simple method for determining cadmium in urine, *Mikrochim. Acta,* 2, 189-98, 1982.

74. **Hasan, M. Z., Kumar, A., and Pande, S. P.,** Determination of cadmium in water by flameless atomic absorption spectrophotometry, *Res. Ind.,* 27, 8-10, 1982.

75. **Bangia., T. R., Kartha, K. N. K., Varghese, M., Dhawale, B. A., and Joshi, B. D.,** Chemical separation and electrothermal atomic absorption spectrophotometric determination of cadmium, cobalt, copper and nickel in high-purity uranium, *Z. Anal. Chem.,* 310, 410, 1982.

76. **Pan, D. and Ling, J.,** Determination of cadmium and lead in cereals and tea using graphite furnace atomic absorption spectroscopy, *Fenxi Huaxue,* 10, 190-1, 1982 (in Chinese).

77. **Lieser, K. H., Sondermeyer, S., and Kliemchen, A.,** Precision and accuracy of analytical results in determining the elements cadmium, chromium, copper, iron, manganese and zinc by flameless atomic absorption spectroscopy, *Fresnius' Z. Anal. Chem.,* 312, 517, 1982.

78. **Atsuya, I. and Itoh, K.,** Direct determination of cadmium in NBS bovine liver by Zeeman atomic absorption spectroscopy with a miniature graphite cup, *Bunseki Kagaku,* 31, 713, 1982.

79. **Gambashidze, L. M.,** Electrothermal atomic absorption analysis of complexes of cadmium with organic ligands, *Issled. Metallsoderzh. Organ. Soedin. Mettodom Atomno-absorbtsion. Spektrometrii, M.* 73, 1982. From Ref. *Zh. Khim.,* Abstr. No. 16G82, 1982.

80. **Sperling, K. R.,** Determination of heavy metals in seawater and in marine organisms by flameless atomic absorption spectrophotometry. XIV. Comments on the usefulness of organohalides as solvents for the extraction of heavy metal (cadmium) complexes, *Fresnius' Z. Anal. Khim.,* 310, 254, 1982.

81. **Cammann, K. and Andersson, J. T.,** Increased sensitivity and reproducibility through signal averaging in ranges near an instrument limit of detection. Thallium and cadmium trace determinations in rock samples, *Fresnius' Z. Anal. Chem.,* 310, 45-50, 1982.

82. **Lamathe, J., Magurno, C., and Equel, J. C.,** An interlaboratory study of the determination of cadmium, copper and lead in seawater by electrothermal atomic absorption spectroscopy, *Anal. Chim. Acta,* 142, 183, 1982 (in French).

83. **Claeys-Thoreau, F.,** Determination of low levels of cadmium and lead in biological fluids with simple dilution by atomic absorption spectrophotometry using Zeeman effect background correction and the L'vov platform, *At. Spectrosc.,* 3, 188, 1982.

84. **Ueda, T., Nishida, S., and Yoshihara, T.,** Hygienic chemical studies on dental treatment materials. II. Determination of arsenic, lead and cadmium in dental cements by flameless atomic absorption spectrophotometry, *Tokyo-toritsu Eisei Kenkyusho Kenkyu Nempo,* 33, 110, 1982 (in Japanese).

85. **Li, R.,** Determination of trace silver and cadmium in geochemical reference samples by flameless atomic absorption spectroscopy, *Yankuang Ceshi,* 11, 60, 1982 (in Chinese).

86. **Knezevic, G.,** The heavy metal content of biscuits. Determination of lead and cadmium by the flameless AAS method, *CCB,* 7, 10, 1982.

87. **Kreuzer, W., Bunzl, K., and Kracke, W.,** Studies on lead and cadmium contents in livers and kidneys of slaughter cattle. III. Cattle from the vicinity of regional and spot sources of emission, *Fleischwirtschaft,* 62, 1479, 1982.

88. **Buckenhueskes, H., Hartman, B., and Gierschner, K.,** Possibilities for analytical detection of environmentally deposited lead and cadmium compounds. I., *Ind. Obst-Gemueseverwert,* 67, 538-43, 1982 (in German).

89. **Obukhov, A. I. and Plekhanova, I. O.,** Effect of sample composites on the determination of environmental lead and cadmium by atomic absorption, *Biol. Nauki* (Moscow), 10, 99, 1982 (in Russian).

90. **Kuhnert, P. M., Erhard, P., and Kuhnert, B. R.,** Analysis of cadmium in whole blood and placental tissue by graphite furnace atomic absorption spectroscopy, *Trace Substr. Environ. Health,* 16, 370, 1982.

91. **Bourcier, D. R., Sharma, R. P., Bracken, W. M., and Taylor, M. J.,** Cadmium-copper interaction: effect of copper pretreatment and cadmium-copper chronic exposure on the distribution and accumulation of cadmium, copper, zinc and iron in mice, *Trace Substr. Environ. Health,* 16, 273, 1982.

92. **Subramanian, K. S., Meranger, J. C., and Mackeen, J. E.,** Graphite furnace atomic absorption spectroscopy with matrix modification for determination of cadmium and lead in human urine, *Anal. Chem.,* 55, 1064, 1983.

93. **Wang, S. T., Strunc, G., and Peter, F.,** Determination of cadmium and lead in whole blood and urine by Zeeman flameless atomic absorption spectroscopy, *Proc. 2nd Int. Conf. Chem. Toxicol. Clin. Chem. Met.,* 1983, 57-60.

94. **Ali, S. L.,** Determination of pesticide residues and other critical impurities such as toxic trace metals in medicinal plants. II. Determination of toxic trace metals in drugs, *Pharm. Ind.,* 45, 1294, 1983 (in German).

95. **Wei, F., Chen, J., and Yin, F.,** Simultaneous determination of ultratrace cadmium and lead in beverages by flameless atomic absorption analysis with automatic atomizing introduction of sample, *Shipin Kexue* (Beijing), 46, 39-43, 1983 (in Chinese).

96. **Subramanian, K. S., Meranger, J. C., and Connor, J.,** The effect of container material, storage time and temperature on cadmium and lead levels in heparinized human whole blood, *J. Anal. Toxicol.,* 7, 15, 1983.

97. **Bourcier, D. R., Sharma, R. P., Bracken, W. M., and Taylor, M. J.,** Cadmium-copper interaction in intestinal mucosal cell cytosol of mice, *Biol. Trace Elem. Res.,* 5, 195-204, 1983.

98. **Steiner, J. W. and Kramer, H. L.,** *In situ* gaseous pretreatment of liver extracts in modified carbon rod atomizer during determination of cadmium and lead, *Analyst,* 108, 1051, 1983.

99. **Aurand, K., Drews, M., and Seifert, B.,** Passive samples for determination of heavy metal burden of indoor environments, *Environ. Technol. Lett.,* 4, 433, 1983.

100. **Guevremont, R. and Whitman, J.,** Automation of graphite furnace atomic absorption spectrometry system with Z80-based microcomputer, *Anal. Chim. Acta,* 154, 295, 1983.

101. **Voellkopf, U., Grobenski, Z., and Welz, B.,** Stabilized temperature platform furnace with zeeman effect background correction for trace analysis in wastewater, *At. Spectrosc.,* 4, 165, 1983.

102. **Legret, M., Demare, D., Marchandise, P., and Robbe, D.,** Interferences of major elements in the determination of lead, copper, cadmium, chromium and nickel in river sediments and sewer sludges by electrothermal atomic absorption spectroscopy, *Anal. Chim. Acta,* 149, 107, 1983.

103. **Brune, D., Gjerdet, N., and Paulsen, G.,** Gastrointestinal and *in vitro* release of copper, cadmium, indium, mercury and zinc from conventional and copper-rich amalgams, *Scand. J. Dent. Res.,* 91, 66, 1983.

104. **Simon, J. and Liesse, T.,** Trace determination of lead and cadmium in bones by electrothermal AAS, *Fresnius' Z. Anal. Chem.,* 314, 483, 1983.

105. **Marletta, G. P. and Favretto, L. G.,** Preliminary investigation on the balance of lead and cadmium content in milk and its by-products, *Z. Lebensm. Unters. Forsch.,* 176, 32, 1983.

106. **Khavezov, I. and Ivanova, E.,** Electrothermal atomization using the L'vov platform. Determination of traces of lead, cadmium, manganese and copper in a sodium chloride matrix, *Fresnius' Z. Anal. Chem.,* 315, 34-7, 1983.

107. **Boiteau, H. L., Metayer, C., Ferre, R., and Pineau, A.,** Automated determination of lead, cadmium, manganese and chromium in whole blood by Zeeman atomic absorption spectrometry, *Analysis,* 11(5), 234-42, 1983 (in French).

108. **Mueller, J. and Kallischnigg, G.,** A ring test for the determination of lead, cadmium and mercury in biological material, *ZEBS-Ber.,* 1, 69, 1983.

109. **Schindler, E.,** Compound procedure to determine lead, cadmium and copper in cereals and flour, *Dtsch. Lebensm. Rundsch,* 79(10), 334-7, 1983.

110. **Janssen, E.,** Determination of lead and cadmium in feeds after ashing under pressure with the Zeeman graphite tube technique, *Landwirtsch. Forsch.,* 36(1-2), 161-71, 1983 (in German).

111. **Takada, K. and Hirokawa, K.,** Determination of traces of lead and cadmium in high-purity tin by polarized Zeeman atomic absorption spectroscopy with direct atomization of solid sample in a graphite-cup cuvette, *Talanta,* 30, 329, 1983.

112. **Zhu, Y.,** Determination of lead and cadmium contents in traditional Chinese medicines by graphite furnace atomic absorption method, *Yaown Fenxi Zazhi,* 3, 175-7, 1983 (in Chinese).

113. **Koops, J. and Westerbeek, D.,** Flameless atomic absorption spectrometric determination of the lead and cadmium contents of Gouda and Edam cheese produced in the Netherlands, *Neth. Milk Dairy J.,* 37, 21-5, 1983.

114. **Postel, W., Meier, B., and Markert, R.,** Determination of lead, cadmium, aluminum and tin in beer using atomic absorption spectrophotometry, *Monatsschr. Brauwiss,* 36, 300, 1983 (in German).

115. **Postel, W., Meier, B., and Markert, R.,** Lead, cadmium, aluminum, tin, zinc, iron and copper in bottled and canned beer, *Monatsschr. Brauwiss,* 36, 360-7, 1983 (in German).

116. **Oliver, W. K., Reeve, S., Hammond, K., and Basketter, F. B.,** Determination of lead and cadmium by electrothermal atomization atomic absorption utilizing the L'vov platform and matrix modification, *J. Inst. Water Eng. Sci.,* 37, 460, 1983.

117. **Van Deijck, W. and Herber, R. F. M.,** Some observations on direct determination of cadmium in urine by electrothermal atomization atomic absorption spectroscopy, *Clin. Chim. Acta,* 128, 379, 1983.

118. **Nakashima, S. and Yagi, M.,** Determination of nanogram amounts of cadmium in water by electrothermal atomic absorption spectroscopy after flotation separation, *Anal. Chim. Acta,* 147, 213, 1983.

119. **Pruszkowska, E., Carnrick, G. R., and Slavin, W.,** Direct determination of cadmium in coastal seawater by atomic absorption spectroscopy with the stabilized temperature platform furnace and Zeeman background correction, *Anal. Chem.,* 55, 182, 1983.

120. **Sharda, B.,** Flameless atomic absorption spectrophotometry of cadmium in plasma, liver tissue, urine, hair and nails, *Curr. Med. Pract.,* 27, 293, 1983.

121. **Vuori, E., Vetter, M., Kuitunen, P., and Salmela, S.,** Cadmium in Finnish breast milk, a longitudinal study, *Arch. Toxicol.,* 53, 207, 1983.

122. **Jawaid, M., Lind, B., and Elinder, C. G.,** Determination of cadmium in urine by extraction and flameless atomic absorption spectrophotometry. Comparison of urine from smokers and non-smokers of different sex and age, *Talanta,* 30, 509, 1983.

123. **Sperling, K. R.,** Experience with cadmium determination in human blood samples, *Proc. 2nd Int. Workshop Trace Elem. Anal. Chem. Med. Biol.,* Braetter, P. and Schramel, P., Eds., Berlin, 1983, 969-74.

124. **Heinrich, R. and Angerer, J.,** Determination of cadmium in urine by ETAAS following extraction, *Fresnius' Z. Anal. Chem.,* 315, 528-33, 1983 (in German).

125. **Hasan, M. Z. and Kumar, A.,** Interference due to calcium and magnesium in flameless atomic absorption spectrophotometric determination of cadmium in water; its suppression by matrix modification, *Indian J. Environ. Health,* 25(3), 161-8, 1983.

126. **Bertocchi, G. and Benfenati, L.,** Determination of cadmium at trace levels in raw materials for pharmaceutical and cosmetic uses, *Relata Tech.,* 15(36), 81-2, 1983 (in Italian).

127. **Slavin, W., Manning, D. C., Carnrick, G., and Pruszkowska, E.,** Properties of the cadmium determination with the platform furnace and zeeman background correction, *Spectrochim. Acta,* 38B(8), 1157-70, 1983.

128. **Benson, W. H., Francis, P. C., Birge, W. J., and Black, J. A.,** A simple method for acid extraction of cadmium from fish eggs or fish tissues, *At. Spectrosc.,* 4(6), 212, 1983.

129. **Black, M., Ottaway, J. M., Fell, G. S., and Aughey, E.,** Model for cadmium toxicity, *Anal. Proc. (London),* 20(12), 592, 1983.

130. **Bermejo Barrera, P., Cochode Juan, J. A., and Bermejo Martinez, F.,** Determination of cadmium in foods, *An. Bromatol.,* 33, 211, 1983.

131. **Falk, H., Hoffman, E., Luedke, C., Carnrick, J. M., and Giri, S. K.,** Furnace atomization with non-thermal extraction-experimental evaluation of detection based on high-resolution echelle monochromator incorporating automatic background correction, *Analyst,* 108, 1459, 1983.

132. **Subramanian, S. K. and Meranger, J. C.,** Blood levels of cadmium, copper, lead and zinc in children in a British Columbia community, *Sci. Total Environ.,* 30, 231-44, 1983.

133. **Steiner, J. W.,** Direct removal of interfering species from the AAS-electrothermal atomizer during the analysis of cadmium and molybdenum in biological materials, *Spurenelem. Symp.,* 411-17, 1983.

134. **Muys, T.,** Quantitative determination of lead and cadmium in foods by programmed dry ashing and atomic absorption spectrophotometry with electrothermal atomization, *Analyst,* 109, 119, 1984.

135. **El-Yazizi, A., Al-Saleh, J., and Al-Mefty, O.,** Concentrations of Ag, Al, Au, Bi, Cd, Cu, Pb, Sb and Se in cerebrospinal fluid of patients with cerebral neoplasma, *Clin. Chem.,* 30, 1358, 1984.

136. **Hasan, M. Z., Kumar, A., and Pande, S. P.,** Occurrence of cadmium in water: its determination by flameless AAS, *J. Indian Inst. Sci.,* 65(5), 105-10, 1984.

137. **Koide, M., Lee, D. S., and Stallard, M. W.,** Concentration and separation of trace metals from seawater using a single anion exchange bead, *Anal. Chem.,* 56, 1956, 1984.

138. Release of metals (lead, cadmium, copper, zinc, nickel and chromium) from kitchen blenders, Statens Levnedsmiddelinstitut Report 1984, PUB-88, Order No. PB84-193416, 26 pp (in Danish). Avail. NTIS. From *Gov. Rep. Announce. Index (U.S.),* 84, 67, 1984.

139. **Hocquellet, P.,** Use of atomic absorption spectrometry with electrothermal atomization for direct determination of trace elements in oils: cadmium, lead, arsenic and tin, *Rev. Fr. Corps Gras,* 31, 117, 1984 (in French).

140. **Sun, H. J., Lin, H. M., and Yang, M. H.,** Determination of lead and cadmium in organic and biological samples by low-temperature plasma ashing combined with graphite furnace atomic absorption spectrometric method, *Hua Hsueh,* 42, 51, 1984 (in Chinese).

141. **Maubach, G.,** Nondestructive simultaneous determination of lead and cadmium by AAS, *Labor Praxis,* 8, 764, 1984.

142. **Voellkopf, U. and Grobenski, Z.,** Interferences in analysis of biological samples using stabilized temperature platform furnace and zeeman background correction, *At. Spectrosc.,* 5, 115, 1984.

143. **Slavin, W. and Carnrick, G. R.,** Possibility of standardless furnace atomic absorption spectroscopy, *Spectrochim. Acta,* 39B, 271, 1984.

144. **Wennrich, R., Dittrich, K., and Bonity, U.,** Matrix interference in laser atomic absorption spectrometry, *Spectrochim. Acta,* 39B, 657, 1984.

145. **Voinovitch, I. A., Druon, M., and Louvrier, J.,** Effect of absence of injection orifice and length of graphite furnaces on sensitivity of determinations by atomic absorption spectrometry with electrothermal atomization, *Analysis,* 12, 256, 1984 (in French).

146. **Loumond, F., Copin-Montegut, G., Courau, P., and Nicholas, E.,** Cadmium, copper and lead in Western Mediterranean sea, *Mar. Chem.,* 15, 251, 1984.

147. **Danielsson, L. G. and Westerlund, S.,** Short term variations in trace metal concentrations in Baltic, *Mar. Chem.,* 15, 273, 1984.

148. **Muys, T.,** Quantitative determination of lead and cadmium in foods by programmed dry ashing and atomic absorption spectrophotometry with electrothermal atomization, *Analyst,* 109, 119, 1984.

149. **Halls, D. J.,** Speeding up determinations by electrothermal atomic absorption spectrometry, *Analyst,* 109, 1081, 1984.

150. **Sharda, B.,** Flameless atomic absorption spectrophotometry of cadmium in plasma, liver tissue, urine, hair and nails, *Curr. Med. Pract.,* 27, 293, 1984.

151. **Jaeger, H.,** Determination of cadmium in plastics by graphite furnace AAS, *Labor Praxis,* 8, 345, 1984 (in German).

152. **Feitsman, K. G., Franke, J. P., and de Zeeuw, R. A.,** Comparison of some matrix modifiers for determination of cadmium in urine by atomic absorption spectrometry with electrothermal atomization, *Analyst,* 109, 789, 1984.

153. **Chakrabarti, C. L., Wu, S., Karwowska, R., Chang, S. B., and Bertels, P. C.,** Studies in atomic absorption spectrometry with laboratory-made graphite furnace, *At. Spectrosc.,* 5, 69, 1984.

154. **Backstrom, K., Danielsson, L. G., and Nord, L.,** Sample work-up for graphite furnace atomic absorption spectrometry using continuous flow extraction, *Analyst,* 109, 323, 1984.

155. **Littlejohn, D., Cook, S., Durie, D., and Ottaway, J. M.,** Investigation of working conditions for graphite probe atomization in electrothermal atomic absorption spectrometry, *Spectrochim. Acta,* 39B, 295, 1984.

156. **Sperling, K. R.,** Tube-in-tube technique in electrothermal atomic absorption spectrometry. III. Zn, Cd, Pb, Cu and Ni, *Spectrochim. Acta,* 39B, 371, 1984.

157. **Favretto, L. G., Marletta, G. P., and Favretto, L.,** Determination of cadmium and lead in natural waters, *Fresnius' Z. Anal. Chem.,* 318, 434, 1984.

158. **Baffi, F., Frache, R., and Dadone, A.,** Comparison between preconcentration methods for determination of chemical forms of cadmium in inshore seawater by atomic absorption spectrophotometry, *Ann. Chim.,* 74, 385, 1984.

159. **Carrondo, M. J. T., Reboredo, F., Ganho, R. M. B., and Oliveira, J. F. S.,** Analysis of sediments for heavy metals by rapid electrothermal atomic absorption procedure, *Talanta,* 31, 561, 1984.

160. **Dungs, K. and Neidhart, B.,** Analysis of urine samples by electrothermal atomization-atomic absorption spectrometry: comparison of natural and control material, *Analyst,* 109, 877, 1984.

161. **McAughey, J. J. and Smith, N. J.,** Direct determination of cadmium in urine by electrothermal atomic absorption spectrometry with L'vov platform, *Anal. Chim. Acta,* 156, 129, 1984.

162. **Nilsson, T. and Berggren, P. D.,** Determination of cadmium in microgram amounts of pancreatic tissue by electrothermal atomic absorption spectrometry, *Anal. Chim. Acta,* 159, 381, 1984.

163. **Niwa, M., Kajimoto, M., Iwai, T., Hirasawa, K., and Inoue, M.,** Analysis of trace metals in human teeth. III. Cadmium concentration in human teeth by flameless atomic absorption spectroscopy, *Shigaku,* 71, 1083, 1984.

164. **Kanipayor, R., Naranjit, D. A., Radziuk, B. H., Van Loon, J. C., and Thomassen, Y.,** Direct analysis of solids for trace elements of combined electrothermal furnace/quartz T-tube/flame atomic absorption spectrometry, *Anal. Chim. Acta,* 166, 39, 1984.

165. **Kantor, T., Bezur, L., and Fazakas, J.,** Volatilization studies of cadmium compounds by a quartz furnace-flame combined atomic absorption method. Effect of magnesium chloride and ascorbic acid as additives, *Magy. Kem. Foly.,* 90, 305, 1984.

166. **Ma, J. and Chen, J.,** Tungsten filament flameless atomic absorption spectrometric determination of cadmium in wastewater, *Guangpuxue Yu Guangpu Fenxi,* 4, 39, 1984 (in Chinese).

167. **Kido, T., Tsuritani, I., Honda, R., Ishizaki, M., Yamada, Y., and Nogawa, K.,** A direct determination of urinary cadmium by graphite-furnace atomic absorption spectroscopy using Zeeman effect, *Kanazawa Ika Daigaku Zasshi,* 9, 70, 1984 (in Japanese).

168. **Chen, Z. and Fan, C.,** Determination of trace cadmium in foods by closed digestion system and graphite furnace atomic absorption spectroscopy, *Shipin Yu Fajiao Gongye,* 4, 8, 1984.

169. **Batz, L., Ganz, S., Hermann, G., Scharmann, A., and Wirz, P.,** Measurement of stable isotope distribution using Zeeman atomic absorption spectroscopy, *Spectrochim. Acta,* 39B, 993, 1984.

170. **Feustel, A. and Wennrich, R.,** Zinc and cadmium in cell fractions of prostatic cancer tissues of different histological grading in comparison to BPH and normal prostate, *Urol. Res.,* 12, 147, 1984.

171. **Jing, S., Wang, R., Yu, C., Zhang, D., Yan, Y., and Ma, Y.,** Design of model ZM-1 zeeman effect atomic absorption spectrometer, *Huanjing Kexue,* 5, 41, 1984 (in Chinese).

172. **Janssen, E. and Bruene, H.,** Determination of mercury, lead and cadmium in fish from the Rhine and Main by flameless atomic absorption, *Z. Lebensm. Unters. Forsch.,* 178, 168, 1984 (in German).

173. **Favretto, L., Marletta, G. P., and Favretto, L. G.,** Nonlinear standard additions in AAS determination of lead and cadmium with electrothermal atomization in a graphite furnace, *At. Spectrosc.,* 5, 51, 1984.

174. **L'vov, B. V. and Yatsenko, L. F.,** Carbothermal reduction of zinc, cadmium, lead and bismuth oxides in graphite furnaces for atomic absorption analysis in the presence of organic substances, *J. Phys. C.,* 17(35), 6415-34, 1984.

175. **Nakaki, K., Fukabori, S., and Masuda, T.,** Lead and cadmium levels in blood samples in general population of urban areas, *Rado Kagaku,* 60(12), 577-83, 1984 (in Japanese).

176. **Mohl, C., Ostapczuk, P., and Stoeppler, M.,** Direct determination of lead and cadmium in urine using graphite furnace AAS and the L'vov platform, *Fortschr. Atomspektrom. Spurenanal.,* 1, 317, 1984 (in German).

177. **Mishima, M.,** Improved accuracy and precision in the determination of trace metals by graphite furnace atomic absorption spectrometry, *Koshu Eiseiin Kenkyu Hokoku,* 33, 16, 1984 (in Japanese).

178. **Terashima, S.,** Determination of cadmium and lead in seventy-seven geological reference samples by atomic absorption spectroscopy, *Geostand. Newsl.,* 8, 13, 1984.

179. **Ohls, K.,** Determination of small cadmium amounts in various materials by using solid samples in ICP and flameless atomic absorption spectrometry, *Spectrochim. Acta,* 39B, 1105, 1984.

180. **Narres, H. D., Mohl, C., and Stoeppler, M.,** Metal analysis in different materials with platform furnace zeeman-atomic absorption spectrometry, *Int. J. Environ. Anal. Chem.,* 18, 267, 1984.

181. **Kaegler, S. H. and Kotzel, R.,** Determination of cadmium at <10 µg/kg in petroleum products by using graphite furnace AAS, *Fortschr. Atomspektrom. Spurenanal.,* 1, 233, 1984 (in German).

182. **Morisi, G., Patriarca, M., and Macchia, T.,** Method for determining cadmium in whole blood in a general population. Evaluation of reliability characteristics and first results of use, *G. Ital. Chim. Clin.,* 9, 59, 1984 (in Italian).

183. **Subramanian, K. S., Meranger, J. C., and Connor, J.,** Effect of container material storage time and temperature on determination of cadmium levels in human urine, *Talanta,* 32, 435, 1985.

184. **Janssen, A., Willmann, K. H., and Simon, F. J.,** Determination of traces of cadmium in zinc by flameless atomic absorption spectrophotometry after extraction, *Z. Anal. Chem.,* 320, 137, 1985 (in German).

185. **Drasch, G., Kauret, G., and Von Meyer, L.,** Cadmium body burden of an occupationally non-burdened population in Southern Bavaria (FRG), *Int. Arch. Occup. Environ. Health,* 55(2), 141-8, 1985.

186. **Baucells, M., Lacort, G., Roura, M., Pascal, M. D., and Felipo, M. T.,** Cadmium determination in soil extracts by furnace atomic absorption, *Int. J. Environ. Anal. Chem.,* 22, 61, 1985.

187. **Wu, J., Zeng, M., and Xu, Q.,** Determination of cadmium in rocks by GFAAS, *Guangpuxue Yu Guangpu Fenxi,* 5(1), 57-8, 1985, CA 20, 103:97999p, 1985.

188. **Clark, D. E., Natiom, J. R., Bourgeois, A. J., Hare, M. F., Baker, D. M., and Hinderberger, E. J.,** The regional distribution of cadmium in the brains of orally exposed adult rats, *Neurotoxicology,* 6(3), 109-14, 1985.

189. **Matsushita, S.,** Pretreatment of samples for measurement of cadmium in blood by flameless atomic absorption method, *Nippon Koshu Eisei Zasshi,* 32(5), 247-51, 1985.

190. **Herber, R. F. M., Roelfsen, A. M., Hazelhoff, R., and Peereboom-Stegeman, J. H. J. C.,** Direct determination of cadmium in placenta. Comparison with a destructive atomic absorption spectrometric method, *Fresnius' Z. Anal. Chem.,* 322(7), 743-6, 1985.

191. **Ruehl, W. J.,** Cadmium in polymers — product control in the automobile industry, *Fresnius' Z. Anal. Chem.,* 322(7), 710-2, 1985.

192. **Fudagawa, N. and Kawase, A.,** Determination of cadmium in coal by metal furnace atomic absorption spectrometry, *Bunseki Kagaku,* 34, 228, 1985 (in Japanese).

193. **Hulanicki, A., Bulska, E., and Wrobel, K.,** Effect of inorganic matrices in determination of cadmium by atomic absorption spectrometry with electrothermal atomization, *Analyst,* 110, 1141, 1985.

194. **Inoue, M. and Nicoa, M.,** Effects of dietary cadmium on the concentration of cadmium in the hard tissues of growing rats, *Shigaku,* 73(5), 1282-94, 1985.

195. **Piperaki, E. A.,** Determination of cadmium in blood by FAAS, *Chem. Chron.,* 14(1), 57-60, 1985.

196. **Erb, R.,** Quality control by Zeeman-AAS in a chemical factory. Direct determination of copper, manganese, iron, cadmium and lead in adhesive tapes and diverse raw materials, *Fresnius' Z. Anal. Chem.,* 322(7), 719-20, 1985.

197. **Kratochvil, B., Thapa, R. S., and Motkosky, N.,** Evaluation of homogeneity of marine biological tissues for lead and cadmium by GFAAS, *Can. J. Chem.,* 63, 2679, 1985.

198. **Narres, H. D., Mohl, C., and Stoeppler, M.,** Metal analysis in difficult materials with platform furnace zeeman-atomic absorption spectroscopy. II. Direct determination of cadmium and lead in milk, *Z. Lebensm. Unters. Forsch.,* 181, 111, 1985.

199. **Jin, F. and Jing, Y.,** Direct determination of lead, cadmium and nickel in soils by atomic absorption spectrometry, *Fenxi Huaxue,* 13, 386, 1985 (in Chinese).

200. **Horvath, Z., Lasztity, A., Szakacs, O., and Bozsai, G.,** Iminodiacetic and ethylcellulose as chelating ion exchanger. 1. Determination of trace metals by atomic absorption spectrometry and collection of uranium, *Anal. Chim. Acta,* 173, 273, 1985.

201. **Hinds, M. W., Jackson, K. W., and Newman, A. P.,** Electrothermal atomization atomic absorption spectrometry with direct introduction of slurries determination of trace metals in soil, *Analyst,* 110, 947, 1985.

202. **Wennrich, R. and Feustel, A.,** Determination of cadmium and zinc in human prostatic tissues by flameless AAS, *Z. Med. Laboratoriumsdiagn.,* 26(7), 365-9, 1985.

203. **Lang, H., Gen, X., Zhu, M., and Bai, G.,** Determination of cadmium, lead and copper in natural water and industrial wastewater by ion-exchange-three-hole ring oven technique, *Xibei Daxue Xuebao Ziran Kexueban,* 14(4), 38-42, 1984, CA 23, 103:146866d, 1985.

204. **Rantala, R. T. T. and Loring, D. H.,** Partition and determination of cadmium, copper, lead and zinc in marine suspended particulate matter, *Int. J. Environ. Anal. Chem.,* 19(3), 165-73, 1985.

205. **Subramanian, K. S., Meranger, J. C., Wan, C. C., and Corsini, A.,** Preconcentration of cadmium, chromium, copper and lead in drinking water on the polyacrylic ester resin, XAD-7, *Int. J. Environ. Anal. Chem.,* 19(4), 261-72, 1985.

206. **Statham, P. J.,** Determination of dissolved manganese and cadmium in seawater at low nmol per liter concentrations by chelation and extraction followed by and extraction followed by electrothermal atomic absorption spectrometry, *Anal. Chim. Acta,* 169, 149, 1985.

207. **Baucells, M., Lacort, G., and Roura, M.,** Determination of cadmium and molybdenum in soil extracts by graphite furnace atomic absorption and ICP, *Analyst (London),* 110(12), 1423-9, 1985.

208. **Kluessendorf, B., Rosopulo, A., and Kreuzer, W.,** Study of the distribution and rapid determination of lead, cadmium and zinc in livers of slaughtered pigs by Zeeman AAS of solid samples, *Fresnius' Z. Anal. Chem.,* 322(7), 721-7, 1985.

209. **Sneddon, J.,** Use of impaction-electrothermal atomization AA spectrometric system for the direct determination of cadmium, copper and manganese in the lab atmosphere, *Anal. Lett.,* 18(A10), 1261-80, 1985.

210. **Criand, A. and Fouillac, C.,** Use of L'vov platform and molybdenum coating for determination of volatile elements in thermomineral waters by atomic absorption spectrometry, *Anal. Chim. Acta,* 167, 257, 1985.

211. **Schindler, E.,** Determination of silver, aluminum, arsenic, cadmium, chromium, copper, iron, manganese, lead, selenium and zinc in drinking water. Procedure for study of drinking water by graphite tube AAS. I, *Dtsch. Lebensm. Rundsch.,* 81, 1, 1985 (in German).

212. **Suzuki, M. and Isobe, K.,** Mechanism of interference elimination by thiourea in electrothermal atomic absorption spectrometry, *Anal. Chim. Acta,* 173, 321, 1985.

213. **Suyama, Y. and Nishimura, M.,** Microdetermination of cadmium in saliva, *Koku Eisei Gakkai Zasshi,* 36, 338, 1986.

214. **Stoeppler, M.,** Recent methodological progress in cadmium determination, *Int. J. Environ. Anal. Chem.,* 27, 231-9, 1986.

215. **Marletta, G. P., Favretto, L. G., and Favretto, L.,** Cadmium in roadside grapes, *J. Sci. Food Agric.,* 37, 1091, 1986.

216. **Lin, S.,** Direct determination of trace cadmium in geochemical materials by Zeeman flameless atomic absorption spectroscopy, *Fenxi Huaxue,* 14(8), 611-15, 1986 (in Chinese).

217. **Black, M. M., Fell, G. S., and Ottaway, J. M.,** Determination of cadmium in blood plasma by graphite furnace atomic absorption spectroscopy, *J. Anal. At. Spectrom.,* 1, 369-72, 1986.

218. **Starkey, B. J., Taylor, A., and Walker, A. W.,** Blood cadmium determination — resin of an external quality assessment scheme, *J. Anal. At. Spectrom.,* 1, 397-400, 1986.

219. **Pan, Z., Guo, W., and Li, P.,** Zeeman effect flameless atomic absorption spectroscopy for blood cadmium, *Zhanghua Yufangyixue Zazhi,* 20, 224-5, 1986 (in Chinese).

220. **Kantor, T. and Bezur, L.,** Volatilization studies of cadmium compounds by the combined quartz furnace and flame AA method: effects of magnesium chloride and ascorbic acid additives, *J. Anal. At. Spectrom.,* 1(1), 9-17, 1986.

221. **Narres, H. D., Mohl, C., and Stoeppler, M.,** Direct determination of cadmium in crude oil and oil products by Zeeman AAS in a graphite cuvette with a L'vov platform, *Erdoel Kohle Erdgras Petrochem.,* 39(4), 193-4, 1986 (in German).

222. **Liang, L.,** Direct determination of cadmium in urine by stable temperature platform flameless Zeeman AAS, *Fenxi Ceshi Tongbao,* 4(5), 19-22, 1985, CA 7, 104:83164w, 1986.

223. **Falk, H., Hoffman, E., Ludke, C., Ottaway, J. M., and Littlejohn, D.,** Studies on the determination of cadmium in blood by furnace atomic nonthermal excitation spectrometry, *Analyst (London),* 111(3), 285-90, 1986.

224. **Fell, G. S. and Ottaway, J. M.,** Determination of cadmium in blood plasma by graphite furnace atomic absorption spectroscopy, *J. Anal. At. Spectrom.,* 1, 369, 1986.

225. **Roberts, C. A. and Clark, J. M.,** Improved determination of cadmium in blood and plasma by flameless atomic absorption spectroscopy, *Bull. Environ. Contam. Toxicol.,* 36, 496, 1986.

226. **Shaojun, L.,** Direct determination of trace cadmium in geochemical materials by Zeeman flameless atomic absorption spectroscopy, *Yankuang Ceshi,* 5, 29, 1986 (in Chinese).

227. **Debus, H., Hermann, G., and Scharmann, A.,** Background correction in combination with spectrometry of optical forward scattering and graphite furnace atomization, *Fresnius' Z. Anal. Chem.,* 323(7), 451, 1986.

228. **Nakano, A., Saito, H., and Wakisaka, I.,** Studies on urinary cadmium and ß2-microglobulin of residents in cadmium-polluted areas, *Kokuritsu Kogai Kenkusho Kenkyu Hokoku,* 84, 13-30, 1985. CA 17, 105:37099b, 1986.

229. **Van Loenen, D. C. and Weers, C. A.,** The determination of arsenic, cadmium and lead in large series of samples from pulverized coal fly ash (PFA) by Zeeman graphite furnace AAS and microliter injection of ultrasonic agitated PFA suspensions, *Fortschr. Atomspektrom. Spurenanal.,* 2, 635-47, 1986.

230. **Stoeppler, M., Mohl, C., Novak, L., and Gardiner, P. E.,** Application of the STPF concept to the determination of aluminum, cadmium and lead in biological and environmental materials, *Fortschr. Atomspektrom. Spurenanal.,* 2, 419, 1986.

231. **Heinz, H., Hoffman, E., Ludke, C., Ottaway, J. M., and Littlejohn, D.,** Studies on the determination of cadmium in blood by furnace atomic nonthermal excitation spectroscopy, *Analyst (London),* 111, 285, 1986.

232. **Clark, J. R.,** Electrothermal atomization atomic absorption conditions and matrix modification for determining Sb, As, Bi, Cd, Ga, Au, In, Pb, Mo, Pd, Pt, Se, Ag, Te, Tl and Sn following back-extraction by organic aminohalide extracts, *J. Anal. At. Spectrom.,* 1, 301, 1986.

233. **Hirokawa, K., Namiki, M., and Kimura, J.,** Determination of silver, copper, lead and cadmium in some metals by graphite furnace coherent forward scattering spectroscopy, *Bunseki Kagaku,* 35, 701, 1986.

234. **Rosopulo, A. and Kreuzer, W.,** Comparison of methods for determination of lead and cadmium in untreated liver tissue of slaughtered animals by solid-sample Zeeman AAS (atomic absorption spectroscopy) and chemical digestion in graphite-tube furnace AAS, *Firtschr. Atomspektrom. Spurenanal.,* 2, 455, 1986.

235. **Mohl, C., Narres, H. D., and Stoeppler, M.,** Oxygen ashing in the determination of lead and cadmium in hard-to-analyze materials by graphite-tube furnace AAS (atomic absorption spectroscopy), *Fortschr. Atomspektrom. Spurenanal.,* 2, 439, 1986.

236. **Kharlamov, I. P., Lebedev, V. I., and Persits, V. Y.,** Complex application of atomic and molecular spectra to the study of matrix effects on the atomization of zinc, cadmium, lead, tin, bismuth and antimony, *Zh. Anal. Khim.,* 41, 1965, 1986.

237. **Fudagawa, N., Hioki, A., Kubota, M., and Kawase, A.,** Determination of copper, lead and cadmium in high-purity zinc by metal furnace AAS (atomic absorption spectroscopy), *Bunseki Kagaku*, 35, T62, 1986 (in Japanese).

238. **Kharlamov, I. P., Lebedev, V. I., Persits, V. Y., and Eremina, G. V.,** Influence of cations and anions on the analytical signals of zinc, cadmium, lead, tin, bismuth and antimony introduced with solutions of steels and alloys into electrothermal atomizers, *Zh. Anal. Khim.*, 41, 1004, 1986.

239. **Hu, Z., Wei, Y., Zhou, Z., and Lou, Q.,** Simultaneous determination of manganese, cobalt, nickel, lead, cadmium and tin in canned foods and infant foods by stabilized temperature platform furnace-atomic absorption spectrophotometry, *Shipin Yu Fajiao Gongye*, 4, 1-17, 1986 (in Chinese).

240. **Hunt, D. T. E. and Winnard, D. A.,** Appraisal of selected techniques for the determination of lead and cadmium in waters by graphite furnace atomic absorption spectroscopy, *Analyst (London)*, 111, 785, 1986.

241. **Zhou, D. and Sun, H.,** Determination of antimony, lead and cadmium released from enameled tableware, *Boli Yu Tangci*, 14(1), 17-20, 1986 (in Chinese).

242. **Dabeka, R. W. and McKenzie, A. D.,** Graphite-furnace atomic absorption spectrometric determination of lead and cadmium in food after nitric-perchloric acid digestion and coprecipitation with amm. pyrrolidine dithiocarbamate, *Can. J. Spectrosc.*, 31(2), 44-52, 1986.

243. **Baasner, J., Berndt, H., and Eiermann, R.,** Use of microcomputer for time resolution in graphite-tube furnace AAS- analytical use for lead and cadmium determination in urine, *Fortschr. Atomspektrom. Spurenanal.*, 2, 387, 1986.

244. **Ostapczuk, P., Valenta, P., and Nuernberg, H. W.,** Square wave voltammetry. A rapid and reliable determination method for zinc, cadmium, lead, copper, nickel and cobalt in biological and environmental samples, *J. Electroanal. Chem. Interfacial Electrochem.*, 214, 51-64, 1986.

245. **Medina, J., Hernandez, F., Pastor, A., Beferull, J. B., and Barbera, J. C.,** Determination of mercury, cadmium, chromium and lead in marine organisms by flameless atomic absorption spectrophotometry, *Mar. Pollut. Bull.*, 17(1), 41-4, 1986.

246. **Welz, B. and Schlemmer, G.,** Determination of arsenic, selenium and cadmium in marine biological tissue samples using a stabilized temperature platform furnace and comparing deuterium arc with Zeeman-effect background correction AAS, *J. Anal. At. Spectrom.*, 1(2), 119-24, 1986.

247. **Xiao, H.,** Determination of silver and cadmium in sediments by extraction and atomic absorption spectroscopy with a graphite furnace, *Fenxi Ceshi Tongbao*, 4(5), 17-18, 1985 (in Chinese), CA 10, 104:16125a, 1986.

248. **Fagioli, F., Landi, S., Locatelli, C., and Bighi, C.,** Determination of lead and cadmium in small amounts of biological material by graphite furnace atomic absorption spectroscopy with sampling of carbonaceous slurry, *At. Spectrosc.*, 7(2), 49-51, 1986.

249. **Stein, K. and Umland, F.,** Trace analysis of lead, cadmium and manganese in honey and sugar, *Z. Anal. Chem.*, 323, 176, 1986.

250. **Radisch, B., Luck, W., and Nau, H.,** Cadmium concentrations in milk and blood of smoking mothers, *Toxicol. Lett.*, 36, 147, 1987.

251. **Roberts, C. A. and Clark, J. M.,** Improved determination of cadmium in blood and plasma by flameless AAS, *Bull. Environ. Contam. Toxicol.*, 36(4), 496-9, 1986.

252. **Ołayinka, K. O., Haswell, S. J., and Grzeskowiak, R.,** Development of a slurry technique for the determination of cadmium in dried foods by electrothermal atomization atomic absorption spectroscopy, *J. Anal. At. Spectrom.*, 1, 297, 1986.

253. **Zhijing, P., Wenbin, G., and Pingjian, L.,** Zeeman effect flameless atomic absorption spectroscopy for blood cadmium, *Zhonghua Yufangyixue Zazhi*, 20, 224, 1986 (in Chinese).

254. **Ueda, J. and Yamazaki, N.,** Flameless atomic absorption spectrometric and differential-pulse polarographic determination of cadmium after coprecipitation with hafnium hydroxide, *Bull. Chem. Soc. Jpn.*, 59, 1845, 1986.

255. **Lum, K. R. and Callaghan, M.,** Direct determination of cadmium in natural waters by electrothermal atomic absorption spectroscopy without matrix modification, *Anal. Chim. Acta*, 187, 157-62, 1986.

256. **Zhang, Z., Liu, J., Lin, R., Yang, X., and He, H.,** Direct graphite furnace atomic absorption determination of cadmium in seawater using an organic matrix modifier and Zeeman background correction, *Zhongshan Daxue Huebao Ziran Kexueban*, 3, 109, 1986 (in Chinese).

257. **Welz, B. and Schlemmer, G.,** Determination of As, Se and Cd in marine biological tissue samples using a stabilized temperature platform furnace and comparing deuterium arc with Zeeman effect background correction atomic absorption spectroscopy, *Spectrochim. Acta*, 41B, 567, 1986.

258. **Zhengzhi, H., Yuzhi, W., Zhiquiang, Z., and Quinzhi, L.,** Simultaneous determination of manganese, cobalt, nickel, lead, cadmium and tin in canned foods and infant foods by stabilized temperature platform furnace-atomic absorption spectrophotometry, *Shipin Yu Fajiao Gongye*, 4, 1, 1986 (in Chinese).

259. **Molin, C. J. and Milling, P. L.,** Enzymic digestion of whole blood for improved determination of cadmium, nickel and chromium by electrothermal atomic absorption spectrophotometry: measurements in patients with rheumatoid arthritis and in normal humans, *Acta Pharmacol. Toxicol. Suppl.*, 59, 399, 1986.

260. **Favretto, L. G. and Favretto, L.,** Cadmium in roadside grapes, *J. Sci. Food Agric.,* 37, 1091, 1986.

261. **Diaz-Mayans, J., Hernandez, F., Medina, J., Del Ramo, J., and Torreblanca, A.,** Cadmium accumulation in the crayfish, *Procambarus clarkii,* using graphite furnace atomic absorption spectroscopy, *Bull. Environ. Contam. Toxicol.,* 37, 722, 1986.

262. **Fudagawa, N., Hioki, A., Kubota, M., and Kawase, A.,** Determination of copper, lead and cadmium in high-purity zinc by metal furnace AAS (atomic absorption spectroscopy), *Bunseki Kagaku,* 35, T62, 1986 (in Japanese).

263. **Cullaj, A.,** A thermodynamic evaluation of the mechanism of electrothermal atomization of cadmium, nickel and lead in chloride solution, *Bul. Shkencave Nat.,* 40(1), 79-83, 1986.

264. **Mohl, C., Narres, H. D., and Stoeppler, M.,** Oxygen ashing in the determination of lead and cadmium in hard-to-analyze materials by graphite-tube furnace AAS (atomic absorption spectroscopy), *Fortschr. Atomspektrom. Spurenanal.,* 2, 439, 1986.

265. **Baasner, J., Berndt, H., and Eiermann, R.,** Use of a microcomputer for time resolution in graphite-tube furnace AAS (atomic absorption spectroscpy)-analytical use for lead and cadmium determination in urine, *Fortschr. Atomspektrom. Spurenanal.,* 2, 387-95, 1986 (in German).

266. **Feustel, A., Wennrich, R., and Dittrich, M.,** Studies of cadmium, zinc and copper levels in human kidney tumors and normal kidney, *Urol. Res.,* 14(2), 105-8, 1986.

267. **Rosopulo, Z. and Kreuzer, W.,** Comparison of methods for determination of lead and cadmium in untreated liver tissue of slaughtered animals by solid-sample Zeeman AAS and chemical digestion in graphite-tube furnace AAS, *Fortschr. Atomspektrom. Spurenanal.,* 2, 455, 1986.

268. **Han, H., Le, X., and Ni, Z.,** Determination of cadmium in seawater by graphite furnace atomic absorption spectroscopy using sodium phosphate as a matrix modifier, *Huanjing Huaxue,* 5, 34, 1986.

269. **Maroof, F. B. A., Hadi, D. A., Khan, A. H., and Chowdhury, A.,** Cadmium and zinc concentrations in drinking water supplies of Dhaka City, Bangladesh, *Sci. Total Environ.,* 53, 233, 1986.

270. **Kharlamov, I. P., Lebedev, V. I., and Persits, V. Y.,** Complex application of atomic and molecular spectra to the study of matrix effects on the atomization of zinc, cadmium, lead, tin, bismuth and antimony, *Zh. Anal. Khim.,* 41, 1965, 1986.

271. **Kirakawa, K., Namiki, M., and Kimura, J.,** Determination of silver, copper, lead and cadmium in some metals by graphite furnace coherent forward scattering spectrometry, *Bunseki Kagaku,* 35, 701, 1986.

272. **Henze, W. and Umland, F.,** Speciation of cadmium and copper in lettuce leaves, *Proc. 4th Int. Workshop Trace Elem. Anal. Chem. Med. Biol.,* Braetter, P. and Schramel, P., Eds., Berlin, 1987, 501-7.

273. **Erler, W., Lehman, R., and Voellkopf, U.,** Determination of cadmium and lead in animal liver and freeze-dried blood by graphite-furnace Zeeman atomic absorption spectroscopy and solid sampling, *Proc. 4th Int. Workshop Trace Elem. Anal. Chem. Med. Biol.,* Braetter, P. and Shraemel, P., Eds., Berlin, 1987, 385-91.

274. **Halls, D. J., Black, M. M., Fell, G. S., and Ottaway, J. M.,** Direct determination of cadmium in urine by electrothermal atomization atomic absorption spectroscopy, *J. Anal. At. Spectrom.,* 2, 305, 1987.

275. **Yin, X., Schlemmer, B., and Welz, B.,** Cadmium determination in biological materials using graphite-furnace atomic absorption spectroscopy with palladium nitrate-amm.nitrate modifier, *Anal. Chem.,* 59, 1462, 1987.

AUTHOR INDEX

SUBJECT INDEX

CALCIUM

CALCIUM, Ca   (ATOMIC WEIGHT 40.08)

Instrumental Parameters:

Wavelength	422.7 nm
Slit width	320 μm
Bandpass	1 nm
Light source	Hollow Cathode Lamp
Lamp current	7 mA
Purge gas	Argon or Nitrogen
Sample size	25 μL
Furnace	Cylindrical cuvette/Pyrolytic graphite

Standard Operating Conditions:

Optimum char temperature	1200°C
Optimum atomization temperature	2400°C
Sensitivity	1 pg/1% absorption
Sensitivity check	0.05-0.04 μg/mL
Working range	0.01-0.4 μg/mL
Background correction	Required only to correct light scattering or non-specific absorption from the sample containing high dissolved solids.

General Note:
1. Use commercial standard or a previously analyzed sample as a working standard.
2. Use ultra clean glass and plastic ware, soaked in 1:5 nitric acid solution and rinsed thoroughly with deionized distilled water.
3. All analytical solutions should be at least 0.2-0.5% (v/v) in nitric acid.
4. Check for blank values on all reagents including water.
5. Dilution is recommended when sample exhibits greater than 0.5 absorbance units.
6. Use of nitrogen as purge gas will suppress the sensitivity.
7. Sensitivity may be enhanced by the use of hydrochloric acid.

REFERENCES

1. **Montaser, A. and Mehrabzadeh, A. A.,** Determination of Ca, Eb, Eu, Ga, In, K, Na, Mo and W by atomic absorption spectroscopy with an electrothermal graphite braid atomizer, *Anal. Chim. Acta,* 111, 297, 1979.
2. **Katskov, D. A. and Grinshtein, I. L.,** Study of the evaporation of beryllium, magnesium, calcium, strontium, barium and aluminum from a graphite surface by an atomic absorption method, *Zh. Prikl. Spektrosk.,* 33, 1004, 1980.
3. **Marumo, Y. and Seta, S.,** Studies on trace elements in human head hair. I. Flameless atomic absorption analysis of calcium in hair and its variation in content, *Eisei Kagaku,* 27, 381-7, 1981 (in Japanese).
4. **Smith, M. R. and Cochran, H. B.,** Determination of calcium and magnesium in saturated sodium chloride brines by graphite furnace atomic absorption spectrophotometry, *At. Spectrosc.,* 2, 97, 1981.
5. **Suzuki, M. and Ohta, K.,** Electrothermal atomization of calcium and strontium in a molybdenum microtube, *Talanta,* 28, 177, 1981.
6. **Katskov, D. A. and Grinshtein, I. L.,** Formation of copper, silver and calcium acetylenides in graphite furnaces for atomic absorption analysis, *Zh. Prikl. Spektrosk.,* 36, 181, 1982.
7. **Powell, L. A. and Tease, R. L.,** Determination of calcium, magnesium, strontium and silicon in brines by graphite furnace atomic absorption spectroscopy, *Anal. Chem.,* 54, 2154, 1982.
8. **Boehmer, R. G., Nieuwenhuis, J. J., and Theron, J. J.,** Determination of calcium in small samples from certain brain areas by furnace atomic absorption spectroscopy, *S. Afr. J. Chem.,* 36, 27-31, 1983.
9. **Berggren, P. O., Andersson, T., and Hellman, B.,** The interaction between barium and calcium in ß-cell-rich pancreatic islets, *Bromed. Res.,* 4, 129, 1983.
10. **Murnane, M. M.,** Analysis of brine by atomic absorption with graphite furnace using direct sample injection, *Chem. N. Z.,* 48, 39, 1984.
11. **Ying, Y.,** Content of zinc, calcium, magnesium, copper and manganese in the hair of Huangshi (China) inhabitants, *Xuebao,* 6(3), 298, 1984 (in Chinese).
12. **DeAlbinati, J. F. P., Troccoli, O. E., and Daraio, M. E.,** Microdetermination of calcium in serum, *Acta Bioquim. Clin. Latinoam.,* 18, 269, 1984 (in Spanish).
13. **Machida, K., Matsubara, J., and Sugawara, K.,** Effect of millimolar level of calcium intake on calcium, phosphorus and cadmium content in rats, *Nutr. Rep. Int.,* 29, 1145, 1984.
14. **Bergsten, P. and Hellman, B.,** Differentiation between the short and long term effects of glucose on the intracellular calcium content of the pancreatic ß-cell, *Endocrinology,* 114, 1854, 1984.
15. **Alvarado, J., Campos, F., and Ottaway, J. M.,** Determination of trace levels of calcium in steels by carbon-furnace AA and AES, *Talanta,* 33(1), 61-5, 1986.
16. **Gerlach, W., Krivan, V., and Sprenger, K.,** Determination of calcium in human erythrocytes by graphite-tube furnace AAS (atomic absorption spectroscopy), *Fortschr. Atomspektrom. Spurenanal.,* 2, 217, 1986.
17. **Barton, I. K., Mansell, M. A., and Hilton, P. J.,** A simple method for the measurement of red-cell calcium concentration, *Ann. Clin. Biochem.,* 23(5), 610, 1986.

AUTHOR INDEX

SUBJECT INDEX

CHROMIUM

CHROMIUM, Cr   (ATOMIC WEIGHT 51.996)

Instrumental Parameters:

Wavelength	357.9 nm
Slit width	320 μm
Bandpass	1 nm
Light source	Hollow Cathode Lamp
Lamp current	6 mA
Purge gas	Argon or Nitrogen
Sample size	25 μL
Furnace	Cylindrical cuvette/Pyrolytic graphite

Standard Operating Conditions:

Optimum char temperature	1200°C
Optimum atomization temperature	2500°C/2200°C
Sensitivity	4 pg/1% absorption
Sensitivity check	0.02-0.05 μg/mL
Working range	0.004-0.05 μg/mL
Background correction	Required only to correct light scattering or non-specific absorption from the sample containing high dissolved solids.

General Note:
1. Use commercial standard or a previously analyzed sample as a working standard.
2. Use ultra clean glass and plastic ware, soaked in 1:5 nitric acid solution and rinsed thoroughly with deionized distilled water.
3. All analytical solutions should be at least 0.2-0.5% (v/v) in nitric acid.
4. Check for blank values on all reagents including water.
5. Dilution is recommended when sample exhibits greater than 0.5 absorbance units.
6. Use step atomization, if increased sensitivity is required.
7. The analytical precision is about 6% RSD.

REFERENCES

1. **Davidson, J. W. F. and Secrest, W. L.,** Determination of chromium in biological materials by atomic absorption spectroscopy using a graphite furnace atomizer, *Anal. Chem.,* 44, 1808, 1972.
2. **Maruta, T. and Takeuchi, T.,** Interferences of chromium (VI) in atomic absorption spectrophotometry with a tantalum filament atomizer, *Bunseki Kagaku,* 22, 602, 1973.
3. **Maruta, T. and Takeuchi, T.,** Determination of trace amounts of chromium by atomic absorption spectroscopy with a tantalum filament atomizer, *Anal. Chim. Acta,* 66, 5, 1973.
4. **Pekarek, R. S., Hauer, E. C., Wannemacher, R. W., Jr., and Beisel, W. R.,** The direct determination of serum chromium by an atomic absorption spectrophotometer with a heated graphite atomizer, *Anal. Biochem.,* 59, 283, 1974.
5. **Schweizer, V. B.,** Determination of cobalt, chromium, copper, molybdenum, nickel and vanadium in carbonate rocks with the HGA-70 graphite furnace, *At. Ab. Newsl.,* 14, 137, 1975.
6. **Maruta, T., Minegishi, K., and Sudoh, G.,** Atomic absorption spectrometric determination of trace amounts of copper, manganese, lead and chromium in cements by direct atomization in a carbon furnace, *Yogyo Kyokai Shi,* 86, 532, 1978.
7. **Guthrie, B. E., Wolf, W. R., and Veillon, C.,** Background correction and related problems in the determination of chromium in urine by graphite furnace atomic absorption spectroscopy, *Anal. Chem.,* 50, 1900, 1978.
8. **Page, A. G., Godbole, S. V., Kulkarni, M. J., Shelar, S. S., and Joshi, B. D.,** Direct AAS determination of cobalt, chromium, manganese and nickel in uranium oxide (U_3O_3) by electrothermal atomization, *Z. Anal. Chem.,* 296, 40, 1979.
9. **Nise, G. and Vesterberg, O.,** Direct determination of chromium in urine by electrothermal atomic absorption spectroscopy, *J. Work. Environ. Health,* 5, 404, 1979.
10. **Shimizu, T., Hiyama, T., Shijo, Y., and Sakai, K.,** Determination of total chromium in human urine by graphite furnace atomic absorption spectroscopy after coprecipitation treatment, *Bunseki Kagaku,* 29, 680, 1980.
11. **Batley, G. E. and Matousek, J. P.,** Determination of chromium speciation in natural waters by electrodeposition on graphite tubes for electrothermal atomization, *Anal. Chem.,* 52, 1570, 1980.
12. **Veillon, C., Guthrie, B. E., and Wolf, W. R.,** Retention of chromium by graphite furnace tubes, *Anal. Chem.,* 52, 457, 1980.
13. **Routh, M. W.,** Analytical parameters for determination of chromium in urine by electrothermal atomic absorption spectroscopy, *Anal. Chem.,* 52, 182, 1980.
14. **Matsusaki, K., Yoshino, T., and Yamamoto, Y.,** The removal of chloride interference in determination of chromate ion by atomic absorption spectroscopy with electrothermal atomization, *Anal. Chim. Acta,* 113, 247, 1980.
15. **Kumpulainen, J.,** Determination of chromium in human milk and urine by graphite furnace atomic absorption spectroscopy, *Anal. Chim. Acta,* 113, 355, 1980.
16. **Chao, S. S. and Pickett, E. E.,** Trace chromium determination by furnace atomic absorption spectroscopy following enrichment by extraction, *Anal. Chem.,* 52, 335, 1980.
17. **Matsusaki, K., Yoshino, T., and Yamamoto, Y.,** Removal of chloride interference in the determination of chromium by atomic absorption spectroscopy with electrothermal atomization, *Anal. Chim. Acta,* 124, 163, 1981.
18. **Genc, O., Akman, S., Ozdural, A. R., Ates, S., and Balkis, T.,** Theoretical analysis of atom formation-time curves for the HGA-74 furnace. II. Evaluation of the atomization mechanisms for managanese, chromium and lead, *Spectrochim. Acta,* 36B, 163, 1981.
19. **Garbett, K., Goodfellow, G. I., and Marshall, G. B.,** Application of atomic absorption techniques in the analysis of metallic sodium. II. Determination of iron, nickel and chromium in sodium salt solutions by electrothermal atomization, *Anal. Chim. Acta,* 126, 147, 1981.
20. **Arafat, N. M. and Glooschenko, W. A.,** Method for the simultaneous determination of arsenic, aluminum, iron, zinc, chromium and copper in plant tissue without the use of perchloric acid, *Analyst* (London), 106, 1174-8, 1981.
21. **Zolotovitakaya, E. S. and Fidel'man, B. M.,** Use of electrothermal atomic absorption spectroscopy for determining iron, nickel and chromium in raw materials and single crystals of corundum, *Zh. Anal. Khim.,* 36, 1564, 1981 (in Russian).
22. **Ranchet, S., Menissier, F., Lamathe, J., and Voinovitch, I.,** Interlaboratory comparison of the determinations of cadmium, chromium, copper and lead by flameless atomic absorption spectrometry, *Bull. Liaison Lab. Ponts Chaussees,* 114, 81-6, 1981 (in French).
23. **Minoia, C., Colli, M., and Pozzoli, L.,** Determination of hexavalent chromium in urine by flameless atomic absorption spectrophotometry, *At. Spectrosc.,* 2, 163, 1981.

24. **Lieser, K. H., Sondermeyer, S., and Kliemchen, A.,** Precision and accuracy of analytical results in determining the elements cadmium, chromium, copper, iron, manganese and zinc by flameless AAS, *Fresnius' Z. Anal. Chem.*, 312, 517, 1982 (in German).

25. **Veillon, C., Patterson, K. Y., and Bryden, N. A.,** Direct determination of chromium in human urine by electrothermal atomic absorption spectroscopy, *Anal. Chim. Acta,* 136, 233, 1982.

26. **Kumpulainen, J., Salmela, S., Vuori, E., and Lehto, J.,** Effects of various washing procedures on the chromium content of human scalp hair, *Anal. Chim. Acta,* 138, 361-4, 1982.

27. **Beninschek, H. I. and Benischek, F.,** Electrothermal atomic absorption spectrometric determination of chromium, iron and nickel in lithium metal, *Anal. Chim. Acta,* 140, 205-12, 1982.

28. **Guenais, B., Poudoulec, A., and Minier, M.,** Flameless atomic absorption analysis of chromium in gallium arsenide, *Analysis,* 10, 78-82, 1982 (in French).

29. **Veillon, C., Patterson, K. Y., and Bryden, N. A.,** Chromium in urine as measured by atomic absorption spectroscopy, *Clin. Chem.,* 28, 2309, 1982.

30. **Nakaaki, K., Tado, O., and Masuda, T.,** Urinary excretion of chromium, manganese, copper and nickel, *Rado Kagaku,* 58, 529, 1982 (in Japanese).

31. **Arnold, D. and Kuennecke, A.,** Determination of 3d elements in high lead-containing glasses, *Silikattechnik,* 33, 312, 1982.

32. **Zhou, B. and Zeng, S.,** Simultaneous determination of chromium, manganese and copper in serum by flameless atomic absorption spectrophotometry, *Yingyang Xuebao,* 4, 323, 1982 (in Chinese).

33. **Leiser, K. H., Sondermeyer, S., and Kliemchen, A.,** Effect of admixtures on the reproducibility and accuracy of analytical results in determining the elements cadmium, chromium, copper, iron, manganese and zinc by flameless AAS, *Fresnius' Z. Anal. Chem.,* 312, 520, 1982.

34. **Ranchet, J., Menissier, F., Lamathe, J., and Voinovitch, I.,** Interlaboratory comparison: The determinations of cadmium, chromium, copper and lead in standard solutions by flameless atomic absorption spectroscopy, *Analysis,* 10, 71-7, 1982.

35. **Martin, T. D. and Riley, J. K.,** Determining dissolved hexavalent chromium in water and wastewater by electrothermal atomization, *At. Spectrosc.,* 3, 174, 1982.

36. **Ping, L., Matsumoto, K., and Fuwa, K.,** Determination of urinary chromium levels for healthy men and diabetic patients by electrothermal atomic absorption spectroscopy, *Anal. Chim. Acta,* 147, 205, 1983.

37. **Verch, R. L., Chu, R., Wallach, S., Peabody, R. A., Jain, J., and Hannan, E.,** Tissue chromium in the rat, *Nutr. Rep. Int.,* 27, 531, 1983.

38. **Willie, S. N., Sturgeon, R. E., and Berman, S. S.,** Determination of total chromium in seawater by graphite furnace atomic absorption spectroscopy, *Anal. Chem.,* 55, 981, 1983.

39. **Offlenbacher, E. G. and Pi-Sunyer, F. X.,** Temperature and pH effects on the release of chromium from stainless steel into water and fruit juices, *J. Agric. Food Chem.,* 31, 89, 1983.

40. **Flatt, P. R. and Berggren, P. O.,** Distribution of chromium in the tissues of normal and genetically diabetic mice, *Biochem. Soc. Trans.,* 11(6), 722, 1983.

41. **Dumitru, R. and Botha, N.,** Determination of urinary chromium concentrations by electrothermal atomic absorption spectrophotometry, *Rev. Ig. Bacteriol. Virusol. Parazitol., Epidemiol. Pneumoftiziol. Ig.,* 32(2), 125-8, 1983 (in Romanian).

42. **Mazzucotelli, A., Minoia, G., Pozzoli, L., and Ariati, L.,** Selective determination of hexavalent chromium in groundwaters by electrothermal atomic absorption and Amberlite LA-1 liquid anion-exchanger separation, *At. Spectrosc.,* 4(5), 182, 1983.

43. **Isozaki, A., Kumagai, K., and Utsumi, S.,** An atomic absorption spectrometric method for the individual determination of chromium (III) and chromium (VI) by atomization of chromium from a chelating resin in a graphite tube, *Anal. Chim. Acta,* 153, 15-22, 1983.

44. **Kumpulainen, J., Lehto, J., Koivistoinen, P., Usitupa, M., and Vuori, E.,** Determination of chromium in human milk, serum and urine by electrothermal atomic absorption spectrometry without preliminary ashing, *Sci. Total Environ.,* 31(1), 71-80, 1983.

45. **Halls, D. J. and Fell, G. S.,** Determination of chromium in urine by graphite furnace atomic absorption spectroscopy, *Proc. 2nd Int. Workshop, Trace Elem. Anal. Chem. Med. Biol.,* Braetter, P. and Schramel, P., Eds., Berlin, 1983, 667-73.

46. **Minoia, C., Mazzucotelli, A., Cavalleri, A., and Mineganti, V.,** Electrothermal atomization atomic absorption spectrophotometric determination of chromium (VI) in urine by solvent extraction separation with liquid anion exchangers, *Analyst* (London), 108, 481, 1983.

47. **Cary, E. E. and Rutzke, M.,** Electrothermal atomic absorption spectroscopic determination of chromium in plant tissues, *J. Assoc. Off. Anal. Chem.,* 66, 850, 1983.

48. **Rossi, G., Omenetto, N., Pigozzi, G., Vivian, R., Mattinez, U., Mousty, F., and Crabi, G.,** Analysis of radioactive waste solutions by atomic absorption spectrometry with electrothermal atomization, *At. Spectrosc.,* 4, 113, 1983.

49. **Campbell, D. E. and Comperat, M.,** Ultratrace analysis of high purity glasses for double crucible low-attention optical waveguides, *Glastech. Ber.,* 56, 898, 1983.

50. **Kumpulainen, J., Lehto, J., Koivistoinen, P., and Vuori, E.,** Direct electrothermal atomic absorption spectrometric determination of selenium and chromium in biological fluids, *Proc. 2nd Int. Workshop, Trace Elem. Anal. Chem. Med. Biol.,* Braetter, P., and Schramel, P., Eds., Berlin, 1983, 951-67.

51. **Voellkopf, U., Grobenski, Z., and Welz, B.,** Stabilized temperature platform furnace with zeeman effect background correction for trace analysis in wastewater, *At. Spectrosc.,* 4, 165, 1983.

52. **Flatt, P. R. and Berggren, P. O.,** Distribution of chromium in tissues of normal and genetically diabetic mice, *Biochem. Soc. Trans.,* 11, 722, 1983.

53. **Legret, M., Demare, D., Marchandise, P., and Robbe, D.,** Interferences of major elements in the determination of lead, copper, cadmium, chromium and nickel in river sediments and sewer sludges by electrothermal atomic absorption spectroscopy, *Anal. Chim. Acta,* 149, 107, 1983.

54. **Burakov, V. S., Verenik, V. N., Malashonok, V. A., Nechaev, S. V., and Puko, R. A.,** Atomic absorption determination of trace iron and chromium in aqueous solutions by using intracavity spectroscopy, *Zh. Anal. Khim.,* 38, 90-3, 1983 (in Russian).

55. **Boiteau, H. L., Metayer, C., Ferre, R., and Pineau, A.,** Automated determination of lead, cadmium, manganese and chromium in whole blood by Zeeman atomic absorption spectrometry, *Analusis,* 11(5), 234-42, 1983 (in French).

56. **Marecek, J., Braunova, I., and Vachtova, I.,** Determination of molybdenum, vanadium and chromium in brines by atomic absorption spectroscopy, *Chem. Prum.,* 33, 429-35, 1983 (in Czech).

57. **Harnly, J. M. and Kane, J. S.,** Optimization of electrothermal atomization parameters for simultaneous multielement atomic absorption spectrometry, *Anal. Chem.,* 56, 48, 1984.

58. **Wendl, W. and Mueller-Vogt, G.,** Chemical reaction in the graphite tube for some carbide and oxide forming elements, *Spectrochim. Acta,* 39B, 237, 1984.

59. **Minoia, C., Mazzucotelli, A., Richelmi, P., Baldi, C., Cavalleri, A., and Micoli, G.,** Determination of hexavalent chromium in bile by electrothermal atomization atomic absorption spectrophotometry and liquid anion-exchange separation, *Mikrochim. Acta,* 1, 353, 1984.

60. **Itoh, K., Akatsuka, K., and Atsuya, I.,** Direct determination of chromium in environmental samples by zeeman atomic absorption spectrometry with graphite miniature cup, *Bunseki Kagaku,* 33, 301, 1984 (in Japanese).

61. **Vallerand, A. L., Cuerrier, J. P., Shapcott, D., Vallerand, R. J., and Gardiner, P. F.,** Influence of excercise training on tissue chromium concentrations in the rat, *Am. J. Clin. Nutr.,* 39, 402, 1984.

62. **Taddia, M. and Lanza, P.,** Determination of chromium in gallium arsenide by electrothermal atomic absorption spectrometry, *Anal. Chim. Acta,* 159, 375, 1984.

63. **Saner, G., Yuzbasiyan, V., and Cigden, S.,** Hair chromium concentration and chromium excretion in tannery workers, *Br. J. Ind. Med.,* 41, 263, 1984.

64. **Zober, A., Kick, A., Schaller, K. H., Schellmann, B., and Valentin, H.,** Studies of normal values of chromium and nickel in human lung, kidney, blood and urine samples, *Zentralbl. Bakteriol., Mikrobiol. Hyg. Abt. 1 Orig. B,* 179, 80, 1984.

65. **Matsusaki, K. and Yoshino, T.,** Electrothermal atomic absorption spectrometric determination of traces of chromium, nickel, iron and beryllium in aluminum and its alloys without preliminary separation, *Anal. Chim. Acta,* 157(1), 193-7, 1984.

66. **Pruszkowska, E. and Barrett, P.,** Determination of arsenic, selenium, chromium, cobalt and nickel in geochemical samples using the stabilized temperature platform furnace and zeeman background correction, *Spectrochim. Acta,* 39B, 485, 1984.

67. **Hoenig, M., Scokart, P. O., and Van Hoeyweghen, P.,** Efficiency of L'vov platform and ascorbic acid modifier for reduction of interferences in analysis of plant samples for lead, thallium, antimony, cadmium, nickel and chromium by ETAAS, *Anal. Lett.,* 17A, 1947, 1984.

68. **Voinovitch, I. A., Druon, M., and Louvrier, J.,** Effect of absence of injection orifice and length of graphite furnaces on sensitivity of determinations by atomic absorption spectrometry with electrothermal atomization, *Analysis,* 12, 256, 1984 (in French).

69. **Boniforti, R., Ferraroli, R., Frigieri, P., Hettai, D., and Queirazza, G.,** Intercomparison of five methods for determination of trace metals in seawater, *Anal. Chim. Acta,* 162, 33, 1984.

70. **Grobenski, L., Lehmann, R., Radzuik, B., and Voellkopf, U.,** Determination of trace metals in seawater using zeeman graphite furnace AAS, *At. Spectrosc.,* 5, 87, 1984.

71. **Carrondo, M. J. T., Reborredo, F., Ganho, R. M. B., and Oliveira, J. F. S.,** Analysis of sediments for heavy metals by rapid electrothermal atomic absorption procedure, *Talanta,* 31, 561, 1984.

72. **Brodie, K. G. and Routh, M. W.,** Trace analysis of lead in blood, aluminum and manganese in serum and chromium in urine by graphite furnace atomic absorption spectrometry, *Clin. Biochem.,* 17, 19, 1984.

73. **Phelan, V. J. and Powell, R. J. W.,** Combined reagent purification and sample dissolution (Corpad) applied to trace analysis of silicon, silica and quartz, *Analyst,* 109, 1269, 1984.

74. Release of metals (lead, cadmium, copper, zinc, nickel and chromium) from kitchen blenders, Statens Levnedsmiddelinstitut REPORT 1984. PUB-88, Order No. PB84-193416, 26 pp (in Danish). Avail. NTIS. From Gov. Rep. Announce. Index (U.S.), 84, 67, 1984.

75. **Littlejohn, D., Cook, S., Durie, D., and Ottaway, J. M.,** Investigation of working conditions for graphite probe atomization in electrothermal atomic absorption spectrometry, *Spectrochim. Acta,* 39B, 295, 1984.

76. **Veillon, C., Patterson, K. Y., and Bryden, N. A.,** Determination of chromium in human serum by electrothermal atomic absorption spectrometry, *Anal. Chim. Acta,* 164, 67, 1984.

77. **Minoia, C., Mazzucotelli, A., Richelmi, P., Baldi, C., Cavalleri, A., and Micoli, G.,** Determination of hexavalent chromium in bile by electrothermal atomization atomic absorption spectrophotometry and liquid anion exchange separation, *Mikrochim. Acta,* 1, 353, 1984.

78. **Taddia, M. and Lanza, P.,** Determination of chromium in gallium arsenide by electrothermal atomic absorption spectroscopy, *Anal. Chim. Acta,* 159, 375, 1984.

79. **Itoh, K., Akatsuka, K., and Atsuya, J.,** Direct determination of chromium in environmental samples by zeeman absorption spectrometry with graphite miniature cup, *Bunseki Kagaku,* 33, 301, 1984.

80. **Vallerand, A. L., Cuerrier, J. P., Shapcott, D., Vallerand, R. J., and Gardiner, P. F.,** Influence of excercise training on tissue chromium concentrations in the rat, *Am. J. Clin. Nutr.,* 39(3), 402-9, 1984.

81. **Yabuta, J., Konishi, Y., Takata, T., and Funato, Y.,** Determination of chromium in urine by flameless AAS, *Shimadzu Hyoron,* 41(4), 279-83, 1984 (in Japanese).

82. **Vos, G.,** Determination of dissolved hexavalent chromium in river water seawater and wastewater, *Z. Anal. Chem.,* 320, 556, 1985.

83. **Morris, B. W. and Kemp, G. J.,** Chromium in plasma and urine measured by electrothermal atomic absorption spectroscopy, *Clin. Chem.,* 31, 171, 1985.

84. **Berggren, P. O. and Flatt, P. R.,** Effects of trivalent chromium administration on endogenous chromium stores in lean and obese hyperglycemic (OB/OB) mice, *Nutr. Rep. Int.,* 31, 213, 1985.

85. **Horio, T. and Arakawa, A.,** Determination of trace amount of chromium in foods using flameless atomic absorption spectrometry, *Kenkyu Hokokusho-Toyo Shokukin Kogyo Tanki Daigaku Toyo Shokuhin Kenkyusho,* 15, 100-11, 1983 (in Japanese), CA 14: 102:219713t, 1985.

86. **Anderson, R. A., Bryden, N. A., and Polansky, M. M.,** Serum chromium of human subjects: effects of chromium supplementation and glucose, *Am. J. Clin. Nutr.,* 41(3), 571, 1985.

87. **Schermaier, A. J., O'Conner, L. H., and Pearson, K. H.,** Semiautomated determination of chromium in whole blood and serum by Zeeman electrothermal atomic absorption spectrophotometry, *Clin. Chim. Acta,* 152, 123, 1985.

88. **Schindler, E.,** Composite procedure to determine chromium in muscle meat, liver and kidneys of slaughtered animals, *Dtsch. Lebensm. Rundsch.,* 81(8), 250-2, 1985.

89. **Cary, E. E.,** Electrothermal atomic absorption spectroscopic determination of chromium in plant tissues: interlaboratory study, *J. Assoc. Off. Anal. Chem.,* 68, 495, 1985.

90. **Morris, B. W., Hardisty, C. A., McCann, J. F., Kemp, G. J., and May, T. W.,** Evidence of chromium toxicity in group of stainless steel welders, *At. Spectrosc.,* 6, 149, 1985.

91. **Wiegand, H. J., Ottenwaeider, H., and Bolt, H. M.,** The formation of glutathione-chromium complexes and their possible role in chromium disposition, *Arch. Toxicol. Suppl.,* 8, 319-21, 1985.

92. **Van Schoor, O., Deelstra, H., and De Leeaw, J.,** Nail chromium concentrations in normal and diabetic subjects, *1st Int. Congr. Adv. Diet Nutr.,* 1985, 56-8.

93. **Dungs, K., Fleischhauer, H., and Neidhart, B.,** Methodical developments for the speciation of chromium (III)/chromium (VI) by electrothermal AAS, *Fresnius' Z. Anal. Chem.,* 322(3), 280-9, 1985.

94. **Hoenig, M., Dehairs, F., and Dekersabiec, A. M.,** Effects of aging of pyrolytically coated tubes on the determination of refractory elements by electrothermal atomization atomic absorption spectroscopy, *J. Anal. At. Spectrom.,* 1, 449, 1986.

95. **Brovko, I. A., Davirov, A., and Koval'chuk, V. L.,** Electrothermal atomizer for atomic absorption determination of elements forming volatile hydrides, *Uzb. Khim. Zh.,* 3, 7-9, 1986 (in Russian).

96. **Matsunaga, H., Hirate, N., and Okada, A.,** Device and method for sample decomposition in the determination of minute amounts of impurities in a semiconductor crystal, Jpn. Kokai Tokkyo Koho JP 61 38, 547 (86 38,547) (Cl. G01N21/31), 24 Feb. 1986, Appl. 84/159,042, 4pp, 31 Jul. 1984.

97. **Barba, M. F. and Valle Fuentes, F. J.,** Chemical analysis of silicon carbide-based materials, *Biol. Soc. Esp. Ceram. Vidrio,* 25, 237-41, 1986 (in Spanish).

98. **Falk, H., Hoffmann, E., Ludke, C., and Schmidt, K. P.,** Direct analysis of solid plant materials by FANES (Furnace atomic nonthermal excitation spectrometry), *Spectrochim. Acta,* 41B, 853-7, 1986 (in German).

99. **Krivan, V. and Egger, K. P.,** Multielement analysis of airborne dust of the city of Ulm and comparison with other regions, *Fresnius' Z. Anal. Chem.,* 325, 41-9, 1986 (in German).

100. **Kimura, M., Yamashita, H., and Komada, J.,** Use of green tea as adsorbent of several metal ions in water, *Bunseki Kagaku,* 35(4), 400-5, 1986 (in Japanese).

101. **Sneddon, J.,** Direct and near real time determination of metallic compounds in the atmosphere by atomic absorption, *Am. Lab.,* 18, 43, 1986.

102. **Okamoto, K. and Fuwa, K.,** Preparation and certification of tea leaves reference material, *Kankyo Kenkyu,* 62, 167, 1986.

103. **Xu, H.,** Analysis of corundum, Changchun Dizhi, *Xueyuan Xuebao,* 2, 62, 1986.

104. **Voellkopf, U. and Gorbenski, Z.,** Use of direct solid sample feed in graphite tube furnace AAS (atomic absorption spectroscopy) for screening analysis of marine animals, *Fortschr. Atomspektrom. Spurenanal.,* 2, 465-77, 1986.

105. **McCamey, D. A., Iannelli, D. P., Bryson, L. J., and Thorpe, T. M.,** Determination of silicone in fats and oils by electrothermal atomic absorption spectroscopy with in-furnace air oxidation, *Anal. Chim. Acta,* 188, 119, 1986.

106. **Li, J., Johnson, W. K., and Wong, C. S.,** Atomic absorption determination of 15 minor and major elements in marine and estuary sediments, *Fenxi Huaxue,* 14(6), 460-3, 1986.

107. **D'Innocenzio, F.,** Annual trend of the deposition of metals and dusts in an urban area, *Acqua Aria,* 4, 361, 1986 (in Italian), CA select 18, 105:65521b, 1986.

108. **Boiteau, H. L., Metayer, C., Ferre, R., and Pineau, A.,** Application of Zeeman atomic absorption spectrometry for the determination of toxic metals in the viscera, *J. Toxicol. Clin. Exp.,* 6(2), 95-106, 1986 (in French), CA 17, 105:36968x, 1986.

109. **Medina, J., Hernandez, F., Pastor, A., Beferull, J. B., and Barbera, J. C.,** Determination of mercury, cadmium, chromium and lead in marine organisms by flameless atomic absorption spectrophotometry, *Mar. Pollut. Bull.,* 17(1), 41-4, 1986.

110. **Frech, W., Lindberg, A. O., Lundberg, E., and Cedergren, A.,** Atomization mechanisms and gas phase reactions in graphite furnace atomic absorption spectrometry, *Fresnius' Z. Anal. Chem.,* 323(7), 716-25, 1986.

111. **De Benzo, Z. A., Castor, G., Carrion, N., and Flores, J.,** Preliminary study on the recuperation of exhausted graphite tubes for the determination of copper and chromium by ETAAS, *Mikrochim. Acta,* 1, 311, 1986.

112. **Christensen, J. M. and Pedersen, L. M.,** Enzymic digestion of whole blood for improved determination of cadmium, nickel and chromium by ETA: Measurements in patients with rheumatoid arthritis and in normal humans, *Acta Pharmacol. Toxicol. Suppl.,* 59, 399, 1986.

113. **Arpadjan, S. and Krivan, V.,** Preatomization separation of Cr(III) from Cr (IV) in the graphite furnace, *Anal. Chem.,* 58, 2611, 1986.

114. **Long, S. and Han, L.,** Determination of chromium (III) and chromium (VI) in human urine by graphite furnace atomic absorption spectroscopy after coprecipitation with aluminum hydroxide, *Fenxi Huaxue,* 14, 529-31, 1986 (in Chinese).

115. **Ferrer, R. S., Garcia, H. G., and Perez, A. T.,** Determination of chromium by flameless AAS, *Rev. Cubana Hig. Epidemiol.,* 23(1), 57-65, 1986.

116. **Johnson, D., Headridge, J. B., McLeod, C. W., Jackson, K. W., and Roberts, J. A.,** Direct determination of chromium in gallium arsenide by electrothermal atomization AAS with Smith-Hieftje background correction, *Anal. Proc. (London),* 23, 8, 1986.

117. **Offenbacher, E. G., Dowling, H. J., Rinko, C. J., and Pi-Sunyer, F.,** Rapid enzymic pretreatment of samples before determining chromium in serum or plasma, *Clin. Chem.,* 32(7), 1383, 1986.

118. **Dungs, K. and Neidhart, B.,** Behavior of chromium in a graphite tube and on a platform, *Fortschr. Atomspektrom. Spurenanal.,* 2, 25-36, 1986.

119. **Halls, D. J. and Fell, G. S.,** The problem of background correction in the determination of chromium in urine by atomic absorption spectroscopy with electrothermal atomization, *J. Anal. At. Spectrom.,* 1, 135, 1986.

120. **Hernandez, F., Diaz, J., Medina, J., Del Ramo, J., and Pastor, A.,** Determination of chromium in treated crayfish, *Procamburus clarkii* by electrothermal AAS-study of chromium accumulation in different tissues, *Bull. Environ. Contam. Toxicol.,* 36, 851, 1986.

121. **Sihua, L. and Lilu, H.,** Determination of chromium (III) and chromium (VI) in human urine by graphite furnace atomic absorption spectroscopy after coprecipitation with aluminum hydroxide, *Fenxi Huaxue,* 14, 529, 1986 (in Chinese).

122. **Van Schoor, O. and Deelstra, H.,** Analytical aspects of the determination of chromium by graphite furnace atomic absorption spectroscopy (GFAAS), *Bull. Soc. Chim. Belg.,* 95, 373, 1986.

123. **Ebdon, L., Hill, S., and Jones, P.,** Application of directly coupled flame atomic absorption spectroscopy-fast protein liquid chromatography to the determination of protein-bound metals, *Analyst (London),* 112, 437, 1987.

124. **Irgolic, K. J.,** Analytical procedures for the determination of organic compounds of metals and metalloids in environmental samples, *Sci. Total Envron.,* 64, 61-73, 1987.

125. **Munro, J. L. and Sneddon, J.,** Determination of six elements in infant formula by flame and furnace atomic absorption spectroscopy, *At. Spectrosc.,* 8, 92, 1987.

126. **Tsalev, D. and Mandzhukov, P.,** Electrothermal AAS determination of hydride-forming elements after simultaneous preconcentration by hydride generation and trapping hydrides in cerium(IV)-potassium iodide absorbing solution, *Microchem. J.*, 35, 83, 1987.
127. **Hall, S., D. J., Mohl, C., and Stoeppler, M.,** Application of rapid furnace programs in atomic absorption spectroscopy to the determination of lead, chromium and copper in digests of plant materials, *Analyst (London)*, 112, 185, 1987.
128. **Lugowski, S., Smith, D. C., and Van Loon, J. C.,** The determination of aluminum, chromium, cobalt, iron and nickel in whole blood by electrothermal atomic absorption spectrophotometry, *J. Biol. Mater. Res.*, 21, 657, 1987.
129. **Mohammed, D. A.,** Determination of chromium by electrothermal atomization atomic absorption spectroscopy, *Analyst* (London), 112, 209, 1987.
130. **Randall, J. A. and Gibson, R. S.,** Serum and urine chromium as indexes of chromium status in tannery workers, *Proc. Soc. Exp. Biol. Med.*, 185, 16, 1987.

AUTHOR INDEX

SUBJECT INDEX

COBALT

COBALT, Co   (ATOMIC WEIGHT 58.9332)

Instrumental Parameters:

Wavelength	240.7 nm
Slit width	80 μm
Bandpass	0.3 nm
Light source	Hollow Cathode Lamp
Lamp current	8 mA
Purge gas	Argon or Nitrogen
Sample size	25 μL
Furnace	Cylindrical cuvette/Pyrolytic graphite

Standard Operating Conditions:

Optimum char temperature	1000°C
Optimum atomization temperature	2400°C
Sensitivity	10 pg/1% absorption
Sensitivity check	0.04 μg/mL
Working range	0.008-0.03 μg/mL
Background correction	Required only to correct light scattering or non-specific absorption from the sample containing high dissolved solids.

General Note:
1. Use commercial standard or a previously analyzed sample as a working standard.
2. Use ultra clean glass and plastic ware, soaked in 1:5 nitric acid solution and rinsed thoroughly with deionized distilled water.
3. All analytical solutions should be at least 0.2-0.5% (v/v) in nitric acid.
4. Check for blank values on all reagents including water.
5. Dilution is recommended when sample exhibits greater than 0.5 absorbance units.
6. Use of nitrogen as purge gas will suppress the sensitivity.

REFERENCES

1. **Schweizer, V. B.,** Determination of cobalt, chromium, copper, molybdenum, nickel and vanadium in carbonate rocks with the HGA-70 graphite furnace, *At. Ab. Newsl.,* 14, 137, 1975.
2. **Mizoguchi, T. and Ishii, H.,** Study of extracting agents for the determination of cobalt and nickel by solvent extraction-carbon furnace atomic absorption spectroscopy, *Bunseki Kagaku,* 26, 839, 1977.
3. **Page, A. G., Godbole, S. V., Kulkarni, M. J., Shelar, S. S., and Joshi, B. D.,** Direct AAS determination of cobalt, chromium, copper, manganese and nickel in uranium oxide (U_3O_3) by electrothermal atomization, *Z. Anal. Chem.,* 296, 40, 1979.
4. **Patel, B. M., Bhatt, P. M., Gupta, N., Pawar, M. M., and Joshi, B. D.,** Electrothermal atomic absorption spectrometric determination of cadmium, chromium and cobalt in uranium without preliminary separation, *Anal. Chim. Acta,* 104, 113, 1979.
5. **Lidums, V. V.,** Determination of cobalt in blood and urine by electrothermal atomic absorption spectroscopy, *At. Ab. Newsl.,* 18, 71, 1979.
6. **Matsuo, H., Kumamara, T., and Hara, S.,** Graphite furnace atomic absorption spectroscopy of cobalt by ion-pair ammonium, *Bunseki Kagaku,* 29, 337, 1980.
7. **Sedykh, E. M., Belyaev, Y. I., and Sorokina, E. V.,** Elimination of matrix effects in electrothermal atomic absorption determination of silver, lead, cobalt, nickel and tellurium in samples of complicated composition, *Zh. Anal. Khim.,* 35, 2348, 1980 (in Russian).
8. **Sedykh, E. M., Belyaev, Y. I., and Sorokina, E. V.,** Matrix effect during electrothermal atomic absorption determination of silver, tellurium, lead, cobalt and nickel in materials of complex composition, *Zh. Anal. Khim.,* 35, 2162, 1980 (in Russian).
9. **Sterritt, R. M. and Lester, J. N.,** Determination of cobalt, manganese and tin in sewage sludge by a rapid electrothermal atomic absorption spectroscopic method, *Analyst,* 105, 616, 1980.
10. **Chakrabarti, C. L., Wan, C. C., and Li, W. C.,** Atomic absorption spectrometric determination of cadmium, lead, zinc, copper, cobalt and iron in oyster tissue by direct atomization from the solid state using graphite furnace platform technique, *Spectrochim. Acta,* 35B, 547, 1980.
11. **Pederson, B., Willems, M., and Stargaard-Joergensen, S.,** Determination of copper, lead, cadmium, nickel and cobalt in EDTA extracts of soil by solvent extraction and graphite furnace atomic absorption spectrophotometry, *Analyst,* 105, 119, 1980.
12. **Lerner, L. A. and Igoshina, E. V.,** Atomic absorption determination with a graphite furnace of copper, cobalt and nickel extracted from soil with ammonium acetate buffer solutions, *Pochvovedenie,* 3, 106, 1980.
13. **Hydes, D. J.,** Reduction of matrix effects with a soluble organic acid in the carbon furnace atomic absorption spectrometric determination of cobalt, copper and manganese in seawater, *Anal. Chem.,* 52, 959, 1980.
14. **Fudagawa, N. and Kawase, A.,** Determination of cobalt in plant materials by graphite furnace atomic absorption spectroscopy after solvent extraction, *Bunseki Kagaku,* 29, 6, 1980.
15. **Hulanicki, A., Karwowska, R., and Sowinski, J.,** Influence of some matrix elements on atomization of cobalt in graphite furnace atomic absorption spectrometry, *Talanta,* 28, 455, 1981.
16. **Oradonskii, S. G. and Kuzimichev, A. N.,** Flameless extraction-atomic absorption method for determination of ultramicro amounts of toxic metals (lead, cadmium, copper, cobalt and nickel) in seawaters and ices in labile form, *Tr. Gos Okeanogr. Inst.,* 162, 22, 1981.
17. **Carelli, G., Cecchetti, G. La Bua, R., Berganaschi, A., and Iannaccone, A.,** Determination of selenium, cadmium, zinc, cobalt, copper and silver using atomic absorption spectrophotometry in aerosols from the workplace in the color television electronic industry, *Ann. Ist. Super Sanita,* 17, 505, 1981.
18. **Park, M. K., Choi, J. K., and Paik, N. H.,** Determination of metals and vitamin B 12 in multi-vitamin preparations by atomic absorption spectrophotometry, *Soul Taehakkyo Yakhak Nonmunjip,* 6, 28, 1981 (in Korean).
19. **Bangia, T. R., Kartha, K. N. K., Varghese, M., Dhawala, B. A., and Joshi, B. D.,** Chemical separation and electrothermal atomic absorption spectrophotometric determination of cadmium, cobalt, copper and nickel in high-purity uranium, *Z. Anal. Chem.,* 310, 410, 1982.
20. **Lerner, L. A., Kakhnovich, J. N., and Igoshina, E. V.,** Electrothermal atomic absorption determination of cobalt extractable from soils with a 1.0 N nitric acid solution, *Pochvovedenie,* 2, 104-9, 1982 (in Russian).
21. **Borggard, O. K., Christensen, H. E. M., Nielsen, T. K., and Willems, M.,** Comparison of four ligands for the determination of cobalt in trace amounts by solvent extraction and graphite furnace atomic absorption spectrophotometry, *Analyst (London),* 107, 1479, 1982.
22. **Stenberg, T.,** Release of cobalt from cobalt chromium alloy constructions in the oral cavity of man, *Scand. J. Dent. Res.,* 90, 472, 1982.
23. **Arnold, D. and Kuennecke, A.,** Determination of 3rd elements in high lead-containing glasses, *Silikaltechnik,* 33, 312, 1982.
24. **Pyatnitskii, I. V. and Pilipyuk, Y. S.,** Atomic absorption determination of nickel and cobalt in chloroform extracts, *Ukr. Khim. Zh.,* 48, 962, 1982 (in Russian).

25. **Gretzinger, K., Kotz, L., Tschoepel, P., and Poelg, G.,** Causes and elimination of systematic errors in the determination of iron and cobalt in aqueous solutions in the ng/mL and pg/mL range, *Talanta,* 29, 1011, 1982.

26. **Perdrix, A., Pellet, F., Vincent, M., De Gaudemaris, R., and Mallion, J. M.,** Cobalt and sintered metal carbides. Value of the determination of cobalt as a tracer for exposure to hard metals, *Toxicol. Eur. Res.,* 5, 233, 1983.

27. **Delves, H. T., Mensikov, R., and Hinks, L.,** Direct determination of cobalt in whole blood by electrothermal atomization and atomic absorption spectroscopy, *Proc. 2nd Int. Workshop Trace Elem. Anal. Chem. Med. Biol.,* Braether, P. and Schramel, P., Eds., Berlin, 1983, 1123.

28. **Castledine, C. G. and Robbins, J. C.,** Practical field-portable atomic absorption analyzer, *J. Geochem. Explor.,* 19, 689, 1983.

29. **Nakeshima, R., Kamata, E., and Shibata, S.,** Determination of cobalt, copper, iron, nickel and lead in bone by graphite furnace atomic absorption spectroscopy, *Kogai,* 18, 37-44, 1983.

30. **Backstrom, K., Danielsson, L. G., and Nord, L.,** Sample work-up for graphite furnace atomic absorption spectrometry using continuous flow extraction, *Analyst,* 109, 323, 1984.

31. **Boniforti, R., Ferraroli, R., Frigieri, P., Heltai, D., and Queirazza, G.,** Intercomparison of five methods for determination of trace metals in seawater, *Anal. Chim. Acta,* 162, 33, 1984.

32. **Blackmore, D. J. and Stanier, P.,** Methods for measurement of trace elements in equine blood by electrothermal atomic absorption spectrophotometry, *At. Spectrosc.,* 5, 215, 1984.

33. **Shijo, Y., Shimizu, T., and Sakai, K.,** Determination of cobalt in seawater by graphite furnace atomic absorption spectrometry after preconcentration with diethyldithiocarbamate extraction and vacuum evaporation of solvent, *Bunseki Kagaku,* 33, E215, 1984.

34. **Schumacher, W.,** Characterization of cobalt-binding proteins in occupational cobalt exposure, *Toxicol. Environ. Chem.,* 8, 185, 1984.

35. **Heinrich, R. and Angerer, J.,** Determination of cobalt in biological materials by voltammetry and electrothermal atomic absorption spectrometry, *Int. J. Environ. Anal. Chem.,* 16, 305, 1984.

36. **Borggaard, O. K., Christensen, H. E. M., and Lund, S. P.,** Determination of cobalt in feeding stuffs by solvent extraction and graphite furnace atomic absorption spectrophotometry, *Analyst,* 109, 1179, 1984.

37. **Angerer, J. and Heinrich, R.,** Determination of cobalt in blood by electrothermal atomic absorption spectrometry, *Z. Anal. Chem.,* 318, 37, 1984.

38. **Jing, S., Wang, R., Yu, C., Yan, Y., and Ma, Y.,** Design of model ZM-1 zeeman effect atomic absorption spectrometer, *Huanjing Kuxue,* 5, 41, 1984 (in Chinese).

39. **Pruszkowska, E. and Barrett, P.,** Determination of arsenic, selenium, chromium, cobalt and nickel in geochemical samples using the stabilized temperature platform furnace and zeeman background correction, *Spectrochim. Acta,* 39B, 485, 1984.

40. **Torrance, K.,** Determination of nickel and cobalt in simulated PWR coolant by differential-pulse polarography, *Analyst (London),* 109(8), 1035, 1984.

41. **Andersen, I. and Hogetveit, A. C.,** Analysis of cobalt in plasma by electrothermal atomic absorption spectrometry, *Z. Anal. Chem.,* 318, 41, 1984 (in German).

42. **Backstrom, K., Danielsson, L. G., and Nord, L.,** Sample work-up for graphite furnace atomic absorption spectrometry using continuous flow extraction, *Analyst,* 109, 323, 1984.

43. **Chakrabarti, C. L., Wu, S., Karwowska, R., Chang, S. B., and Bertels, P. C.,** Studies in atomic absorption spectrometry with laboratory-made graphite furnace, *At. Spectrosc.,* 5, 69, 1984.

44. **Slavin, W. and Carnrick, G. R.,** Possibility of standardless furnace atomic absorption spectroscopy, *Spectrochim. Acta,* 39B, 271, 1984.

45. **Mo, S. and Zhang, Z.,** Determination of cobalt and cadmium in high-purity rare earth oxides by graphite furnace atomic absorption spectroscopy, *Fenxi Huaxue,* 12, 408, 1984.

46. **Pellet, F., Perdrix, A., Vincent, M., and Mallion, J. M.,** Biological determination of urinary cobalt. Significance in occupational medicine in the monitoring of exposures to sintered metallic carbides, *Arch. Mal. Prof. Med. Trav. Secur. Soc.,* 45, 81-5, 1984 (in French).

47. **Lerner, L. A.,** Electrothermal atomic absorption determination of cobalt in a nitrate medium, *Metody Spektr. Anal. Miner. Syr'ya (Mater. Vses. Konf. Nov. Metodam Spektr. Anal. Ikh Primen.),* 2nd, Lontsikh, S. V., Ed., Nauka, S. b. otd.: Novosibirsk, U.S.S.R., 1984, 120-3.

48. **Pellet, F., Perdrix, A., Vincent, M., De Gaudemaris, R., Debrru, J. L., Cau, G., and Mallion, J. M.,** Biological determination of urinary cobalt. Value, methodology and preliminary results, *Collect. Med. Leg. Toxicol. Med.,* 125, 71-7, 1984.

49. **Liang, D., Liang, Z., and Wang, W.,** Determination of trace cobalt in sediments by flameless atomic absorption spectrophotometry — application of modified simplex optimization method, *Redai Haiyang,* 3, 7-15, 1984 (in Chinese).

50. **Molin, C. J. and Mikkelsen, S.,** Cobalt concentration in whole blood and urine from pottery plate painters exposed to cobalt paint, *5th Int. Conf. Heavy Met. Environ.,* Lekkas, T. D., Ed., CEP, Consult, Edinburgh, U.K., 1985, 86-8.

51. **Barbera, R., Farre, R., and Montoro, R.,** AA spectrophotometric determination of cobalt in foods, *J. Assoc. Off. Anal. Chem.,* 68(3), 511, 1985.

52. **Heinrich, R. and Angerer, J.,** Electrothermal AAS (ETAAS) determination of cobalt in whole blood. Comparison of direct and deproteinization procedures, *Fresnius' Z. Anal. Chem.,* 322(8), 772-4, 1985.

53. **Taylor, P., Desmet, B., and Dams, R.,** The determination of cobalt in high purity nickel for reactor neutron dosimetry by means of ion exchange combined with ICP-AES and ion exchange combined with GF-AAS, *Anal. Lett.,* 18(A19), 2477-87, 1985.

54. **Firriolo, J. M. and Kutzman, R. S.,** Determination of cobalt in samples containing cobalt and tungsten carbide by electrothermal AAS, *Am. Ind. Hyg. Assoc. J.,* 46(9), 476-80, 1985.

55. **Wang, X., Lu, P., and Zhang, G.,** Determination of cobalt, chromium and nickel in doped gadolinium gallium garnets by GFAAS, *Guangpuxue Yu Guangpu Fenxi,* 5(1), 63-5, 1985.

56. **Raichevski, G., Gancheva, Y., Kuncheva, M., and Tomov, I.,** Anodic behavior and passivation of electroplated and electroless deposited nickel and cobalt, *Zashch. Met.,* 21(3), 449-52, 1985.

57. **Kleijburg, M. R. and Pijpers, F. W.,** Calibration graphs in atomic-absorption spectrophotometry, *Analyst,* 110, 147, 1985.

58. **Ma, Y. and Cheng, J.,** Comparative study on several tubular atomizers for determination of cobalt with graphite furnace atomic absorption spectroscopy, *Fenxi Huaxue,* 14, 746, 1986.

59. **Anderson, J. and Victor, A. H.,** Determination of cobalt in South African primary and secondary reference samples by ion exchange chromatography and electrothermal AAS, *Geostand. Newsl.,* 10(1), 27-8, 1986.

60. **Shimizu, T., Koyanagi, H., Shijo, Y., and Sakai, K.,** Determination of cobalt in natural waters by graphite furnace AAS after micro solvent extraction with capriquat, *Chem. Lett.,* 3, 319-22, 1986.

61. **Bouman, A. A., Platenkamp, A. J., and Posma, F. D.,** Determination of cobalt in urine by flameless atomic absorption spectroscopy. Comparison of direct analysis using Zeeman background correction and indirect analysis using extraction in organic solution, *Ann. Clin. Biochem.,* 23(3), 346-50, 1986.

62. **Christensen, J. M. and Mikkelsen, S.,** Cobalt concentration in whole blood and urine from pottery plate painters exposed to cobalt paint, *5th Int. Conf. Heavy Met. Environ.,* 2, 86, 1986.

63. **Alary, J., Bourbon, P., and Vandaele, J.,** Determination of cobalt in forages by Zeeman electrothermal atomic absorption spectrophotometry without preliminary extraction, *Sci. Total Environ.,* 46, 181, 1986.

64. **McMahon, J. W., Doherty, A. E., Judd, J. M. A., and Gentner, S. R.,** Determination of ultratrace amounts of cobalt in fish by graphite furnace Zeeman effect atomic absorption spectrometry, *Int. J. Environ. Anal. Chem.,* 24(4), 297-303, 1986.

65. **Zhengli, H., Yuzhi, W., Zhiquiang, Z., and Quinzhi, L.,** Simultaneous determination of manganese, cobalt, nickel, lead, cadmium and tin in canned foods and infant foods by stabilized temperature platform furnace-atomic absorption spectrophotometry, *Shipin Yu Fajiao Gongye,* 4, 1, 1986 (in Chinese).

66. **Kuroda, R., Nakano, T., Miura, Y., and Oguma, K.,** Determination of trace amounts of nickel and cobalt in silicate rocks by graphite furnace atomic absorption spectroscopy. Elimination of matrix effects with an ammonium fluoride modifier, *J. Anal. At. Spectrom.,* 1, 429, 1986.

67. **Mitchell, M. C., Berrow, M. L., and Shand, C. A.,** Direct determination of cobalt in acetic acid extracts of soils by graphite furnace atomic absorption spectrometry, *J. Anal. At. Spectrom.,* 2, 261, 1987.

68. **Kimberly, M. M., Bailey, G. G., and Paschal, D. C.,** Determination of urinary cobalt using matrix modification and graphite furnace atomic absorption spectroscopy with zeeman background correction, *Analyst (London),* 112, 287, 1987.

AUTHOR INDEX

SUBJECT INDEX

COPPER

COPPER, Cu   (ATOMIC WEIGHT 63.54)

Instrumental Parameters:

Wavelength	324.7 nm
Slit width	320 μm
Bandpass	1 nm
Light source	Hollow Cathode Lamp
Lamp current	5 mA
Purge gas	Argon or Nitrogen
Sample size	25 μL
Furnace	Cylindrical cuvette/Pyrolytic graphite

Standard Operating Conditions:

Optimum char temperature	900°C
Optimum atomization temperature	2200°C/2000°C
Sensitivity	4 pg/1% absorption
Sensitivity check	0.05 μg/mL
Working range	0.004-0.08 μg/mL
Background correction	Required only to correct light scattering or non-specific absorption from the sample containing high dissolved solids.

General Note:

1. Use commercial standard or a previously analyzed sample as a working standard.
2. Use ultra clean glass and plastic ware, soaked in 1:5 nitric acid solution and rinsed thoroughly with deionized distilled water.
3. All analytical solutions should be at least 0.2-0.5% (v/v) in nitric acid.
4. Check for blank values on all reagents including water.
5. Dilution is recommended when sample exhibits greater than 0.5 absorbance units.
6. Use of nitrogen as a purge gas will suppress the sensitivity.
7. For increased sensitivity, step atomization is recommended.
8. For analyzing oil or highly viscous samples, use a maximum 5 μL or 5 μg sample size.

REFERENCES

1. **Glenn, M., Savory, J., Hart, L., Glenn, T., and Winefordner, J. D.,** Determination of copper in serum with a graphite rod atomizer for atomic absorption spectrophotometry, *Anal. Chim. Acta,* 57, 263, 1971.
2. **Barnett, W. B. and Kahn, H. L.,** Determination of copper in fingernails by atomic absorption with the graphite furnace, *Clin. Chem.,* 18, 923, 1972.
3. **Maurer, L.,** Rapid, simple procedure for the determination of copper in cheese with a graphite furnace, *Z. Lebensm. Unters. Forsch.,* 156, 284, 1974.
4. **Muzzarelli, R. A. A. and Rocchetti, A.,** Determination of copper in seawater by atomic absorption spectroscopy with a graphite atomizer after elution from Chitosan, *Anal. Chim. Acta,* 69, 35, 1974.
5. **Downes, T. E. H. and Labuschagne, J. H.,** Determination of copper in milk and butter with the graphite furnace atomic absorption apparatus, *S. Afr. J. Dairy Technol.,* 7, 167, 1975.
6. **Simmons, W. and Loneragan, J. F.,** Determination of copper in small amounts of plant material by atomic absorption spectrophotometry using a heated graphite atomizer, *Anal. Chem.,* 47, 566, 1975.
7. **Evenson, M. A. and Anderson, C. T., Jr.,** Ultramicro analysis for copper, cadmium and zinc in human liver tissue by use of atomic absorption spectrophotometry and the heated graphite tube atomizer, *Clin. Chem.,* 21, 537, 1975.
8. **Vondenhoff, T.,** Determination of lead, cadmium, copper and zinc in plant and animal material by atomic absorption in a flame and in a graphite tube after sample decomposition by the Schoniger technique, *MHBl. Gdch. Fachgr. Lebensmittelchem. Gerichtl. Chem.,* 29, 341, 1975 (in German).
9. **Schweizer, V. B.,** Determination of copper, chromium, cobalt, molybdenum, nickel and vanadium in carbonate rocks with the HGA-70 graphite furnace, *At. Ab. Newsl.,* 14, 137, 1975.
10. **Evenson, M. A. and Warren, B. L.,** Determination of serum copper by atomic absorption with use of the graphite cuvette, *Clin. Chem.,* 21, 619, 1975.
11. **Smeyers-Verbeke, J., Michotte, Y., Van den Winkel, P., and Massart, D. L.,** Matrix effects in the determination of copper and manganese in biological materials using carbon furnace atomic absorption spectroscopy, *Anal. Chem.,* 48, 125, 1976.
12. **Janssen, A., Melchior, H., and Scholz, D.,** Application of flameless atomic absorption (heated graphite atomizer) to the determination of traces of copper in steel and other metal alloys after extraction, *Z. Anal. Chem.,* 283, 1 1977.
13. **Del-Castiho, P. and Herber, R. F.,** The rapid determination of cadmium, lead, copper and zinc in whole blood by atomic absorption spectroscopy with electrothermal atomization. Improvements in precision with a peak-shape monitoring device, *Anal. Chim. Acta,* 94, 269, 1977.
14. **Tanaka, T., Hayashi, Y., and Ishizawa, M.,** Simultaneous determination of cadmium and copper in water by a graphite furnace dual channel atomic absorption spectrophotometry, *Bunseki Kagaku,* 27, 499, 1978.
15. **Smeyers-Verbeke, J., Michotte, Y., and Massart, D. L.,** Influence of some matrix elements on the determination of copper and manganese by furnace atomic absorption spectroscopy, *Anal. Chem.,* 50, 10, 1978.
16. **Maruta, T., Minegishi, K., and Sudoh, G.,** Atomic absorption spectrometric determination of trace amounts of copper, manganese, lead and chromium in cements by direct atomization in a carbon furnace, *Yogyo Kyokai Shi,* 86, 532, 1978.
17. **Potter, N. M.,** Determination of copper in gasoline by atomic absorption spectroscopy with electrothermal atomization, *Anal. Chim. Acta,* 102, 201, 1978.
18. **Nagahoro, T., Fujino, O., Matsui, M., and Shigematsu, T.,** Determination of cadmium in individual organs and divided shells of seawater clam by atomic absorption spectroscopy with a carbon tube atomizer, *Bull. Inst. Chem. Res. Kyoto Univ.,* 56, 274, 1978.
19. **Kamel, H., Teape, J., Brown, D. H., Ottaway, J. M., and Smith, W. E.,** Determination of copper in plasma ultrafiltrate by atomic absorption spectroscopy using carbon furnace atomization, *Analyst,* 103, 921, 1978.
20. **Norval, E.,** A tungsten carbide-coated crucible for electrothermal atomization. Determination of copper in some biological standards, *Anal. Chim. Acta,* 97, 399, 1978.
21. **Churella, D. J. and Copeland, T. R.,** Interference of salt matrices in the determination of copper by atomic absorption spectroscopy with electrothermal atomization, *Anal. Chem.,* 50, 309, 1978.
22. **Katskov, D. A. and Grinshtein, I. L.,** Study of the chemical interaction of copper, gold and silver with carbon by an atomic absorption method using an electrothermal atomizer, *Zh. Prikl. Spektrosk,* 30, 787, 1979.
23. **Teape, J., Kamel, H., Brown, D. H., Ottaway, J. M., and Smith, W. E.,** An evaluation of the use of electrophoresis and carbon furnace atomic absorption spectroscopy to determine the copper level in separated serum protein fractions, *Clin. Chim. Acta,* 94, 1, 1979.
24. **Isozaki, A., Soeda, N., Okutani, T., and Utsumi, S.,** Flameless atomic absorption spectrophotometry of copper by the introduction of chelating resin into a carbon tube atomizer, *Nippon Kagaku Kaishi,* 4, 549, 1979.
25. **Patel, B. M., Gupta, N., Purohit, P., and Joshi, B. D.,** Electrothermal atomic absorption spectrometric determination of lithium, sodium, potassium and copper in uranium without preliminary chemical separation, *Anal. Chim. Acta,* 118, 163, 1980.

26. **Chakrabarti, C. L., Wan, C. C., and Li, W. C.,** Atomic absorption spectrometric determination of cadmium, lead, zinc, copper, cobalt and iron in oyster tissue by direct atomization from the solid state using the graphite furnace platform technique, *Spectrochim. Acta,* 35B, 547, 1980.

27. **Hydes, D. J.,** Reduction of matrix effects with a soluble organic acid in the carbon furnace atomic absorption spectrometric determination of cobalt, copper and manganese in sea water, *Anal. Chem.,* 52, 954, 1980.

28. **Pederson, B., Willems, M., and Stargaard-Joergensen, S.,** Determination of copper, lead, cadmium, nickel and cobalt in EDTA extracts of soil by solvent extraction and graphite furnace atomic absorption spectrophotometry, *Analyst,* 105, 119, 1980.

29. **Lerner, L. A. and Igoshina, E. V.,** Atomic absorption determination with a graphite furnace of copper, cobalt and nickel extracted from soil with amm. acetate buffer solutions, *Pochvovedenie,* 3, 106, 1980.

30. **Kuga, K.,** Rapid determination of copper, iron and manganese in polyimide resins by atomic absorption spectroscopy using a graphite furnace atomizer, *Bunseki Kagaku,* 29, 342, 1980.

31. **Tsushida, T. and Takeo, T.,** Determination of copper, lead and cadmium in tea by graphite furnace atomic absorption spectrophotometry, *Nippon Shokuhin Kogyo Gakkaishi,* 27, 585, 1980 (in Japanese).

32. **Montgomery, J. R. and Peterson, G. N.,** Effects of amm. nitrate on sensitivity for determination of copper, iron and manganese in seawater by atomic absorption spectroscopy with pyrolytically coated graphite tubes, *Anal. Chim. Acta,* 117, 397, 1980.

33. **Battistoni, P., Bruni, P., Cardellini, L., Fava, G., and Gobbi, G.,** Trace element determination. I. Use of 2,9-dimethyl-1,10- phenanthroline in determination of copper in heavy matrixes by carbon furnace atomic absorption spectroscopy, *Talanta,* 27, 623, 1980.

34. **Baker, A. A. and Headridge, J. B.,** Determination of bismuth, lead and tellurium in copper by atomic absorption spectroscopy with introduction of solid samples into an induction furnace, *Anal. Chim. Acta,* 125, 93, 1981.

35. **Oradovskii, S. G. and Kuzimichev, A. N.,** Flameless extraction-atomic absorption method for determination of ultramicro amounts of toxic metals (lead, cadmium, copper, cobalt and nickel) in seawaters and ices in labile form, *Tr. Gos Okeanogr. Inst.,* 162, 22, 1981.

36. **Arafat, N. M. and Glooschenko, W. A.,** Method for the simultaneous determination of arsenic, aluminum, iron, zinc, chromium and copper in plant tissue without the use of perchloric acid, *Analyst (London),* 106, 1174-8, 1981.

37. **Farmer, J. G. and Gibson, M. J.,** Direct determination of cadmium, chromium, copper and lead in siliceous standard reference materials from a fluoboric acid matrix by graphite furnace atomic absorption spectroscopy, *At. Spectrosc.,* 2, 176-8, 1981.

38. **Ranchet, J., Menissier, F., Lamathe, J., and Voinovitch, I.,** Interlaboratory comparison of the determinations of cadmium, chromium, copper and lead by flameless atomic absorption spectrometry, *Bull. Liaison Lab. Ponts Chaussees,* 114, 81-6, 1981 (in French).

39. **Suzuki, M., Ohta, K., and Yamakita, T.,** Elimination of alkali chloride interference with thiourea in electrothermal atomic absorption spectroscopy of copper and manganese, *Anal. Chem.,* 53, 9, 1981.

40. **Sekiya, T., Tanimura, H., and Hkiasa, Y.,** Simplified determination of copper, zinc and manganese in plasma and bile by flameless atomic absorption spectroscopy, *Archib. Jpn. Chir.,* 50, 729-39, 1981.

41. **Bouzanne, M.,** Determination of traces of inorganic pollutants (copper, lead, cadmium and zinc) in waters by differential pulse anodic stripping voltammetry at a rotating mercury film electrode, *Analusis,* 9, 461-7, 1981 (in French).

42. **Isozaki, A., Soeda, N., and Utsumi, S.,** Sensitive atomic absorption spectrometric method for copper employing the direct introduction of chelating resin into a carbon tube atomizer, *Bull. Chem. Soc. Jpn,* 54, 1364, 1981.

43. **Halls, D. J., Fell, G. S., and Dunbar, P. M.,** Determination of copper in urine by graphite furnace atomic absorption spectroscopy, *Clin. Chim. Acta,* 114, 21, 1981.

44. **Takada, T., Okano, H., Koide, T., Fujita, K., and Nakano, K.,** Determination of trace copper by electrothermal atomic absorption spectroscopy with direct heating of metal-adsorbed ion exchange resin, *Nippon Kagaku Kaishi,* 1, 13, 1981.

45. **Masuda, N.,** Determination of total dissolved copper in seawater by carbon furnace atomic absorption spectroscopy, *Hokkaido Daigaku Suisangakubu Kenkyu Iho,* 32, 425-33, 1981.

46. **Velghe, N., Campe, A., and Claeys, A.,** Determination of copper in undiluted serum and whole blood by atomic absorption spectrophotometry with graphite furnace, *At. Spectrosc.,* 3, 48, 1982.

47. **Matsuzaki, A., Kondo, O., and Saitu, N.,** Determination of trace copper in electrical insulating oil by atomic absorption spectrometry with a graphite furnace atomizer, *Banjasz. Kohasz. Lapok, Koolaj Foldgaz,* 15, 74-8, 1982 (in Hungarian).

48. **Krivan, V. and Lang, M.,** Radiotracer studies on the direct determination of copper in biological matrixes by flameless AAS, *Fresnius' Z. Anal. Chem.,* 312, 324-30, 1982.

49. **Tarui, T., Nakatani, S., and Tokairin, H.,** Determination of copper in petroleum by graphite furnace atomic absorption spectroscopy, *Bunseki Kagaku,* 31, T29-T33, 1982 (in Japanese).

50. **Gonzalez, O. C., Baez, C. M., Ruiz, A. F., and Miranda, L. A.,** Method for the determination of copper in plant fibers by flameless atomic absorption spectrophotometry, *Bol. Soc. Clin. Quim.,* 27, 259-61, 1982.

51. **Fagioli, F., Landi, S., and Lucci, G.,** Determination of manganese and copper in small amounts of maize roots by graphite tube furnace atomic absorption spectroscopy with liquid and solid sampling techniques, *Ann. Chim.,* 72, 63, 1982.

52. **Bangia, T. R., Kartha, N. K., Varghese, M., Dhawale, B. A., and Joshi, B. D.,** Chemical separation and electrothermal atomic absorption spectrophotometric determination of cadmium, cobalt, copper and nickel in high-purity uranium, *Z. Anal. Chem.,* 310, 410, 1982.

53. **Lieser, K. H., Sondermeyer, S., and Kliemchen, A.,** Precision and accuracy of analytical results in determining the elements cadmium, chromium, copper, iron, manganese and zinc by flameless AAS, *Fresnius' Z. Anal. Chem.,* 312, 517, 1982 (in German).

54. **Suzuki, M., Ohta, K., and Katsuno, T.,** Determination of traces of lead and copper in foods by electrothermal atomic absorption spectroscopy with metal atomizer, *Mikrochim. Acta,* 2, 225, 1982.

55. **Katskov, D. A. and Grinshtein, I. L.,** Formation of copper, silver and calcium acetylenides in graphite furnaces for atomic absorption analysis, *Zh. Prikl. Spektrosk.,* 36, 181, 1982.

56. **Gardiner, P. E., Roesick, E., Roesick, U., Braetter, P., and Kynast, G.,** Application of gel filtration, immunonephelometry and electrothermal atomic absorption spectroscopy to the study of the distribution of copper, iron and zinc bound constituents in human amniotic fluid, *Clin. Chim. Acta,* 120, 103, 1982.

57. **Prokof'ev, A. K., Oradorskii, S. G., and Georgievskii, V. V.,** Flameless atomic-absorption method for the determination of copper, lead and cadmium in marine bottom sediments, *Tr. Gos. Okeanogr. In-ta,* 162, 51-5, 1981 (in Russian). From *Ref. Zh. Khim.,* Abstr. No. 11G258, 1982.

58. **Johnsen, A. C., Wibetoe, G., Langmyhr, F. J., and Aaseth, J.,** Atomic absorption spectrometric determination of the total content and distribution of copper and gold in synovial fluid from patients with rheumatoid arthritis, *Anal. Chim. Acta,* 135, 243-8, 1982.

59. **Kobayashi, T., Ohya, K., and Murakoshi, T.,** Determination of copper in lead pipes by flameless atomic absorption spectroscopy using a graphite tube, *Aichi-ken Kogyo Gijutsu Senta Hokoku,* 17, 61, 1982 (in Japanese).

60. **Wawschinek, O. and Beyer, W.,** Determination of copper in liver biopsy samples for diagnosis of Wilson's disease, *J. Clin. Chem. Clin. Biochem.,* 20, 929, 1982 (in German).

61. **Arnold, D. and Kuennecke, A.,** Determination of 3rd elements in high lead-containing glasses, *Silikattechnik,* 33, 312, 1982.

62. **Nakaaki, K., Tada, O., and Masuda, T.,** Urinary excretion of chromium, manganese, copper and nickel, *Rodo Kagaku,* 58, 529, 1982 (in Japanese).

63. **Zhou, B. and Zeng, S.,** Simultaneous determination of chromium, manganese and copper in serum by flameless atomic absorption spectrophotometry, *Yingyang Xuebao,* 4, 323, 1982 (in Chinese).

64. **Bourcier, D. R., Sharma, R. P., Bracken, W. M., and Taylor, M. J.,** Cadmium-copper interaction: effect of copper pretreatment and cadmium-copper chronic exposure on the distribution and accumulation of cadmium, copper, zinc and iron in mice, *Trace Substr. Environ. Health,* 16, 273, 1982.

65. **Matsusaki, K., Yoshino, T., and Yamamoto, Y.,** Electrothermal atomic absorption spectrometric determination of trace lead, copper and manganese in aluminum and its alloys without preliminary separation, *Anal. Chim. Acta,* 144, 189, 1982.

66. **Lamathe, J., Magurno, C., and Equel, J. C.,** An interlaboratory study of the determination of cadmium, copper and lead in seawater by electrothermal atomic absorption spectroscopy, *Anal. Chim. Acta,* 142, 183, 1982 (in French).

67. **Ke, C. H., Lo, J. M., and Yeh, S. J.,** Determination of trace amount of copper in reactor coolant by graphite furnace atomic absorption spectroscopy, *Hua Hsueh,* 41, 132, 1983 (in Chinese).

68. **Kida, A.,** Efficacy of coprecipitation or acid decomposition as a pretreatment for the determination of total arsenic, *Hirochima-ken Kankyo Senta Kenkyu Hokoku,* 5, 99, 1983.

69. **Komarek, J., Kolcava, D., and Sommer, L.,** Atomic absorption spectrometry of copper in the presence of organic chelating agents, *Ser. Fac. Sci. Nat. Univ. Purkynianae Brun.,* 13(3-4), 131-6, 1983.

70. **McGahan, M. C. and Bito, L. Z.,** Determination of copper concentration in blood plasma and in ocular and cerebrospinal fluids using GFAAS (graphite furnace atomic absorption spectroscopy), *Anal. Biochem.,* 135(1), 186-92, 1983.

71. **Schindler, E.,** Compound procedure to determine lead, cadmium and copper in cereals and flours, *Dtsch. Lebensm. Rundsch.,* 79, 334, 1983.

72. **Becher, G., Oestvold, G., Paus, P., and Seip, H. M.,** Complexation of copper by aquatic humic matter studied by reversed-phase liquid chromatography and atomic absorption spectroscopy, *Chemosphere,* 12, 1209-15, 1983.

73. **Wei, F., Yang, J. T., and Yin, P.,** Determination of trace copper in cord blood serum from newborn infants by flameless AAS with aerosol-deposition technique, *Anal. Lett.,* 16(B7), 501-8, 1983.

74. **Hayashi, Y., Yabuta, Y., Tanaka, T., and Nose, T.,** Determination of copper, iron and zinc in cutaneous leg lymph of rabbits by atomic absorption spectrophotometry, *Bunseki Kagaku,* 32, 212, 1983.

75. **Favier, A. and Ruffieux, D.,** Physiological variations of serum levels of copper, zinc, iron and manganese, *Biomed. Pharmacother,* 37, 462, 1983.

76. **Brune, D., Gjerdet, N., and Paulsen, G.,** Gastrointestinal and *in vitro* release of copper, cadmium, indium, mercury and zinc from conventional and copper-rich amalgams, *Scand. J. Dent. Res.,* 91, 66, 1983.

77. **Takada, K.,** Determination of small amounts of copper and silver in tin ingot by polarized Zeeman atomic absorption spectroscopy with direct atomization of solid sample in a graphite-cup cuvette, *Bunseki Kagaku,* 32, 197, 1983.

78. **Borggaard, O. K., Christensen, H. E. M., and Ilsoee, C.,** Determination of copper in milk by graphite furnace atomic absorption spectrophotometry, *Milchwissenschaft,* 39(12), 725-7, 1984.

79. **Carelli, G., Bergamaschi, A., and Altavista, M. C.,** Use of stabilized temperature platform furnace in determination of copper in human red blood cells, *At. Spectrosc.,* 5, 46, 1984.

80. **Bahreyni-Toosi, M. H., Dawson, J. B., Chilvers, D. C., and Ellis, D. J.,** An improved electrothermal atomic absorption technique for the determination of copper and zinc in plasma protein fractions, *Proc. 2nd Int. Workshop Trace Elem. Anal. Chem. Med. Biol.,* Braetter, P. and Schramel, P., Eds., Berlin, 1983, 811-18.

81. **Legret, M., Demare, D., Marchandise, P., and Robbe, D.,** Interferences of major elements in the determination of lead, copper, cadmium, chromium and nickel in river sediments and sewer sludges by electrothermal atomic absorption spectroscopy, *Anal. Chim. Acta,* 149, 107, 1983.

82. **Castledine, C. G. and Robbins, J. C.,** Practical field-portable atomic absorption analyzer, *J. Geochem. Explor.,* 19, 689, 1983.

83. **Nakashima, R., Kamata, E., and Shibata, S.,** Determination of cobalt, copper, iron, nickel and lead in bone by graphite furnace atomic absorption spectroscopy, *Kogai,* 18, 37-44, 1983.

84. **Campbell, D. E. and Comperat, M.,** Ultratrace analyses of high purity glasses for double-crucible low-attention optical waveguides, *Glastech. Ber.,* 56, 898, 1983.

85. **Khavezov, I. and Ivanova, E.,** Electrothermal atomization using the L'vov platform. Determination of traces of lead, cadmium, manganese and copper in a sodium chloride matrix, *Fresnius' Z. Anal. Chem.,* 315, 34-7, 1983.

86. **Stevens, B. J.,** Determination of aluminum, copper and zinc in human hair, *At. Spectrosc.,* 4(5), 176-8, 1983.

87. **Kawano, Y.,** Determination of plasma zinc, magnesium and copper by atomic absorption spectrometry with a flame or flameless system, *Kyushu Yakugakkai Kaiho,* 37, 43-52, 1983.

88. **Subramanian, K. S. and Meranger, J. C.,** Blood levels of cadmium, copper, lead and zinc in children in a British Columbia community, *Sci. Total Environ.,* 30, 231-44, 1983.

89. **Bourcier, D. R., Sharma, R. P., Bracken, W. M., and Taylor, M. J.,** Cadmium-copper interaction in intestinal mucosal cell cetosol of mice, *Biol. Trace Elem. Res.,* 5, 195-204, 1983.

90. **Hinks, L. J., Colmese, M., and Delves, H. T.,** Measurement of zinc and copper in leukocytes, *Proc. 2nd Int. Workshop Trace Elem. Anal. Chem. Med. Biol.,* Braetter, P. and Schramel, P., Eds., Berlin, 1982, 885-92.

91. **Postel, W., Meier, B., and Markert, R.,** Lead, cadmium, aluminum, tin, zinc, iron and copper in bottled and canned beer, *Moatsschr. Brauwiss,* 36, 360-7, 1983.

92. **Grobenski, Z., Lehmann, R., Radzuik, B., and Voelkopf, U.,** Determination of trace metals in seawater using Zeeman graphite furnace AAS, *At. Spectrosc.,* 5, 87, 1984.

93. **Backstrom, K., Danielsson, L. G., and Nord, L.,** Sample work-up for graphite furnace atomic absorption spectrometry using continuous flow extraction, *Analyst,* 109, 323, 1984.

94. **Kanipayor, R., Naranjit, D. A., Radziuk, B. H., Van Loon, J. C., and Thomassen, Y.,** Direct analysis of solids for trace elements by combined electrothermal furnace/quartz T-tube/flame atomic absorption spectrometry, *Anal. Chim. Acta,* 166, 39, 1984.

95. **Kerven, G. L., Edwards, D. G., and Asher, C. J.,** Determination of native ionic copper concentrations and copper complexation in peat soil extracts, *Soil Sci.,* 137, 91, 1984.

96. **Freedman, J. H. and Peisach, J.,** Determination of copper in biological materials by atomic absorption spectroscopy: reevaluation of extinction coefficients for azurin and stellacyanin, *Anal. Biochem.,* 1412, 301, 1984.

97. **Bahreyni-Toosi, M. H., Dawson, J. B., Ellis, D. J., and Duffield, R. J.,** Examination of instrumental systems for reducing cycle time in atomic absorption spectroscopy with electrothermal atomization, *Analyst,* 109, 1607, 1984.

98. **Lewis, S. A., O'Haver, T. C., and Harnly, J. M.,** Simultaneous multielement analysis of microliter quantities of serum for copper, iron and zinc by graphite furnace atomic absorption spectroscopy, *Anal. Chem.,* 56, 1651, 1984.

99. **Chung, C.,** Atomization mechanism with arrhenius plots taking the dissipation function into account in graphite furnace atomic absorption spectrometry, *Anal. Chem.,* 56, 2714, 1984.

100. **Boniforti, R., Ferraroli, R., Frigieri, P., Hettai, D., and Queirazza, G.,** Intercomparison of five methods for determination of trace metals in seawater, *Anal. Chim. Acta,* 162, 33, 1984.

101. **Mazzucotelli, A., Collecchi, P., Esposito, M., and Frache, R.,** Interactions among metals in determination of trace amounts of copper in environmental samples by electrothermal atomic absorption spectrophotometry, *Ann. Chim.,* 74, 289, 1984.

102. **Hulanicki, A. and Bulska, E.,** Effect of organic solvents on atomization of nickel, copper and lead in graphite furnaces in atomic absorption spectrometry, *Can. J. Spectrosc.,* 29, 148, 1984.

103. **Carrondo, M. J. T., Reboredo, F., Ganho, R. M. B., and Oliiveira, J. F. S.,** Analysis of sediments for heavy metals by rapid electrothermal atomic absorption procedure, *Talanta,* 31, 561, 1984.

104. **El-Yazigi, A., Al-Saleh, I., and Al-Mefty, O.,** Concentrations of Ag, Al, Au, Bi, Cd, Cu, Pb, Sb and Se in cerebrospinal fluid of patients with cerebral neoplasma, *Clin. Chem.,* 30, 1358, 1984.

105. **Halls, D. J.,** Speeding up determinations by electrothermal atomic absorption spectrometry, *Analyst,* 109, 1081, 1984.

106. **Phelan, V. J. and Powell, R. J. W.,** Combined reagent purification and sample dissolution (corpad) applied to trace analysis of silicon, silica and quartz, *Analyst,* 109, 1269, 1984.

107. Release of metals (lead, cadmium, copper, zinc, nickel and chromium) from kitchen blenders, Statens Levnedsmiddelinstitut Report, PUB-88, Order No. PB84-193416, 26 pp (in Danish), 1984. Avail. NTIS. From Gov. Rep. Announce. Index (U.S.), 84, 67, 1984.

108. **Blakemore, D. J. and Stanier, P.,** Methods for measurement of trace elements in equine blood by electrothermal atomic absorption spectrophotometry, *At. Spectrosc.,* 5, 215, 1984.

109. **Sperling, K. R.,** Tube-in-tube technique in electrothermal atomic absorption spectrometry. III. Atomization behavior of Zn, Cd, Pb, Cu and Ni, *Spectrochim. Acta,* 39B, 371, 1984.

110. **Littlejohn, D., Cook, S., Durie, D., and Ottaway, J. M.,** Investigation of working conditions for graphite probe atomization in electrothermal atomic absorption spectrometry, *Spectrochim. Acta,* 39B, 295, 1984.

111. **Danielsson, L. G. and Westerlund, S.,** Short term variations in trace metal concentrations in Baltic, *Mar. Chem.,* 15, 273, 1984.

112. **Voinovitch, I. A., Druon, M., and Louvrier, J.,** Effect of absence of injection orifice and length of graphite furnaces on sensitivity of determinations by atomic absorption spectrometry with electrothermal atomization, *Analysis,* 12, 256, 1984 (in French).

113. **Carelli, G., Bergamaschi, A., and Altavista, M. C.,** Use of the stabilized temperature platform furnace in the determination of copper in human red blood cells, *At. Spectrosc.,* 5, 46, 1984.

114. **Feinberg, M. and Wirth, P.,** General introduction to optimization in analytical chemistry, *Analysis,* 12, 490, 1984.

115. **Laumond, F., Copin-Montegut, G., Courau, P., and Nicholas, E.,** Cadmium, copper and lead in Western Mediterranean sea, *Mar. Chem.,* 15, 251, 1984.

116. **Ding, Y., Yao, M., Che, K., and Yang, W.,** Content of Zn, Ca, Mg, Cu and Mn in the hair of Huangshi (China) inhabitants, *Yingyang Xuebao,* 6(3), 298, 1984.

117. **Mittelman, M. W. and Geesey, G. G.,** Copper-binding characteristics of exopolymers from a freshwater-sediment bacterium, *Appl. Environ. Microbiol.,* 49(4), 846, 1985.

118. **Liska, S. K., Kerkay, J., and Pearson, K. H.,** Determination of copper in whole blood, plasma and serum using zeeman effect atomic absorption spectroscopy, *Clin. Chim. Acta,* 150, 11, 1985.

119. **Subramanian, K. S., Meranger, J. C., Wan, C. C., and Corsini, A.,** Preconcentration of cadmium, chromium, copper and lead in drinking water on polyacrylic ester resin XAD-7, *Int. J. Environ. Anal. Chem.,* 19, 261, 1985.

120. **Miura, H.,** Measurement of magnesium and copper in normal human serum by flameless AAS, *Eisei Kensa,* 33(11), 1425-8, 1984, CA 21,103:101226d, 1985.

121. **Hutchinson, D. J., Disinski, F. J., and Nardelli, C. A.,** Determination of copper in infant formula by graphite furnace atomic absorption spectroscopy with a L'vov platform, *J. Assoc. Off. Anal. Chem.,* 69, 60, 1985.

122. **Wirth, P. L. and Linder, M. C.,** Distribution of copper among components of human serum, *J. Natl. Cancer Inst.,* 75(2), 277-84, 1985.

123. **Horvath, Z., Lasztity, A., Szakacs, O., and Bozsai, G.,** Iminodiacetic acid/ethylcellulose as chelating ion exchanger. I. Determination of trace metals by atomic absorption spectrometry and collection of uranium, *Anal. Chim. Acta,* 173, 273, 1985.

124. **Schindler, E.,** Determination of silver, aluminum, arsenic, cadmium, chromium, copper, iron, manganese, lead, selenium and zinc in drinking water by graphite tube AAS. I, *Dtsch. Lebensm. Rundsch.,* 81, 1, 1985.

125. **Lang, H., Gen, X., Zhu, M., and Bai, G.,** Determination of cadmium, lead and copper in natural water and industrial wastewater by ion-exchange-three-hole ring oven technique, *Xibei Daxue Xuebao Ziran Kexueban,* 14(4), 38-42, 1984, CA 23, 103:146866d, 1985.

126. **Jin, L. and Wu, D.,** Determination of lead, copper, cadmium chromium in river sediments and coal fly ash by graphite FAAS, *Huanjing Huaxue,* 2(5), 13-19, 1983. CA 22, 103:134062a, 1985.

127. **Khammas, Z. A., Marshall, J., Littlejohn, D., Ottaway, J. M., and Stephen, S. C.,** Determination of copper in milk powder by electrothermal atomic absorption and atomic emission spectrometry, *Mikrochim. Acta,* 1, 333, 1985.

128. **Erb, R.,** Quality control by Zeeman-AAS in a chemical factory. Direct determination of copper, manganese, iron, cadmium and lead in adhesive tapes and diverse raw materials, *Fresnius' Z. Anal. Chem.,* 322(7), 719, 1985.

129. **Kozak, E., Hudnik, V., and Gomiscek, S.,** Determination of copper and gold in human tissues by electrothermal AAS, *Vestn. Slov. Kem. Drus,* 32(3), 249-62, 1985.

130. **Suzuki, M. and Isobe, K.,** Mechanism of interference elimination by thiourea in electrothermal atomic absorption spectrometry, *Anal. Chim. Acta,* 173, 321, 1985.

131. **Kleijburg, M. R. and Pijpers, F. W.,** Calibration graphs in atomic-absorption spectrophotometry, *Analyst,* 110, 147, 1985.

132. **Sneddon, J.,** Use of a impaction-electrothermal atomization atomic absorption spectrometric system for the direct determination of cadmium, copper and manganese in the laboratory atmosphere, *Anal. Lett.,* 18(A10), 1261-80, 1985.

133. **Honjo, T. and Makino, T.,** The preconcentration and separation of a trace amount of copper (II) in water as its silica-immobilized salicylaldimine chelate by means of column chromatography, *Bull. Chem. Soc. Jpn.,* 59, 3273, 1986.

134. **Kuroda, R., Takekawa, F., and Oguma, K.,** Determination of copper in standard silicate rocks by graphite furnace AAS after lithium carbonate-boric acid fusion, *Bunseki Kagaku,* 35, T86, 1986 (in Japanese).

135. **Game, I., Balabanoff, L., Valdebenito, R., and Vivaldi, L.,** Use of matrix modifier and L'vov platform in the determination of copper in pooled human saliva by electrothermal atomic absorption spectroscopy, *Analyst (London),* 111, 1139-41, 1986.

136. **Nishioka, H., Maeda, Y., and Azumi, T.,** Environmental analysis. XXII. Graphite furnace atomic absorption spectroscopy of copper in seawater by using copper concentration method, *Nippon Kaishi Gakkaishi,* 40, 100, 1986 (in Japanese).

137. **Manning, D. C. and Slavin. W.,** Test of copper wavelengths for Zeeman furnaces, *At. Spectrosc.,* 7, 179, 1986.

138. **Black, S. S., Riddle, M. R., and Holcombe, J. A.,** A Monte Carlo simulation for graphite furnace atomization of copper, *Appl. Spectrosc.,* 40, 925, 1986.

139. **Darlington, S. T., Gower, A. M., and Ebdon, L.,** The measurement of copper in individual aquatic insect larvas, *Environ. Technol. Lett.,* 7(3), 141-6, 1986.

140. **Hutchinson, D. J., Disinski, F. J., and Nardelli, C. A.,** Determination of copper in infants formula by graphite FAAS with a L'vov platform, *J. Assoc. Off. Anal. Chem.,* 69(1), 60-4, 1986.

141. **Ostapozuk, P., Valenta, P., and Nuernberg, H. W.,** Square wave voltammetry. A rapid and reliable determination method for zinc, cadmium, lead, copper, nickel and cobalt in biological and environmental samples, *J. Electroanal. Chem. Interfacial Electrochem.,* 214, 51-64, 1986.

142. **Feustel, A., Wennrich, R., and Dittrich, M.,** Studies of cadmium, zinc and copper levels in human kidney tumors and normal kidney, *Urol. Res.,* 14(2), 105-8, 1986.

143. **Komarek, J., Kacirova, A., and Sommer, L.,** A contribution to the determination of lead and copper using electrothermal AAS, *Scr. Fac. Sci. Nat. Univ. Purkynianas Brun.,* 15(6), 329-34, 1985, CA 3, 104:28124h, 1986.

144. **Frech, W., Lindberg, A. O., Lundberg, E., and Cedergren, A.,** Atomization mechanisms and gas phase reactions in graphite furnace atomic absorption spectrometry, *Fresnius' Z. Anal. Chem.,* 323(7), 716-25, 1986.

145. **Schmidt. K. P. and Falk, H.,** Direct determination of silver, copper and nickel in solid materials by graphite furnace atomic absorption spectroscopy using a specially designed graphite tube, *Spectrochim. Acta,* 42B, 431, 1987.

146. **Halls, D. J., Mohl, C., and Stoeppler, M.,** Application of rapid furnace programs in atomic absorption spectroscopy to the determination of lead, chromium and copper in digests of plant materials, *Analyst (London),* 112, 185, 1987.

147. **Gonzalez, M. C., Rodriguez, A. R., and Gonzalez, V.,** Determination of vanadium, nickel, iron, copper and lead in petroleum fractions by atomic absorption spectrophotometry with a graphite furnace, *Microchem. J.,* 35, 94-106, 1987.

148. **Smith, A. J. and Collins, C.,** Determination of trace copper in naphtha at levels below 50μg/L, *At. Spectrosc.,* 8, 96, 1987.

149. **Ebdon, L. and Evans, E. H.,** Determination of copper in biological microsamples by direct solid sampling graphite furnace atomic absorption spectroscopy, *J. Anal. At. Spectrom.,* 2, 317, 1987.

150. **Ueda, J. and Yamaguchi, N.,** Determination of copper by electrothermal atomization atomic absorption spectroscopy following coprecipitation with hafnium hydroxide, *Analyst,* 112, 283, 1987.

AUTHOR INDEX

Altavista, M. C., 79, 113
Anderson, C. T., 7
Arafat, N. M., 36
Arnold, D., 61
Asher, C. J., 95
Azumi, T., 136
Backstrom, K., 93
Baez, C. M., 50
Bahreyni-Toosi, M. H., 80, 97
Bai, G., 125
Baker, A. A., 34
Balabanoff, L., 135
Bangia, T. R., 52
Barnett, W. B., 2
Battistoni, P., 33
Becher, G., 72
Bergamaschi, A., 79, 113
Beyer, W., 60
Bito, L. Z., 70
Black, S. S., 138
Blakemore, D. J., 108
Boniforti, R., 100
Borggaard, O. K., 78
Bourcier, D. R., 64, 89
Bouzanne, M., 41
Bozsai, G., 123
Bracken, W. M., 64, 89
Braetter, P., 56
Brown, D. H., 19, 23
Brune, D., 76
Bruni, P., 33
Bulska, E., 102
Campbell, D. E., 84
Campe, A., 46
Cardellini, L., 33
Carelli, G., 79, 113
Carrondo, M. J. T., 103
Castledine, C. G., 82
Cedergren, A., 144
Chakrabarti, C. L., 26
Che, K., 116
Chilvers, D. C., 80
Christensen, H. E. M., 78
Chung, C., 99
Churella, D. J., 21
Claeys, A., 46
Collecchi, P., 101
Collins, C., 148
Colmese, M., 90
Comperat, M., 84
Cook, S., 110
Copeland, T. R., 21
Copin-Montegut, G., 115
Corsini, A., 119
Courau, P., 115
Danielsson, L. G., 93, 111
Darlington, S. T., 139
Dawson, J. B., 80, 97
Del-Castiho, P., 13

Delves, H. T., 90
Demare, D., 81
Dhawale, B. A., 52
Ding, Y., 116
Disinski, F. J., 121, 140
Dittrich, M., 142
Downes, T. E. H., 5
Druon, M., 112
Duffield, R. J., 97
Dunbar, P. M., 43
Durie, D., 110
Ebdon, L., 139, 149
Edwards, D. G., 95
El-Yazigi, A., 104
Ellis, D. J., 80, 97
Equel, J. C., 66
Erb, R., 128
Esposito, M., 101
Evans, E. H., 149
Evenson, M. A., 7, 10
Fagioli, F., 51
Falk, H., 145
Farmer, J. G., 37
Fava, G., 33
Favier, A., 75
Feinberg, M., 114
Fell, G. S., 43
Ferraroli, R., 100
Feustel, A., 142
Frache, R., 101
Frech, W., 144
Freedman, J. H., 96
Frigieri, P., 100
Fujino, O., 18
Fujita, K., 44
Game, I., 135
Ganho, R. M. B., 103
Gardiner, P. E., 56
Geesey, G. G., 117
Gen, X., 125
Gibson, M. J., 37
Gjerdet, N., 76
Glenn, M., 1
Glenn, T., 1
Glooschenko, W. A., 36
Gobbi, G., 33
Gomiscek, S., 129
Gonzalez, M. C., 147
Gonzalez, O. C., 50
Gonzalez, V., 147
Gower, A. M., 139
Grinshtein, I. L., 22, 55
Grobenski, Z., 92
Gupta, N., 25
Halls, D. J., 43, 105, 146
Harnly, J. M., 98
Hart, L., 1
Hayashi, Y., 14, 74
Headridge, J. B., 34

SUBJECT INDEX

GALLIUM

GALLIUM, Ga   ATOMIC WEIGHT 69.72

Instrumental Parameters:

Wavelength	287.4 nm
Slit width	80 μm
Bandpass	0.3 nm
Light source	Hollow Cathode Lamp
Lamp current	5 mA
Purge gas	Argon
Sample size	25 μL
Furnace	Cylindrical cuvette/Pyrolytic graphite

Standard Operating Conditions:

Optimum char temperature	1000°C
Optimum atomization temperature	2700°C
Sensitivity	5 pg/1% absorption
Sensitivity check	0.1-1.0μg/mL
Working range	0.15-1.0 μg/mL
Background correction	Required only to correct light scattering or non-specific absorption from the sample containing high dissolved solids.

General Note:
1.   Use commercial standard or a previously analyzed sample as a working standard.
2.   Use ultra clean glass and plastic ware, soaked in 1:5 nitric acid solution and rinsed thoroughly with deionized distilled water.
3.   All analytical solutions should be prepared in 1% (v/v) nitric acid and hydrogen peroxide (1 mL of 30% hydrogen peroxide in 100 mL water).
4.   Check for blank values on all reagents including water.
5.   Dilution is recommended when sample exhibits greater than 0.5 absorbance units.

REFERENCES

1. **Langmyhr, F. J. and Rasmussen, S.,** Atomic absorption spectrometric determination of gallium and indium in inorganic materials by direct atomization from the solid state in a graphite furnace, *Anal. Chim. Acta,* 72, 79, 1974.
2. **Dittrich, K.,** Atomic spectroscopic trace analysis in $A^{III}B^V$ semiconductor microsamples. II. Determination of trace gallium arsenide and phosphide by atomic absorption spectroscopy with electrothermal atomization, *Talanta,* 24, 735, 1977.
3. **Sukhoveeva, L. N., Spivakov, B. Y., Karyakin, A. V., and Zolotov, Y. A.,** Atomic absorption determination of gallium in flame and graphite furnaces, *Zh. Anal. Khim.,* 34, 693, 1979.
4. **Tsunoda, K., Haragachi, H., and Fuwa, K.,** Studies on the occurrence of atoms and molecules of aluminum, gallium, indium and their monohalides in an electrothermal carbon furnace, *Spectrochim. Acta,* 35B, 715, 1980.
5. **Katskov, D. A., Grinshtein, I. L., and Kruglikova, L. P.,** Study of the evaporation of the metals In, Ga, Tl, Ge, Sn, Pb, Sb, Bi, Se and Te from a graphite surface by the atomic absorption method, *Zh. Prikl. Spektrosk.,* 33, 804, 1980.
6. **Sukhoveeva, L. N., Butrimenko, G. G., and Spivakov, B. Y.,** Atomic absorption determination of gallium and indium in a graphite furnace by evaporation from a substrate, *Zh. Anal. Khim.,* 35, 649, 1980.
7. **Kuga, K.,** Determination of gallium by graphite furnace atomic absorption spectroscopy using a zirconium impregnated graphite tube, *Bunseki Kagaku,* 30, 529, 1981.
8. **Nakamura, K., Fujimori, M., Tsuchiya, H., and Orii, H.,** Determination of gallium in biological materials by electrothermal atomic absorption spectroscopy, *Anal. Chim. Acta,* 138, 129-36, 1982.
9. **Wu, J., Luo, Z., and Yung, B.,** Atomic absorption determination of gallium in bauxite by using a flameless atomizer, *Yankuang Ceshi,* 1, 73, 1982 (in Chinese).
10. **Han, H. and Ni, Z.,** Atomic absorption spectrophotometric determination of gallium in environmental samples using graphite furnace graphite platform technique, *Fenxi Huaxue,* 11(8), 571, 1983.
11. **Kuga, K., Ooyu, S., Kitazume, E., and Tsujii, K.,** Determination of gallium depth profiles in semiconductor silicon by chemical etching and graphite furnace atomic absorption spectrometry, *Bunseki Kagaku,* 33, 29E, 1984.
12. **Botha, P. V. and Fazakas, J.,** Some observations on atomic absorption spectrometry of gallium with electrothermal atomizers, *Anal. Chim. Acta,* 162, 413, 1984.
13. **Collery, P., Millart, H., Simoneau, J. P., Pluot, M., Halpern, S., Pechery, C., Choisy, H., and Etienne, J. C.,** Experimental treatment of mammary carcinomas by gallium chloride after oral administration: intratumor concentrations of gallium, anatomopathologic study and intracellular microanalysis, *Trace Elem. Med.,* 1(4), 159, 1984.
14. **Quin, Z. and Zhou, J.,** Determination of gallium in rocks by flameless atomic absorption spectroscopy using matrix correction, *Yanshi Kuangwu Ji Ceshi,* 3(4), 367, 1984.
15. **Shan, X., Yuan, Z., and Ni, Z.,** Determination of gallium in sediment, coal, coal fly ash and botanical samples by graphite surface atomic absorption spectrometry using nickel matrix modification, *Anal. Chem.,* 57, 857, 1985.
16. **Anderson, J., Van der Watt, T. N., and Strelow, F. W. E.,** Determination of gallium at the ppb level in South African primary and secondary reference samples by ion exchange chromatography and electrothermal atomic absorption spectrometry, *Geostand. Newsl.,* 9, 17, 1985.
17. **Quin, Z., Wu, Z., Yang, S., and Li, Q.,** Determination of traces of gallium, indium and thallium in geological samples by flameless AAS using vanadium as matrix modifier, *Yanshi Kuangwu Ji Ceshi,* 4(2), 160, 1985.
18. **Kobayashi, T., Hirose, F., Hasegawa, S., and Okuchi, H.,** Determination of trace tellurium and gallium in nickel-base heat-resisting alloys by GFAAS, *Nippon Kinzoku Gakkaishi,* 49(8), 656, 1985.
19. **Barron, D. C. and Haynes, B. W.,** Determination of gallium in phosphorus flue dust and other materials by graphite furnace AAS, *Analyst (London),* 111(1), 19, 1986.

AUTHOR INDEX

SUBJECT INDEX

GERMANIUM

GERMANIUM, Ge   (ATOMIC WEIGHT 72.59)

Instrumental Parameters:

Wavelength	265.1 nm
Slit width	320 μm
Bandpass	1 nm
Light source	Hollow Cathode Lamp
Lamp current	5 mA
Purge gas	Argon
Sample size	25 μL
Furnace	Cylindrical cuvette/Pyrolytic graphite

Standard Operating Conditions:

Optimum char temperature	1000°C
Optimum atomization temperature	2300°C
Sensitivity	40 pg/1% absorption
Sensitivity check	0.2-2.0 μg/mL
Working range	0.1-4.0 μg/mL
Background correction	Required only to correct light scattering or non-specific absorption from the sample containing high dissolved solids.

General Note:
1. Use commercial standard or a previously analyzed sample as a working standard.
2. Use ultra clean glass and plastic ware, soaked in 1:5 nitric acid solution and rinsed thoroughly with deionized distilled water.
3. Prepare all analytical solutions in 1% (v/v) nitric acid.
4. Check for blank values on all reagents including water.
5. Dilution is recommended when sample exhibits greater than 0.5 absorbance units.
6. Since germanium is not very stable in very dilute solutions, use solutions within 4 h of preparation.

REFERENCES

1. **Johnson, D. J., West, T. S., and Dagnall, R. M.,** Determination of germanium by atomic absorption spectroscopy with a graphite tube atomizer, *Anal. Chim. Acta,* 67, 79, 1973.
2. **Mino, Y., Shimonura, S., and Ota, N.,** Determination of germanium in different media by atomic absorption spectroscopy with electrothermal atomization, *Anal. Chim. Acta,* 107, 395, 1979.
3. **Katskov, D. A., Grinshtein, I. L., and Kruglikova, L. P.,** Study of the evaporation of the metals In, Ga, Tl, Ge, Sn, Pb, Sb, Bi, Se and Te from a graphite surface by the atomic absorption method, *Zh. Prikl. Spektrosk.,* 33, 804, 1980.
4. **Studnicki, M.,** Determination of germanium, vanadium and titanium by carbon furnace atomic absorption spectroscopy, *Anal. Chem.,* 52, 1762, 1980.
5. **Mino, Y., Ota, N., Sakao, S., and Shimonura, S.,** Determination of germanium in medicinal plants by atomic absorption spectroscopy with electrothermal atomization, *Chem. Pharm. Bull.,* 28, 2687, 1980.
6. **Andreae, M. O. and Froelich, P. N.,** Determination of germanium in natural waters by graphite furnace atomic absorption spectroscopy with hydride generation, *Anal. Chem.,* 53, 287, 1981.
7. **Gao, Y. and Ni, Z.,** Determination of germanium by graphite furnace atomic absorption spectroscopy using a zirconium- coated tube, *Huaxue Xuebao,* 40, 1021, 1982 (in Chinese).
8. **Gao, Y. and Ni, Z.,** Determination of germanium by atomic absorption spectroscopy using the graphite furnace platform technique, *Xiyou Jinshu,* 1, 57-60, 1982 (in Chinese).
9. **Inui, T., Terada, S., Tamura, H., and Ichinose, N.,** Determination of germanium by hydride generation with reducing tube followed by graphite furnace atomic absorption spectroscopy using methane/argon as sweeper gas, *Fresnius' Z. Anal. Chem.,* 315, 598-601, 1983.
10. **Hambrick, G. A., III, Froelich, P. N., Jr., Andreae, M. O., and Lewis, B. L.,** Determination of methylgermanium species in natural waters by graphite furnace atomic absorption spectrometry with hydride generation, *Anal. Chem.,* 56, 421, 1984.
11. **Pelieva, L. A. and Martynenko, K. P.,** Atomic absorption determination of germanium in graphite furnace, *Zh. Prikl. Spektrosk.,* 40, 33, 1984.
12. **Wendl, W. and Mueller-Vogt, G.,** Chemical reactions in the graphite tube for some carbide and oxide forming elements, *Spectrochim. Acta,* 39B, 237, 1984.
13. **Carnrick, G. R. and Barnett, W. B.,** Determination of germanium by zeeman atomic absorption spectrophotometry, *At. Spectrosc.,* 5, 213, 1984.
14. **Pelieva, L. A. and Martynenko, K. P.,** Atomic absorption determination of germanium by using a graphite furnace, *Zh. Prikl. Spektrosk.,* 40(1), 33-7, 1984.
15. **Hambrick, G. A., III, Froelich, P. N., Jr., Andrease, M. O., Lewis, M. O., and Lewis, B. L.,** Determination of methylgermanium species in natural waters by graphite furnace atomic absorption spectrometry with hydride generation, *Anal. Chem.,* 56(3), 421-4, 1984.
16. **Criand, A. and Fouillac, C.,** Use of L'vov platform and molybdenum coating for determination of volatile elements in thermomineral waters by atomic absorption spectrometry, *Anal. Chim. Acta,* 167, 257, 1985.
17. **Dittrich, K., Mandry, R., Mothes, W., and Judelevich, J. G.,** Investigation of effects of matrix modification on atomization of germanium in atomic absorption spectrometry with electrothermal atomization and its application to determination of germanium in $A^{III}B^{V}$ semiconductor microsamples, *Analyst,* 110, 169, 1985.
18. **Papadoyannis, I. N., Matis, K. A., and Zoumboulis, A. I.,** Extraction and flameless AAS determination of germanium in lignite and fly ash, *Anal. Lett.,* 18(A 19), 2467, 1985.
19. **Sohrin, Y., Isshiki, K., Kuwamoto, T., and Nakayama, E.,** Determination of germanium by graphite furnace atomic absorption spectroscopy, *Talanta,* 341, 341, 1987.

AUTHOR INDEX

SUBJECT INDEX

GOLD

GOLD, Au   (ATOMIC WEIGHT 196.967)

Instrumental Parameters:

Wavelength	253.7 nm
Slit width	320 μm
Bandpass	1 nm
Light source	Hollow Cathode Lamp
Lamp current	5 mA
Purge gas	Argon or Nitrogen
Sample size	25 μL
Furnace	Cylindrical cuvette/Pyrolytic graphite

Standard Operating Conditions:

Optimum char temperature	600°C
Optimum atomization temperature	1800°C
Sensitivity	5 pg/1% absorption
Sensitivity check	0.05-0.1μg/mL
Working range	0.04-0.1 μg/mL
Background correction	Required only to correct light scattering or non-specific absorption from the sample containing high dissolved solids.

General Note:
1. Use commercial standard or a previously analyzed sample as a working standard.
2. Use ultra clean glass and plastic ware, soaked in 1:5 nitric acid solution and rinsed thoroughly with deionized distilled water.
3. Prepare all analytical solutions in 1% (v/v) nitric acid or aqua regia.
4. Check for blank values on all reagents including water.
5. Dilution is recommended when sample exhibits greater than 0.5 absorbance units.
6. Do not use solutions containing gold stored in plastic containers for a long time. Gold can be absorbed into some plastic surfaces.
7. The analytical precision of this method is 6% RSD.

REFERENCES

1. **Kerber, J. D.,** The direct determination of gold in polyester fibers with the HGA-70 graphite furnace, *At. Ab. Newsl.,* 10, 104, 1971.
2. **Martin, L.,** Determination of gold in minerals by atomic absorption spectrophotometry using the Massman graphite furnace, *An. Quim.,* 72, 217, 1976 (in Spanish).
3. **Bea-Barredo, F., Polo-Polo, C., and Polo Diaz, C.,** The simultaneous determination of gold, silver and cadmium at ppb levels in silicate rocks by atomic absorption spectroscopy with electrothermal atomization, *Anal. Chim. Acta,* 94, 783, 1977.
4. **Torgov, V. G. and Khlebnikova, A. A.,** Atomic absorption determination of gold in a flame and a flameless graphite atomizer with prior solvent extraction with petroleum sulfides, *Zh. Anal. Khim.,* 32, 960, 1977.
5. **Kamel, H., Brown, D. H., Ottaway, J. M., and Smith, W. E.,** Determination of gold in separate protein fractions of blood serum by carbon furnace atomic absorption spectroscopy, *Analyst,* 102, 645, 1977.
6. **Fishkova, N. L. and Vilenkin, V. A.,** Application of an HGA-74 graphite atomizer to atomic absorption determination of gold, silver, platinum and palladium in solutions of complicated composition, *Zh. Anal. Khim.,* 33, 897, 1978 (in Russian).
7. **Katskov, D. A. and Grinshtein, I. L.,** Study of the chemical interaction of copper, gold and silver with carbon by an atomic absorption method using electrothermal atomizer, *Zh. Prikl. Spektrosk.,* 30, 787, 1979.
8. **Korda, T. M., Zelentsova, L. V., and Yudelevich, G. I.,** Solvent selection for organic sulfides for extraction — atomic absorption determination of gold and palladium with electrothermal atomization, *Zh. Anal. Khim.,* 36, 86, 1981 (in Russian).
9. **Zhang, Z. and Lin, J.,** Determination of trace gold in geological samples by graphite furnace atomic absorption spectroscopy, *Fenxi Huaxue,* 9, 703-5, 1981 (in Chinese).
10. **Sharma, R. P.,** A microanalytical method for the analysis of gold in biological media by flameless atomic absorption spectroscopy, *Ther. Drug Monit.,* 4, 219-24, 1982.
11. **Rodgers, A. I. A., Brown, D. H., Smith, W. E., Lewis, D., and Capell, H. A.,** Distribution of gold in blood following administration of myocrisin and auranofin, *Anal. Proc. (London),* 19, 87-8, 1982.
12. **Huang, M.,** Application of Zephiramine in the analysis of noble metals. I. Determination of microamounts of gold, platinum and palladium in ores by graphite -furnace -atomic-absorption spectroscopy after preconcentration with Zephiramine, *Fenxi Huaxue,* 10, 661, 1982 (in Chinese).
13. **Robert, R. V. D. and Ormrod, G. T. W.,** The development of an on-line gold analyzer, *Rep. MINTEK* M50, 26, 1982.
14. **Fazakas, J.,** Determination of gold by graphite furnace atomic absorption spectroscopy with atomization under pressure, *Rev. Roum. Chim.,* 27, 685, 1982.
15. **McHugh, J. B.,** Determination of gold in water in the nanogram range by electrothermal atomization after coprecipitation with tellurium, *At. Spectrosc.,* 4, 66, 1983.
16. **Aleksenko, A. N., Guletskii, N. N., and Korennoi, E. P.,** Flameless atomic-absorption analysis of solid rock samples for gold, *Probl. Sovrem. Anal. Khim.,* 4, 101-4, 1983 (in Russian).
17. **Lang, J., Wawschinek, O., Rainer, F., and Lanzer, G.,** Determination of gold in tissues and blood-cells by a combination of cold ashing and flameless atomic absorption in a graphite furnace, *Mikrochim. Acta,* 3, 53-60, 1983 (in German).
18. **Koshima, H. and Ohnishi, H.,** Adsorption of silver, gold and platinum from aqueous solutions by carbonaceous materials, *Bunseki Kagaku,* 32, E149, 1983.
19. **Vasilikiotis, G. S., Papadoyannis, I. N., and Kouuimtizs, T. A.,** Analytical applications of crown ethers. IV. Extraction and determination of gold, *Microchim. J.,* 29, 356, 1984.
20. **El-Azizi, A., Al-Saleh, I., and Al-Mefty, O.,** Concentrations of Ag, Al, Au, Bi, Cd, Cu, Pb, Sb and Se in cerebrospinal fluid of patients with cerebral neoplasma, *Clin. Chem.,* 30, 1358, 1984.
21. **Koide, M., Lee, D. S., and Stallard, M. W.,** Concentration and separation of trace metals from seawater using a single anion exchange bead, *Anal. Chem.,* 56, 1956, 1984.
22. **Shimizu, H., Onoue, T., and Murayama, T.,** Determination of ultratrace gold in ores by graphite furnace zeeman atomic absorption spectrometry, *Bunseki Kagaku,* 33(12), T123-T126, 1984 (in Japanese).
23. **Sighinolfi, G. P., Gorgoni, C., and Mohamed, A. H.,** Comprehensive analysis of precious metals in some geological standards by flameless atomic absorption spectroscopy, *Geostand. Newsl.,* 8, 25-9, 1984.
24. **Pchelintseva, N. F.,** Extraction-atomic absorption determination of gold in rocks, *Zh. Anal. Khim.,* 39, 462, 1984.
25. **McHugh, J. B.,** Gold in natural water: A method of determination by solvent extraction and electrothermal atomization, *J. Geochem. Explor.,* 20, 303, 1984.
26. **Hu, J. and Lin, Y.,** Graphite furnace atomic absorption determination of trace gold in rocks, *Yanshi Kuangwu Ji Ceshi,* 3(2), 158-61, 1984.
27. **Matsuno, K., Iwao, S., Kodama, Y., and Kaizu, K.,** Gold metabolism in rats, *Nippon Eiseigaku Zasshi,* 39(2), 535-40, 1984 (in Japanese).

185

28. **Zhang, Z., Gan, S., and Ying, P.,** Solvent extraction-graphite furnace atomic absorption determination of trace gold in geological samples, *Fenxi Huaxue,* 12(6), 528-30, 1984.
29. **Hahn, R. and Ikramuddin, M.,** New method for determination of gold in natural waters by electrothermal atomic absorption spectrophotometry, *At. Spectrosc.,* 6(3), 77, 1985.
30. **Aleksenko, A. N., Turkin, Y. I., Guletskii, N. N., and Korennoi, E. P.,** Flameless atomic absorption analyzer for determining gold in rocks, *Zavod. Lab.,* 51(6), 30, 1985.
31. **Yukhin, Y. M., Udalova, T. A., and Tsimbalist, V. G.,** Flameless atomic absorption determination of noble metals after liquid-liquid extraction by mixture of bis(2 ethylhexyl) dithiophosphate and p-octylaniline, *Zh. Anal. Khim.,* 40, 850, 1985 (in Russian).
32. **Amosse, J., Fischer, W. Allibert, M., and Piboule, M.,** Determination of ultratraces of Pt, Pd, Rh and Au in silicate rocks by electrothermal atomic absorption spectrometry, *Analysis,* 14, 26, 1986 (in French).
33. **Li, L., Sun, Y., Han, Z., Gao, M., Li, H., Zhang, S., and Qu, R.,** Flameless atomic absorption spectrometric determination of trace amounts of gold in ores after separation by extraction chromatographhy with PTSO (di-p-tolyl-sulfoxides), *Lihua Jianyan Huaxue Fence,* 22, 81, 1986.
34. **Bazhov, A. S., Zekki, L. N., and Klimenkova, N. B.,** Atomic absorption determination of low gold contents by using a pulsed tungsten atomizer, *Zh. Anal. Khim.,* 41, 2190, 1986.
35. **Myasoedova, G. V., Antokol'skaya, I. I., Kubrakova, I. V., Belova, E. V., Mezhirova, M. S., Varshal, G. M., Girshina, O. N., Zhukova, N. G., Danilova, F. I., and Savvin, S. B.,** Preconcentration of platinum group metals and gold by sorption on polyorgs XI-H and their atomic absorption determination in sorbent suspension, *Zh. Anal. Khim.,* 41, 1816, 1986.
36. **Hall, G. E. M., Vaive, J. E., and Ballantyne, S. B.,** Field and laboratory procedures for determining gold in natural waters. Relative merits of preconcentration with activated charcoal, *J. Geochem. Explor.,* 26, 191, 1986.
37. **Hara, S., Matsuo, H., and Kumamaru, T.,** Simultaneous determination of gold (III) and platinum (IV) by graphite furnace atomic absorption spectroscopy after ion-pair extraction with Zephiramine, *Bunseki Kagaku,* 35(6), 503-7, 1986.
38. **Bratjer, K. and Slonawska, K.,** Determination of gold in the presence of platinum and palladium by electrothermal atomization atomic absorption spectroscopy, *J. Anal. At. Spectrom.,* 2, 167, 1987.
39. **Shan, X., Egila, J., Littlejohn, D., and Ottaway, J. M.,** Direct determination of gold in whole blood and plasma by electrothermal atomization atomic-absorption spectroscopy using Zeeman-effect background correction and matrix modification, *J. Anal. At. Spectrom.,* 2, 299, 1987.
40. **Stafilov, T. and Todorovski, T.,** Determination of gold in arsenic-antimony ore by flameless atomic absorption spectroscopy, *At. Spectrosc.,* 8, 12, 1987.
41. **Egila, J., Littlejohn, D., Ottaway, J. M., and Shan, X.,** Comparison of interferences and matrix modifiers in the determination of gold by electrothermal atomization atomic absorption spectroscopy with Zeeman-effect background correction, *J. Anal. At. Spectrom.,* 2, 293, 1987.

AUTHOR INDEX

SUBJECT INDEX

INDIUM

INDIUM, In   (ATOMIC WEIGHT 114.82)

Instrumental Parameters:

Wavelength	303.9 nm
Slit width	160 μm
Bandpass	0.5 nm
Light source	Hollow Cathode Lamp
Lamp current	5 mA
Purge gas	Argon
Sample size	25 μL
Furnace	Cylindrical cuvette/Pyrolytic graphite

Standard Operating Conditions:

Optimum char temperature	800°C
Optimum atomization temperature	2000°C
Sensitivity	100 pg/1% absorption
Sensitivity check	0.25-0.5 μg/mL
Working range	0.5-1.0 μg/mL
Background correction	Required only to correct light scattering or non-specific absorption from the sample containing high dissolved solids.

General Note:
1. Use commercial standard or a previously analyzed sample as a working standard.
2. Use ultra clean glass and plastic ware, soaked in 1:5 nitric acid solution and rinsed thoroughly with deionized distilled water.
3. Prepare all analytical solutions in 0.25-0.5% (v/v) nitric acid
4. Check for blank values on all reagents including water.
5. Dilution is recommended when sample exhibits greater than 0.5 absorbance units.
6. If higher pyrolysis temperatures are required, prepare all analytical solutions in 1% (w/v) ammonium fluoride (NH_4F).

REFERENCES

1. **Langmyhr, F. J. and Rasmussen, S.,** Atomic absorption spectrometric determination of gallium and indium in inorganic materials by direct atomization from the solid state in a graphite furnace, *Anal. Chim. Acta,* 72, 79, 1974.
2. **Dittrich, K.,** Molecular absorption spectroscopy by electrothermal volatilization with graphite furnace. I. Principle of the method and studies of the molecular absorption of gallium and indium halides, *Anal. Chim. Acta,* 97, 59, 1978.
3. **Spivakov, B. Y., Sukhoveeva, L. N., Dittrich, K., Karyakin, A. V., and Zolotov, Y. A.,** Atomic absorption determination of indium in extracts and aqueous solutions using a graphite furnace and flame, *Zh. Anal. Khim.,* 34, 1947, 1979.
4. **Tsunoda, K., Haraguchi, H., and Fuwa, K.,** Studies on the occurrence of atoms and molecules of aluminum, gallium, indium and their monohalides in an electrothermal carbon furnace, *Spectrochim, Acta,* 35B, 715, 1980.
5. **Sukhoveeva, L. N., Butrimenko, G. G., and Spivakov, B. Y.,** Atomic absorption determination of gallium and indium in a graphite furnace by evaporation from a substrate, *Zh. Anal. Khim.,* 35, 649, 1980.
6. **Katskov, D. A., Grinshtein, I. L., and Kruglikova, L. P.,** Study of the evaporation of the metals In, Ga, Tl, Ge, Sn, Pb, Sb, Bi, Se and Te from a graphite surface by the atomic absorption method, *Zh. Prikl. Spektrosk.,* 33, 804, 1980.
7. **Busheiva, I. S. and Headridge, J. B.,** Determination of cadmium, indium and zinc in nickel-base alloys by atomic absorption spectroscopy with introduction of solid samples into furnaces, *Anal. Chim. Acta,* 142, 197, 1982.
8. **Brune, D., Gjerdet, N., and Paulsen, G.,** Gastrointestinal and *in vitro* release of copper, cadmium, indium, mercury and zinc from conventional and copper-rich amalgams, *Scand. J. Dent. Res.,* 91, 66, 1983.
9. **Botha, P. V. and Fazakas, J.,** Use of nonresonance lines for determination of lead, indium and thallium by electrothermal atomic absorption spectrometry, *Spectrochim. Acta,* 39B, 379, 1984.
10. **Zhou, L., Chao, T. T., and Meier, A. L.,** Determination of indium in geological materials by electrothermal atomization atomic absorption spectrometry with tungsten-impregnated graphite furnace, *Anal. Chim. Acta,* 161, 369, 1984.
11. **Qin, Z., Wu, Z., Yang, S., and Li, Q.,** Determination of traces of gallium, indium and thallium in geological samples by flameless AAS using vanadium as matrix modifier, *Yanshi Kuangwu Ji Ceshi,* 4(2), 160-3, 1985.
12. **Samchuk, A. I.,** Behavior of extracts of thallium, indium, beryllium and molybdenum chelates in a graphite furnace during AA analysis, *Ukr. Khim. Zh.,* 51(3), 287-91, 1985.
13. **Kobayoshi, T., Yokota, F., and Ohnishi, Y.,** Determination of indium in metal by graphite furnace atomic absorption spectroscopy, *Aichi-ken Kogyo Gijutsu Senta Hokoku,* 21, 79, 1985 (in Japanese).
14. **Shan, X., Ni, Z., and Yuan, Z.,** Determination of indium in minerals, river sediments and coal fly ash by electrothermal atomic absorption spectrometry with palladium as matrix modifier, *Anal. Chim. Acta,* 171, 269, 1985.
15. **Brajter, K. and Olbrych-Sleszynska, E.,** Application of electrothermal atomic absorption spectroscopy to the determination of trace amounts of indium in metallic zinc and lead, *Analyst (London),* 111(9), 1023-7, 1986.
16. **Donard, O. F. X., Rapsomanikis, S., and Weber, J. H.,** Speciation of inorganic tin and alkyltin compounds by AAS using electrothermal quartz furnace after hydride generation, *Anal. Chem.,* 58(4), 772-7, 1986.

AUTHOR INDEX

SUBJECT INDEX

IRON

IRON, Fe   (ATOMIC WEIGHT 55.84)

Instrumental Parameters:

Wavelength	248.3 nm
Slit width	80 μm
Bandpass	0.3 nm
Light source	Hollow Cathode Lamp
Lamp current	10 mA
Purge gas	Argon or Nitrogen
Sample size	25 μL
Furnace	Cylindrical cuvette/Pyrolytic graphite

Standard Operating Conditions:

Optimum char temperature	1000°C
Optimum atomization temperature	2400°C
Sensitivity	3 pg/1% absorption
Sensitivity check	0.02-0.04 μg/mL
Working range	0.003-0.03 μg/mL
Background correction	Required only to correct light scattering or non-specific absorption from the sample containing high dissolved solids.

General Note:

1. Use commercial standard or a previously analyzed sample as a working standard.
2. Use ultra clean glass and plastic ware, soaked in 1:5 nitric acid solution and rinsed thoroughly with deionized distilled water.
3. Prepare all analytical solutions in 0.5% (v/v) nitric acid
4. Check for blank values on all reagents including water.
5. Dilution is recommended when sample exhibits greater than 0.5 absorbance units.
6. For increased sensitivity, use step atomization.

REFERENCES

1. **Kragten, J. and Reynaert, A. P.,** Application of the graphite furnace to the determination of trace iron in gold and silver, *Talanta,* 21, 618, 1974.
2. **La Brecque, J. J., Mendelovici, E., Villallba, R. E., and Bellorin, C. C.,** The determination of total iron in Venezuelan Laterites: The investigation of interferences of aluminum and silicon on the determination of iron in the fluoboric-boric acid matrix by atomic absorption, *Appl. Spectrosc.,* 32(1), 57, 1978.
3. **Allen, P. D., Hampson, N. A., Moore, D. C. A., and Willars, M. J.,** The determination of iron in aqueous perchlorate solutions by atomic absorption spectroscopy with electrothermal atomization, *Anal. Chim. Acta,* 101, 401, 1978.
4. **Sturgeon, R. E., Berman S. S., Desaulniers, A., and Russell, D. S.,** Determination of iron, manganese and zinc in seawater by graphite furnace atomic absorption spectroscopy, *Anal. Chem.,* 51, 2364, 1979.
5. **Sturgeon, R. E., Berman, S. S., Desaulniers, A., and Russell, D. S.,** Determination of iron, manganese and zinc by graphite furnace atomic absorption spectroscopy. Reply to comments, *Anal. Chem.,* 52, 1767, 1980.
6. **Montgomery, J. R. and Peterson, G. N.,** Effects of ammonium nitrate on sensitivity for determination of copper, iron and manganese in seawater by atomic absorption spectroscopy with pyrolytically coated graphite tubes, *Anal. Chim. Acta,* 117, 397, 1980.
7. **Chakrabarti, C. L., Wan, C. C., and Li, W. C.,** Atomic absorption spectrometric determination of cadmium, lead, zinc, copper, cobalt and iron in oyster tissue by direct atomization from the solid state using the graphite furnace platform technique, *Spectrochim. Acta,* 35B, 547, 1980.
8. **Kuga, K.,** Rapid determination of copper, iron and manganese in poly imide resins by atomic absorption spectroscopy using a graphite furnace atomizer, *Bunseki Kagaku,* 29, 342, 1980.
9. **Segar, D. A. and Cantillo, A. Y.,** Determination of iron, manganese and zinc in sea water by graphite furnace atomic absorption spectroscopy. Comments, *Anal. Chem.,* 52, 1766, 1980.
10. **Zolotovitskaya, E. S. and Fidel'man, B. M.,** Use of electrothermal atomic absorption spectroscopy for determining iron, nickel and chromium in raw materials and single crystals of corundum, *Zh. Anal. Khim.,* 36, 1564, 1981.
11. **Garbett, K., Goosdfellow, G. I., and Marshall, G. B.,** Application of atomic absorption techniques in the analysis of metallic sodium. II. Determination of iron, nickel and chromium in sodium salt solutions by electrothermal atomization, *Anal. Chim. Acta,* 126, 147, 1981.
12. **Tanaka, T., Hayashi, Y., Funakawa, K., and Ishizawa, M.,** Simultaneous determination of iron and manganese in human hair by graphite furnace-two channel atomic absorption spectrophotometry, *Nippon Kagaku Kaishi,* 1, 169, 1981.
13. **Arafat, N. M. and Glooschenko, W. A.,** Method for the simultaneous determination of arsenic, aluminum, iron, zinc, chromium and copper in plant tissue without the use of perchloric acid, *Analyst (London),* 106, 1174-8, 1981.
14. **Gardiner, P. E., Roesick, E., Roesick, U., Braetter, P., and Kynast, G.,** Application of gel filtration, immunonephelometry and electrothermal atomic absorption spectroscopy to the study of the distribution of copper, iron and zinc bound constituents in human amniotic fluid, *Clin. Chim. Acta,* 120, 103, 1982.
15. **Beninschek, F. and Huber, I.,** Electrothermal atomic absorption spectrometric determination of chromium, iron and nickel in lithium metal, *Anal. Chim. Acta,* 140, 205-12, 1982.
16. **Lieser, K. H., Sondermeyer, S., and Kliemchen, A.,** Precision and accuracy of analytical results in determining the elements cadmium, chromium, copper, iron, manganese and zinc by flameless AAS, *Fresnius' Z. Anal. Chem.,* 312, 517, 1982 (in German).
17. **Arnold, D. and Kuennecke, A.,** Determination of 3rd elements in high lead-containing glasses, *Silikattechnik,* 33, 312, 1982.
18. **Gretzinger, K., Kotz, L., Tschoepel, P., and Toelg, G.,** Causes and elimination of systematic errors in the determination of iron and cobalt in aqueous solutions in the ng/mL and pg/mL range, *Talanta,* 29, 1011, 1982.
19. **Bourcier, D. R., Sharma, R. P., Bracken, W. M., and Taylor, M. J.,** Cadmium-copper interactions: effect of copper pretreatment and cadmium-copper chromic exposure on the distribution and accumulation of cadmium, copper, zinc and iron in mice, *Trace Subst. Environ. Health,* 16, 273, 1982.
20. **Favier, A., Maljournal, B., Decoix, G., and Ruffieux, D.,** Microassay of serum iron by atomic absorption spectrophotometry in a graphite furnace: the improvement and evaluation of this technique, *Ann. Biol. Clin. (Paris),* 41, 45, 1983.
22. **Stewart, D. A. and Newton, D. C.,** Determination of iron in semiconductor-grade silicon by furnace atomic absorption spectrometry, *Analyst (London),* 108(1293), 1450-8, 1983.
23. **Yu, Z., Zhang, Z., and Liu, S.,** Chemical concentration-flameless atomic absorption spectrometric determination of trace iron in high-purity boric acid, *Huaxue Shiji,* 5(5), 315-19, 1983.
24. **Burakov, V. S., Verenik, V. N., Malashonok, V. A., Nechaev, S. V., and Puko, R. A.,** Atomic absorption determination of trace iron and chromium in aqueous solutions by using intracavity spectroscopy, *Zh. Anal. Khim.,* 38, 90-3, 1983 (in Russian).

25. **Hayashi, Y., Yabuta, Y., Tanaka, T., and Nose, T.,** Determination of copper, iron and zinc in cutaneous leg lymph of rabbits by atomic absorption spectrophotometry, *Bunseki Kagaku,* 32, 212, 1983.

26. **Nakashima, R., Kamata, E., and Shibata, S.,** Determination of cobalt, copper, iron, nickel and lead in bone by graphite furnace atomic absorption spectroscopy, *Kogai,* 18, 37-44, 1983.

27. **Favier, A. and Ruffieux, D.,** Physiological variations of serum levels of copper, zinc, iron and manganese, *Biomed. Pharmacother.,* 37, 462, 1983.

28. **Rossi, G., Omenetto, N., Pigozzi, G., Vivian, R., Mattonez, U., Mousty, F., and Crabi, G.,** Analysis of radioactive waste solutions by atomic absorption spectrometry with electrothermal atomization, *At. Spectrosc.,* 4, 113, 1983.

29. **Postel, W., Meier, B., and Markert, R.,** Lead, cadmium, aluminum, tin, zinc, iron and copper in bottled and canned beer, *Monatsschr. Brauwiss,* 36, 360-7, 1983 (in German).

30. **Campbell, D. E. and Comperat, M.,** Ultratrace analyses of high purity glasses for double-crucible low-attention optical waveguides, *Glastech. Ber.,* 56, 898, 1983.

31. **Stewart, D. A. and Newton, D. C.,** Determination of iron in semiconductor-grade silicon by furnace atomic absorption spectrometry, *Analyst,* 108, 1450, 1983.

32. **Zhong, Y., Wang, C., Liang, S., Hu, Y., and Zhang, M.,** Ultramicro determination of serum iron and total iron binding capacity by electrothermal atomic absorption spectrophotometry, *Gaodeng Xuexiao Huaxue Xuebao,* 5, 643, 1984 (in Chinese).

33. **Chung, C.,** Atomization mechanism with arrhenius plots taking the dissipation function into account in graphite furnace atomic absorption spectrometry, *Anal. Chem.,* 56, 2714, 1984.

34. **Boniforti, R., Ferraroli, R., Frigieri, P., Heltai, D., and Queirazza, G.,** Intercomparison of five methods for determination of trace metals in seawater, *Anal. Chim. Acta,* 162, 33, 1984.

35. **Phelan, V. J. and Powell, R. J. W.,** Combined reagent purification and sample dissolution (corpad) applied to trace analysis of silicon, silica and quartz, *Analyst,* 109, 1269, 1984.

36. **Murnane, M. M.,** Analysis of brine by atomic absorption with graphite furnace using direct sample injection, *Chem. N. Z.,* 48, 39, 1984.

37. **Lewis, S. A., O'Haver, T. C., and Harnly, J. M.,** Simultaneous multielement analysis of microliter quantities of serum for copper, iron and zinc by graphite furnace atomic absorption spectrometry, *Anal. Chem.,* 56, 1651, 1984.

38. **Backstrom, K., Danielsson, L. G., and Nord, L.,** Sample work-up for graphite furnace atomic absorption spectrometry using continuous flow extraction, *Analyst,* 109, 323, 1984.

39. **Matsusaki, K. and Yoshino, T.,** Electrothermal atomic absorption spectrometric determination of traces of chromium, nickel, iron and beryllium in aluminum and its alloys without preliminary separation, *Anal. Chim. Acta,* 157, 193, 1984.

40. **Slavin, W. and Carnrick, G. R.,** Possibility of standardless furnace atomic absorption spectroscopy, *Spectrochim. Acta,* 39B, 271, 1984.

41. **Brossier, P. and Moise, C.,** Determination of iron in ferrocene and some derivatives by flameless atomic absorption spectrometry, *Analysis,* 12(4), 223-4, 1984.

42. **Kreeftenberg, H. G., Koopman, B. J., Huizenga, J. R., Van Vilsteren, T., Wolthers, B. G., and Gips, C. H.,** Measurement of iron in liver biopsies — a comparison of three analytical methods, *Clin. Chim. Acta,* 144(2-3), 255-62, 1984.

43. **Danielsson, L. G. and Westerlund, S.,** Short term variations in trace metal concentrations in Baltic, *Mar. Chem.,* 15, 273, 1984.

44. **Kleijburg, M. R. and Pijpers, F. W.,** Calibration graphs in atomic absorption spectrophotometry, *Analyst,* 110, 147, 1985.

45. **Erb, R.,** Quality control by Zeeman -AAS in a chemical factory. Direct determination of copper, manganese, iron, cadmium and lead in adhesive tapes and diverse raw materials, *Fresnius' Z. Anal. Chem.,* 322(7), 719, 1985.

46. **Montgomery, J. R., Hucks, M., and Peterson, G. N.,** A portable noncontaminating sampling system for iron and manganese in sediment pore water, *Fla. Sci.,* 48(1), 46-9, 1985.

47. **Ito, M., Sato, M., Shibata, N., and Suzuki, S.,** Application of zirconium hydroxide coprecipitation for the determination of iron in the standard reference material mussel, *Annu. Rep. Tohoku Coll. Pharm.,* 32, 187-90, 1985 (in Japanese).

48. **He, J. and Tang, Y.,** Direct determination of trace iron in phosphate by polarized Zeeman AAS, *Gaodeng Xuexiao Huaxue Xuebao,* 6(8), 689, 1985.

49. **McGahan, M. C. and Fleisher, L. N.,** A micro method for the determination of iron and total iron-binding capacity in intraocular fluids and plasma using electrothermal atomic absorption spectroscopy, *Anal. Biochem.,* 156, 397, 1986.

50. **Nakamura, S. and Kubota, M.,** Comparison of furnace materials for electrothermal atomic absorption spectrometry on the basis of chemical thermodynamic calculation, *Bunseki Kagaku,* 35(6), 548, 1986.

51. **Clark, J. R.,** Electrothermal atomization atomic absorption conditions and matrix modifications for determining Sb, As, Bi, Cd, Ga, Au, Pb, Mo, Pd, Pt, Se, Ag, Te, Tl and Sn following back-extraction of organic aminohalide extracts, *J. Anal. At. Spectrom.*, 1, 301-8, 1986.
52. **El-Yazigi, A., Al-Saleh, J., and Al-Mefty, G.,** Concentrations of zinc, iron, molybdenum, arsenic and lithium in cerebrospinal fluid of patients with brain tumors, *Clin. Chem.*, 32, 2187, 1986.
53. **Gonzalez, M., Rodriguez, A. R., and Gonzalez, V.,** Determination of vanadium, nickel, iron, copper and lead in petroleum fractions by atomic absorption spectrophotometry with a graphite furnace, *Microchem. J.*, 35, 94-106, 1987.
54. **Lugowski, S., Smith, D. C., and Van Loon, J. C.,** The determination of aluminum, chromium, cobalt, iron and nickel in whole blood by electrothermal atomic absorption spectrophotometry, *J. Biomed. Mater. Res.*, 21, 657, 1987.

AUTHOR INDEX

SUBJECT INDEX

LEAD

LEAD, Pb   (ATOMIC WEIGHT 207.19)

Instrumental Parameters:

Wavelength	283.3 nm
Slit width	160 μm
Bandpass	0.5 nm
Light source	Hollow Cathode Lamp
Lamp current	5 mA
Purge gas	Argon or Nitrogen
Sample size	25 μL
Furnace	Cylindrical cuvette/Pyrolytic graphite

Standard Operating Conditions:

Optimum char temperature	500°C
Optimum atomization temperature	2000°C
Sensitivity	5 pg/1% absorption
Sensitivity check	0.1 μg/mL
Working range	0.005-0.05 μg/mL
Background correction	Required only to correct light scattering or non-specific absorption from the sample containing high dissolved solids.

General Note:

1. Use commercial standard or a previously analyzed sample as a working standard.
2. Use ultra clean glass and plastic ware, soaked in 1:5 nitric acid solution and rinsed thoroughly with deionized distilled water.
3. Prepare all analytical solutions in 0.5% (v/v) nitric acid.
4. Check for blank values on all reagents including water.
5. Dilution is recommended when sample exhibits greater than 0.5 absorbance units.
6. Prepare dilute solutions (1:25) every 4 h, when needed.
7. For increased sensitivity, use step atomization.
8. The analytical precision for this method is approximately 5% RSD.

REFERENCES

1. **Kubasik, N. P., Volosin, M. T., and Murray, M. H.,** Carbon rod atomizer applied to measurement of lead in whole blood by atomic absorption spectrophotometry, *Clin. Chem.,* 18, 410, 1972.
2. **Norval, E. and Butler, L. R. P.,** Determination of lead by atomic absorption with the high temperature graphite tube, *Anal. Chim. Acta,* 58, 47, 1972.
3. **Kubasik, N. P. and Volosin, M. T.,** Simplified determination of urinary cadmium, lead and thallium with use of carbon rod atomization and atomic absorption spectrophotometry, *Clin. Chem.,* 19, 954, 1973.
4. **Chauvin, J. V., Newton, M. P., and Davis, D. G.,** The determination of lead and nickel by atomic absorption spectroscopy with a flameless wire loop atomizer, *Anal. Chim. Acta,* 65, 291, 1973.
5. **Shigematsu, T., Matsui, M., Fujino, O., and Kinoshita, K.,** Determination of lead by atomic absorption spectroscopy with a carbon tube atomizer, *Nippon Kagaku Kaishi,* 2123, 1973.
6. **Shaw, F. and Ottaway, J. M.,** Determination of trace amounts of lead in high purity copper and copper alloys by atomic absorption spectroscopy with graphite furnace atomization, *At. Absorp. Newsl.,* 13, 77, 1974.
7. **Jensen, F. O., Dolezal, J., and Langmyhr, F. J.,** Atomic absorption spectrometric determination of cadmium, lead and zinc in salts or salt solutions by hanging mercury drop electrodeposition and atomization in a graphite furnace, *Anal. Chim. Acta,* 72, 245, 1974.
8. **Lech, J. F., Siemer, D., and Woodriff, R.,** Determination of lead in atmospheric particulates by furnace atomic absorption, *Environ. Sci. Technol.,* 8, 890, 1974.
9. **Brady, D. V., Montalvo, J. G., Jr., Jung, J., and Curran, R. A.,** Direct determination of lead in plant leaves via graphite furnace atomic absorption, *At. Ab. Newsl.,* 13, 118, 1974.
10. **Shaw, F. and Ottaway, J. M.,** Determination of trace amounts of lead in steel and cast iron by atomic absorption spectroscopy with the use of carbon furnace atomization, *Analyst,* 99, 184, 1974.
11. **Evenson, M. A. and Pendergast, D. D.,** Rapid ultramicro direct determination of erythrocyte lead concentrations by atomic absorption spectrophotometry with use of a graphite tube furnace, *Clin. Chem.,* 20, 163, 1974.
12. **Vondenhoff, T.,** Determination of lead, cadmium, copper and zinc in plant and animal material by atomic absorption in a flame and in a graphite tube after sample decomposition by the Schoniger technique, *Mitteilungsbl. GDch (Gas. Dtsh. Chem.) Fachgr. Lebensmittelchem. Gerichtl. Chem.,* 29, 341, 1975.
13. **Nakahara, T. and Musha, S.,** Chemical interference effects in the atomic absorption spectrometric determination of lead with premixed inert gas (entrained air)-hydrogen flames, *Appl. Spectrosc.,* 29, 352, 1975. *Mittielungs GDCh. Fachgr. Lebensmittelchem. Gerichtl. Chem.,* 29, 341, 1975 (in German).
14. **Campbell, W. C. and Ottaway, J. M.,** Determination of lead in carbonate rocks by carbon furnace atomic absorption spectroscopy after dissolution in nitric acid, *Talanta,* 22, 72a, 1975.
15. **Barlow, P. J. and Khera, A. K.,** Sample preparation using tissue solubilization by solvene-350TM for lead determinations by graphite furnace atomic absorption spectrophotometry, *At. Ab. Newsl.,* 14, 149, 1975.
16. **Fernandez, F. J.,** Micromethod for lead determination in whole blood by atomic absorption with use of the graphite furnace, *Clin. Chem.,* 21, 558, 1975.
17. **Kilroe-Smith, T. A.,** Linearization of calibration curves with the HGA-72 flameless cuvette for the determination of lead in blood, *Anal. Chim. Acta,* 82, 421, 1976.
18. **Sabet, S., Ottaway, J. M., and Fell, G. S.,** Comparison of the Delves cup and carbon furnace atomization used in atomic absorption spectroscopy for the determination of lead in blood, *Proc. Anal. Div. Chem. Soc.,* 14, 300, 1977.
19. **Robinson, J. W., Kiesel, E. L., Goodbread, J. P., Bliss, R., and Marshall, R.,** The development of a gas chromatography-furnace atomic absorption combination for the determination of organic lead compounds. Atomization processes in furnace atomizers, *Anal. Chim. Acta,* 92, 321, 1977.
20. **McLaren, J. W. and Wheeler, R. C.,** Double peaks in the atomic absorption determination of lead using electrothermal atomization, *Analyst,* 102, 542, 1977.
21. **Thompson, K. C., Wagstaff, K., and Wheatstone, K. C.,** Method for the minimization of matrix interferences in the determination of lead and cadmium in non-saline waters by atomic absorption spectrophotometry using electrothermal atomization, *Analyst,* 102, 310, 1977.
22. **Kujirai, O., Kobayashi, T., and Sudo, E.,** Determination of trace quantities of lead and bismuth in heat-resisting alloys by atomic absorption spectroscopy with heated graphite atomizer, *Trans. Jpn. Inst. Met.,* 18, 775, 1977.
23. **Brodie, K. G. and Stevens, B. J.,** Measurement of whole blood lead and cadmium at low levels using an automatic sample dispenser and furnace atomic absorption, *J. Anal. Toxicol.,* 1, 282, 1977.
24. **Wegscheider, W., Knapp, G., and Spitzy, H.,** Statistical investigations of interferences in graphite furnace atomic absorption spectroscopy. III. Lead, *Z. Anal. Chem.,* 283, 183, 1977.
25. **Fuller, C. W.,** The effect of graphite tube condition on the determination of lead in the presence of magnesium chloride by electrothermal atomic absorption spectroscopy, *At. Ab. Newsl.,* 16, 106, 1977.
26. **Del-Castiho, P. and Herber, R. F.,** The rapid determination of cadmium, lead, copper and zinc in whole blood by atomic absorption spectroscopy with electrothermal atomization. Improvements in precision with a peak-shape monitoring device, *Anal. Chim. Acta,* 94, 269, 1977.

27. **Murata, T., Minegishi, K., and Sudoh, G.,** Atomic absorption spectrometric determination of trace amounts of Cu, Mn, Pb and Cr in cements by direct atomization in a carbon furnace, *Yogyo Kyokai Shi,* 86, 532, 1978.

28. **Stoeppler, M., Brandt, K., and Rains, T. C.,** Contributions to automated trace analysis. II. Rapid method for the automated determination of lead in whole blood by electrothermal atomic absorption spectrophotometry, *Analyst,* 103, 714, 1978.

29. **Manning, D. C. and Slavin, W.,** Determination of lead in a chloride matrix with the graphite furnace, *Anal. Chem.,* 50, 1234, 1978.

30. **Bye, R. and Paus, P. E.,** Atomic absorption spectroscopy used as a specific gas chromatography detector. Comparison of flame and graphite furnace techniques in the determination of tetraalkyllead compounds, *At. Ab. Newsl.,* 17, 131, 1978.

31. **Andrews, D. G., Aziz-Alrahman, A. M., and Headridge, J. G.,** Determination of lead in irons and steels by atomic absorption spectrophotometry with the introduction of solid samples into an induction furnace, *Analyst,* 103, 909, 1978.

32. **Siemer, D. D. and Wei, H.,** Determination of lead in rocks and glasses by temperature controlled graphite cup atomic absorption spectroscopy, *Anal. Chem.,* 50, 147, 1978.

33. **Pickford, C. J. and Rossi, G.,** Determination of lead in atmospheric particulates using an automated atomic absorption spectrophotometric system with electrothermal atomization, *Analyst,* 103, 341, 1978.

34. **Ohta, K. and Suzuki, M.,** Determination of lead in waters by atomic absorption spectroscopy with electrothermal atomization, *Z. Anal. Chem.,* 298, 140, 1979.

35. **Aruscavage, P. J. and Campbell, E. Y.,** The determination of lead in 13 USGS rocks, *Talanta,* 26, 1052, 1979.

36. **Baeckman, S. and Karlsson, R. W.,** Determination of lead, bismuth, zinc, silver and antimony in steel and nickel-base alloys by atomic absorption spectrophotometry using direct atomization of solid samples in a graphite furnace, *Analyst,* 104, 1017, 1979.

37. **Tominaga, M. and Umezaki, Y.,** Suppression of interferences in the determination of lead and cadmium by graphite furnace atomic absorption spectroscopy, *Bunseki Kagaku,* 28, 347, 1979.

38. **Dabeka, R. W.,** Graphite furnace atomic absorption spectrometric determination of lead and cadmium in foods after solvent extraction and stripping, *Anal. Chem.,* 51, 902, 1979.

39. **Slavin, W. and Manning, D. C.,** Reduction of matrix interferences for lead determination with the L'vov platform and the graphite furnace, *Anal. Chem.,* 51, 261, 1979.

40. **Lundberg, E. and Frech, W.,** Direct determination of trace metals in solid samples by atomic absorption spectroscopy with electrothermal atomizers. I. Investigations of homogeneity for lead and antimony in metallurgical materials, *Anal. Chim. Acta,* 104, 67, 1979.

41. **Lundberg, E. and Frech, W.,** Direct determination of trace metals in solid samples by atomic absorption spectroscopy with electrothermal atomizers. II. Determination of lead in steels and nickel-base alloys, *Anal. Chim. Acta,* 104, 75, 1979.

42. **Bengtsson, M., Danielsson, L. G., and Magnusson, B.,** Determination of cadmium and lead in seawater after extraction using electrothermal atomization. Minimization of interferences from coextracted sea salts, *Anal. Lett.,* 12, 1367, 1979.

43. **Sedykh, E. M., Belyaev, Y. I., and Ozhegov, P. I.,** Study of the mechanism of atomization of lead compounds during atomic absorption determination of lead in a graphite furnace, *Zh. Anal. Khim.,* 34, 1984, 1979.

44. **Bailey, P., Norval, E., Kilroe-Smith, T. A., Skikne, M. I., and Rollin, H. B.,** The application of metal-coated graphite tubes to the determination of trace metals in biological materials. I. The determination of lead in blood using a tungsten-coated graphite tube, *Microchem. J.,* 24, 107, 1979.

45. **Hirao, Y., Fukumoto, K., Sugisaki, H., and Kimura, K.,** Determination of lead in seawater by furnace atomic absorption spectroscopy after concentrations with yield tracer, *Anal. Chem.,* 51, 651, 1979.

46. **Allain, P. and Mauras, Y.,** Microdetermination of lead and cadmium in blood and urine by graphite furnace atomic absorption, *Clin. Chim. Acta,* 91, 41, 1979.

47. **Vickrey, T. M., Harrison, G. V., and Ramelow, G. J.,** Zirconium surface treatment of graphite furnace tubes for the analysis of organo-lead compounds by atomic absorption spectroscopy, *At. Spectrosc.,* 1, 116, 1980.

48. **Tarui, T., Hasegawa, T., and Tokairin, H.,** Analysis of trace metals in crude oil. I. Determination of lead in petroleum by graphite furnace atomic absorption spectroscopy after thorium hydroxide coprecipitation, *Bunseki Kagaku,* 29, 423, 1980.

49. **Sefflova, A. and Komarek, J.,** Determination of lead by atomic absorption spectroscopy with electrothermal atomization, *Chem. Listy,* 74, 971, 1980.

50. **Halliday, M. C., Houghton, C., and Ottaway, J. M.,** Direct determination of lead in polluted sea water by carbon furnace atomic absorption spectroscopy, *Anal. Chim. Acta,* 119, 67, 1980.

51. **Shan, X. and Ni, Z.,** Matrix modification for the determination of lead in seawater using graphite furnace atomic absorption spectroscopy, *Environ. Sci.,* 1, 24, 1980 (in Chinese).

52. **Kruse, R.,** Multiple determination of lead and cadmium in fish by electrothermal AAS after wet-ashing in commercially available teflon beakers, *Z. Lebensm. Unters. Forsch.,* 171, 261, 1980 (in German).

53. **Mitcham, R. P.,** Determination of lead in drinking water by atomic absorption spectrophotometry using an electrically heated graphite furnace and an amm. tetramethlene dithiocarbamate extraction technique, *Analyst*, 105, 43, 1980.

54. **DeJonghe, W., Chakraborti, D., and Adams, F.,** Graphite furnace atomic absorption spectroscopy as a metal specific detection system for tetraalkyl lead compounds separated by gas-liquid chromatography, *Anal. Chim. Acta*, 115, 89, 1980.

55. **Tsushida, T. and Takeo, T.,** Determination of copper, lead and cadmium in tea by graphite furnace atomic absorption spectrophotometry, *Nippon Shokuhin Kogyo Gakkaishi*, 27, 585, 1980 (in Japanese).

56. **Sedykh, E. M., Belyaev, Y. I., and Sorokina, E. V.,** Elimination of matrix effects in electrothermal atomic absorption determination of Ag, Pb, Co, Ni, and Te in samples of complicated composition, *Zh. Anal. Khim.*, 35, 2348, 1980 (in Russian).

57. **Katskov, D. A., Grinshtein, I. L., and Kruglikova, L. P.,** Study of the evaporation of the metals In, Ga, Tl, Ge, Sn, Pb, Sb, Bi, Se and Te from a graphite surface by the atomic absorption method, *Zh. Prikl. Spektrosk.*, 33, 804, 1980.

58. **Glenc, T., Jurczyk, J., and Robosz-Kabza, A.,** Determination of tin and lead in transformer, carbon and alloy steels in the range from 0.0005 to 0.04% by atomic absorption method with electrothermal atomization, *Chem. Anal.*, 25, 513, 1980.

59. **Chakrabarti, C. L., Wan, C. C., and Li, W. C.,** Atomic absorption spectrometric determination of Cd, Pb, Zn, Cu, Co and Fe in oyster tissue by direct atomization from the solid state using the graphite furnace platform technique, *Spectrochim. Acta*, 35B, 547, 1980.

60. **Pederson, B., Willems, M., and Storgaard-Joergensen, S.,** Determination of Cu, Pb, Cd, Ni., and Co in EDTA extracts of soil by solvent extraction and graphite furnace atomic absorption spectrophotometry, *Analyst*, 105, 119, 1980.

61. **Legotte, P. A., Rosa, W. C., and Sutton, D. C.,** Determination of cadmium and lead in urine and other biological samples by graphite furnace atomic absorption spectroscopy, *Talanta*, 27, 39, 1980.

62. **Vickrey, T. M., Howell, H. E., Harrison, G. V., and Ramelow, G. J.,** Post column digestion methods for liquid chromatography-graphite furnace atomic absorption speciation of organo lead and organo tin compounds, *Anal. Chem.*, 52, 1743, 1980.

63. **Knutti, R.,** Matrix effects and matrix modifications in graphite furnace atomic absorption spectrophotometry as exemplified by determination of lead in urine, *Mitt. Geb. Lebensmittelunters. Hyg.*, 72, 183, 1981 (in German).

64. **Frech, W., Lundberg, E., and Barbooti, M. M.,** Direct determination of trace metals in solid samples by atomic absorption spectrometry with electrothermal atomizers. IV. Interference effects in the determination of lead and bismuth in steels, *Anal. Chim. Acta*, 131, 45, 1981.

65. **Chakraborti, D., Jiang, S. G., Surkijn, P., DeJonghe, W., and Adams, F.,** Determination of tetraalkyl lead compounds in environmental samples by gas chromatography-graphite furnace atomic absorption spectroscopy, *Anal. Proc. (London)*, 18, 347, 1981.

66. **Simon, J. and Liese, T.,** Trace determination of lead and cadmium in bones. Comparison of atomic absorption spectroscopy and inverse polarography, *Fresnius' Z. Anal. Chem.*, 309, 383, 1981 (in German).

67. **Paschal, D. C. and Bell, C. J.,** Improved accuracy in the determination of blood lead by electrothermal atomic absorption, *At. Spectrosc.*, 2, 146, 1981.

68. **Subramanian, K. S. and Meranger, J. C.,** Rapid electrothermal atomic absorption spectrophotometric method for cadmium and lead in human whole blood, *Clin. Chem.*, 27, 1866, 1981.

69. **Genc, O., Akman, S., Ozdural, A. R., Ates, S., and Balkis, T.,** Theoretical analysis of atom formation-time curves for the HGA-74 furnace. II. Evaluation of the atomization mechanisms for manganese, chromium and lead, *Spectrochim. Acta*, 36B, 663, 1981.

70. **Baker, A. A. and Headridge, J. B.,** Determination of Bi, Pb and Te in copper by atomic absorption spectroscopy with introduction of solid samples into an induction furnace, *Anal. Chim. Acta*, 125, 93, 1981

71. **Brodie, K. G. and Rowland, J. J.,** Trace analysis of arsenic, lead and tin, *Eur. Spectrosc. News*, 36, 41-44, 1981.

72. **Wittmers, L. W., Jr., Alich, A., and Aufderheide, A. C.,** Lead in bone. I. Direct analysis for lead in milligram quantities of bone ash by graphite furnace atomic absorption spectroscopy, *Am. J. Clin. Pathol.*, 75, 80, 1981.

73. **Behari, J. R.,** Determination of lead in blood, *Int. J. Environ. Anal. Chem.*, 10, 149, 1981.

74. **Subramanian, K. S. and Meranger, J. C.,** A rapid electrothermal atomic absorption spectrophotometric method for cadmium and lead in human whole blood, *Clin. Chem. (Winston-Salem, N.C.)*, 27, 1866-1871, 1981.

75. **Farmer, J. G. and Gibson, M. J.,** Direct determination of cadmium, copper and lead in siliceous standard reference materials from a fluoboric acid matrix by graphite-furnace atomic absorption spectroscopy, *At. Spectrosc.*, 2, 176-178, 1981.

76. **Ranchet, I., Monssier, F., Lamathe, J., and Voinovitch, I.,** Interlaboratory comparison of the determinations of cadmium, chromium, copper and lead by flameless atomic absorption spectrometry, *Bull. Liaison Lab. Ponts Chaussees*, 114, 81-86, 1981 (in French).

77. **Bouzanne, M.,** Determination of traces of inorganic pollutants (copper, lead, cadmium, zinc) in waters, by differential pulse anodic stripping voltammetry at a rotating mercury film electrode, *Analysis,* 9, 461-467, 1981 (in French).

78. **Jackson, K. W., Ebdon, L., Webb, D. C., and Cox, A. G.,** Determination of lead in vegetation by a rapid microsampling cup atomic absorption procedure with solid sample introduction, *Anal. Chim. Acta,* 128, 67, 1981.

79. **Torsi, G., Desimone, E., Palmisano, F., and Sabbatini, L.,** Determination of lead in seawater by electrothermal atomic absorption after electrolytic accumulation on a glassy carbon furnace, *Anal. Chim. Acta,* 124, 143, 1981.

80. **Sugisaki, H., Nakamura, H., Hirao, Y., and Kimura, K.,** Determination of lead in geostandard rocks by electrothermal atomic absorption spectroscopy after isolation of lead with yield monitoring, *Anal. Chim. Acta,* 125, 203, 1981.

81. **Hodges, D. J. and Skelding, D.,** Determination of lead in urine by atomic absorption spectroscopy with electrothermal atomization, *Analyst,* 106, 299, 1981.

82. **Bertenshaw, M. P., Gelsthorpe, D., and Wheatstone, K. C.,** Determination of lead in drinking water by atomic absorption spectrophotometry with electrothermal atomization, *Analyst,* 106, 23, 1981.

83. **Ikeda, S., Nagashima, C., and Ishikawa, T.,** A rapid determination of lead in blood by flameless atomic absorption spectrophotometer, *Tokyo-Toritsu Eisei Kenkyusho Kenkyyu Nempo,* 32-1, 227-231, 1981 (in Japanese).

84. **Kanikawa, M., Kojima, S., and Nakamura, A.,** Analysis of lead in paint scrapings and stationaries used by children, *Eisei Kagaku,* 27, 391-398, 1981 (in Japanese).

85. **Suddendorf, R. F., Gajan, R. J., and Boyer, K. W.,** Utilization of atomic spectroscopy for the determination of lead and cadmium in zinc salts used as food additives, *Dev. At. Plasma Spectrochem. Anal. Proc. Int. Winter Conf., 1980,* 1981, 706-12.

86. **Minoia, C., Caballeri, A., Colli, M., and Baruffini, A.,** Plasma lead determination by flameless atomic absorption spectroscopy, *Ann. Ist. Super Sanita,* 17, 641, 1981 (in Italian).

87. **Knutti, R.,** Matrix effects and matrix modification in graphite-furnace AAS as illustrated by direct determination of lead in urine, *Mitt. Geb. Lebensmittelunters. Hyg.,* 72, 183, 1981.

88. **Kodama, K., Tsuchiya, H., and Nakata, T.,** Iron inteference in blood lead determination, *Nagoyo-shi Eisei Kenkyusho Ho,* 28, 39, 1981 (in Japanese).

89. **Oradovskii, S. G. and Kuzimichev, A. N.,** Flameless extraction-atomic absorption method for determination of ultramicro amounts of toxic metals (lead, cadmium, copper, cobalt and nickel) in seawaters and ices in labile form, *Tr. Gos. Okeanogr. Inst.,* 162, 22, 1981.

90. **Chakraborti, D., Jiang, S. G., Surkijn, P., DeJonghe, W., and Adams, F.,** Determination of tetraalkyllead compounds in environmental samples by gas chromatography-graphite furnace atomic absorption spectroscopy, *Anal. Proc. (London),* 18, 347-50, 1981.

91. **Amakawa, E., Ohonishi, K., Seki, H., and Matsumuto, M.,** Determination of lead in soft drinks by flameless atomic absorption spectroscopy using a method of copper with zirconium hydroxide, *Tokyo-toritsu Eisei Kenkyusho Kenkyu Nempo,* 32-1, 199-201, 1981.

92. **Smith, B. M. and Griffiths, M. A.,** Determination of lead and antimony in urine by atomic absorption spectroscopy with electrothermal atomization, *Analyst (London),* 107, 253-9, 1982.

93. **Sakata, M. and Shimoda, O.,** Simple and rapid method for determination of lead and cadmium in sediment by graphite furnace atomic absorption spectroscopy, *Water Res.,* 16, 231-5, 1982.

94. **Camara Rica, C. and Kirkbright, G. F.,** Determination of trace concentrations of lead and nickel in human milk by electrothermal atomization atomic absorption spectrophotometry and inductively coupled plasma emission spectroscopy, *Sci. Total Environ.,* 22, 193, 1982.

95. **Suzuki, M., Ohta, K., and Katsuno, T.,** Determination of traces of lead and copper in foods by electrothermal atomic absorption spectroscopy with metal atomizer, *Mikrochim. Acta,* 2, 225, 1982.

96. **Erspamer, J. P. and Niemczyk, T. M.,** Effect of graphite surface type on determination of lead and nickel in a magnesium chloride matrix by furnace atomic absorption spectroscopy, *Anal. Chem.,* 54, 2150, 1982.

97. **Croce, A.,** Atomic absorption spectroscopy in the determination of lead, *Tinctoria,* 79, 381, 1982 (in Italian).

98. **Harms, U.,** Experiences with atomic absorption spectroscopy in the trace analysis of lead in biological matrixes, *Atomspektrom. Spurenanal., Vortr. Kolloq.,* Welz, B., Ed., Springer-Verlag, Weinheim, 1982, 91-96.

99. **Fleischer, M. and Schaller, K. H.,** Experiences in the routine use of the Triton X100 method for blood lead determination, *Atomspektrom. Spurenanal., Vortr. Kolloq.,* 47-55, 1981, Welz, B., Ed., Springer-Verlag, Weinheim, 1982, 47-55.

100. **Reeves, R., Kjellstroem, T., and Dallow, M.,** Analysis of lead in blood, paint, soil and house dust for the assessment of human lead exposure in Aukland (New Zealand), *N.Z. J. Sci.,* 25, 221, 1982.

101. **Clayes-Thoreau, F.,** Determination of low levels of cadmium and lead in biological fluids with simple dilution by atomic absorption spectrophotometry using Zeeman effect background correction and the L'vov platform, *At. Spectrosc.,* 3, 108, 1982.

102. **Ueta, T., Nishida, S., and Yoshimura, T.,** Hygienic chemical studies on dental treatment materials. II. Determination of arsenic, lead and cadmium in dental cements by flameless atomic-absorption spectrophotometry, *Tokyo-Toritsu Eisei Kenkyusho Kenkyu Nempo*, 33, 110, 1982 (in Japanese).

103. **Dumitru, R., Niculescu, T., and Botha, C.,** Determination of lead concentration in hair by electrothermal atomic absoroton spectrophotometry as a test of occupational exposure, *Rev. Ig. Bacterial. Virusol. Parazitol. Epidemiol. Pneumoftoziol. Ig.*, 31, 227-234, 1982.

104. **Ranchet, J., Menissier, F., Lamathe, J., and Voinovitch, I.,** Interlaboratory comparison: the determination of Cd, Cr, Cu and lead in standard solutions by flameless atomic absorption spectroscopy, *Analysis*, 10, 71-77, 1982.

105. **Kijewski, H. and Lowitz, H. D.,** The estimation of lead in hydride form from biopsy specimens from the iliac crests of patients who had contracted lead poisoning many years previously, *Arch. Toxicol.*, 50, 301-311, 1982.

106. **Shan, Q. and Ni, Z.,** Matrix modification for the determination of lead in urine by graphite furnace atomic absorption spectroscopy, *Can. J. Spectrosc.*, 27, 75-81, 1982.

107. **Isozaki, A., Fukuda, Y., and Utsumi, S.,** Electrothermal atomic absorption spectroscopy for lead by a direct atomization of lead adsorbed chelating resin, *Bunseki Kagaku*, 31, 404-406, 1982 (in Japanese).

108. **Dabeka, R. W.,** Workshop on the determination of lead in foods by graphite furnace atomic absorption spectroscopy: preliminary report, *J. Assoc. Off. Anal. Chem.*, 65, 1005, 1982.

109. **Fukuya, Y., Gotoh, M., Matsumoto, T., and Okutani, H.,** Automated microdetermination of lead in capillary blood from ear lobes by flameless atomic absorption spectrophotometry, *Sangyo Igaku*, 24, 126-132, 1982.

110. **Pan, D. and Ling, J.,** Determination of cadmium and lead in cereals and tea using graphite furnace atomic absorption spectroscopy, *Fenxi Huaxue*, 10, 190, 1982.

111. **Miles, J. P.,** Analytical methods used by industry for lead in infant formula, *J. Assoc. Off. Anal. Chem.*, 65, 1016-1024, 1982.

112. **Koirtyohann, S. R., Kaiser, M. L., and Hinderberger, E. J.,** Food analysis for lead using furnace atomic absorption and a L'vov platform, *J. Assoc. Off. Anal. Chem.*, 65, 999-1004, 1982.

113. **Rains, T. C., Rush, T. A., and Butler, T. A.,** Innovations in atomic absorption spectrophotometry with electrothermal atomization for determining lead in foods, *J. Assoc. Off. Anal. Chem.*, 65, 994-998, 1982.

114. **Fernandez, F. J. and Hillgoss, D.,** An improved graphite furnace method for the determination of lead in blood using matrix modifications and the L'vov platform, *At. Spectrosc.*, 3, 130, 1982.

115. **Ma, J. and Chen, J.,** Tungsten filament electrolytic concentration flameless atomic absorption spectrometric determination of lead in seawater, *Huaxue Tongbao*, 5, 276, 1982.

116. **Croce, A. and Brizzi, P.,** Paper and card board used in food packaging. Comparison between flame- and flameless systems for the determination of lead using atomic absorption spectroscopy in extracts from food-packaging papers and card boards, *Ind. Carta*, 20, 89, 1982 (in Italian).

117. **May, T. W. and Brumbaugh, W. G.,** Matrix modifier and L'vov platform for elimination of matrix interferences in the analysis of fish tissues for lead by graphite furnace atomic absorption spectroscopy, *Anal. Chem.*, 54, 1032, 1982.

118. **Bertenshaw, M. P., Gelsthorpe, D., and Wheatstone, K. C.,** Reduction of matrix interferences in the determination of lead in aqueous samples by atomic absorption spectrophotometry with electrothermal atomization with lanthanum pretreatment, *Analyst (London)*, 107, 163-171, 1982.

119. **Prokof'ov, A. K., Oradovskii, S. G., and Geogievskii, V. V.,** Flameless atomic absorption method for the determination of copper, lead and cadmium in marine bottom sediments, *Tr. Gos. Okeanogr. In-ta*, 162, 51-55, 1981 (in Russian). From Ref. *Zh. Khim.*, Abstr. No. 11G258, 1982.

120. **Fazakas, J.,** Determination of lead in mineral waters by graphite furnace-atomic absorption spectroscopy, *Rev. Chim. (Bucharest)*, 33, 173-176, 1982 (in Romanian).

121. **Bruhn, F. C., Rodriguez, E. A., Navarrete, A. G., and Bravo, C.,** Method for the direct determination of lead in blood by graphite furnace atomic absorption spectroscopy, *Bol. Soc. Chil. Quim.*, 27, 346-348, 1982 (in Spanish).

122. **Herber, R. F. M. and Van Deyck, W.,** On the optimization of blood lead standards in electrothermal atomization atomic absorption spectroscopy, *Clin. Chim. Acta*, 120, 313-320, 1982.

123. **Herber, R. F. and Van Deyck, W.,** On the optimization of blood lead standards in electrothermal atomization atomic absorption spectroscopy, *Clin. Chim. Acta*, 120, 313, 1982.

124. **Rohbock, E.,** Determination of gaseous alkylleads in the air of large cities, *Atomspektrom. Spurenanal., Vortr. Kolloq., 1981*, 267-274, 1982.

125. **Knutti, R.,** Matrix effects and matrix modification in graphite-tube AAS. Determination of lead in urine, *Atomspektrom. Spurenanal., Vortr. Kolloq., 1981*, 57-66, 1982.

126. **Wittmann, Z.,** Determination of lead in oils by atomic absorption spectroscopy using a modified Delves cup technique, *Acta Chim. Acad. Sci. Hung.*, 111, 85-88, 1982.

127. **Moattar, F. and Rahimi, H.,** Determination of lead in atmospheric particulates using flameless atomic absorption techniques, *Int. J. Environ. Stud.*, 20, 67, 1982.

128. **Yamaguchi, M., Okuba, N., Yokoyama, H., Ui, T., and Asada, E.,** Determination of lead in airborne particulates using direct introduction of filter paper in graphite furnace atomic absorption spectroscopy, *Bunseki Kagaku,* 31, 561, 1982.

129. **Collins, M. F., Hrdina, P. D., Whittle, E., and Singhal, R. L.,** Lead in blood and brain regions of rats chronically exposed to low doses of the metal, *Toxicol. Appl. Pharmacol.,* 65, 314, 1982.

130. **Marletta, G. P., Favretto, L. G., and Calzolari, C.,** A statistical approach to the balance of lead in milk and some dairy products, *Proc. 1st Eur. Conf. Food Chem. Recent Dev. Food Anal. 1981,* 1982, 377-382

131. **Jin, J. and Li, F.,** Determination of trace lead in gasoline by tantalum ribbon flameless atomic absorption spectrometry, *Fenxi Huaxue,* 10, 487, 1982.

132. **Del Rosario, A. R., Guirguis, G. N., Perez, G. P., Matias, V. C., Li, T. H., and Flessel, C. P.,** A rapid and precise system for lead determination in whole blood, *Int. J. Environ. Anal. Chem.,* 12, 223, 1982.

133. **Aznarez, J., Palacios, F., and Vidal, J. C.,** Determination of lead in environmental samples by electrothermal atomic absorption spectrophotometry after extraction with pyrrolidene-dithiocarbamate, *At. Spectrosc.,* 3, 192, 1982.

134. **Takada, K. and Hirokawa, K.,** Origin of double-peak signals for trace lead, bismuth, silver and zinc in a micro amount of steel in atomic absorption spectroscopy with direct electrothermal atomization of a solid sample in a graphite-cup cuvette, *Talanta,* 29, 849, 1982.

135. **Atsuya, I. and Itoh, K.,** Direct determination of lead in NBS bovine liver by Zeeman atomic absorption spectroscopy with a miniature graphite cup, *Bunseki Kagaku,* 31, 708, 1982 (in Japanese).

136. **Jin, K. and Taga, M.,** Determination of lead by continuous flow hydride generation and atomic absorption spectroscopy. Comparison of malic acid-dichromate, nitric acid-hydrogen peroxide and nitric acid-peroxodisulfate reaction matrixes in combination with sodium tetrahydroborate, *Anal. Chim. Acta,* 143, 229, 1982.

137. **Chen, Y.,** Study on the determination of lead in soy sauce using graphite furnace-atomic absorption spectroscopy, *Zhongguo Niangzao,* 1, 21, 1982 (in Chinese).

138. **Kuga, K.,** Determination of exhausting gas from flameless atomizer in atomic absorption spectroscopy. An estimation of chloride interference phenomenon in lead determination, *Bunseki Kagaku,* 37, 27-32, 1982.

139. **Frigieri, P., Croce, E., Pilotini, F., Buratti, M., Cambiaghi, G., and Bertelli, G.,** Application of Zeeman effect atomic absorption spectrophotometric method for determination of lead in plasma, *Med. Lav.,* 74, 231-238, 1983.

140. **Hodges, D. J. and Skelding, D.,** Determination of lead in blood by atomic absorption spectroscopy with electrothermal atomization, *Analyst (London),* 108, 813-820, 1983.

141. **Huat, L. H., Zakariya, D., and Eng, K. H.,** Lead concentrations in breast milk of Malayasian urban and rural mothers, *Arch. Environ. Health,* 38(4), 205-209, 1983.

142. **Fletcher, I. J.,** A comparison of methods for the determination of lead in potable waters by electrothermal atomic absorption spectrometry with lanthanum matrix modification and standard addition procedures, *Anal. Chim. Acta,* 154, 235-249, 1983.

143. **Torsi, G. and Palmisano, F.,** Electrostatic capture of gaseous tetraalkyllead compounds and their determination by electrothermal atomic absorption spectrometry, *Analyst (London),* 108(1292), 1318, 1983.

144. **Fukushi, K. and Hiiro, K.,** Determination of lead in seawater by graphite furnace atomic absorption spectrometry after electrolytic preconcentration on a platinum electrode, *Bunseki Kagaku,* 32(11), 688, 1983.

145. **Kubo, T., Takano, Y., Kumazawa, J., and Komoike, Y.,** Determination of lead in blood by graphite furnace atomic absorption spectrophotometry with micro ashing and matrix modification, *Sumitomo Sangyo Eisei,* 19, 121-129, 1983.

146. **Giri, S. K., Shields, C. K., Littlejohn, D., and Ottaway, J. M.,** Determination of lead in whole blood by electrothermal atomic absorption spectroscopy using graphite probe atomization, *Analyst (London),* 108, 244, 1983.

147. **Pruszkowska, E., Carnrick, G. R., and Slavin, W.,** Blood lead determination with the platform furnace technique, *At. Spectrosc.,* 4, 59, 1983.

148. **Epstein, M. S.,** Determination of ultratrace levels of lead in reference fuels by graphite furnace atomic absorption, *At. Spectrosc.,* 4, 62, 1983.

149. **Diehl, K. H., Rosopulo, A., and Krenzer, W.,** Determination of lead from tetraalkyllead compounds using AAS, *Fresnius' Z. Anal. Chem.,* 314, 755, 1983.

150. **Shamberger, R. J.,** Effects of blood and urine on lead analyzed by flameless atomic absorption, *J. Clin. Chem. Clin. Biochem.,* 21, 107, 1983.

151. **Niculescu, T., Dumitru, R., Botha, V., Alexandrescu, R., and Mavolescu, N.,** Relationship between the lead concentrations in hair and occupational exposure, *Br. J. Ind. Med.,* 40, 67, 1983.

152. **Jackson, K. W. and Newman, A. P.,** Determination of lead in soil by graphite furnace atomic absorption spectroscopy with the direct introduction of slurries, *Analyst (London),* 108, 261, 1983.

153. **Sturgeon, R. E., Mitchell, D. F., and Berman, S. S.,** Atomization of lead in graphite furnace atomic absorption spectroscopy, *Anal. Chem.,* 55, 1059, 1983.

154. **DeJonghe, W. R. A., Van Mol, W. E., and Adams, F. C.,** Determination of trialkyllead compounds in water by extraction and graphite furnace atomic absorption spectroscopy, *Anal. Chem.,* 55, 1050, 1983.

155. **Eaton, D. K. and Holcombe, J. A.,** Oxygen ashing and matrix modifiers in graphite furnace atomic absorption spectrometric determination of lead in whole blood, *Anal. Chem.,* 55, 946, 1983.

156. **Zakhariya, A. N., Dolgushina, L. E., and Olenovich, N. L.,** Extraction-atomic absorption determination of lead in seawater using a furnace-flame atomizer, *Ukr. Khim. Zh.,* 49, 54, 1983 (in Russian).

157. **Haswell, S. J., O'Neill, P., and Banecroft, K. C. C.,** Association of lead with polar dissolved organic compounds in soil pore waters, *4th Int. Conf. Heavy Met. Environ.,* CEP Consult, Edinburgh, U.K., 1983, 1215-1218.

158. **Chen, Z. and Fan, C.,** Determination of lead in preserved duck egg by electrothermal atomic absorption spectroscopy, *Shipin Yu Fajiao Gongye,* 5, 38-41, 1983.

159. **Kosaka, H. and Miyajima, K.,** Studies on the determination of the lead in whole blood by flameless atomic absorption spectrophotometry, *Osaku-furitsu Koshu Eisei Kenkyusho Kenkyu Hokoku, Rodo Eisei Hen,* 21, 17, 1983 (in Japanese).

160. **Hasan, M. Z. and Kumar, A.,** A graphite furnace atomic absorption spectrophotometric method for the determination of lead in water by chemical modification, *Proc. Indian Natl. Sci. Acad. Part A,* 49(6), 654, 1983.

161. **Hasan, M. Z. and Kumar, A.,** Elimination of sodium interference in flameless atomic absorption spectrophotometric determination of lead in potable water by matrix modification, *J. Indian Int. Sci.,* 64(5), 127-131, 1983.

162. **Hasan, M. Z. and Kumar, A.,** Flameless AAS method for determination of lead in blood, *Indian J. Environ. Health,* 25(3), 191-199, 1983.

163. **Maeda, T., Kawakatsu, M., Tanimoto, Y., and Tanabe, M.,** Atomization of lead compounds in graphite furnace atomic absorption spectroscopy, *Shimadzu Hyoron,* 40(2/3), 165-170, 1983.

164. **Panetti, M. and Ferrero, F.,** Direct determination of lead in atmospheric particulates by flameless atomic absorption with a carbon rod atomizer, *Rass. Chim.,* 35(5), 329, 1983.

165. **Scott, D. R., Holboke, L. E., and Hadeishi, T.,** Determination of lead in gasoline by Zeeman atomic absorption spectroscopy, *Anal. Chem.,* 55, 2006, 1983.

166. **Grobenski, Z. and Lehmann, R.,** Determination of lead in rock samples by using solid sampling with the stabilized temperature platform furnace and zeeman background correction, *At. Spectrosc.,* 4, 111, 1983.

167. **Dela Guardia, M., Durrieu, F., Voinovitch, I. A., and Louvrier, J.,** The use of surface active agents in improving the sensitivity and repeatablity of lead determinations by atomic absorption spectroscopy with electrothermal atomization, *Spectrochim. Acta,* 38B, 617-624, 1983 (in French).

168. **Butrimovitz, G. P., Sherlip, I., and Lo, R.,** Extremely low seminal lead concentrations and male fertility, *Clin. Chim. Acta,* 135, 229, 1983.

169. **Ellis, B. H., Bourcier, D. R., and Galke, W. A.,** Canned fruit juices and orange drink as a source of lead in the human environment, *Trace Subst. Environ. Health,* 17, 358-364, 1983.

170. **Alder, J. F. and Batoreu, M. C. C.,** Practical aspects of lead and nickel in epithelial tissue by electrothermal atomic absorption spectrometry, *Anal. Chim. Acta,* 155, 199, 1983.

171. **Legret, M., Demare, D., Marchandise, P., and Robbe, D.,** Interferences of major elements in the determination of lead, copper, cadmium, chromium and nickel in river sediment and sewer sludges by electrothermal atomic absorption spectroscopy, *Anal. Chem.,* 149, 107, 1983.

172. **Simon, J. and Liese, T.,** Trace determination of lead and cadmium in bones by electrothermal atomic absorption spectroscopy, *Fresnius' Z. Anal. Chem.,* 314, 483, 1983.

173. **Marletta, G. P. and Favretto, L. G.,** Preliminary investigation on the balance of lead and cadmium content in milk and its by-products, *Z. Lebensm. Unters. Forsch.,* 176, 32, 1983.

174. **Khavezov, I. and Ivanova, E.,** Electrothermal atomization using the L'vov platform. Determination of traces of lead, cadmium, manganese and copper in a sodium chloride matrix, *Fresnius' Z. Anal. Chem.,* 315, 34-37, 1983.

175. **Boiteau, H. L., Metayer, C., Ferre, R., and Pineau, A.,** Automated determination of lead, cadmium, manganese and chromium in whole blood by Zeeman atomic absorption spectrometry, *Analysis,* 11(5), 234-42, 1983 (in French).

176. **Mueller, J. and Kallischnigg, G.,** A ring test for the determination of lead, cadmium and mercury in biological material, *ZEBS-Ber.,* 1, 69, 1983.

177. **Schindler, E.,** Compound procedure to determine lead, cadmium and copper in cereals and flour, *Dtsch. Lebensm. Rundsch.,* 79(10), 33-37, 1983.

178. **Janssen, E.,** Determination of lead and cadmium in feeds after ashing under pressure with Zeeman graphite tube technique, *Landwirtsch. Forsch.,* 36(1-2), 161-167, 1983 (in German).

179. **Ali, S. L.,** Determination of pesticide residues and other critical impurities such toxic trace metals in medicinal plants. II. Determination of toxic trace metals in drugs, *Pharm. Ind.,* 45, 1294, 1983 (in German).

180. **Takada, K. and Hirokawa, K.,** Determination of traces of lead and cadmium in high-purity tin by polarized Zeeman atomic absorption spectroscopy with direct atomization of solid sample in a graphite-cup cuvette, *Talanta,* 30, 329, 1983.

181. **Zhu, Y.,** Determination of lead and cadmium contents in traditional Chinese medicines by graphite furnace atomic absorption method, *Yaowu Fenxi Zazhi,* 3, 175-177, 1983 (in Chinese).

182. **Vanloo, B., Dams, R., and Hoste, J.,** Determination of lead and bismuth in steel and cast iron by Zeeman atomic absorption spectroscopy and hydride generation, *Fonderie Belge,* 53, 7-16, 1983.

183. **Postel, W., Meier, B., and Markert, R.,** Determination of lead, cadmium, aluminum and tin in beer using atomic absorption spectrophotometry, *Monatsschr. Brauwiss,* 36, 300, 1983 (in German).

184. **Postel, W., Meier, B., and Markert, R.,** Lead, cadmium, aluminum, tin, zinc, iron and copper in bottled and canned beer, *Monatsschr. Brauwiss,* 36, 360-367, 1983 (in German).

185. **Janssen, E.,** Determination of lead and cadmium in feeds after ashing under pressure with Zeeman graphite tube technique, *Landwirtsch. Forsch.,* 36, 161, 1983.

186. **Castledine, C. G. and Robbins, J. C.,** Practical field-portable atomic absorption analyzer, *J. Geochem. Explor.,* 19, 689, 1983.

187. **Subramanian, K. S., Meranger, J. C., and Mackeen, J. E.,** Graphite atomic absorption spectroscopy with matrix modification for determination of cadmium and lead in human urine, *Anal. Chem.,* 55, 1064, 1983.

188. **Subramanian, K. S., Meranger, J. C., and Connor, J.,** The effect of container material storage time and temperature as cadmium and lead levels in heparinized human whole blood, *J. Anal. Toxicol.,* 7, 15, 1983.

189. **Nakashima, R., Kamata, E., and Shibata, S.,** Determination of cobalt, copper, iron, nickel and lead in bone by graphite furnace atomic absorption spectroscopy, *Kogai,* 18, 37-44, 1983.

190. **Wang, S. T., Strunc, G., and Peter, F.,** Determination of cadmium and lead in whole blood and urine by Zeeman flameless atomic absorption spectroscopy, *Proc. 2nd Int. Conf. Chem. Toxicol. Clin. Chem. Met.,* 1983, 57-60.

191. **Torsi, G. and Palmisano, F.,** Electrostatic capture of gaseous tetraalkyllead compounds and their determination by electrothermal atomic absorption spectrometry, *Analyst,* 108, 1318, 1983.

192. **Panetti, M. and Ferrero, F.,** Direct determination of lead in atmospheric particulates by flameless atomic absorption with carbon rod atomizer, *Rass. Chim.,* 35, 329, 1983 (in Italian).

193. **Wei, F., Chen, J., and Yin, F.,** Simultaneous determination of ultratrace cadmium and lead in beverages by flameless atomic absorption analysis with automatic atomizing introduction of sample, *Shipin Kexue (Beijing),* 46, 39-43, 1983 (in Chinese).

194. **Steiner, J. W. and Kramer, H. L.,** *In situ* gaseous pretreatmenmt of liver extracts in modified carbon rod atomizer during determination of cadmium and lead, *Analyst,* 108, 1051, 1983.

195. **Voellkopf, U., Grobenski, Z., and Welz, B.,** Stabilized temperature platform furnace with Zeeman effect background correction for trace analysis in wastewater, *At. Spectrosc.,* 4, 165, 1983.

196. **Aurand, K., Drews, M., and Seifert, B.,** Passive sampler for determination of heavy metal burden of indoor environments, *Environ. Technol. Lett.,* 4, 433, 1983.

197. **Subramanian, K. S. and Meranger, J. C.,** Blood levels of cadmium, copper, lead and zinc in children in a British Columbia community, *Sci. Total Environ.,* 30, 231-244, 1983.

198. **Mazzucotelli, A., Minoia, C., and Frache, R.,** Electrothermal atomic absorption spectrophotometry of aluminum, lead and zinc in CSF samples, *Proc. 2nd Int. Workshop Trace Elem. Anal. Chem. Med. Biol.,* 1983, 975-980.

199. **Oliver, W. K., Reeve, S., Hammond, K., and Basketter, F. B.,** Determination of lead and cadmium by electrothermal atomization atomic absorption utilizing the L'vov platform and matrix modification, *J. Inst. Water Eng. Sci.,* 37, 460, 1983.

200. **Falk, H., Hoffmann, E., Luedke, C., Ottaway, J. M., and Giri, S. K.,** Furnace atomization with nonthermal excitation-experimental evaluation of detection based on high-resolution echelle monochromator incorporating automatic background correction, *Analyst,* 108, 1459, 1983.

201. **Zerezghi, M., Mulligan, K. J., and Caruso, J. A.,** Simultaneous multielement determination in microliter samples by rapid-scanning spectrometry coupled to microwave-induced plasma, *Anal. Chim. Acta,* 154, 219, 1983.

202. **Schindler, E.,** Compound procedure to determine lead, cadmium and copper in cereals and flours, *Dtsch. Lebensm. Rundsch.,* 79, 334, 1983.

203. **Janssen, E. and Bruene, H.,** Determination of mercury, lead and cadmium in fish from the Rhine and Main by flameless atomic absorption, *Z. Lebensm. Unters. Forsch.,* 178, 168, 1984 (in German).

204. **Hulanicki, A. and Bulska, E.,** Effect of organic solvents in atomization of nickel, copper and lead in graphite furnace in atomic absorption spectrometry, *Can. J. Spectrosc.,* 29, 148, 1984.

205. **Laumond, F., Copin-Montegut, G., Courau, P., and Nicholas, E.,** Cadmium, copper and lead in the Western Mediterranean sea, *Mar. Chem.,* 15(3), 251-261, 1984.

206. **Hoenig, M., Scokart, P. O., and Van Hoeyweghen, P.,** Efficiency of L'vov platform and ascorbic acid modifier for reduction of interferences in analysis of plant samples for lead, thallium, antimony, cadmium, nickel and chromium by ETAAS, *Anal. Lett.,* 17A, 1947, 1984.

207. **Danielsson, L. G. and Westerlund, S.,** Short term variations in trace metal concentrations in Baltic, *Mar. Chem.,* 15, 273, 1984.

208. **Voellkopf, U. and Grobenski, Z.,** Interference in analysis of biological samples using stabilized temperature platform furnace and Zeeman background correction, *At. Spectrosc.,* 5, 115, 1984.

209. **Voinovitch, I. A., Druon, M., and Louvrier, J.,** Effect of absence of injection orifice and length of graphite furnaces on sensitivity of determination by atomic absorption spectrometry with electrothermal atomization, *Analysis,* 12, 256, 1984 (in French).

210. **L'vov, B. V. and Yatsenko, L. F.,** Carbothermal reduction of zinc, cadmium, lead and bismuth oxides in graphite furnaces for atomic absorption analysis in presence of organic substances, *Zh. Anal. Khim.,* 39, 1773, 1984 (in Russian).

211. **Favretto, L., Marletta, G. P., and Favretto, L. G.,** Nonlinear standard additions in AAS determination of lead and cadmium with electrothermal atomization in a graphite furnace, *At. Spectrosc.,* 5, 51, 1984.

212. **Muys, T.,** Quantitative determination of lead and cadmium in foods by programmed dry ashing and atomic absorption spectrophotometry with electrothermal atomization, *Analyst (London),* 109, 119, 1984.

213. **Headridge, J. B. and Riddington, I. M.,** Determination of silver, lead and bismuth in glasses by atomic absorption spectroscopy with introduction of solid samples into furnaces, *Analyst (London),* 109, 113, 1984.

214. **L'vov, B. V. and Yatsenko, L. F.,** Carbothermal reduction of zinc, cadmium, lead and bismuth oxides in graphite furnace atomic absorption analysis in the presence of organic substances, *J. Phys. C.,* 17(35), 6415-6434, 1984.

215. **Hulanicki, A. and Bulska, E.,** Effect of organic solvents on atomization of nickel, copper and lead in graphite furnace in atomic absorption spectrometry, *Can. J. Spectrosc.,* 29(6), 148-152, 1984.

216. **Andersson, K., Nilsson, C. A., and Nygren, O.,** A new method for the analysis of tetramethyllead in blood, *Scand. J. Work Environ. Health,* 10, 51, 1984.

217. **Dabeka, R. W.,** Collaborative study of a graphite furnace atomic absorption screening method for the determination of lead in infant formulas, *Analyst (London),* 109(10), 1259-1263, 1984.

218. **Edmonstone, G. and Van Loon, J. C.,** Determination of lead in blood and precipitation using a tungsten probe and graphite furnace, *Spectrosc. Lett.,* 17(10), 591-602, 1984.

219. **Webster, J. and Wood, A.,** Evaluation of an electrothermal atomization procedure for the determination of lead in potable water, *Analyst (London),* 109(10), 1255-1258, 1984.

220. **Brams, E., Anthony, W., Lyons, P., and Head, P.,** Lead in tissues of goats fed hay from adultrated soil: an assessment of an agricultural food chain, *Trace Subst. Environ. Health,* 18, 442, 1984.

221. **Sthapit, P. R., Ottaway, J. M., Halls, D. J., and Fell, G. S.,** Suppression of interferences in the determination of lead in natural and drinking waters by graphite furnace atomic absorpotion spectroscopy, *Anal. Chim. Acta,* 165, 121, 1984.

222. **Lin, S. M., Tsai, J. L., Chiang, C. H., Tzeng, J. L., and Yang, M. H.,** Comparative determination of lead in whole blood by flame atomic absorption spectrophotometry following extraction and graphite furnace atomic absorption spectrophotometry, *Huan Ching Pao Hu (Taipei),* 7(2), 59-70, 1984.

223. **Troccoli, O. E. and Daraio, M. E.,** Microdetermination of lead in whole lead by electrothermal atomization, *Acta Bioquim. Clin. Latinoam.,* 18(4), 639, 1984 (in Spanish).

224. **Larkins, P. L.,** Effect of pressure broadening and line shift on determination of lead isotopic ratios by atomic absorption and fluorescence spectroscopy, *Spectrochim. Acta,* 39B, 1365, 1984.

225. **Siemer, D. D., Lundberg, E., and Frech, W.,** Gas phase temperatures in massman furnaces equipped with L'vov platforms, *Appl. Spectrosc.,* 38, 389, 1984.

226. **Sneddon, J.,** Direct collection of lead in the atmosphere by impaction for determination by electrothermal atomization atomic absorption spectrometry, *Anal. Chem.,* 56, 1982, 1984.

227. **Fudagawa, N.,** Determination of lead by atomic absorption spectrometry with tungsten ribbon atomizer, *Bunseki Kagaku,* 33, 328, 1984 (in Japanese).

228. **Fazakas, J.,** Effect of atomization temperature on sensitivity of lead in atomic absorption spectrometry under electrothermal atomization and tube wall or platform vaporization, *Rev. Roum. Chim.,* 29, 207, 1984.

229. **de Benzo, Z. A. and Fraile, V. R.,** Behavior of low salt content aqueous solutions of lead in hafnium treated graphite tubes for atomic absorption spectrometry, *At. Spectrosc.,* 5, 204, 1984.

230. **Andersson, K., Nilsson, C. A., and Nygren, O.,** New method for analysis of tetramethyllead, *Scand. J. Work Environ. Health,* 10, 51, 1984.

231. **Lundberg, E., Frech, W., and Lindberg, I.,** Determination of lead in biological materials by constant temperature electrothermal atomic absorption spectrometry, *Anal. Chim. Acta,* 160, 205, 1984.

232. **Matousek, J. P.,** Removal of sample vapor in electrothermal atomization studied at constant-temperature conditions, *Spectrochim. Acta,* 39B, 205, 1984.

233. **Mohl, C., Ostapczuk, P., and Stoeppler, M.,** Direct determination of lead and cadmium in urine using graphite furnace atomic absorption spectroscopy and the L'vov platform, *Fortschr. Atomspektrom. Spurenanal.,* 1, 317, 1984 (in German).

234. **Sumamur, P. K. and Heryuni, S.,** Relationship of blood lead and manganese in occupationally unexposed subjects, *Arh. Hig. Rada Toksikol.,* 35(2), 169-172, 1984.

235. **Nakaaki, K., Fukobari, S., and Masuda, T.,** Lead and cadmium levels in blood samples in general population of urban areas, *Rado Kagaku,* 60(12), 577-583, 1984 (in Japanese).

236. **Botha, P. V. and Fazakas, J.,** The use of nonresonance lines for the determination of lead, indium and thallium by electrothermal atomic absorption spectrometry, *Spectrochim. Acta,* 39B, 379, 1984.

237. **Koyama, T., Miyamoto, S., and Maeda, S.,** Effect of long-term feeding of lead into milk and the accumulation in internal organs, *Chikusan Shikenjo Kenkyu Hokoku,* 42, 15-19, 1984 (in Japanese).

238. **Mobarak, N. and P'an, A. Y. S.,** Lead distribution in the saliva and blood fractions of rats after intraperitoneal injections, *Toxicology,* 32, 67-74, 1984.

239. **Feinberg, M. and Ducauze, C.,** Presentation of analytical results: statistical and computer-based approach, *Analysis,* 12, 26, 1984.

240. **Stoeppler, M., Mohl, C., Ostapczuk, P., Goedde, M., Roth, M., and Waidmann, E.,** Rapid and reliable determination of elevated blood lead levels, *Fresnius' Z. Anal. Chem.,* 317, 486-490, 1984.

241. **Slavin, W. and Carnrick, G. R.,** Possibility of standardless furnace atomic absorption spectroscopy, *Spectrochim. Acta,* 39B, 271, 1984.

242. **Slavin, W., Carnrick, G. R., and Manning, D. C.,** Chloride interferences in graphite furnace atomic absorption spectrometry, *Anal. Chem.,* 56, 163, 1984.

243. **Brodie, K. G. and Routh, M. W.,** Trace analysis of lead in blood, aluminum and manganese in serum and chromium in urine by graphite furnace atomic absorption spectroscopy, *Clin. Biochem.,* 17, 19-26, 1984.

244. **Backstrom, K., Danielsson, L. G., and Nord, L.,** Sample work-up for graphite furnace atomic absorption spectrometry using continuous flow extraction, *Analyst,* 109, 323, 1984.

245. **Littlejohn, D., Cook, S., Durie, D., and Ottaway, J. M.,** Investigation of working conditions for graphite probe atomization in electrothermal atomic absorption spectrometry, *Spectrochim. Acta,* 39B, 295, 1984.

246. **Littlejohn, D., Duncan, I., Marshall, J., and Ottaway, J. M.,** Analytical evaluation of totally pyrolytic graphite cuvettes for electrothermal atomic absorption spectrometry, *Anal. Chim. Acta,* 157, 291, 1984.

247. **Headridsge, J. B. and Riddington, J. M.,** Determination of silver, lead and bismuth in glasses by atomic absorption spectrometry with introduction of solid samples into furnaces, *Analyst,* 109, 113, 1984.

248. **Chakrabarti, C. L., Wu, S., Karwowska, R., Chang, S. B., and Bertels, P.,** Studies in atomic absorption spectrometry with laboratory-made graphite furnace, *At. Spectrosc.,* 5, 69, 1984.

249. **Carrondo, M. J. T., Roboredo, F., Ganho, R. M. B., and Oliveira, J. F. S.,** Analysis of sediments for heavy metals by rapid electrothermal atomic absorption procedure, *Talanta,* 31, 561, 1984.

250. **Sperling, K. R.,** Tube-in-tube technique in electrothermal atomic absorption spectrometry. III. Atomization behavior of Zn, Cd, Pb, Cu and Ni, *Spectrochim. Acta,* 39B, 371, 1984.

251. **El-Azizi, A., Al-Saleh, I., and Al-Mefty, O.,** Concentrations of Ag, Al, Au, Bi, Cd, Cu, Pb, Sb and Se in cerebrospinal fluid of patients with cerebral neoplasma, *Clin. Chem.,* 30, 1358, 1984.

252. **Hocquellet, P.,** Use of atomic absorption spectrometry with electrothermal atomization for direct determination of trace elements in oils: cadmium, lead, arsenic and tin, *Rev. Fr. Corps Gras,* 31, 117, 1984 (in French).

253. **Halls, D. J.,** Speeding up determinations by electrothermal atomic absorption spectrometry, *Analyst,* 109, 1081, 1984.

254. **Phelan, V. J. and Powell, R. J. W.,** Combined reagent purification and sample dissolution (CORPAD) applied to trace analysis of silicon, silica and quartz, *Analysis,* 109, 1269, 1984.

255. **Mausbach, G.,** Nondestructive simultaneous determination of lead and cadmium by AAS, *Labor Praxis,* 8, 764, 1984.

256. **Diehl, K. H., Rosopulo, A., and Krenzer, W.,** Possibilities for determination of lead tetraalkyls in different matrixes, *Z. Anal. Chem.,* 317, 469, 1984 (in German).

257. Flameless atomic absorption spectroscopy of lead in aqueous media, Mitsushita Electric Industrial Co. Ltd. Jpn. Tokkyo Koho JP59 13,695 (84 13, 695) (Cl.G01N21/31), 31 Mar. 1984, 6pp. Appl. 76/98,510, 17 Aug. 1976.

258. **Kuntsevich, I. C., Golubev, I. R., Dmitriev, M. T., Dubrovskaya, G. N., and Tereshchenko, O. V.,** Determination of lead in environmental samples and biological materials by flameless atomic absorption spectrophotometry, *Zdravookhr. Beloruss.,* 7, 41, 1984 (in Russian).

259. **Hinds, M. W., Jackson, K. W., and Newman, A. P.,** Electrothermal atomization atomic absorption spectrometry with direct introduction of slurries determination of trace metals in soil, *Analyst,* 110, 947, 1985.

260. **Horvath, Z., Lasztity, A., Szakacs, O., and Bozsai, G.,** Iminodiacetic acid/ethylcellulose as chelating ion exchanger. I. Determination of trace metals by atomic absorption spectrometry and collection of uranium, *Anal. Chim. Acta,* 173, 273, 1985.

261. **Erb, R.,** Quality control by Zeeman-atomic absorption spectroscopy in a chemical factory. Direct determination of copper, manganese, iron, cadmium and lead in adhesive tapes and diverse raw materials, *Fresnius' Z. Anal. Chem.,* 322(7), 719-720, 1985.

262. **Lang, H., Gen, X., Zhu, M., and Bai, G.,** Determination of cadmium, lead and copper in natural water and industrial wastewater by ion-exchange-three-hole ring oven technique, *Xibei Daxue Xuebao Ziran Kexueban,* 14(4), 38-42, 1984, CA 23, 103:146866d, 1985.

263. **Narres, H. D., Mohl, C., and Stoeppler, M.,** Metal analysis in difficult materials with platform furnace Zeeman-AAS. II. Direct determination of cadmium and lead in milk, *Z. Lebensm. Unters. Forsch.,* 18(2), 111-116, 1985.

264. **Tominaga, M., Bansho, K., and Umazaki, Y.,** Electrothermal atomic absorption spectrometric determination of lead, manganese, vanadium and molybdenum in seawater with ascorbic acid to reduce in matrix effects, *Anal. Chim. Acta,* 169, 171, 1985.

265. **Fengming, J. and Yingchun, J.,** Direct determination of lead, cadmium and nickel in soils by atomic absorption spectrometry, *Fenxi Huaxue,* 13, 386, 1985 (in Chinese).

266. **Kratochvil, B., Thapa, R. S., and Matkosky, N.,** Evaluation of homogeneity of marine biological tissues for lead and cadmium by GFAAS, *Can. J. Chem.,* 63(60), 2679-2682, 1985.

267. **Kluessendorf, B., Rosopulo, A., and Krenzer, W.,** Study of the distribution and rapid determination of lead, cadmium and zinc in livers of slaughtered pigs by Zeeman atomic absorption spectroscopy of solid samples, *Fresnius' Z. Anal. Chem.,* 322(7), 721-727, 1985.

268. **Rozanska, B., Skorko-Trybula, Z., and Slodownik, A.,** Precipitation and extraction separation of lead, iron and zinc prior to the determination of cobalt, nickel and manganese by atomic absorption in dust from electric furnace, *Chem. Anal. (Warsaw),* 30(3), 471-479, 1985.

269. **Jin, L. and Wu, D.,** Determination of lead, copper, cadmium and chromium in river sediment and coal fly ash by graphite furnace atomic absorption spectroscopy, *Huanjing Huaxue,* 2(5), 13-19, 1983, CA 22, 103:1340629, 1985.

270. **Aneva, Z. and Iancheva, M.,** Simultaneous extraction and determination of traces of lead and arsenic in petrol by electrothermal atomic absorption spectrometry, *Anal. Chim. Acta,* 167, 371, 1985.

271. **Shan, X. and Wang, D.,** X-ray photoelectron spectroscopic study of mechanism of palladium matrix modification in electrothermal atomic absorption spectrometric determination of lead and bismuth, *Anal. Chim. Acta,* 173, 315, 1985.

272. **Subramanian, K. S., Meranger, J. C., Wan, C. C., and Corsini, A.,** Preconcentration of cadmium, chromium, copper and lead in drinking water on polyacrylic ester resin XAD-7, *Int. J. Environ. Anal. Chem.,* 19, 261, 1985.

273. **Suzuki, M. and Isobe, K.,** Mechanism of interference elimination by thiourea in electrothermal atomic absorption spectrometry, *Anal. Chim. Acta,* 173, 321, 1985.

274. **Wang, S. T. and Peter, F.,** The stability of human blood lead in storage, *J. Anal. Toxicol.,* 9(2), 85, 1985.

275. **Maeda, T. and Tanimoto, Y.,** Determination of lead in whole blood by atomic absorption spectroscopy with a graphite furnace, *Shimadzu Hyoron,* 40(4), 197-201, 1983, published 1984, CA 25, 103:173498s, 1985.

276. **Koga, N., Hirai, Y., and Tomokuni, K.,** Direct determination of lead in blood by FAAS, *Nippon Eiseigaku Zasshi,* 40(2), 636-640, 1985.

277. **Harms, U.,** Possibilities of improving the determination of extremely low lead concentrations in marine fish by GFAAS, *Fresnius' Z. Anal. Chem.,* 322(1), 53-56, 1985.

278. **Telisman, S., Pongracic, J., Azaric, J., and Prpic-Majic, D.,** Sensitive method for the determination of lead in human and cow milk by the use of ET-AAS technique, *5th Int. Conf. Heavy Met. Environ.,* Lekkas, T. D., Ed., CEP Consult., Edinburgh, U.K., 1985, 490.

279. **Pongracic, J., Azaric, J., Telisman, S., and Prpic-Majic, D.,** The possible effect of different anticoagulants on the results of blood lead determination, *5th Int. Conf. Heavy Met. Environ.,* Lekkas, T. D., Ed., CEP Consult., Edinburgh, U.K., 1985, 430.

280. **Turlakiewicz, Z., Jakubowski, M., and Chmielnicke, J.,** Determination of diethyllead in urine by flameless atomic absorption spectrometry, *Br. J. Ind. Med.,* 42, 63, 1985.

281. **Schmid, W. and Krivan, V.,** Radio tracer study of preatomization behavior of lead in graphite furnace, *Anal. Chem.,* 57, 30, 1985.

282. **Andersson, J. R.,** Zeeman-corrected graphite furnace atomic absorption spectrometric screening method for determination of lead in infant formulas and powdered milk, *Analyst,* 110(3), 315, 1985.

283. **Hee, S. S. Q., Macdonald, T. J., and Bornschein, R. L.,** Blood lead by furnace-zeeman atomic absorption spectrophotometry, *Microchem. J.,* 32, 55, 1985.

284. **Yansheng, Z. and Yansheng, L.,** Influence of chloride matrix on lead in GA-3 graphite furnace atomizer, *Kexue Tongbao,* 30, 239, 1985.

285. **Miller, R. G., Doerger, J. U., Kopfler, F. C., Stober, J., and Roberson, P.,** Influence of the time of acidification after sample collection on the preservation of drinking water for lead determination, *Anal. Chem.,* 57(6), 1020, 1985.

286. **Soulin, L., Yi, Z., Guizhen, S., and Huajie, T.,** Mechanism of double-peak formation and the atomization of lead in graphite furnace atomic absorption determination of lead, *Fenxi Huaxue,* 13(11), 847-849, 1985 (in Chinese).

287. **Othman, I.,** A preliminary investigation of the lead level in whole blood of normal and occupationally exposed populations in Damascus City, *Sci. Total Environ.,* 43(1-2), 141-148, 1985.

288. **Luo, Y. and Mu, C.,** Treatment of lead pollution, *Huanjing Huaxue,* 2(4), 47-51, 1983, CA 18, 103:58446m, 1985.

289. **Hozman, R.,** Determination of trace lead concentrations in gasoline by flameless AAS (atomic absorption spectrophotometry), *Sb. Pr. Vyzk. Chem. Vyuziti Uhli Dehtu Ropy,* 17, 55-69, 1984 (in Czech), CA 18, 103:56381n, 1985.

290. **Ma, Y., Su, W., and Sun, D.,** Determination of lead in water sediments and coal fly ashes by graphite-furnace AAS using a temperature stabilized pyrographite tube lined with tungsten and tantalum, *Fenxi Huaxue,* 13(5), 379-82, 1985.

291. **Ong, C. N., Phoon, W. O., Law, H. Y., Tye, C. Y., and Lim, H. H.,** Concentrations of lead in maternal blood, cord blood and breast milk, *Arch. Dis. Child,* 60(8), 756-759, 1985.

292. **Turlakiewicz, Z. and Chmielnicke, J.,** Diethyl as a specific indicator of occupational exposure to tetraethyllead, *Br. J. Ind. Med.,* 42(10), 682-685, 1985.

293. **Jahr, K.,** Clinical significance and technique of blood-lead determination with Zeeman AAS, *Fresnius' Z. Anal. Chem.,* 322(7), 736-738, 1985 (in Chinese).

294. **Orren, D. K., Caldwell-Kenkel, J. C., and Mushak, P.,** Quantitative analysis of total and trimethyl lead in mammalian tissues using ion exchange HPLC and AAS detection, *J. Anal. Toxicol.,* 9(6), 258-261, 1985.

295. **Funato, Y., Iwanaga, Y., and Ito, N.,** Determination of trace amounts of lead in nickel-base alloy by graphite furnace atomic absorption spectrophotometry, *Shimadzu Hyoron,* 42(3), 269-272, 1985 (in Japanese).

296. **Stephen, S. C., Littlejohn, D., and Ottaway, J. M.,** Evaluation of slurry technique for determination of lead in spinach by electrothermal atomic absorption spectrometry, *Analyst,* 110, 1147, 1985.

297. **Paschal, D. C. and Kimberly, M. M.,** Determination of urinary lead by electrothermal atomic absorption with the stabilized temperature platform furnace and matrix modification, *At. Spectrosc.,* 6, 134, 1985.

298. **Hurst, P. L. and Hughes, P. M.,** Poor correlation between erythrocytic 5-aminolevulinic acid dehydratase activity and blood lead concentration in a group of eleven-year-old children, *Clin. Chem.,* 31, 1244, 1985.

299. **Osberghaus, U., Kurfuerst, U., and Stoeppler, M.,** Determination of lead in whole blood by three versions of AAS, *Fresnius' Z. Anal. Chem.,* 322(7), 739-742, 1985.

300. **Kotani, T.,** Determination of lead in straw, chaff and unpolished rice by graphite cup furnace AAS with addition of phosphoric acid as a matrix modifier, *Bunseki Kagaku,* 35, 880, 1986.

301. **Ebdon, L. and Lechotycki, A.,** The determination of lead in environmental samples by slurry atomization-graphite-furnace-atomic absorption spectrophotometry using matrix modification, *Microchem. J.,* 34, 340-348, 1986.

302. **Zheng, Y. and Liu, Y.,** Direct determination of lead in soils by graphite furnace atomic absorption spectroscopy with a modifier, *Fenxi Huaxue,* 14, 605, 1986 (in Chinese).

303. **Arai, F.,** Determination of triethyllead, diethyllead and inorganic lead in urine by atomic absorption spectroscopy, *Ind. Health,* 24, 139, 1986.

304. **Lian, L.,** Direct determination of lead in urine by Zeeman atomic absorption spectroscopy with the stabilized temperature platform furnace, *Spectrochim. Acta,* 411B, 1131-1135, 1986.

305. **Lee, R. Y. and Tsai, W. C.,** Determination of lead in canned foods with graphite furnace atomic absorption spectrophotometry, *Shih Pin K'o Hsueh (Taipei),* 13(1-2), 93-102, 1986 (in Chinese).

306. **Xu, B., Xu, T., Shen, M., and Fang, Y.,** Determination of subnanogram lead in water by electrolytic deposition on mercury film tungsten electrodes with flameless atomic absorption spectrophotometry, *Fenxi Huaxue,* 14, 623-626, 1986.

307. **Chakraborti, D., Van Cleuvenberger, R. J. A., and Adams, F. C.,** Determination of total ionic alkyllead in water by electrothermal atomization AAS, *J. Anal. At. Spectrom.,* 1(4), 293-295, 1986.

308. **Shuttler, I.L. and Delves, H. T.,** Determination of lead in blood by AAS with electrothermal atomization, *Analyst (London),* 111(6), 651-660, 1986.

309. **Ni, Z., Han, H., and Le, X.,** Determination of lead by graphite furnace AAS with argon-hydrogen as the purge gas using low-temperature atomization, *J. Anal. At. Spectrosc.,* 1(2), 131-134, 1986.

310. **Brodie, K. G., Routh, M. W., and Pohl, B.,** Trace element analysis with graphite tube atomic absorption spectroscopy. Lead in whole blood, manganese and aluminum in serum and chromium in wine, *Labor Med.,* 9(4), 218, 1986 (in German).

311. **Kumamaru, T., Matsuo, H., Riordan, J. F., and Vallee, B. L.,** Determination of lead in whole blood by automated atomic absorption spectrometry using a graphite furnace atomizer, *Hiroshima Daigaku Sogo Kagakubu Kiyo,* 4, 11, 17-24, 1986.

312. **Cui, D.,** The orthogonal test on the optimization of instrumental condition for determination of lead by graphite furnace atomic absorption method, *Shipin Yu Fajiao Gongye,* 6, 55-66, 1985, CA 8, 104: 122123d, 1986.

313. **Bester, W. and Venter, B. G.,** Determination of the lead content of dairy products by flameless atomic absorption spectroscopy, *S. Afr. Tydskr. Suiwelkd.,* 18, 105, 1986.

314. **Yansheng, Z. and Yansheng, L.,** Direct determination of lead in soils by graphite furnace atomic absorption spectroscopy with a modifier, *Fenxi Huaxue,* 14, 605, 1986.

315. **Clark, J. R.,** Electrothermal atomization atomic absorption conditions and matrix modifications for determining Sb, As, Bi, Cd, Ga, Au, In, Pb, Mo, Pd, Pt, Se, Ag, Te, Tl and Sn following back-extraction of organic aminohalide extracts, *J. Anal. At. Spectrom.,* 1, 301-308, 1986.

316. **Tsukahara, I.,** Determination of nonferrous metals: copper, lead and zinc, *Bunseki Kagaku,* 8, 561-567, 1986 (in Japanese).

317. **Xu, M., Wang, Y., and Yie, A.,** Direct determination of copper and lead in river sediments by Zeeman effect atomic absorption, *Huanjin Wuran Yu Fangzhi,* 7(2), 29-32, 1985. CA 3, 104:23965, 1986.

318. **Medina, J., Hernandez, F., Pastor, A., Beferull, J. B., and Barbera, J. C.,** Determination of mercury, cadmium, chromium and lead in marine organisms by flameless atomic absorption spectrophotometry, *Mar. Pollut. Bull.,* 17(1), 41-44, 1986.

319. **Cullaj, A.,** A thermodynamic evaluation of the mechanism of electrothermal atomization of cadmium, nickel and lead in chloride solution, *Bull. Shkencave Nat.,* 40(1), 79-83, 1986.

320. **Frech, W., Lindberg, A. O., Lundberg, E., and Cedergren, A.,** Atomization mechanisms and gas phase reactions in graphite furnace atomic absorption spectrometry, *Fresnius' Z. Anal. Chem.,* 323(7), 716-725, 1986.

321. **Van Loenen, D. C. and Weers, C. A.,** The determination of arsenic, cadmium and lead in large series of samples from pulverized coal fly ash (PFA) by Zeeman graphite furnace AAS and microliter injection of ultrasonic agitated PFA suspension, *Fortschr. Atomspektrom. Spurenanal.,* 2, 63-47, 1986.

322. **Stoeppler, M., Mohl, C., Novak, L., and Gardiner, P. E.,** Application of the STPF concept to the determination of aluminum, cadmium and lead in biological and environmental materials, *Fortschr. Atomspektrom. Spurenanal.,* 2, 419, 1986.

323. **Hirokawa, K., Namiki, M., and Kimura, J.,** Determination of silver, copper, lead and cadmium in same metals by graphite furnace coherent forward scattering spectrometry, *Bunseki Kagaku,* 35, 701, 1986.

324. **Kharlamov, I. P., Lebedev, V. I., and Persits, V. Y.,** Complex applications of atomic and molecular spectra to the study of matrix effects on the atomization of zinc, cadmium, lead, tin, bismuth and antimony, *Zh. Anal. Khim.,* 41, 1965, 1986.

325. **Fudagawa, N., Hioki, A., Kubota, M., and Kawase, A.,** Determination of copper, lead and cadmium in high-purity zinc by metal furnace AAS (atomic absorption spectroscopy), *Bunseki Kagaklu,* 35, T62, 1986 (in Japanese).

326. **Kharlamov, I. P., Lebedev, V. I., Persits, V. Y., and Eremina, G. V.,** Influence of cations and anions on the analytical signals of zinc, cadmium, lead, tin, bismuth and antimony introduced with solutions of steels and alloys into electrothermal atomizers, *Zh. Anal. Khim.,* 41, 1004, 1986.

327. **Zengzhi, H., Yuzhi, W., Zhiquiang, Z., and Qinzhi, L.,** Simultaneous determination of manganese, cobalt, nickel, lead, cadmium and tin in canned foods and infant foods by stabilized temperature platform furnace-atomic absorption spectrophotometry, *Sgipin Yu Fajiao Gongye,* 4, 1, 1986 (in Chinese).

328. **Zhou, D. and Sun, H.,** Determination of antimony, lead and cadmium released from enameled tableware, *Boli Yu Tangei,* 14(1), 17-20, 1986 (in Chinese).

329. **Dams, R., Alluyn, F., Van Loon, B., Wauters, G., and Vandecasteele, C.,** Gray cast iron reference material certified for antimony, arsenic, bismuth and lead, *Fresnius' Z. Anal. Chem.,* 325, 163-167, 1986.

330. **Donard, O. F. X., Randall, L., Papsomanikis, S., and Weber, J. H.,** Developments in the speciation and determination of alkylmetals (Sn, Pb) using volatilization technique and chromatography atomic absorption spectroscopy, *Int. J. Environ. Anal. Chem.,* 22, 55-67, 1986.

331. **Wang, J. and Tang, F.,** Enrichment of trace lead in water by manganese dioxide for determination with graphite-furnace atomic absorption spectroscopy, *Huanjing Kexue,* 7, 80, 1986 (in Chinese).

332. **Vermaak, H., Kujirai, O., Hanamura, S., and Winefordner, J. D.,** A potential method for determination of gaseous and particulate lead in exhaust gas by microwave-induced air-plasma emission spectrometry and Zeeman furnace atomic absorption spectroscopy, *Can. J. Spectrosc.,* 31, 95-99, 1986.

333. **Boxing, X., Tongming, X., Mingneng, S., and Yuzhi, F.,** Determination of submicrogram lead in water by electrolytic deposition on mercury film tungsten electrodes with flameless atomic absorption spectrometry, *Fenxi Huaxue,* 14, 623, 1986.

334. **Koreckova, J. and Pavelka, J.,** Determination of lead in milk by atomic absorption spectroscopy using a graphite cell and a platform technique, *Chem. Listy,* 80, 1309, 1986.

335. **Roeyset, O. and Thomassen, Y.,** Activated carbon as adsorbent for alkyllead in air, *Anal. Chim. Acta,* 188, 247, 1986.

336. **Hewitt, C. N., Harrison, R. M., and Radojevic, M.,** The determination of individual gaseous ionic alkyllead species in the atmospheere, *Anal. Chim. Acta,* 188, 229, 1986.

337. **Harms, U.,** Determination of reliable analytical data on the lead content of marine fish, *Fortschr. Atomspektrom. Spurenanal.,* 2, 479, 1986.

338. **Yang, L. R. and Chong, T. W.,** Determination of lead in canned foods with graphite furnace atomic absorption spectrophotometry, *Shih P'in K'o Hsueh (Taipei),* 13, 93, 1986.

339. **Karwowska, R. and Jackson, K. W.,** Lead atomization in the presence of aluminum matrices by electrothermal atomic absorption spectroscopy. A comparative study of slurry vs solution sample introduction, *Spectrochim. Acta,* 41B, 947, 1986.

340. **Novak, L. and Stoeppler, M.,** Use of hydrogen for the elimination of matrix interferences in the determination of lead by graphite furnace atomic absorption spectrometry, *Fresnius' Z. Anal. Chem.,* 323(7), 737-741, 1986.

341. **Holcombe, J. A. and Droessler, M. S.,** Surface reactions and the effects of oxygen on lead atomization in graphite furnace atomic absorption spectrometry, *Fresnius' Z. Anal. Chem.,* 323(7), 689-696, 1986.

342. **Ostapazuk, P., Valenta, P., and Nuerrnberg, H. W.,** Square wave voltammetry, a rapid and reliable determination method for zinc, cadmium, lead, copper, nickel and cobalt in biological and environmental samples, *J. Electroanal. Chem. Interfacial Electrochem.,* 214, 51-64, 1986.

343. **Tominaga, M. and Bansho, K.,** Determination of Cu, Pb, V and Co by GFAAS after a simple and rapid extraction, *Kogai Shigen Kenkyusho Iho,* 15(2), 97-106, 1985. CA 7, 104:95119y, 1986.

344. **Stein, K. and Umland, F.,** Trace analysis of lead, cadmium and manganese in honey and sugar, *Z. Anal. Chem.,* 323, 176, 1986.

345. **Gonzalez, M. C., Rodriguez, A. R., and Gonzalez, V.,** Determination of vanadium, nickel, iron, copper and lead in petroleum fractions by atomic absorption spectrophotometry with a graphite furnace, *Mikrochem. J.,* 35, 94-106, 1987.

346. **Erler, W., Lehman, R., and Voellkopf, U.,** Determination of cadmium and lead in animal liver and freeze dried blood by graphite furnace Zeeman atomic absorption spectroscopy and solid sampling, *Proc. 4th Int. Workshop Trace Elem. Anal. Chem. Med. Biol.,* Braetter, P. and Shoemel, P., Eds. , Berlin, 1987, 385-391.

347. **Nagourney, S. J., Heit, M., and Bogen, D. C.,** Electrothermal atomic absorption spectrometric analysis of lake waters for manganese, iron, lead and cadmium, *Talanta,* 34, 465, 1987.

348. **Subramanian, K. S.,** Determination of lead in blood: comparison of two GFAAS methods, *At. Spectrosc.,* 8, 7-11, 1987.

349. **Drasch, G. A., Brehm, J., and Baur, C.,** Lead in human bones. Investigations on an occupationally unexposed population in Southern Bavaria (F.R.G.). I. Adults, *Sci. Total Environ.,* 64, 303, 1987.

350. **McDonald, C. and Tessmer, D.,** Stability of blood-lead analysis by graphite furnace-atomic absorption spectrophotometry, *Microchem. J.,* 35, 227, 1987.

351. **Medina, J., Hernandez , F., Casanova, M., and Pastor, A.,** Study of the effect of matrix modifiers on the determination of lead in several marine organisms by flameless atomic absorption spectroscopy, *Analysis,* 15, 47, 1987.

352. **Costantini, S., Giordano, R., and Rubbiani, M.,** Comparison of flameless atomic absorption spectrophotometry and anodic stripping voltammetry for the determination of blood lead, *Microchem. J.,* 35, 70, 1987.

353. **Hansson, H., Ekholm, A. P., and Ross, H. B.,** Rainwater analysis: a comparison between proton-induced X-ray emission and graphite furnace atomic absorption spectroscopy, *Environ. Sci. Technol.,* 22, 527-531, 1988.

AUTHOR INDEX

SUBJECT INDEX

MAGNESIUM

MAGNESIUM, Mg   (ATOMIC WEIGHT 24.312)

Instrumental Parameters:

Wavelength	309.2 nm
Slit width	320 μm
Bandpass	1 nm
Light source	Hollow Cathode Lamp
Lamp current	5 mA
Purge gas	Argon or Nitrogen
Sample size	25 μL
Furnace	Cylindrical cuvette/Pyrolytic graphite

Standard Operating Conditions:

Optimum char temperature	1000°C
Optimum atomization temperature	2200°C
Sensitivity	0.1 pg/1% absorption
Sensitivity check	0.3-0.5 μg/mL
Working range	0.001-0.05 μg/mL
Background correction	Required only to correct light scattering or non-specific absorption from the sample containing high dissolved solids.

General Note:

1. Use commercial standard or a previously analyzed sample as a working standard.
2. Use ultra clean glass and plastic ware, soaked in 1:5 nitric acid solution and rinsed thoroughly with deionized distilled water.
3. Prepare all analytical solutions in 0.5% (v/v) nitric acid.
4. Check for blank values on all reagents including water.
5. Dilution is recommended when sample exhibits greater than 0.5 absorbance units. Dilutions can be made with deionized distilled water.
6. Prepare dilute solutions (1:25) every 4 h, when needed.
7. For increased sensitivity, use step atomization.
8. The analytical precision for this method is approximately 6% RSD.

REFERENCES

1. **Katskov, D. A. and Grinshtein, J. L.,** Study of the evaporation of Be, Mg, Ca, Sr, Ba and Al from a graphite surface by an atomic absorption method, *Zh. Prikl. Spectrosk.,* 33, 1004, 1980.
2. **Berggren, P. O.,** Determination of Ba, La and Mg in pancreatic islets by electrothermal atomic absorption spectroscopy, *Anal. Chim. Acta,* 119, 161, 1980.
3. **Smith, M. R. and Cochran, H. B.,** Determination of calcium and magnesium in saturated sodium chloride brines by graphite furnace atomic absorption spectrophotometry, *At. Spectrosc.,* 2, 97, 1981.
4. **Powell, L. A. and Tease, R. L.,** Determination of calcium, magnesium, strontium and silicon in brines by graphite furnace atomic absorption spectroscopy, *Anal. Chem.,* 54, 2154, 1982.
5. **Kantor, T., Bezur, L., Pungor, E., and Winefordner, J. D.,** Volatilization studies of magnesium compound by a graphite furnace and flame combined atomic absorption method. The use of halogenating atmosphere, *Spectrochim. Acta,* 38B, 581-607, 1983.
6. **Patel, B. M., Goyal, N., Purohit, P., Dhobale, A. R., and Joshi, B. D.,** Direct determination of magnesium, manganese, nickel and zinc in uranium by electrothermal atomic absorption spectroscopy, *Fresnius' Z. Anal. Chem.,* 315, 42-46, 1983.
7. **Kawani, Y.,** Determination of plasma zinc, magnesium and copper by atomic absorption spectrometry with flame or flameless system, *Kyushu Yakagakkai Kaiho,* 37, 43, 1983.
8. **Chung, C.,** Atomization mechanism with arrhenius plots taking the dissipation function into account in graphite furnace atomic absorption spectrometry, *Anal. Chem.,* 56, 2714, 1984.
9. **Ding, Y., Yao, M., Che, K., and Yang, W.,** Content of zinc, calcium, magnesium, copper and manganese in the hair of Huangshi (China) inhabitants, *Yingyang Xuebao,* 6(3), 298, 1984 (in Chinese).
10. **Miura, H.,** Measurement of magnesium and copper in normal human serum by flameless atomic absorption spectroscopy, *Eisei Kensa,* 33(11), 1425-1428, 1984, CA 21, 103:101226d, 1985.
11. **Tsai, J. L. and Lin, S. M.,** Method of standard addition: optimal conditions of analyzing serum manganese by graphite furnace AAS, *Hua Hsueh,* 44(1), 19-28, 1986 (in Chinese).

MANGANESE

MANGANESE, Mn   (ATOMIC WEIGHT 54.938)

Instrumental Parameters:

Wavelength	279.5 nm
Slit width	80 μm
Bandpass	0.3 nm
Light source	Hollow Cathode Lamp
Lamp current	5 mA
Purge gas	Argon or Nitrogen
Sample size	25 μL
Furnace	Cylindrical cuvette/Pyrolytic graphite

Standard Operating Conditions:

Optimum char temperature	1000°C
Optimum atomization temperature	2000°C
Sensitivity	1 pg/1% absorption
Sensitivity check	3.0-5.0 μg/mL
Working range	1.0-10.0 μg/mL
Background correction	Required only to correct light scattering or non-specific absorption from the sample containing high dissolved solids.

General Note:
1. Use commercial standard or a previously analyzed sample as a working standard.
2. Use ultra clean glass and plastic ware, soaked in 1:5 nitric acid solution and rinsed thoroughly with deionized distilled water.
3. Prepare all analytical solutions in 0.2% (v/v) nitric acid.
4. Check for blank values on all reagents including water.
5. Dilution is recommended when sample exhibits greater than 0.5 absorbance units. Dilutions can be made with deionized distilled water.
6. The analytical precision for this method is approximately 5% RSD.

REFERENCES

1. **Ross, R. T. and Gonzalez, J. G.,** Direct determination of trace quantities of manganese in blood and serum samples using selective volatilization and graphite tube reservoir atomic absorption spectrophotometry, *Bull. Environ. Contam. Toxicol.,* 12, 470, 1974.

2. **Bek, F., Janousek, J. and Molden, B.,** Determination of manganese and strontium in blood serum using the Perkin-Elmer HGA-70 graphite furnace, *At. Absorp. Newsl.,* 13, 47, 1974.

3. **Smeyers-Verbeke, J., Michotte, Y., Van den Winkel, P., and Massart, D. L.,** Matrix effects in the determination of copper and manganese in biological materials using carbon furnace atomic absorbtion spectroscopy, *Anal. Chem.,* 48, 125, 1976.

4. **Maruta, T., Minegishi, K., and Sudoh, G.,** Atomic absorption spectrometric determination of trace amounts of copper, manganese, lead and chromium in cements by direct atomization in a carbon furnace, *Yogyo Kyokou Shi,* 86, 532, 1978.

5. **Hageman, L., Mubarak, A., and Woodriff, R.,** Comparison of interference effects for manganese in constant temperature versus pulse-type electrothermal atomization, *Appl. Spectrosc.,* 33(3), 226, 1979.

6. **Sturgeon, R. E., Berman, S. S., Desaulniers, A., and Russell, B. S.,** Determination of iron, manganese and zinc in seawater by graphite furnace atomic absorption spectroscopy, *Anal. Chem.,* 51, 2364, 1979.

7. **Page, A. G., Godbole, S. V., Kulkarni, M. J., Shelar, S. S., and Joshi, B. D.,** Direct atomic absorption spectroscopy determination of cobalt, chromium, copper, manganese and nickel in uranium oxide (U_3O_3) by electrothermal atomization, *Z. Anal. Chem.,* 296, 40, 1979.

8. **Geladi, P. and Adams, F.,** The determination of beryllium and manganese in aerosols by atomic absorption spectroscopy with electrothermal atomization, *Anal. Chim. Acta,* 105, 219, 1979.

9. **Manning, D. C. and Slavin, W.,** Determination of manganese by electrothermal atomic absorption spectroscopy with a graphite furnace at constant temperature, *Anal. Chim. Acta,* 118, 301, 1980.

10. **Kuga, K.,** Rapid determination of copper, iron and manganese in polyimide resins by atomic absorption spectroscopy using a graphite furnace atomizer, *Bunseki Kagaku,* 29, 342, 1980.

11. **Sterritt, R. M. and Lester, J. N.,** Determination of silver, cobalt, manganese, molybdenum and tin in sewage sludge by a rapid electrothermal atomic absorption spectroscopic method, *Analyst,* 105, 616, 1980.

12. **Hydes, D. J.,** Reduction of matrix effects with a soluble organic acid in the carbon furnace atomic absorption spectrometric determination of cobalt, copper and manganese in seawater, *Anal. Chem.,* 52, 959, 1980.

13. **Segar, D. A. and Cantillo, A. Y.,** Determination of iron, manganese and zinc in seawater by graphite furnace atomic absorption spectroscopy. Comments, *Anal. Chem.,* 52, 1766, 1980.

14. **Sturgeon, R. E., Berman, S. S., Desaulniers, A., and Russell, D. S.,** Determination of iron, manganese and zinc by graphite furnace atomic absorption spectroscopy. Reply to comments, *Anal. Chem.,* 52, 1767, 1980.

15. **Montgomery, J. R. and Peterson, G. N.,** Effects of amm. nitrate on sensitivity for determination of copper, iron and manganese in seawater by atomic absorption spectroscopy with pyrolytically coated graphite tubes, *Anal. Chim. Acta,* 117, 397, 1980.

16. **Sekiya, T., Tanimura, H., and Hkiasa, Y.,** Simplified determination of copper, zinc and manganese in plasma and bile by flameless atomic absorption spectroscopy, *Archiv. Jpn. Chir.,* 50, 729-739, 1981.

17. **Halls, D. J. and Fell, G. S.,** Determination of manganese in serum and urine by electrothermal atomic absorption spectroscopy, *Anal. Chim. Acta,* 129, 205, 1981.

18. **Tominaga, M. and Umezaki, Y.,** Determination of submicrogram amounts of manganese in seawater by graphite furnace atomic absorption spectroscopy, *Nippon Kagaku Kaishi,* 1, 7, 1981.

19. **Carnrick, G. R., Slavin, W., and Manning, D. C.,** Direct determination of manganese in seawater with the L'vov platform and Zeeman background correction in the graphite furnace, *Anal. Chem.,* 53, 1866, 1981.

20. **Genc, O., Akman, S., Ozdural, A. R., Ates, S., and Balkis, T.,** Theoretical analysis of atom formation-time curves for the HGA-74 furnace. II. Evaluation of the atomization mechanisms for manganese, chromium and lead, *Spectrochim. Acta,* 36B, 663, 1981.

21. **Matsusaki, K., Yoshino, T., and Yamamoto, Y.,** Electrothermal atomic absorption spectrometric determination of trace lead, copper and manganese in aluminum and its alloys without preliminary separation, *Anal. Chim. Acta,* 144, 189, 1982.

22. **Arnold, D. and Kuennecke, A.,** Determination of 3d elements in high lead-containing glasses, *Silikattechnik,* 33, 312, 1982.

23. **Favier, A., Ruffieux, D., Alcaraz, A., and Maljournal, B.,** Flameless atomic absorption assay of serum manganese, *Clin. Chim. Acta,* 124, 239-244, 1982.

24. **Okuda, M., Ikeda, M., Hayashi, K., Suzuki, H., and Wakabayashi, J.,** Flameless atomic absorption spectrophotometric determination of manganese and blood and serum manganese values in healthy subjects, *Rinsho Byori,* 30, 93-96, 1982 (in Japanese).

25. **Wei, F. S., Qu, W. Q., and Yin, F.,** Direct determination of ultratrace manganese in serum by controlled-temperature graphite furnace method of atomic absorption, *Anal. Lett.,* 15, 721, 1982.

26. **Fagioli, F., Landi, S., and Lucci, G.,** Determination of manganese and copper in small amounts of maize roots by graphite-tube furnace atomic-absorption spectroscopy with liquid and solid sampling technique, *Ann. Chim. (Rome),* 72, 63-71, 1982.

27. **Rorsman, P. and Berggren, P. O.,** Direct determination of manganese in microgram amounts of pancreatic tissue by electrothermal atomic absorption spectroscopy, *Anal. Chim. Acta,* 140, 325-329, 1982.

28. **Clegg, M. S., Keen, C. L., Lonnerdal, B., and Hurly, L. S.,** Analysis of trace elements in animal tissues. III. Determination of manganese by graphite furnace atomic absorption spectrophotometry, *Biol. Trace Elem. Res.,* 4, 145-156, 1982.

29. **Zhou, B. and Zeng, S.,** Simultaneous determination of chromium, manganese and copper in serum by flameless atomic absorption spectrometry, *Yingyang Xuebao,* 4, 323, 1982 (in Chinese).

30. **Nakaaki, K., Tada, O., and Masuda, T.,** Urinary excretion of chromium, manganese, copper and nickel, *Rodo Kagaku,* 58, 529, 1982 (in Japanese).

31. **Lieser, K. H., Sondermeyer, S., and Kliemchen, A.,** Precision and accuracy of analytical results in determining the elements cadmium, chromium, copper, iron, manganese and zinc by flameless atomic absorption spectroscopy, *Fresnius' Z. Anal. Chem.,* 312, 517, 1982 (in German).

32. **Chernykh, O. A.,** Determination of iron and manganese in natural water by atomic absorption, *Gidrokhim. Mater.,* 85, 76, 1983.

33. **Patel, B. M., Goyal, N., Purohit, P., Dhobale, A. R., and Joshi, B. D.,** Direct determination of magnesium, manganese, nickel and zinc in uranium by electrothermal atomic absorption spectroscopy, *Fresnius' Z. Anal. Chem.,* 315, 42-46, 1983.

34. **Khavezov, I. and Ivanova, E.,** Electrothermal atomization using the L'vov platform. Determination of traces of lead, cadmium, manganese and copper in a sodium chloride matrix, *Fresnius' Z. Anal. Chem.,* 315, 34-37, 1983.

35. **Boiteau, H. L., Metayer, C., Ferre, R., and Pineau, A.,** Automated determination of lead, cadmium, manganese and chromium in whole blood by Zeeman atomic absorption spectrometry, *Analysis,* 11(5), 234-242, 1983 (in French).

36. **Favier, A. and Ruffieux, D.,** Physiological variations of serum levels of copper, zinc, iron and manganese, *Biomed. Pharmacother,* 37, 462, 1983.

37. **Rossi, G., Omenetto, N., Pigozzi, G., Vivian, R., Mattinez, U., Mousty, F., and Crabi, G.,** Analysis of radioactive waste solutions by atomic absorption spectrometry with electrothermal atomization, *At. Spectrosc.,* 4, 113, 1983.

38. **Chen, G. H. and Risby, T. H.,** Determination of free silica and manganese in airborne particles by flameless atomic absorption spectroscopy, *Anal. Chem.,* 55, 943, 1983.

39. **Hatano, S., Nishi, Y., and Usui, T.,** Erythrocyte manganese concentration in healthy Japanese children, adults and the elderly and in cord blood, *Am. J. Clin. Nutr.,* 37, 457, 1983.

40. **Koen, E. and Razboinikova, F.,** Flameless atomic-absorption method with graphite tubes for the determination of manganese traces in whole blood and urine, *Khig. Zdraveopaz.,* 26, 422, 1983 (in Bulgarian).

41. **Wei, F. and Yin, F.,** Determination of ultratrace manganese in drinking water by flameless atomic absorption spectrophotometry with the aerosol deposition technique, *Zhongguo Kexue Jishu Daxue Xuebao,* 13, 194, 1983 (in Chinese).

42. **Littlejohn, D., Cook, S., Durie, D., and Ottaway, J. M.,** Investigation of working conditions for graphite probe atomization in electrothermal atomic absorption spectrometry, *Spectrochim. Acta,* 39B, 295, 1984.

43. **Littlejohn, D., Duncan, I., Marshall, J., and Ottaway, J. M.,** Analytical evaluation of totally pyrolytic graphite cuvettes for electrothermal atomic absorption spectrometry, *Anal. Chim. Acta,* 157, 291, 1984.

44. **Grobenski, Z., Lehmann, R., Rudziuk, B., and Voellkpf, U.,** Determination of trace metals in seawater using Zeeman graphite furnace AAS, *At. Spectrosc.,* 5, 87, 1984.

45. **Sumamur, P. K. and Heryuni, S.,** Relationship of blood lead and manganese in occupationally unexposed subjects, *Arh. Hig. Rada Toksikol.,* 35(2), 169-172, 1984.

46. **Ding, Y., Yao, M., Che, K., and Yang, W.,** Content of zinc, calcium, magnesium, copper and manganese in the hair of Huangshi (China) inhabitants, *Yingyang Xuebao,* 6(3), 298, 1984 (in Chinese).

47. **Slavin, W. and Carnrick, G. R.,** Possibility of standardless atomic absorption spectroscopy, *Spectrochim. Acta,* 39B, 271, 1984.

48. **Slavin, W., Carnrick, G. R., and Manning, D. C.,** Chloride interferences in graphite furnace atomic absorption spectrometry, *Anal. Chem.,* 56, 163, 1984.

49. **Nagouney, S. J., Negro, V. C., Bogen, D. C., Latner, N., and Cassidy, M. E.,** Electrothermal atomic absorption spectroscopic analysis using computer, *Comput. Appl. Lab.,* 2, 49, 1984.

50. **Tapia, T. A., Combs, P. A., and Sneddon, J.,** Investigation of an aerosol deposition technique for sample introduction for the determination of manganese by electrothermal atomization atomic absorption spectrometry, *Anal. Lett.,* 17(A20), 2333-2347, 1984.

51. **Kurkus, J., Alcock, N. W., and Shils, M. E.,** Manganese content of large-volume parenteral solutions and of nutrient additives, *J. Parenter. Enteral Nutr.,* 8, 254, 1984.
52. **Brodie, K. G. and Routh, M. W.,** Trace analysis of lead in blood, aluminum and manganese in serum and chromium in urine by graphite furnace atomic absorption spectrometry, *Clin. Biochem.,* 17, 19, 1984.
53. **Boniforti, R., Ferraroli, R., Frigieri, P., Heltai, D., and Queirazza, G.,** Intercomparison of five methods for determination of trace metals in seawater, *Anal. Chim. Acta,* 162, 33, 1984.
54. **Vannini, P.,** Determination of manganese in milk: method for detection of milk adulteration with integrated mixed feed for calves based on skim milk powder reconstituted with water, *Latte,* 9, 395, 1984.
55. **Blackmore, D. J. and Stanier, P.,** Methods for measurement of trace elements in equine blood by electrothermal atomic absorption spectrophotometry, *At. Spectrosc.,* 5, 215, 1984.
56. **Guillard, O., Brugier, J. C., Piriou, A., Menard, M., Gombert, J., and Reiss, D.,** Improved determination of manganese in hair by use of a mini-autoclave and flameless atomic absorption spectroscopy with Zeeman background correction: an evaluation in unexposed subjects, *Clin. Chem. (Winston-Salem, N.C.),* 30, 1642, 1984.
57. **Guillard, O., Gombert, J., Barriere, M., Reiss, D., and Piriou, A.,** Manganese concentration in hair of greying ("salt and pepper") men reconsidered, *Clin. Chem.,* 31, 1251, 1985.
58. **Tominaga, M., Bansho, K., and Umazaki, Y.,** Electrothermal atomic absorption spectrometric determination of lead, manganese, vanadium and molybdenum in seawater with ascorbic acid to reduce matrix effects, *Anal. Chim. Acta,* 169, 171, 1985.
59. **Statham, P. J.,** Determination of dissolved manganese and cadmium in seawater at low nmol per liter concentrations by chelation and extraction followed by electrothermal atomic absorption spectrometry, *Anal. Chim. Acta,* 169, 149, 1985.
60. **Lewis, S. A., O'Haver, T. C., and Harnly, J. M.,** Determination of metals at microgram-per-liter level in blood serum by simultaneous multielement atomic absorption spectrometry with graphite furnace atomization, *Anal. Chem.,* 57, 2, 1985.
61. **Montgomery, J. R., Hucks, M., and Peterson, G. N.,** A portable non-contaminating sampling system for iron and manganese in sediment pore water, *Fla. Sci.,* 48(1), 46-49, 1985.
62. **Kitagishi, K. and Obata, H.,** Extra situ quantitative histochemical determination of zinc and manganese in plant tissues by flameless atomic absorption spectrophotometry, *Mie Daigaku Kankyo Kagaku Kenkyu Kiyo,* 10, 171-179, 1985.
63. **Rozanska, B., Skorko-Trybula, Z., and Slodownik, A.,** Precipitation and extraction separation of lead, iron and zinc prior to the determination of cobalt, nickel and manganese by atomic absorption in dust from electric furnace, *Chem. Anal. (Warsaw),* 30(3), 471-479, 1985.
64. **Subramanian, K. S. and Meranger, J. C.,** GFAAS with nitric acid deproteinization for determination of manganese in human plasma, *Anal. Chem.,* 57(13), 2478-2481, 1985.
65. **Giannissis, D., Dorange, G., and Martin, G.,** Effect of humic substances on the elimination of manganese in a water treatment plant, *Rev. Fr. Sci. Eau,* 4(2-3), 149-162, 1985 (in French).
66. **Swovick, M.,** Detection of manganese in gunshot residue, *At. Spectrosc.,* 6(3), 79-80, 1985.
67. **Dougherty, J. P., Michel, R. G., and Slavin, W.,** Precision considerations in the determination of manganese in mouse brains by FAA with Zeeman background correction, *Anal. Lett.,* 18(A10), 1231-1244, 1985.
68. **Saner, G. and Dagoglu, T.,** Hair manganese concentrations in newborns and their mothers, *Am. J. Clin. Nutr.,* 41(5), 1042, 1985.
69. **Martin, M. L. and Danzerick, K.,** Graphite furnace atomic absorption determination of trace amounts of manganese in lead monoxide, Wiss. Z. Friedrich-Schiller Univ., *Jena, Naturwiss Reihe,* 34(5-6), 803, 1985.
70. **Erb, R.,** Quality control by Zeeman-AAS in a chemical factory. Direct determination of copper, manganese, iron, cadmium and materials, *Fresnius' Z. Anal. Chem.,* 322(7), 719-720, 1985.
71. **Xinhua, Y., Shaoquan, L., and Yanzi, Z.,** Determination of manganese in serum and blood by platform atomic absorption spectroscopy, *Fenxi Ceshi Tongbao,* 4, 14, 1985 (in Chinese).
72. **Sneddon, J.,** Use of a impaction-electrothermal atomization atomic absorption spectrometric system for the direct determination of cadmium, copper and manganese in the laboratory atmosphere, *Anal. Lett.,* 18(A10), 1261-1280, 1985.
73. **Itoh, K., Itoh, T., Akatsuka, K., and Atsuya, I.,** Direct determination of manganese in several biological samples by polarized Zeeman atomic absorption spectrometry with a graphite miniature cup, *Bunseki Kagaku,* 35(2), 122-127, 1986 (in Japanese).
74. **Kubo, T., Takano, Y., Komoike, Y., and Haraguchi, S.,** Synergistic extraction of manganese in blood with dithizone and o-phenanthroline and determination by atomic absorption spectrophotometry, *Sumitomo Sangyo Eisei,* 21, 101-112, 1985, CA 8, 104:103653k, 1986.
75. **Deano, P. and Robinson, J. W.,** Direct determination of manganese in perspiration and urine using atomic absorption spectroscopy, *Spectrosc. Lett.,* 19(1), 11-19, 1986.
76. **Clegg, M. S., Lonnerdal, B., Hurly, L. S., and Keen, C. L.,** Analysis of whole blood manganese by flameless atomic absorption spectrophotometry and its use as an indicator of manganese status in animals, *Anal. Biochem.,* 157, 12, 1986.

77. **Uchida, T. and Vallee, B. L.,** Simple and micro determination of manganese in serum by graphite furnace atomic absorption spectrometry, *Anal. Sci.,* 2(1), 71-75, 1986.
78. **Yang, X., Li, S., and Zhang, Y.,** Determination of manganese in serum, and blood by platform atomic absorption spectroscopy, *Fenxi Ceshi Tongbao,* 4(5), 14-17, 1986.
79. **Bayer, W.,** Manganese determination in whole blood, *Fortschr. Atmospektrom. Spurenanal.,* 2, 197, 1986.
80. **Lian, T. J. and Meei, L. S.,** Method of standard addition: Optimal conditions of analyzing serum manganese by graphite furnace atomic absorption spectrophotometry, *Hua Hsueh,* 44, 19, 1986.
81. **Kumar, A., Hasan, M. Z., and Deshmukh, B. T.,** Interference-free determination of manganese in water by graphite furnace atomic absorption spectrophotometry, *Indian J. Pure Appl. Phys.,* 24, 465-468,1986.
82. **Shijo, Y., Watanabe, J., Akiyama, S., Shimizu, T., and Sakai, K.,** Determination of manganese in seawater by graphite furnace AAS after micro-solvent extraction, *Bunseki Kagaku,* 36, 59, 1987.

AUTHOR INDEX

SUBJECT INDEX

MERCURY

MERCURY, Hg   (ATOMIC WEIGHT 200.59)

Instrumental Parameters:

Wavelength	253.7 nm
Slit width	320 μm
Bandpass	1.0 nm
Light source	Hollow Cathode Lamp
Lamp current	3 mA
Purge gas	Argon or Nitrogen
Sample size	25 μL
Furnace	Cylindrical cuvette/Pyrolytic graphite

Standard Operating Conditions:

Optimum char temperature	250°C
Optimum atomization temperature	2000°C
Sensitivity	40 pg/1% absorption
Sensitivity check	2.0 μg/mL
Working range	40.0-1000.0 μg/mL
Background correction	Required only to correct light scattering or non-specific absorption from the sample containing high dissolved solids.

General Note:
1. Use commercial standard or a previously analyzed sample as a working standard.
2. Use ultra clean glass and plastic ware, soaked in 1:5 nitric acid solution and rinsed thoroughly with deionized distilled water.
3. Prepare all analytical solutions in 0.2% (v/v) nitric acid.
4. Check for blank values on all reagents including water.
5. Dilution is recommended when sample exhibits greater than 0.5 absorbance units. Dilutions can be made with deionized distilled water.
6. Use 1000 μg/mL tellurium in 1% (v/v) hydrochloric acid or 0.1% (w/v) potassium dichromate in 0.5% (v/v) nitric acid as diluent.
7. Dilute mercury solutions are unstable. Do not store in plastic containers.
8. Addition of organo-sulfur compounds (such as cysteine), as complexing agents, increases the maximum pyrolysis temperature to 500°C. High char or pyrolysis temperature is necessary to minimize chemical and bulk matrix interference.

REFERENCES

1. **Kamada, T., Hayashi, Y., Kumamaru, T., and Yamamoto, Y.,** Flameless atomic absorption spectrophotometry for determination of inorganic and organic mercury at the parts per billion level in water using the vapor phase equilibrated with the solution, *Bunseki Kagaku,* 22, 1481, 1973.
2. **Siemer, D. and Woodriff, R.,** Application of the carbon rod atomizer to the determination of mercury in the gaseous products of oxygen combustion of solid samples, *Anal. Chem.,* 46, 597, 1974.
3. **Siemer, D., Lech, J., and Woodriff, R.,** Application of carbon rod atomizer for the analysis of mercury in air, *Appl. Spectrosc.,* 28, 68, 1974.
4. **Fujiwara, K., Sato, K., and Fuwa, K.,** Atomic absorption spectroscopy of mercury using a graphite furnace atomizer, *Bunseki Kagaku,* 26, 772, 1977.
5. **Alder, J. F. and Hickman, D. A.,** Determination of mercury by atomic absorption spectrometry with graphite tube atomization, *Anal. Chem.,* 49, 336, 1977.
6. **Hocquellet, P.,** Application of electrothermal atomization to the determination of arsenic, antimony, selenium and mercury by atomic absorption spectroscopy, *Analysis,* 6, 426, 1978.
7. **Shan, X. and Ni, Z.,** Matrix modification for the determination of mercury using electrothermal atomic absorption spectroscopy, *Acta Chim. Sinica,* 37, 261, 1979.
8. **Shum, G. T., Freeman, H. C., and Uthe, J. F.,** Determination of organo (methyl)mercury in fish by graphite furnace atomic absorption spectrophotometry, *Anal. Chem.,* 51, 414, 1979.
9. **Kunert, I., Komarek, J., and Sommer, L.,** Determination of mercury by atomic absorption spectroscopy with cold vapor and electrothermal technique, *Anal. Chim. Acta,* 106, 285, 1979.
10. **Karmanova, N. G. and Pogrebnyak, Y. F.,** Determination of mercury in powdered samples of sulfide ores by an atomic absorption spectral method with a "graphite capsule adapter" atomizer, *Zh. Prikl. Spektrosk.,* 33, 813, 1980.
11. **Siemer, D. D. and Hageman, L.,** Determination of mercury in water by furnace atomic absorption spectroscopy after reduction and aeration, *Anal. Chem.,* 52, 105, 1980.
12. **Kirkbright, G. F., Shan, H. C., and Snook, R. D.,** Evaluation of some matrix modification procedures for use in the determination of mercury and selenium by atomic absorption spectroscopy with a graphite tube electrothermal atomizer, *At. Spectrosc.,* 1, 85, 1980.
13. **Cavallaro, A.,** Mercury pollution: flameless atomic absorption analysis, *Rischi Tossic. Inquin. Met. Cromo Mercurio (Conv. Naz.),* 145-70, 1980 (in Italian).
14. **Halasz, A., Polyak, K., and Gegus, E.,** Processes taking place in the graphite tube: anion and matrix effects in atomic absorption spectroscopy. II. Determination of mercury, *Mikrochim. Acta,* 2, 229, 1981.
15. **Gras, G. and Mondain, J.,** Colorimetric microdetermination of total mercury with di-beta-naphthylthiocarbazone. Application to control in fish, *Ann. Pharm. Fr.,* 39, 529-36, 1981 (in French).
16. **Michalewska, M.,** Determination of mercury compounds in copper ores by flameless atomic absorption spectrophotometry, *Chem. Anal. (Warsaw),* 26, 659, 1981 (in German).
17. **Romanova, I. B. and Kashparova, E. V.,** Determination of mercury content in cultural fluids, *Deposited Doc. VINITI,* 227-81, 13pp, 1981. Avail. VINITI.
18. **Mueller, B., Wenzel, B., and Shroeder, L.,** Quantitative determination of phenylmercury acetate in the presence of inorganic mercury salts using flameless atomic absorption, *Z. Chem.,* 21, 367, 1981 (in German).
19. **Didorenko, T. O., Vitkun, R. A., and Zelyukova, Y. V.,** Mercury determination by flameless atomic absorption method, *Lab. Delo,* 9, 536, 1981 (in Russian).
20. **Miyagawa, H.,** Microdetermination of mercury by carbon tube flameless atomic absorption spectroscopy, *Nenpo-Fukui-ken Kogyo Shikenjo,* 54-6, 1980, published in 1981 (in Japanese).
21. **Torsi, G., Desimoni, E., Palmisano, F., and Sabbatini, L.,** Determination of mercury vapor in air using electrothermal atomic absorption spectroscopy with an electrostatic accumulation furnace, *Analyst (London),* 107, 96-103, 1982.
22. **Katalevskii, N. I., Anikanov, A. M., and Semenov, A. D.,** Determination of mercury in natural waters by a flameless atomic absorption method, *Metodiki Analiza Mor. Vod. Tr. Sov.-Bolg. Sotrudnichestva,* L., 78-82, 1981 (in Russian). From Ref. *Zh. Khim.,* Abstr. No. 7G164, 1982.
23. **Halasz, A., Polyak, K., and Gegus, E.,** Determination of mercury by electrothermal atomization following enrichment by extraction, *Magy. Kem. Foly,* 88, 139-43, 1982.
24. **Flanagan, F. J., Moore, R., and Aruscavage, P. J.,** Mercury in geologic reference samples, *Geostand. Newsl.,* 6(1), 25-46, 1982.
25. **Chen, J. S., Wang, N. S., Wei, J. C., Ke, J. N., Huang, M. H., and Yang, M. H.,** Determination of trace amounts of mercury in biological and environmental samples by cold vapor atomic absorption spectroscopy and neutron activation analysis, *Hua Hsueh,* 40(2), 45-51, 1982 (in Chinese).
26. **Doi, R. and Kobayashi, T.,** Organ distribution and biological half-time of methylmercury in four strains of mice, *Jpn. J. Exp. Med.,* 52, 307, 1982.
27. **Robinson, J. W. and Skelly, E. M.,** The direct determination of mercury in breath and saliva by carbon bed atomic absorption spectroscopy, *Spectrosc. Lett.,* 15, 631, 1982.

28. **Freiman, P. and Schmidt, D.**, Determination of mercury in seawater by cold vapor atomic absorption spectrophotometry, *Fresnius' Z. Anal. Chem.*, 313, 200, 1982.
29. **Roschig, M. and Wuenscher, R. G.**, Proposal to the 1983 GDR pharmacopeia diagnostic laboratory methods: determination of mercury in biological material (flameless atomic absorption spectroscopy), *Zentralbl. Pharm., Pharmakother. Laboratoriumsdiagn.*, 121, 893, 1982.
30. **Kurfuerst, U. and Rues, B.**, Direct mercury determination in solids, *Labor Praxis*, 6, 1094, 1982.
31. **Magyar, B., Vonmont, H., and Cicciarelli, R.**, Determination of mercury in sludge and materials used for mercury eliminations from industrial wastewaters by Zeeman atomic absorption spectroscopy after matrix modification with potassium bromide and bromine, *Mikrochim. Acta*, 2, 407, 1982.
32. **Dumarey, R., Dams, R., and Sandra, P.**, Interfacing an atomic absorption spectrophotometer with a fused silica capillary gas chromatographic system for the determination of trace organomercury (II) compounds, *J. High Resolut. Chromatogr. Chromatogr. Commun.*, 5, 687, 1982.
33. **Zelentsova, L. V., Yudelevich, I. G., and Chanysheva, T. A.**, Flameless atomic absorption determination of mercury in high-purity metals, *Izv. Sib. Otd. Akad. Nauk SSSR, Ser. Khim. Nauk*, 1, 130, 1982.
34. **Bitkun, R. A., Didorenko, T. O., Zelyukova, Y. V., and Poluektov, N. S.**, Determination of mercury using flameless atomic absorption with dihydroxymaleic acid as the reducing agent, *Zh. Anal. Khim.*, 37, 833-6, 1982 (in Russian).
35. **Ni, Z. and Yang, F.**, Determination of mercury in soil by graphite furnace atomic absorption spectroscopy using citric acid as matrix modifier, *Huanjing Huaxue*, 1, 83, 1982 (in Chinese).
36. **Skorzynska, K.**, Application of flameless atomic absorption to mercury determination in pharmaceutical preparations of vegetable origin, *Farm. Pol.*, 38, 199, 1982 (in Polish).
37. **Furuta, R. and Mihara, Y.**, The determination of trace mercury in solid fuel by atomic absorption spectrophotometry, *Bunseki Kagaku*, 31, 367, 1982 (in Japanese).
38. **Torsi, G., Desimoni, E., and Palmisano, F.**, Determination of mercury vapor in air using electrothermal atomic absorption spectroscopy with an electrostatic accumulation furnace, *Analyst*, 107, 96, 1982.
39. **Zelyukova, Y. V., Vitkun, R. A., and Poluektov, N. S.**, Kinetics of the separation of mercury from solution for its determination by flameless atomic absorption, *At. Spectrosc.*, 3, 146, 1982.
40. **Polos, L., Fodor, P., and Pungor, E.**, Direct method for the determination of mercury content of solid samples, *Hung. Sci. Instrum.*, 53, 11-14, 1982.
41. **Metil, N. I., Taushan, M. D., Chagir, T. S., and Shevchuk, I. A.**, Flameless atomic absorption method for the determination of blood mercury content, *Lab. Delo*, 1, 25, 1982 (in Russian).
42. **Mueller, J. and Kallischnigg, G.**, A ring test for the determination of lead, cadmium and mercury in biological material, *ZEBS-Ber.*, 1, 69, 1983.
43. **Aftabi, A. and Azzaria, L. M.**, Distribution of mercury compounds in ore and host rocks at sigma gold mine, Vald'or, Quebec, Canada, *J. Geochem. Explor.*, 19, 447, 1983.
44. **Fitzgerald, W. F., Gill, G. A., and Hewitt, A. D.**, Air-sea exchange of mercury, *NATO Conf. Ser.*, 4, 9, 1983.
45. **Matsumoto, M.**, Studies of behavior of atmospheric mercury in general and mercury deposit areas, *Taiki Osen Gakkaishi*, 18, 66, 1983 (in Japanese).
46. **Robinson, J. W. and Skelly, E. M.**, The direct determination of mercury in water: A more accurate method than cold vapor-AAS, *Spectrosc. Lett.*, 16, 33-58, 1983.
47. **Robinson, J. W. and Skelly, E. M.**, The direct determination of mercury in whole blood and serum, *Spectrosc. Lett.*, 16, 59-76, 1983.
48. **Robinson, J. W. and Skelly, E. M.**, The direct determination of mercury in urine, *Spectrosc. Lett.*, 16, 117, 1983.
49. **Hon, P. K., Lau, O. W., and Wong, M. C.**, Novel static cold vapor atomic absorption method for the determination of mercury, *Analyst (London)*, 108, 64, 1983.
50. **Liu, Y., Gong, B., Sun, J., and Lin, T.**, Development and performance of an automatic hydride/mercury system, *Bunseki Kagaku*, 32, E9, 1983.
51. **Fujita, M.**, Continuous flow reducing vessel in determination of mercuric compounds by liquid chromatography/cold vapor atomic absorption spectrometry, *Anal. Chem.*, 55, 454, 1983.
52. **Tanida, K., Fukuda, H., and Hoshino, M.**, Continuous determination of mercury in air by gold amalgamation and flameless atomic absorption, *Bunseki Kagaku*, 32, 352, 1983.
53. **Ayyadurai, K., Kamalam, N., and Rajagopal, C. K.**, Mercury pollution in water in Madras City, *Indian J. Environ. Health*, 25(1), 15-20, 1983.
54. **Baba, T., Ohmiya, S., Hosokawa, M., and Ishibashi, T.**, An improved pretreatment method for the determination of total mercury, *Seikatsu Eisei*, 27(5), 258-63, 1983 (in Japanese).
55. **Metil, N. I. and Shevchuk, I. A.**, Flameless atomic absorption assay of mercury in the blood, *Lab. Delo*, 9, 25-7, 1983 (in Russian).
56. **Konishi, T. and Takahashi, H.**, Direct determination of inorganic mercury in biological materials after alkali digestion and amalgamation, *Analyst (London)*, 108, 827-34, 1983.
57. **Dmitriev, M. T., Granovskii, E. I., and Slashchev, A. Y.**, Determination of mercury levels in the environment and in biological materials by atomic absorption, *Gig. Sanit.*, 9, 50-3, 1983 (in Russian).

58. **Bertocchi, G., Benfenati, L., and Bertolini, C.,** Determination of mercury in pharmaceutical products using atomic absorption spectroscopy, *Relata Tech.,* 15, 45-8, 1983 (in Italian).
59. **Fukuzaki, N. and Tamura, R.,** Method for the measurement of accumulative amounts of mercury in air, *Bunseki Kagaku,* 32, 391, 1983 (in Japanese).
60. **Tanida, K., Fukuda, H., and Hoshino, M.,** Continuous determination of mercury in air by gold amalgamation and flameless atomic absorption, *Bunseki Kagaku,* 32, 352, 1983.
61. **Zelyukova, Y. V. and Diderenko, T. O.,** Atomic absorption determination in semifinished products and wastes of the lead-zinc industry, *Ukr. Khim. Zh. (Russian Ed.),* 49, 526, 1983 (in Russian).
62. **Diggs, T. H. and Ledbetter, J. O.,** Palladium chloride enhancement of low level mercury analysis, *Am. Ind. Hyg. Assoc. J.,* 44, 606, 1983.
63. **Holak, W.,** Atomic absorption spectrophotometric determination of mercury in mercury-containing drugs: collaborative study, *J. Assoc. Off. Anal. Chem.,* 66, 1203, 1983.
64. **Brune, D., Gjerdet, N., and Paulsen, G.,** Gastrointestinal and *in vitro* release of copper, cadmium, indium, mercury and zinc from conventional and copper-rich amalgams, *Scand. J. Dent. Res.,* 91, 66, 1983.
65. **Karimian-Teherani, D., Kiss, I., Altmann, H., Wallisch, G., and Kapeller, K.,** Accumulation and distribution of elements in plants (paprika), *Acta Aliment.,* 12, 301, 1983.
66. **Filippelli, M.,** Determination of trace amounts of mercury in seawater by graphite furnace atomic absorption spectrophotometry, *Analyst,* 109, 515, 1984.
67. **Protasowicki, M.,** Review of sample preparation methods for determination of mercury in fish tissues by the cold vapor technique, *Mater. Konwersatorium Spektrom. At. Zastrsow. Metod. Emisyjnej Absorpc. Spektrom. At. Naukach Ryabackich Roln.,* 121, 56-60,1983.
68. **Nakagawa, R., Tatsumoto, H., and Horiuchi, N.,** Behavior of mercury in atmosphere. Studies of the distribution of atmospheric mercury in Japan, *Kankyo Kagaku Kenkyu Hokoku,* 9, 16-23, 1984.
69. **Keller, B. J., Peden, M. E., and Rattonetti, A.,** Graphite furnace atomic absorption method for trace-level determination of total mercury, *Anal. Chem.,* 56, 2617, 1984.
70. Apparatus for determination of mercury, Hitachi, Ltd., Jpn. Kokai Tokkyo Koho JP59 79,837 (8479,837) (Cl.G01 N21/31), 09 May 1984, 2pp. Appl. 82/184,051, 29 Oct. 1982.
71. **Krueger, K. E. and Kruse, R.,** Mercury content of canned tuna from Southern Europe, Africa and Asia, *Arch. Lebensmittelhyg.,* 35, 55, 1984.
72. **Luca, C., Tarabic, M., Sarachie, V., and Danet, A. F.,** Method and apparatus for determination of mercury, *Rev. Chim. (Bucharest),* 35, 526, 1984.
73. **Tajima, K., Fujita, M., Kai, F., and Takamatsu, M.,** Differentiation of mercury alkane thiolates by high performance liquid chromatography, *J. Chromatogr. Sci.,* 22, 244, 1984.
74. **Batz, L., Ganz, S., Hermann, G., Scharmann, A., and Wirz, P.,** Measurement of stable isotope distribution using Zeeman atomic absorption spectroscopy, *Spectrochim. Acta,* 39B, 993, 1984.
75. **Keller, B. J., Peden, M. E., and Rattonetti, A.,** Graphite-furnace atomic absorption method for trace-level determination of total mercury, *Anal. Chem.,* 56, 2617, 1984.
76. **Seth, P. C. and Pandey, G. S.,** Mercury as pollutant in dust fallout from steel plant, *Indian J. Environ. Health,* 26(4), 298-303, 1984.
77. **Fukuzaki, N. and Ichikawa, Y.,** Determination of particulate mercury in air, *Bunseki Kagaku,* 33, 178, 1984 (in Japanese).
78. **Didorenko, T. O., Zelyukova, Y. V., and Poluektov, N. S.,** Use of dimethylaminoborane as a reducing agent in flameless atomic absorption determination of mercury, *Zh. Anal. Khim.,* 39(2), 282, 1984.
79. **Jaffar, M. and Athar, M.,** Inorganic mercury determination in local public utility waters by flameless atomic absorption technique, *Pak. J. Sci. Ind. Res.,* 27(3), 121-3, 1984.
80. **Adebeg, V., Struckmann, I., and Kirsch, D.,** Determination of higher mercury concentrations by flameless atomic absorption, *Z. Chem.,* 24(9), 339, 1984 (in German).
81. **Nabrzyski, M. and Kostrzewska, B.,** Comparative determination of total mercury in food by the dithizone method and flameless atomic absorption spectrometry, *Rocz. Panstw. Zakl. Hig.,* 35(2), 125-30, 1984.
82. **Ziegler, E. and Ziegler, B.,** Simple instrumental determination of mercury in urine by using graphite FAAS, *Fortschr. Atomspektrom. Spurenanal.,* 1, 309-15, 1984.
83. **Uchikawa, H., Furata, R., and Mihara, Y.,** Determination of trace mercury in solid fuel by atomic absorption spectrophotometry, Onoda *Kenkyu Hokoku,* 34(107), 19-25, 1982, CA 7, 100:88350f, 1984.
84. **Graeser, K. and Staiger, K.,** Determination of contents of mercury in the nanogram range in biological material and in soil samples by an efficient analyzing procedure-technique in measuring the analyzing procedure, *Z. Gesamte Hyg. Ihre Grenzgeb.,* 29(12), 737-9, 1983, CA 7, 100:80775c, 1984.
85. **Graeser, K. and Staiger, K.,** Determination of the contents of mercury in the nanograms range in biological material and in soil samples by an efficient analyzing procedure-decomposing procedure for organic and soil materials, *Z. Gesamte Hyg. Ihre Grenzgeb.,* 29(12), 734-7, 1983, CA 6, 100:62968, 1984.
86. **Breder, R. and Flucht, R.,** Mercury levels in atmosphere of various regions and locations in Italy, *Sci. Total Environ.,* 40, 231, 1984.

87. **Tomita, T., Kadowaki, S., Takahashi, T., Yamamoto, H., and Morita, Y.**, Determination of atmospheric mercury. On gold amalgamation and flameless atomic absorption spectrometry, *Aichi-ken Kogai Chosa Senta Shoho,* 12, 16-20, 1984.

88. **Nomura, T. and Karasawa, I.**, Behavior of mercury (II) salts in electrothermal atomic absorption spectrometry. Application to determination of thiosulfates, *Anal. Chim. Acta,* 167, 269, 1985.

89. **Gorbenski, Z., Erler, W., and Voellkopf, U.**, Determination of mercury with Zeeman graphite furnace AAS, *At. Spectrosc.,* 6(4), 91-3, 1985.

90. **Michitsuji, H., Ohara, A., Yamaguchi, K., and Fujiki, Y.**, Mercury determination by graphite furnace atomic absorption spectroscopy, *Shojinkai Igakashi,* 24, 253, 1985.

91. **Fox, J. B.**, Distribution of mercury during simulated *in situ* oil shale retorting, *Environ. Sci. Technol.,* 19(4), 316-22, 1985.

92. **Nakagawa, R.**, Particulate mercury concentrations and its behavior in urban ambient air, *Nippon Kagaku Kaishi,* 4, 709, 1985 (in Japanese).

93. **Vackova, M. and Zemberyova, M.**, Determination of mercury in environmental samples by flameless atomic absorption spectroscopy (AAS), *Acta Fac. Rerum Nat. Univ. Comenianae Form. Prot. Nat.,* 9, 223-32, 1984, CA 21, 103:115273d, 1985.

94. **Ping, L., Fuwa, K., and Matsumoto, K.**, Sensitivity enhancement by palladium addition in the electrothermal AAS of mercury, *Anal. Chim. Acta,* 171, 279-84, 1985.

95. **Fleckenstein, J.**, Direct determination of mercury in solid biological samples by Zeeman AAS (ZAAS) with the graphite furnace, *Fresnius' Z. Anal. Chem.,* 322(7), 704-7, 1985.

96. **Rozanska, B. and Lachowicz, E.**, Determination of mercury in industrial materials by atomic absorption spectroscopy after thermal volatilization, *Anal. Chim. Acta,* 175, 211-17, 1985.

97. **Nakagawa, R.**, Particulate mercury concentration and its behavior in urban ambient air, *Anal. Chim. Acta,* 172, 127, 1985.

98. **Xu, B., Xu, T., Shen, M., and Fang, Y.**, Determination of sub-μg/mL levels of mercury in water by electrolytic deposition and electrothermal atomic absorption spectrophotometry, *Talanta,* 32(10), 1016-18, 1985.

99. **Protasovitski, M.**, Determination of mercury by FAAS by the cold-vapor technique, *Probl. Fonovogo Monit. Sostoyaniya Prir. Sredy,* 3, 48-53, 1985.

100. **Komarova, Z. V. and Guletskaya, N. A.**, Determination of mercury in soils by flameless atomic absorption, *Clays Clay Miner.,* 33(6), 563-6, 1985.

101. **Tamura, R., Fukuzaki, N., Hirano, Y., and Mizushima, Y.**, Evaluation of mercury contamination using plant leaves and humus as indicators, *Chemosphere,* 14(11-12), 1687-93, 1985.

102. **Horvat, M., Dermelj, M., and Kosta, L.**, Tentative determination of total mercury, methylmercury (Me-Hg) and selenium in the hair of persons from different parts of Yugoslavia, *5th Int. Conf., Heavy Met. Environ.,* Lekkas, T. D., Ed., CEP Consult., Edinburgh, U.K., 1985, 73-5.

103. **Tatsumoto, H., Nakagawa, R., Suzuki, M., and Suzuki, S.**, Increase of mercury vapor concentrations in the air of a working environment, *Taiki Osen Gakkaishi,* 21, 197-203, 1986 (in Japanese).

104. **Semu, E., Singh, B. R., and Selmer-Olsen, A. R.**, Mercury pollution of a effluent air and soil near a battery factory in Tanzania, *Water, Air, Soil Pollut.,* 27(1-2), 141-6, 1986.

105. **Lin, R., Zhong, C., Zeng, X., Lin, S., Li, J., Huang, G., Zhang, R., and He, H.**, Development of KT-81 Zeeman effect mercury analyzer, *Fenxi Ceshi Tongbao,* 4(5), 51-5, 1985, CA 8, 104:122049j, 1986.

106. **Rincon-Leon, F. and Zurera-Cosano, G.**, Flameless atomic absorption spectrophotometric determination of mercury in mushroom samples using a mercury/hydride system, *At. Spectrosc.,* 7(3), 82-4, 1986.

107. **DeArmas, F. and Diaz, P.**, Flameless atomic absorption (spectroscopy) for the determination of mercury in the urine of exposed workers. Modification to A.O. Rathje method, *Rev. Cubana Farm.,* 20, 35-9, 1986 (in Spanish).

108. **Rozanska, B. and Domanska, M.**, Use of additives in pyrolytic separation of mercury from industrial samples for cold-vapor atomic absorption spectroscopy, *Anal. Chim. Acta,* 187, 317, 1986.

109. **Nakamura, E. and Namiki, H.**, Preconcentration and determination of nanogram per liter levels of mercury in water, *Bunseki Kagaku,* 36, 120, 1987.

110. **Filippelli, M.**, Determination of trace amounts of organic and inorganic mercury in biological material by graphite furnace atomic absorption spectroscopy and organic mercury speciation by gas chromatography, *Anal. Chem.,* 59, 116-18, 1987.

111. **Gill, G. A. and Fitzgerald, W. F.**, Picomolar mercury measurements in seawater and other materials using stannous chloride reduction and two-stage gold amalgamation with gas phase detection, *Mar. Chem.,* 20, 227, 1987.

AUTHOR INDEX

SUBJECT INDEX

MOLYBDENUM

MOLYBDENUM, Mo   (ATOMIC WEIGHT 95.94)

Instrumental Parameters:

Wavelength	313.3 nm
Slit width	160 μm
Bandpass	0.5 nm
Light source	Hollow Cathode Lamp
Lamp current	6 mA
Purge gas	Argon
Sample size	25 μL
Furnace	Cylindrical cuvette/Pyrolytic graphite

Standard Operating Conditions:

Optimum char temperature	1500°C
Optimum atomization temperature	2700°C
Sensitivity	15 pg/1% absorption
Sensitivity check	0.05-0.3 μg/mL
Working range	0.01-0.6 μg/mL
Background correction	Required only to correct light scattering or non-specific absorption from the sample containing high dissolved solids.

General Note:

1. Use commercial standard or a previously analyzed sample as a working standard.
2. Use ultra clean glass and plastic ware, soaked in 1:5 nitric acid solution and rinsed thoroughly with deionized distilled water.
3. Prepare all analytical solutions in 0.2% (v/v) nitric acid.
4. Check for blank values on all reagents including water.
5. Dilution is recommended when sample exhibits greater than 0.5 absorbance units. Dilutions can be made with deionized distilled water
6. Use of nitrogen as purge gas and high acid concentrations (>3% v/v) cause reduced sensitivity.

REFERENCES

1. **Bodrov, N. V. and Nikolaev, G. I.,** Characteristics of the atomic absorption determination of molybdenum in a graphite cell, *Zh. Anal. Khim.,* 24, 1314, 1969.

2. **Muzzarelli, R. A. A. and Rocchetti, R.,** Determination of molybdenum in seawater by hot graphite atomic absorption spectroscopy after concentration on p-amino benzyl cellulose or chitosan, *Anal. Chim. Acta,* 64, 371, 1973.

3. **Schweizer, V. B.,** Determination of Co, Cr, Cu, Mo, Ni and V in carbonate rocks with the HGA-70 graphite furnace, *At. Ab. Newsl.,* 14, 137, 1975.

4. **Ni, Z., Jin, L., and Wu, D.,** Applications of graphite furnace atomic absorption to the determination of molybdenum, *Environ. Sci.,* 6, 25, 1979 (in Chinese).

5. **Sugimae, A.,** Matrix effects in the determination of molybdenum in plants by carbon furnace atomic absorption spectroscopy, *Anal. Chem.,* 51, 1336, 1979.

6. **Nakahara, T. and Chakrabarti, C. L.,** Direct determination of traces of molybdenum in synthetic sea water by atomic absorption spectroscopy with electrothermal atomization and selective volatilization of the salt matrix, *Anal. Chim. Acta,* 104, 173, 1979.

7. **Neuman, D. R. and Munshower, F. F.,** Rapid determination of molybdenum in botanical material by electrothermal atomic absorption spectroscopy, *Anal. Chim. Acta,* 123, 325, 1981.

8. **Barbooti, M. M. and Jasim, F.,** Electrothermal atomic absorption spectrophotometric determination of molybdenum, *Talanta,* 28, 359, 1981.

9. **Bet-Pera, F.,** Electrothermal reduction of perchlorate and nitrate in the presence of molybdenum (VI): analytical applications of molybdenum (VI) for the determination of sub micro amounts of phosphate using spectrophotometric and atomic absorption with graphite furnace, 177pp, 1982. *Avail. Univ. Microfilms Int.,* Order No. DA8212277. From Diss. Abstr. Int. B, 42, 4804, 1982.

10. **Yoshimura, C. and Matsumura, K.,** Effect of carbon black on flameless atomic absorption spectrophotometry of molybdenum and vanadium, *Nippon Kagaku Kaishi,* 8, 1337, 1982 (in Japanese).

11. **Miyai, Y., Sugasaka, K., and Katoh, S.,** Studies on the extraction of soluble resources in seawater. I. Determination of molybdenum and vanadium in seawater by graphite furnace atomic absorption spectrophotometry with oxine-solvent extraction, *Nippon Kaishi Gakkaishi,* 36, 159, 1982 (in Japanese).

12. **Ngah, S. W. S., Sarkissian, L. L., and Tyson, J. F.,** Comparison of electrothermal atomization methods for molybdenum, *Anal. Proc.(London),* 20(12), 597-599, 1983.

13. **Mueller-Vogt, G., Wendl, W., and Pfundstein, P.,** Chemical reactions in the determination of molybdenum by electrothermal atomic absorption spectroscopy, *Fresnius' Z. Anal. Chem.,* 314, 638, 1983.

14. **Ghe, A. M., Carati, D., and Stefanelli, G.,** Determination of trace amounts of molybdenum by flame and flameless atomic absorption spectrometry, *Ann. Chim. (Rome),* 73(11-12), 705-718, 1983.

15. **Steiner, J. W.,** Direct removal of interfering species from the AAS-electrothermal atomizer during the analysis of cadmium and molybdenum in biological materials, *Spurenelem. Symp.,* 411-17, 1983.

16. **Marecek, J., Braunova, I., and Vachtova, I.,** Determination of molybdenum, vanadium and chromium in brines by atomic absorption spectroscopy, *Chem. Prum.,* 33, 429-435, 1983 (in Czech).

17. **Rossi, G., Omenetto, N., Pigozzi, G., Vivian, R., Martinez, V., Mousty, F., and Crabi, G.,** Analysis of radioactive waste solutions by atomic absorption spectrometry with electrothermal atomization, *At. Spectrosc.,* 4, 113, 1983.

18. **Kolcava, D. and Janacek, J.,** Determination of molybdenum in rocks by atomic absorption spectroscopy with electrothermal atomization after its separation with alpha-benzoin oxime, *Collect. Czech. Chem. Commun.,* 49(2), 370-7, 1984.

19. **Gohda, S., Yamazaki, H., and Kataoka, H.,** Determination of molybdenum in environmental materials by polarized Zeeman atomic absorption spectrometry with graphite furnace, *Bunseki Kagaku,* 33, 407, 1984 (in Japanese).

20. **Steiner, J. W. and Ryan, K. M.,** Rapid atomic absorption spectroscopic analysis of molybdenum in plant tissue with modified carbon rod atomizer, *Analyst,* 109, 581, 1984.

21. **Chakrabarti, C. L., Wu, S., Karwowska, R., Chang, S. B., and Bertels, P. C.,** Studies in atomic absorption spectrometry with laboratory-made graphite furnace, *At. Spectrosc.,* 5, 69, 1984.

22. **Wendl, W. and Mueller-Vogt, G.,** Chemical reactions in the graphite tube for some carbide and oxide forming elements, *Spectrochim. Acta,* 39B, 237, 1984.

23. **Niu, S., Qin, Z., Li, Z., Li, S., Wang, T., and Jin, Q.,** Determination of molybdenum and iron in cluster compounds by atomic absorption spectroscopy with a graphite furnace atomizer, *Gaodeng Xuexiao Huaxue Xuebao,* 5, 122, 1984.

24. **Grobenski, Z., Lehmann, R., Radziuk, B., and Voelkopf, U.,** Determination of trace metals in seawater using Zeeman graphite furnace AAS, *At. Spectrosc.,* 5, 87, 1984.

25. **Pilipenko, A. T., Samchuk, A. I., and Zul'figarov, O. S.,** Extraction and atomic absorption determination of molybdenum using N-benzoylphenylhydroxylamine, *Zh. Anal. Khim.,* 39(11), 2051, 1985.

26. **Curtis, P. R. and Grusovin, J.,** Determination of molybdenum in plant tissue by graphite furnace atomic absorption spectrophotometry (GFAAS), *Commun. Soil Sci. Plant Anal.,* 16, 1279, 1985.
27. **Wang, L., Wu, X., and Pan, D.,** Rapid determination of molybdenum in wheat by GFAAS, *Guangpuxue Yu Guangpu Fenxi,* 5(2), 667-668, 1985 (in Chinese), CA 18, 103:52770r, 1985.
28. **Baez, M. E., Gonzalez, C., and Lachica, M.,** Lifetime of pyrolytic graphite-tube atomizers in molybdenum determination, *Analysis,* 13(10), 474-476, 1985.
29. **Yoshimura, C. and Huzino, T.,** Direct determination of molybdenum disulfide by atomic absorption spectrometry in presence of carbon black, *Nippon Kagaku Kaishi,* 5, 882, 1985 (in Japanese).
30. **Samchuk, A. I.,** Behavior of extracts of thallium, beryllium and molybdenum in a graphite furnace during AA analysis, *Ukr. Khim. Zh.,* 51(3), 287-291, 1985.
31. **Baucella, M., Lacort, G., and Roura, M.,** Determination of cadmium and molybdenum in soil extracts by graphite furnace atomic absorption and inductively coupled plasma spectroscopy, *Analyst (London),* 110(12), 1423-1429, 1985.
32. **Tominaga, M., Bansho, K., and Umazaki, Y.,** Electrothermal atomic absorption spectrometric determination of lead, manganese, vanadium and molybdenum in seawater with ascorbic acid to reduce matrix effects, *Anal. Chim. Acta,* 169, 171, 1985.
33. **Sneddon, J. and Fuavao, V. A.,** Effect of atomizer surface type on quantitation of molybdenum and ytterbium by electrothermal atomization atomic absorption spectrometry, *Anal. Chim. Acta,* 167, 317, 1985.
34. **Hoenig, M., Van Elsen, Y., and Van Cauter, R.,** Factors influencing the determination of molybdenum in plant samples by electrothermal AAS, *Anal. Chem.,* 58(4), 777-780, 1986.
35. **Chakrabarti, C. L., Wu, S., Marcantonio, F., and Headrick, K. L.,** Chemical reactions in the atomization of molybdenum in graphite furnace atomic absorption spectrometry, *Fresnius' Z. Anal. Chem.,* 323(7), 730-736, 1986.
36. **Lajunen, L. H. J. and Kubin, A.,** Determination of trace amounts of molybdenum in plant tissue by solvent extraction-atomic absorption and direct-current plasma emission spectrometry, *Talanta,* 33(3), 265-270, 1986.
37. **Gohda, S., Yamazaki, H., and Shigematsu, T.,** Determination and characterization of molybdenum in natural water by solvent extraction-atomic absorption spectrometry using graphite furnace, *Anal. Sci.,* 2(1), 37-42, 1986.
38. **El-Yazigi, A., Al-Saleh, I., and Al-Mefty, O.,** Concentrations of zinc, iron, molybdenum, arsenic and lithium in cerebrospinal fluids of patients with brain tumors, *Clin. Chem.,* 32, 2187, 1986.
39. **Clark, J. R.,** Electrothermal atomization atomic absorption conditions and matrix modifications for determining Sb, As, Bi, Cd, Ga, Au, In, Pb, Mo, Pd, Pt, Se, Ag, Te, Tl and Sn following back-extraction of organic aminohalide extracts, *J. Anal. At. Spectrom.,* 1, 301-308, 1986.
40. **Botha, P. V., Fazakas, J., Harles, A., and Lacatusu, R.,** Critical study of the determination of molybdenum in soils and plants by electrothermal atomic absorption spectroscopy with thermally shielded furnace, *At. Spectrosc.,* 8, 84, 1987.
41. **Emerick, R. J.,** Calcium chloride matrix modification for preventing sulfate interference in the determination of molybdenum by graphite furnace atomic absorption spectroscopy, *At. Spectrosc.,* 8, 69, 1987.

AUTHOR INDEX

SUBJECT INDEX

NICKEL

NICKEL, Ni   (ATOMIC WEIGHT 58.71)

Instrumental Parameters:

Wavelength	232.0 nm
Slit width	40 μm
Bandpass	0.15 nm
Light source	Hollow Cathode Lamp
Lamp current	10 mA
Purge gas	Argon or Nitrogen
Sample size	25 μL (5 μL or mg for oil samples.)
Furnace	Cylindrical cuvette/Pyrolytic graphite

Standard Operating Conditions:

Optimum char temperature	1000°C
Optimum atomization temperature	2700°C
Sensitivity	20 pg/1% absorption
Sensitivity check	0.05-0.2 μg/mL
Working range	0.05-0.3 μg/mL
Background correction	Required only to correct light scattering or non-specific absorption from the sample containing high dissolved solids.

General Note:
1. Use commercial standard or a previously analyzed sample as a working standard.
2. Use ultra clean glass and plastic ware, soaked in 1:5 nitric acid solution and rinsed thoroughly with deionized distilled water.
3. Prepare all analytical solutions in 0.2% (v/v) nitric acid.
4. Check for blank values on all reagents including water.
5. Dilution is recommended when sample exhibits greater than 0.5 absorbance units. Dilutions can be made with deionized distilled water.

REFERENCES

1. **Chauvin, J. V., Newton, M. P., and Davis, D. G.,** The determination of lead and nickel by atomic absorption spectroscopy with a flameless wire loop atomizer, *Anal. Chim. Acta*, 65, 291, 1973.

2. **Schweizer, V. B.,** Determination of cobalt, chromium, copper, molybdenum, nickel and vanadium in carbonate rocks with the HGA-70 graphite furnace, *At. Ab. Newsl.*, 14, 137, 1975.

3. **Mizoguchi, T. and Ishii, H.,** Study of extracting agents for the determination of cobalt and nickel by solvent extraction-carbon furnace atomic absorption spectroscopy, *Bunseki Kagaku*, 26, 839, 1977.

4. **Sutter, E. M. and Loroy, M. J.,** Nature of the interference of nitric acid in the determination of nickel and vanadium by atomic absorption spectroscopy with electrothermal atomization, *Anal. Chim. Acta*, 96, 243, 1978.

5. **Andersen, I., Torjussen, W., and Zachariasen, H.,** Analysis for nickel in plasma and urine by electrothermal atomic absorption spectroscopy with sample preparation by protein precipitation, *Clin. Chem.*, 24, 1198, 1978.

6. **Fudagawa, N. and Kawase, A.,** Determination of nickel by graphite tube furnace atomic absorption spectroscopy. Applications to nickel determination in plant materials, *Bunseki Kagaku*, 27, 37, 1978.

7. **Page, A. G., Godbole, S. V., Kulkarni, M. J., Shelar, S. S., and Joshi, B. D.,** Direct AAS determination of cobalt, copper, manganese and nickel in uranium oxide (U3O8) by electrothermal atomization, *Z. Anal. Chem.*, 296, 40, 1979.

8. **Sedykh, E. M., Belyaev, Y. I., and Sorokina, E. V.,** Elimination of matrix effects in electrothermal atomic absorption determination of Ag, Pb, Co, Ni and Te in samples of complicated composition, *Zh. Anal. Khim.*, 35, 2348, 1980 (in Russian).

9. **Pederson, B., Willems, M., and Storgaard-Joergensen, S.,** Determination of copper, lead, cadmium, nickel and cobalt in EDTA extracts of soil by solvent extraction and graphite furnace atomic absorption spectrophotometry, *Analyst*, 105, 119, 1980.

10. **Lerner, L. A. and Igoshine, E. V.,** Atomic absorption determination with a graphite furnace of copper, cobalt and nickel extracted from soil with ammonium acetate buffer solutions, *Pochvovedenie*, 3, 106, 1980.

11. **Matsuo, H., Kumamaru, T., and Hara, S.,** Graphite furnace atomic absorption spectroscopy of nickel by ion-pair extraction of thiocyanate pyridine nickel (II) with Zephir amine (tetra decyl dimethyl benzyl ammonium), *Bunseki Kagaku*, 29, 801, 1980.

12. **Giacintov, P., Zima, S., Kozman, J., and Kukla, J.,** Nickel content in several organs of farm animals, *Spurenelem. Symp.: Nickel*, 3, 263-268, 1980.

13. **Vollkopf, U., Grobenski, Z., and Welz, B.,** Determination of nickel in serum using graphite furnace atomic absorption, *At. Spectrosc.*, 2, 168, 1981.

14. **Suzuki, M., Ohta, K., and Yamakita, T.,** Elimination of alkali chloride interference with thio urea in electrothermal atomic absorption spectroscopy of copper and manganese, *Anal. Chem.*, 53, 9, 1981.

15. **Oradonskii, S. G. and Kuzimichev, A. N.,** Flameless extraction-atomic absorption method for determination of ultramicro amounts of toxic metals (lead, cadmium, copper, cobalt and nickel) in seawaters and ices in labile form, *Tr. Gos Okeanogr. Inst.*, 162, 22, 1981.

16. **Garbett, K., Goodfellow, G. I., and Marshall, G. B.,** Application of atomic absorption technique in the analysis of metallic sodium. II. Determination of iron, nickel and chromium in sodium salt solutions by electrothermal atomization, *Anal. Chim. Acta*, 126, 147, 1981.

17. **Zolotovitakaya, E. S. and Fidel'man, B. M.,** Use of electrothermal atomic absorption spectroscopy for determining iron, nickel and chromium in raw materials and single crystals of corundum, *Zh. Anal. Khim.*, 36, 1564, 1981 (in Russian).

18. **Brown, S. S., Nomoto, S., Stoeppler, M., and Sunderman, F. W., Jr.,** IUPAC reference method for analysis of nickel in serum and urine by electrothermal atomic absorption spectroscopy, *Clin. Biochem. (Ottawa)*, 14, 295-299, 1981.

19. **Mundt-Becker, W. and Angerer, J.,** Practicable method for the determination of nickel in urine by atomic absorption spectroscopy, *Int. Arch. Occup. Environ. Health*, 49, 187-192, 1981 (in German).

20. **Erspamer, J. P. and Niemczyk, T. M.,** Effect of graphite surface type on determination of lead and nickel in a magnesium chloride matrix by furnace atomic absorption spectroscopy, *Anal. Chem.*, 54, 2150, 1982.

21. **Kuehn, K. and Sunderman, F. W., Jr.,** Dissolution half-times of nickel compounds in water, rat serum and renal cytosol, *J. Inorg. Biochem.*, 17, 29-39, 1982.

22. **Camara Rica, C. and Kirkbright, G. F.,** Determination of trace concentrations of lead and nickel in human milk by electrothermal atomization atomic absorption spectrophotometry and inductively coupled plasma emission spectroscopy, *Sci. Total Environ.*, 22, 193, 1982.

23. **Koops, J., Klomp, H., and Westerbeek, D.,** Spectrophotometric determination of nickel with furildioxime with special reference to milk and milk products and to the release of nickel from stainless steel by acidic dairy products and by acid cleaning, *Neth. Milk Dairy J.*, 36, 333, 1982.

24. **Kosprzak, K. S., Stoeppler, M., Gorski, Z., and Kopczynski, T.,** New oxime extractants for analysis of nickel by electrothermal atomic absorption spectroscopy, *Chem. Anal. (Warsaw)*, 27, 115, 1982.

25. **Bangia, T. R., Kartha, K. N. K., Varghese, M., Dhawale, B. A., and Joshi, B. D.,** Chemical separation and electrothermal atomic absorption spectrophotometric determination of Cd, Co, Cu and Ni in high-purity uranium, *Z. Anal. Chem.,* 310, 410, 1982.

26. **Beninschek, A., Huber, I., and Benischek, F.,** Electrothermal atomic absorption spectrometric determination of chromium, iron and nickel in lithium metal, *Anal. Chim. Acta,* 140, 205-212, 1982.

27. **Nakaaki, K., Tada, O., and Masuda, T.,** Urinary excretion of chromium, manganese, copper and nickel, *Rodo Kagaku,* 58, 529, 1982 (in Japanese).

28. **Arnold, D. and Kuennecke, A.,** Determination of 3d elements in high lead-containing glasses, *Silikattechnik,* 33, 312, 1982.

29. **Pyatnitskii, I. V. and Pilipyuk, Y. S.,** Atomic absorption determination of nickel and cobalt in chloroform extracts, *Ukr. Khim. Zh.,* 48, 962, 1982 (in Russian).

30. **Bruhn, F. C. and Cabalin, G. V.,** Direct determination of nickel in gas oil by atomic absorption spectroscopy with electrothermal atomization, *Anal. Chim. Acta,* 147, 193, 1983.

31. **Alder, J. F. and Batoreu, M. C. C.,** Practical aspects of lead and nickel in epithelial tissue by electrothermal atomic absorption spectrometry, *Anal. Chim. Acta,* 155, 199, 1983.

32. **Legret, M., Demare, D., Marchandise, P., and Robbe, D.,** Interferences of major elements in the determination of lead, copper, cadmium, chromium and nickel in river sediments and sewer sludges by electrothermal atomic absorption spectroscopy, *Anal. Chim. Acta,* 149, 107, 1983.

33. **Nakashima, R., Kamata, E., and Shibata, S.,** Determination of cobalt, copper, iron, nickel and lead in bone by graphite furnace atomic absorption spectroscopy, *Kogai,* 18, 37-44, 1983.

34. **Komarek, J. and Gomiscek, S.,** Electrothermal atomic absorption spectrometry of nickel in the presence of dithiocarbamate, *Vestn. Slov. Kem. Drus.,* 30(4), 443-58, 1983.

35. **Campbell, D. E. and Comperat, M.,** Ultratrace analyses of high-purity glasses for double-crucible low-attention optical waveguides, *Glastech. Ber.,* 56, 898, 1983.

36. **Vorberg, B., Peters, H. J., Koehler, H., Pohle, B., and Haustein, B.,** The possibility of determining the trace element nickel in human serum, *Mongen-Spurenelem., Arbeitstag,* (in German). Anke, M., Ed., Karl-Marx University. Leipzig, East Germany, 1983, 327-330

37. **Patel, B. M., Goyal, N., Purohit, P., Dobale, A. R., and Joshi, B. D.,** Direct determination of magnesium, manganese, nickel and zinc in uranium by electrothermal atomic absorption spectroscopy, *Fresnius' Z. Anal. Chem.,* 315, 42-46, 1983.

38. **Castledine, C. G. and Robbins, J. C.,** Practical field-portable atomic absorption analyzer, *J. Geochem. Explor.,* 19, 689, 1983.

39. **Alder, J. F. and Batoreu, M. C. C.,** Determination of lead and nickel in epithelial tissue by electrothermal atomic absorption spectrometry, *Anal. Chim. Acta,* 155, 199-207, 1983.

40. **Carlin, L. M., Colovos, G., Garland, D., Jamin, M. E., and Klenck, M.,** Analytical methods evaluation and validation: arsenic, nickel, tungsten, vanadium, talc wood dust, Report NIOSH-210-79-0060; Order No. PB83-155325, 223 pp, 1981. Avail. NTIS from Govt. Rep. Announce. Index (U.S.), 83, 2355, 1983.

41. **Senft, V. and Reinvart, M.,** Determination of nickel in urine and serum by atomic absorption spectroscopy, *Prac. Lek.,* 36, 171, 1984.

42. **Green, R. J. and Asher, C. J.,** Measurement of submicrogram amounts of nickel in plant material by electrothermal atomic absorption spectroscopy, *Analyst (London),* 109, 503, 1984.

43. **Sunderman, F. W., Jr., Criostomo, M. C., Reid, M. C., Hopfer, S. M., and Nomoto, S.,** Rapid analysis of nickel in serum and whole blood by electrothermal atomic absorption spectrophotometry, *Ann. Clin. Lab. Sci.,* 14, 232, 1984.

44. **Wei, F. S. and Qi, W. Q.,** Direct determination of ultratrace nickel in serum by controlled temperature graphite furnace atomic absorption spectrophotometry, *Anal. Lett.,* 17(B14), 1607-1616, 1984.

45. **Green, R. J. and Asher, C. J.,** Measurement of submicrogram amounts of nickel in plant material by electrothermal atomic absorption spectroscopy, *Analyst,* 109, 503, 1984.

46. **Torrance, K.,** Determination of ionic nickel and cobalt in simulated PWR coolant by differential pulse polarography, *Analyst (London),* 109(8), 1035, 1984.

47. **Phelan, V. J. and Powell, R. J. W.,** Combined reagent purification and sample dissolution (CORPAD) applied to trace analysis of silicon, silica and quartz, *Analyst,* 109, 1269, 1984.

48. Release of metals (lead, cadmium, copper, zinc, nickel and chromium) from kitchen blenders, Statens Levnedsmiddelinstitut REPORT 1984, PUB-88, Order No. PB84-193416, 26 pp (in Danish). Avail. NTIS From Gov. Rep. Announce. Index (U.S.), 84, 67, 1984.

49. **Sperling, K. R.,** Tube-in-tube technique in electrothermal atomic absorption spectrometry. III. Atomization behavior of Zn, Cd, Pb, Cu and Ni, *Spectrochim. Acta,* 39B, 371, 1984.

50. **Matsusaki, K. and Yoshino, T.,** Electrothermal atomic absorption spectrometric determination of traces of chromium, nickel, iron and beryllium in aluminum and its alloys without preliminary separation, *Anal. Chim. Acta,* 157, 193, 1984.

51. **Carrondo, M. J. T., Reboredo, F., Ganho, R. M. B., and Oliveira, F. S.,** Analysis of sediments for heavy metals by rapid electrothermal atomic absorption procedure, *Talanta*, 31, 561, 1984.

52. **Fish, R. H., Komlenic, J. J., and Wines, B. K.,** Characterization and comparison of vanadyl and nickel compounds in heavy crude petroleums and asphaltenes by reverse-phase and size-exclusion liquid chromatography/graphite furnace atomic absorption spectrometry, *Anal. Chem.*, 56, 2452, 1984.

53. **Sunderman, F. W., Jr., Crisostomo, M. C., Reid, M. C., Hopfer, S. M., and Nomoto, S.,** Rapid analysis of nickel in serum and whole blood by electrothermal atomic absorption spectrophotometry, *Ann. Clin. Lab. Sci.*, 14, 232, 1984.

54. **Chakrabarti, C. L., Wu, S., Karwowska, R., Chang, S. B., and Bertels, P. C.,** Studies in atomic absorption spectrometry with laboratory-made graphite furnace, *At. Spectrosc.*, 5, 69, 1984.

55. **Pruszkowska, E. and Barrett, P.,** Determination of arsenic, selenium, chromium, cobalt and nickel in geochemical samples using the stabilized temperature platform furnace and Zeeman background correction, *Spectrochim. Acta*, 39B, 485, 1984.

56. **Zober, A., Kick, A., Schaller, K. H., Schellman, B., and Valentin, H.,** Studies of normal values of chromium and nickel in human lung, kidney, blood and urine samples, *Zentralbl. Bakterial. Mikrobiol. Hyg. Abt. 1 Orig. B*, 179, 80, 1984.

57. **Abreu, M., and Buenafama, H.,** Matrix effects in the determination of vanadium and nickel in petroleum by FAAS, *Collect. Colloq. Semin. (Inst. Fr. Pet.)*, 40, 179-184, 1984.

58. **Hoenig, M., Scokart, P. O., and Van Hoeyweghen, P.,** Efficiency of L'vov platform and ascorbic acid modifier for reduction of interferences in analysis of plant samples for lead, thallium, antimony, cadmium, nickel and chromium by ETAAS, *Anal. Lett.*, 17A, 1947, 1984.

59. **Danielsson, L. G. and Westerlund, S.,** Short term variations in trace metal concentrations in Baltic, *Mar. Chem.*, 15, 273, 1984.

60. **Hulanicki, A. and Bulska, E.,** Effect of organic solvents on atomization of nickel, copper and lead in graphite furnaces in atomic absorption spectrometry, *Can. J. Spectrosc.*, 29(6), 148-152, 1984.

61. **Zamilova, L. M., Biktimirova, T. G., and Sokolova, V. I.,** Atomic absorption determination of vanadium and nickel in gas-oil fractions of crude oils, *Neftekhimiya*, 25, 159, 1985 (in Russian).

62. **Lewis, S. A., O'Haver, T. C., and Harnly, J. M.,** Determination of metals at microgram-per-liter level in blood serum by simultaneous multielement atomic absorption spectrometry with graphite furnace atomization, *Anal. Chem.*, 57, 2, 1985.

63. **Wang, X., Lu, P., and Zhang, G.,** Determination of cobalt, chromium and nickel in doped gadolinium gallium garnets by GFAAS, *Guangpuxue Yu Guangpu Fenxi*, 5(1), 63-65, 1985.

64. **Rozanska, B., Skorko-Trybula, Z., and Slodownik, A.,** Precipitation and extraction separation of lead, iron and zinc prior to the determination of cobalt, nickel and manganese by atomic absorption in dust from electric furnace, *Chem. Anal.(Warsaw)*, 30(3), 471-479, 1985.

65. **Kollmeier, H., Witting, C., Seemann, J., Wittig, P., and Rothe, R.,** Increased chromium and nickel content in lung tissue, *J. Cancer Res. Clin. Oncol.*, 110(2), 173-176, 1985.

66. **Raichevski, G., Gancheva, Y., Kuncheva, M., and Tomov, I.,** Anodic behavior and passivation of electroplated and electroless deposited nickel and cobalt, *Zashch. Met.*, 21(3), 449-452, 1985.

67. **Jin, F. and Jing, Y.,** Direct determination of lead, cadmium and nickel in soils by atomic absorption spectrometry, *Fenxi Huaxue*, 13, 386, 1985 (in Chinese).

68. **Marzouk, A. and Sunderman, F. W., Jr.,** Biliary excretion of nickel in rats, *Toxicol. Lett.*, 27(1-3), 65-71, 1985.

69. **Shijo, Y., Shimizu, T., and Sakai, K.,** Micro-solvant extraction method for preconcentration. Determination of nickel in seawater by graphite furnace atomic absorption spectroscopy, *Anal. Sci.*, 1(5), 479, 1985.

70. **Sunderman, F. W., Jr., Marzouk, A., Grisostomo, M. C., and Weatherby, D. R.,** Electrothermal atomic absorption spectrophotometry of nickel in tissue homogenates, *Ann. Clin. Lab. Sci.*, 15, 299, 1985.

71. **Jin, L. and Ni, Z.,** Determination of nickel in urine and other biological samples by graphite furnace atomic absorption spectroscopy, *Fresnius' Z. Anal. Chem.*, 321(1), 72-6, 1985.

72. **Drazniowsky, M., Parkinson, I. S., Ward, M. K., Channon, S. M., and Kerr, D. N. S.,** Method for determination of nickel in water and serum by flameless atomic absorption spectrophotometry, *Clin. Chim. Acta*, 145 (2), 219-226, 1985.

73. **Kumar, A., Hasan, M. Z., and Deshmukh, B. T.,** Interference-free determination of nickel in water by graphite furnace atomic absorption spectrophotometry, *Indian J. Pure Appl. Phys.*, 24, 594, 1986.

74. **Filkova, L. and Jager, J.,** Nonoccupational exposure to nickel tetracarbonyl, *Cesk. Hug.*, 31(5), 255-259, 1986 (in Czech).

75. **Sunderman, F. W., Jr., Hopfer, S. M., Crisostomo, M. C., and Stoeppler, M.,** Rapid analysis of nickel in urine by electrothermal atomic absorption spectrophotometry, *Ann. Clin. Lab. Sci.*, 16(3), 219-230, 1986.

76. **Eller, P. M.,** Determination of nickel carbonyl by charcoal tube collection and furnace atomic absorption spectroscopy, *Appl. Ind. Hyg.*, 1(3), 115-118, 1986.

77. **Andersen, J. R., Gammelgaard, B., and Reimert, S.,** Direct determination of nickel in human plasma by Zeeman-corrected atomic absorption spectroscopy, *Analyst (London),* 11(6), 721-727, 1986.
78. **Nishioka, H., Assadamongkol, S., Maeda, Y., and Azumi, T.,** Environmental analysis. XXXII. Graphite furnace atomic absorption spectrometric determination of nickel in seawater by a coprecipitation concentration method, *Nippon Kaisui Gakkkaishi,* 40(227), 286-290, 1987 (in Japanese).
79. **Hopfer, S. M., Linen, J. W., Rezuke, W. N., O'Brien, J. E., Smith, L., Watters, F., and Sunderman, F. W., Jr.,** Increased nickel concentrations in body fluids of patients with chronic alcoholism during disulfiram therapy, *Res. Commun. Chem. Pathol. Pharmacol.,* 55, 101, 1987.
80. **Schmidt, K. P. and Falk, H.,** Direct determination of silver, copper and nickel in solid materials by graphite furnace atomic absorption spectroscopy using a specially designed graphite tube, *Spectrochim. Acta,* 42B, 431, 1987.
81. **Gonzalez, M. C., Rodriguez, A. R., and Gonzalez, V.,** Determination of vanadium, nickel, iron, copper and lead in petroleum fractions by atomic absorption spectrophotometry with a graphite furnace, *Microchem. J.,* 35, 94, 1987.
82. **Rezuke, W. N., Knight, J. A., and Sunderman, F. W., Jr.,** Reference values for nickel concentrations in human tissues and bile, *Am. J. Ind. Med.,* 11, 419, 1987.

AUTHOR INDEX

SUBJECT INDEX

PHOSPHORUS

PHOSPHORUS, P   (ATOMIC WEIGHT 30.9738)

Instrumental Parameters:

Wavelength	213.6 nm
Slit width	320 μm
Bandpass	1 nm
Light source	Hollow Cathode Lamp
Lamp current	8 mA
Purge gas	Argon or Nitrogen
Sample size	25 μL
Furnace	Cylindrical cuvette/Pyrolytic graphite

Standard Operating Conditions:

Optimum char temperature	1500°C
Optimum atomization temperature	2700°C
Sensitivity	25000 pg/1% absorption
Sensitivity check	100.0-200.0 μg/mL
Working range	100.0-500.0 μg/mL
Background correction	Required only to correct light scattering or non-specific absorption from the sample containing high dissolved solids.

General Note:
1. Use commercial standard or a previously analyzed sample as a working standard.
2. Use ultra clean glass and plastic ware, soaked in 1:5 nitric acid solution and rinsed thoroughly with deionized distilled water.
3. Prepare all analytical solutions in 1% (v/v) nitric acid.
4. Check for blank values on all reagents including water.
5. Dilution is recommended when sample exhibits greater than 0.5 absorbance units. Dilutions can be made with deionized distilled water.
6. Use 0.1% lanthanum as lanthanum nitrate for dilutions.
7. Addition of nickel (200 μg/mL as nickel nitrate) allows the use of higher temperature for pyrolysis.

REFERENCES

1. **Ediger, R. D.,** Standard conditions for the determination of phosphorus with the HGA graphite furnace, *At. Ab. Newsl.,* 15, 145, 1976.
2. **Vigler, M. S., Strecker, A., and Varnes, A.,** Investigations of the determination of phosphorus in organic media by atomic absorption heated graphite atomizer, *Appl. Spectrosc.,* 32, 60, 1978.
3. **L'vov, B. V. and Pelieva, L. A.,** Atomic absorption of phosphorus with an HGA atomizer using sample evaporation from a probe introduced into heated furnace, *Zh. Anal. Khim.,* 33, 1572, 1978.
4. **Ediger, R. D., Knott, A. R., Peterson, G. E., and Beaty, R. D.,** The determination of phosphorus by atomic absorption using the graphite furnace, *At. Ab. Newsl.,* 17, 28, 1978.
5. **Khavezov, I., Ruseva, E., and Iordanov, N.,** Flameless atomic absorption determination of phosphorus using zirconium carbide coated graphite atomizer tubes, *Z. Anal. Chem.,* 296, 125, 1979.
6. **Slikkerveer, F. J., Braad, A. A., and Hendrikse, P. W.,** Determination of phosphorus in edible oils by graphite furnace atomic spectroscopy, *At. Spectrosc.,* 1, 30, 1980.
7. **Tittarelli, P. and Mascherpa, E. S.,** Atomic absorption for organophosphorus additives in lubricating oil, *Anal. Chem.,* 53, 1981.
8. **Mimura, T. and Wakisaka, S.,** Indirect determination of small amounts of phosphorus and ATPase by flameless atomic absorption spectroscopy, *Teikyo Igaku Zasshi,* 4, 143-148, 1981 (in Japanese).
9. **Welz, B., Voellkopf, U., and Grobenski, Z.,** Determination of phosphorus in steel with a stabilized temperature graphite furnace and Zeeman-corrected atomic absorption spectroscopy, *Anal. Chim. Acta,* 136, 201-214, 1982.
10. **Persson, K.,** Determination of phosphorus in edible oils by graphite furnace atomic absorption spectrophotometry using a Varian CRA-90, *Proc. 11th Scand. Symp. Lipids* Marcuse, R., Ed., Lipid FORUM, Goetborg, Sweden, 1982, 224-229.
11. **Hogen, M. L.,** Detection of phosphorus on starch by atomic absorption and the graphite furnace, *Cereal Chem.,* 60, 403, 1983.
12. **Ruseva, E., Khavezov, I., Spivakov, B. Y., and Shkinev, V. M.,** Electrothermal atomic absorption determination of traces of arsenic and phosphorus in copper-nickel alloys, *Fresnius' Z. Anal. Chem.,* 315, 499, 1983.
13. **McCarthy, J. P., Nunn, E. B., and Kinard, C.,** Atomic absorption spectrophotometric methods for the determination of phosphorus and silicon in steel, *Anal. Chem. Symp. Ser.,* 19, 349, 1984.
14. **Lin, S. W. and Julshamn, K.,** A comparative study of the determination of phosphorus by electrothermal atomic absorption spectroscopy and solution spectrophotometry, *Anal. Chim. Acta,* 158, 199, 1984.
15. **Kubota, T., Ueda, T., and Okutani, T.,** Determination of phosphorus by atomic absorption spectrometry using a zirconium treated graphite tube, *Bunseki Kagaku,* 33(12), 633, 1984 (in Japanese).
16. **Luo, S.,** Indirect determination of total phosphorus and phosphate in water by GFAAS, *Huanjing Huaxue,* 4(2), 64-6, 1985.
17. **Russeva, E., Havezov, I., Jordanov, N., and Ortner, H.,** Extraction of atomic absorption determination of traces of phosphorus in tungsten trioxide, *Fresnius' Z. Anal. Chem.,* 321(7), 677-678, 1985.
18. **Rao, A. L. J., Gupta, U., and Puri, B. K.,** Determination of phosphorus by graphite furnace atomic absorption spectroscopy. I. Determination in the absence of a modifier, *Analyst (London),* 111, 1401, 1986.
19. **Curtius, A. J., Schlemmer, G., and Welz, B.,** Determination of phosphorus by graphite furnace atomic absorption spectroscopy. I. Analysis of biological reference materials, *J. Anal. At. Spectrom.,* 1, 421-427, 1986.
20. **Casetta, B., Giaretta, A., and Rampazzo, G.,** Determination of phosphorus in siliceous rocks by atomic absorption spectroscopy with a graphite furnace, *At. Spectrosc.,* 7, 155-157, 1986.
21. **Kubota, T., Ueda, T., and Okutani, T.,** Determination of phosphorus in natural water by atomic absorption spectrometry after coprecipitation enrichment, *Bunseki Kagaku,* 35, 75, 1986.
22. **Ruseva, E. and Khavezov, I.,** Electrothermal atomic absorption determination of phosphorus in lubricating oils and related products, *Izv. Khim.,* 19, 422, 1986.
23. **Welz, B., Curtius, A. J., Schlemmer, G., Ortner, H. M., and Birzer, W.,** Scanning electron microscopy studies on surfaces from electrothermal atomic absorption spectroscopy. III. The lanthanum modifier and the determination of phosphorus, *Spectrochim. Acta,* 41B, 1175, 1986.
24. **Curtius, A. J., Schlemmer, G., and Welz, B.,** Determination of phosphorus by graphite furnace atomic absorption spectroscopy. III. Analysis of biological reference materials, *J. Anal. At. Spectrom.,* 2, 371, 1987.

AUTHOR INDEX

SUBJECT INDEX

PLATINUM GROUP METALS
(IRIDIUM, OSMIUM, PALLADIUM, PLATINUM,
RHODIUM, AND RUTHENIUM)

IRIDIUM, Ir   (ATOMIC WEIGHT 192.2)

Instrumental Parameters:

Wavelength	284.9 nm
Slit width	160 μm
Bandpass	0.5 nm
Light source	Hollow Cathode Lamp
Lamp current	15 mA
Purge gas	Argon
Sample size	25 μL
Furnace	Cylindrical cuvette/Pyrolytic graphite

Standard Operating Conditions:

Optimum char temperature	1000°C
Optimum atomization temperature	2600°C
Sensitivity	200 pg/1% absorption
Sensitivity check	1.0-3.0 μg/mL
Working range	5.0-20.0 μg/mL
Background correction	Required only to correct light scattering or non-specific absorption from the sample containing high dissolved solids.

General Note:
1. Use commercial standard or a previously analyzed sample as a working standard.
2. Use ultra clean glass and plastic ware, soaked in 1:5 nitric acid solution and rinsed thoroughly with deionized distilled water.
3. All analytical solutions should be at least 0.2-0.5% (v/v) in nitric acid.
4. Check for blank values on all reagents including water.
5. Dilution is recommended when sample exhibits greater than 0.5 absorbance units.
6. Lower atomization temperature may cause reduced sensitivity.

OSMIUM, Os   (ATOMIC WEIGHT 190.2)

Instrumental Parameters:

Wavelength	290.9 nm
Slit width	80 μm
Bandpass	0.3 nm
Light source	Hollow Cathode Lamp
Lamp current	15 mA
Purge gas	Argon
Sample size	25 μL
Furnace	Cylindrical cuvette/Pyrolytic graphite

Standard Operating Conditions:

Optimum char temperature	550°C
Optimum atomization temperature	2700°C
Sensitivity	270 pg/1% absorption
Sensitivity check	5.0-10.0 μg/mL
Working range	5.0-25.0 μg/mL
Background correction	Required only to correct light scattering or nonspecific absorption from the sample containing high dissolved solids.

General Note:
1. Use commercial standard or a previously analyzed sample as a working standard.
2. Use ultra clean glass and plastic ware, soaked in 1:5 nitric acid solution and rinsed thoroughly with deionized distilled water.
3. All analytical solutions should be prepared in 0.5% (v/v) nitric acid.
4. Check for blank values on all reagents including water.
5. Dilution is recommended when sample exhibits greater than 0.5 absorbance units.
6. Lower atomization temperature may cause reduced sensitivity.
7. Since osmium is highly toxic, take extreme precautions in handling the solutions.

PALLADIUM, Pd   (ATOMIC WEIGHT 106.4)

Instrumental Parameters:

Wavelength	247.6 nm
Slit width	160 μm
Bandpass	0.5 nm
Light source	Hollow Cathode Lamp
Lamp current	5mA
Purge gas	Argon
Sample size	25 μL
Furnace	Cylindrical cuvette/Pyrolytic graphite

Standard Operating Conditions:

Optimum char temperature	1000°C
Optimum atomization temperature	2700°C
Sensitivity	12 pg/1% absorption
Sensitivity check	0.02-1.0 μg/mL
Working range	0.02-2.0 μg/mL
Background correction	Required only to correct light scattering or non-specific absorption from the sample containing high dissolved solids.

General Note:
1. Use commercial standard or a previously analyzed sample as a working standard.
2. Use ultra clean glass and plastic ware, soaked in 1:5 nitric acid solution and rinsed thoroughly with deionized distilled water.
3. All analytical solutions should be prepared in 0.5% (v/v) nitric acid.
4. Check for blank values on all reagents including water.
5. Dilution is recommended when sample exhibits greater than 0.5 absorbance units.
6. Use of nitrogen as purge gas and high concentrations of >3% nitric acid (v/v) will suppress the sensitivity.

PLATINUM, Pt   (ATOMIC WEIGHT 195.09)

Instrumental Parameters:

Wavelength	265.9 nm
Slit width	160 μm
Bandpass	0.5 nm
Light source	Hollow Cathode Lamp
Lamp current	10 mA
Purge gas	Argon
Sample size	25 μL
Furnace	Cylindrical cuvette/Pyrolytic graphite

Standard Operating Conditions:

Optimum char temperature	1200°C (in nitric acid)
	500°C (in hydrochloric acid)
Optimum atomization temperature	2700°C
Sensitivity	200 pg/1% absorption
Sensitivity check	0.5-1.5 μg/mL
Working range	0.08-0.50 μg/mL
Background correction	Required only to correct light scattering or non-specific absorption from the sample containing high dissolved solids.

General Note:
1. Use commercial standard or a previously analyzed sample as a working standard.
2. Use ultra clean glass and plastic ware, soaked in 1:5 nitric acid solution and rinsed thoroughly with deionized distilled water.
3. All analytical solutions should be prepared in 0.5% (v/v) nitric acid or 1% (v/v) hydrochloric acid.
4. Check for blank values on all reagents including water.
5. Dilution is recommended when sample exhibits greater than 0.5 absorbance units.

RHODIUM, Rh   (ATOMIC WEIGHT 102.905)

Instrumental Parameters:

Wavelength	343.5 nm
Slit width	160 μm
Bandpass	0.5 nm
Light source	Hollow Cathode Lamp
Lamp current	5 mA
Purge gas	Argon or Nitrogen
Sample size	25 μL
Furnace	Cylindrical cuvette/Pyrolytic graphite

Standard Operating Conditions:

Optimum char temperature	1100°C
Optimum atomization temperature	2700°C
Sensitivity	20 pg/1% absorption
Sensitivity check	0.05-0.10 μg/mL
Working range	0.02-1.0 μg/mL
Background correction	Required only to correct light scattering or non-specific absorption from the sample containing high dissolved solids.

General Note:
1. Use commercial standard or a previously analyzed sample as a working standard.
2. Use ultra clean glass and plastic ware, soaked in 1:5 nitric acid solution and rinsed thoroughly with deionized distilled water.
3. All analytical solutions should be prepared in 0.5% (v/v) nitric acid.
4. Check for blank values on all reagents including water.
5. Dilution is recommended when sample exhibits greater than 0.5 absorbance units.
6. Use of nitrogen as purge gas and highly concentrated nitric acid (>3% v/v) will suppress the sensitivity.

RUTHENIUM, Ru   (ATOMIC WEIGHT 101.07)

Instrumental Parameters:

Wavelength	349.9 nm
Slit width	160 μm
Bandpass	0.5 nm
Light source	Hollow Cathode Lamp
Lamp current	10 mA
Purge gas	Argon
Sample size	25 μL
Furnace	Cylindrical cuvette/Pyrolytic graphite

Standard Operating Conditions:

Optimum char temperature	1400°C
Optimum atomization temperature	2500°C
Sensitivity	32 pg/1% absorption
Sensitivity check	0.1 μg/mL
Working range	0.05-0.50μg/mL
Background correction	Required only to correct light scattering or non-specific absorption from the sample containing high dissolved solids.

General Note:
1. Use commercial standard or a previously analyzed sample as a working standard.
2. Use ultra clean glass and plastic ware, soaked in 1:5 nitric acid solution and rinsed thoroughly with deionized distilled water.
3. All analytical solutions should be at least 0.2-0.5% (v/v) in nitric acid.
4. Check for blank values on all reagents including water.
5. Dilution is recommended when sample exhibits greater than 0.5 absorbance units.
6. Use of nitrogen as purge gas and high concentrations of >3% nitric acid (v/v) will suppress the sensitivity.

REFERENCES

1. **Guerin, B. D.,** Determination of the noble metals by atomic absorption spectrophotometry with the carbon rod furnace, *J. S. Afr. Chem. Inst.,* 25, 230, 1972.

2. **Janouskova, J., Nehasilova, M., and Sychra, V.,** The determination of platinum with the HGA-70 graphite furnace, *At. Ab. Newsl., 12,* 161, 1973.

3. **Tello, A. and Sepulveda, N.,** Determination of platinum in sands by flameless atomic absorption using the HGA-2100, *At. Ab. Newsl.,* 16, 67, 1977.

4. **Fishkova, N. L. and Vilenkin, V. A.,** Application of an HGA-74 graphite atomizer to atomic absorption determination of gold, silver, platinum and palladium in solutions of complicated composition, *Zh. Anal. Khim.,* 33, 897, 1978 (in Russian).

5. **L'vov, B. V., Pelieva, L. A., Mandrazhi, E. K., and Kalinin, S. K.,** Atomic absorption determination of platinum group elements in an HGA graphite furnace, *Zavod. Lab.,* 45, 1098, 1979.

6. **Rowston, W. B. and Ottaway, J. M.,** Determination of noble metals by carbon furnace atomic absorption spectroscopy. I. Atom formation processes, *Analyst,* 104, 645, 1979.

7. **Potter, N. M. and Waldo, R. A.,** Determination of platinum in alumina-supported automative catalyst material by electrothermal atomic absorption spectroscopy, *Anal. Chim. Acta,* 110, 29, 1979.

8. **Bound, E. A., Norris, J. D., Sanz-Model, A., and West, T. S.,** Atomic absorption spectrometric determination of palladium with a carbon filament electrothermal atomizer, *Anal. Chim. Acta,* 104, 385, 1979.

9. **Ambrosetti, P. and Librici, F.,** Palladium determination in photographic films by furnace atomic absorption after enzymatic digestion and extraction with dibutylsulfide, *At. Ab. Newsl.,* 18, 38, 1979.

10. **Bel'skii, N. K., Nebol'sina, L. A., Shubochkin, L. K., and Verkhoturov, G. N.,** Atomic absorption determination of rhodium in silver with chemical sample preparation directly in a graphite furnace without dissolution, *Zh. Anal. Khim.,* 35, 799, 1980.

11. **Yudelevich, I. G., Startseva, E. A., and Popova, N. M.,** Atomic absorption determination of platinum group metals in gold and silver with electrothermal atomization of the solid phase, *Izv. Sib. Otd. Akad. Nauk SSSR Ser. Khim. Nauk,* 5, 90, 1981.

12. **Smeyers-Verbeke, J., Detaevernier, M. E., Denis, L., and Massart, D. L.,** Determination of platinum in biological fluid by graphite furnace atomic absorption spectroscopy, *Clin. Chim. Acta,* 113, 329, 1981.

13. **Fazakas, J.,** Critical study of the determination of palladium by graphite furnace atomic absorption spectroscopy, *Anal. Lett.,* 14, 535, 1981.

14. **Priesner, D., Sternson, L. A., and Repta, A. J.,** Analysis of total platinum in tissue samples by flameless atomic absorption spectrophotometry. Elimination of the need for sample digestion, *Anal.Lett.,* 14 1255-1268, 1981.

15. **Fazakas, J.,** Determination of palladium by graphite furnace atomic absorption spectroscopy with vaporization from a platform, *Anal. Lett.,* 15, 245-265, 1982.

16. **Bel'skii, N. K., Nebol'sina, L. A., and Shubochkin, L. K.,** Atomic absorption determination of nanogram amounts of Pt, Pd and Rh in a silver alloy using a graphite boat and an electrothermal atomizer, *Zh. Anal. Khim.,* 37, 61-65, 1982 (in Russian).

17. **Huang, M.,** Application of Zephiramine in the analysis of noble metals. I. Determination of microamounts of gold, platinum and palladium in ores by graphite furnace atomic absorption spectroscopy after preconcentration with Zephiramine, *Fenxi Huaxue,* 10, 661, 1982 (in Chinese).

18. **Samchuk, A. I. and Latysh, I. K.,** Platinum and palladium determination in rocks by cupellation and atomic absorption, *Ukr. Khim. Zh.,* 48, 638-640, 1982 (Russian ed.).

19. **Haines, J. and Robert, R. V. D.,** The determination by atomic absorption spectrophotometry using electrothermal atomization of platinum, palladium, rhodium, ruthenium and iridium, Report 1982, Rep. MINTEK, M34, 15pp 1982; Order No. PB82-257908. Avail. NTIS. From Govt. Rep. Announce. Index (U.S.), 82, 5217, 1982.

20. **Fazakas, J.,** Analytical potential of nonresonance lines for the determination of palladium by graphite furnace atomic absorption spectroscopy, *Fresnius' Z. Anal. Chem.,* 312, 227, 1982.

21. **Cano, J. P., Catalin, J., and Bues-Charbit, M.,** Platinum determination in plasma and urine by flameless atomic absorption spectrophotometry, *J. Appl. Toxicol.,* 2, 33-28, 1982.

22. **Shiryaeva, O. A., Kolonina, L. N., Vladimirskaya, I. N., Shestakov, V. A., Malofeeva, G. I., Petrukhin, O. M., Zolotov, Y. A., Marcheva, E. V., Murinov, Y. I., and Nikita, Y. E.,** Atomic absorption determination of platinum metals after their sorption concentration on a thioether polymer, *Zh. Anal. Khim.,* 37, 281-284, 1982 (in Russian).

23. **Shiryaeva, O. A., Kolonina, L. N., Vladimirskaya, I. N., Malofeeva, G. I., Petrukhin, O. M., Zolotov, Y. A., Marcheva, E. V., Murinov, Y. I., and Nikita, Y. E.,** Flameless sorption-atomic absorption determination of platinum-group metals in products of the processing of copper-nickel slimes, *Metody Bydeleniya i Opredeleniya Blagorod. Elementov, M.,* 90-92, 1981 (in Russian). From Ref. *Zh. Metall,* Abstr. No. 11K63, 1982.

24. **Fazakas, J.,** Influence of pressure on the determination of palladium by graphite furnace atomic absorption spectroscopy using resonance and nonresonance lines, *Spectrochim. Acta,* 37B, 921, 1982.

25. **Johnsen, A. C., Wibetoe, G., Langmyhr, F. J., and Aaseth, J.,** Atomic absorption spectrometric determination of the total content and distribution of copper and gold in synovial fluid from patients with rheumatoid arthritis, *Anal. Chim. Acta,* 135, 243-248, 1982.

26. **Farago, M. E. and Parsons, P. J.,** Determination of platinum, palladium and rhodium by atomic-absorption spectroscopy with electrothermal atomization, *Analyst (London),* 107, 1218, 1982.

27. **Belova, V. V., Vasil'eva, A. A., and Androsova, N. V.,** Flameless atomic absorption determination of platinum-group metals in sulfate solutions, *Zavod. Lab.,* 48, 32, 1982.

28. **Kalinina, L. B., Novakovskaya, E. G., and Khalonin, A. S.,** Extraction-atomic absorption determination of platinum, ruthenium and iridium, *Zavod. Lab.,* 48, 6, 1982.

29. **Ito, E. and Kidani, Y.,** Determination of platinum in the environmental samples by graphite furnace atomic absorption spectrophotometry, *Bunseki Kagaku,* 31, E381, 1982.

30. **Gao, Y. and Ni, Z.,** Pyrolytic treatment of graphite tubes and its application to the determination of platinum group metals by graphite furnace atomic absorption spectroscopy, *Xiyou Jinshu,* 1, 99, 1982 (in Chinese).

31. **Brajter, K., Slonawska, K., and Vorbrodt, Z.,** Studies on the conditions for the determination of platinum, palladium, rhodium and iridium by flameless atomic absorption spectroscopy, *Chem. Anal. (Warsaw),* 27, 239, 1982.

32. **Belova, V. V., Vasil'eva, A. A., and Androsova, N. V.,** Extraction-atomic-absorption determination of iridium, ruthenium and rhodium in sulfate solutions, *Izv. Sib. Otd. Akad. Nauk SSSR Ser. Khim. Nauk,* 1, 89, 1983 (in Russian).

33. **Kubrakova, I. V., Vaarshal, G. M., Sedykh, E. M., Myasoedova, G. V., Antokol'skaya, I. I., and Shemarykina, T. P.,** Determination of platinum metals in complex natural samples using electrothermal atomization of the sorbent, *Zh. Anal. Khim.,* 38(12), 2205-2209, 1983.

34. **Chowdhury, A. N. and Pal, J. C.,** Determination of platinum in USGS manganese nodule reference samples by a fire assay-spectrographic method, *Geostand. Newsl.,* 7(2), 279, 1983., 1983.

35. **Sharma, R. P. and Edwards, I. R.,** Microanalysis of platinum in biological media by graphite furnace atomic absorption spectroscopy, *Ther. Drug Monit.,* 367, 1983.

36. **Vinetskaya, T. N., Postnikova, I. S., and Bykovskaya, Y. I.,** Extraction atomic-absorption determination of platinum-group metals in industrial solutions of a complex salt composition, Deposited Doc. VINITI 5511-5583, 12 pp, 1983. Avail VINITI.

37. **Matsumoto, K., Solin, T., and Fuwa, K.,** Elimination of nitrate interference in determination of platinum in some animal tissues using furnace atomic absorption spectrometry, *Spectrochim. Acta,* 39B, 481, 1984.

38. **Suzuki, T. and Ohta, K.,** Atomization of platinum metals from a metal surface, *Spectrochim. Acta,* 39B, 473, 1984.

39. **Reece, P. A., McCall, J. T., Powis, G., and Richardson, R. L.,** Sensitive high performance liquid chromatographic assay for platinum in plasma ultrafiltrate, *J. Chromatogr.,* 306, 417, 1984.

40. **Aruscavage, P. J., Simon, F. O., and Moore, R.,** Flameless atomic absorption determination of platinum, palladium, rhodium in geologic materials, *Geostand. Newsl.,* 8, 3, 1984.

41. **Sighinolfi, G. P., Gorgoni, C., and Mohgamed, A. H.,** Comprehensive analysis of precious metals in some geological standards by flameless atomic absorption spectroscopy, *Geostand. Newsl.,* 8, 25-29, 1984.

42. **Haines, J. and Robert, R. V. D.,** Measurement of platinum, palladium, rhodium, ruthenium and iridium by atomic-absorption spectroscopy using electrothermal atomization, *S. Afr. J. Chem.,* 37, 121, 1984.

43. **Korde, M., Lee, D. S., and Stallard, M. W.,** Concentration and separation of trace metals from seawater using a single anion exchange bead, *Anal. Chem.,* 56, 1956, 1984.

44. **Aruscavage, P. J., Simon, F. O., and Moore, R.,** Flameless atomic absorption determination of platinum, palladium and rhodium in geologic materials, *Geostand. Newsl.,* 8, 3-6, 1984.

45. **Matsumoto, K., Solin, T., and Fuwa, K.,** Elimination of nitrate interference in the determination of platinum in some animal tissues using furnace atomic absorption spectroscopy, *Spectrochim. Acta,* 39B, 481, 1984.

46. **Drummer, O. H., Proudfoot, A., Howes, L., and Louis, W. J.,** High-performance liquid chromatographic determination of platinum (II) in plasma ultrafiltrate and urine: comparison with a flameless atomic spectrometric method, *Clin. Chim. Acta,* 136(1), 65-74, 1984.

47. **Marsh, K. C., Sternson, L. A., and Repta, A. J.,** Postcolumn reaction detector for platinum (II) antineoplastic agents, *Anal. Chem.,* 56(3), 491-497, 1984.

48. **Kolosova, L. P., Novatskaya, N. V., Ryzhov, R. I., and Aladyshkina, A. E.,** Atomic absorption (flame and graphite furnace) determination of noble metals in natural and industrial samples after fire assay preconcentration by lead and incomplete cuppellation, *Zh. Anal. Khim.,* 39, 14775, 1984.

49. **Kizu, R., Hamaki, M., Shimozawa, M., and Miyazaki, M.,** Atomic absorption spectrophotometric determination of platinum in biological materials using nickel as internal standard, *Rinshe Kogaku,* 13, 155, 1984 (in Japanese).

50. **Yukhin, Y. M., Udallova, T. A., and Tsimbalist, V. G.,** Flameless atomic absorption determination of noble metals after liquid-liquid extraction by mixture of bis (2 ethylhexyl) dithiophosphate and p-octylaniline, *Zh. Anal. Khim.,* 40(5), 850-854, 1985 (in Russian).

51. **Kritsotakis, K. and Tobschall, H. J.,** Determination of the precious metals gold, palladium, platinum, rhodium and iridium in rocks and ores by electrothermal atomic absorption spectrometry, *Fresnius' Z. Anal. Chem.,* 320(1), 15-21, 1985.

52. **Hara, S., Matsuo, H., and Kumamaru, T.,** Simultaneous determination of gold (II) and platinum (IV) by graphite furnace atomic absorption spectroscopy after ion-pair extraction with Zephiramine, *Bunseki Kagaku,* 35(6), 503-507 1986 (in Japanese).

53. **Kontas, E., Niskavaara, H., and Virtasalo, J.,** Flameless atomic absorption determination of gold and palladium in geological reference samples, *Geostand. Newsl.,* 10, 169, 1986.

54. **Hodge, V., Stallard, M., Koide, M., and Goldberg, E. D.,** Determination of platinum and iridium in marine waters, sediments and organisms, *Anal. Chem.,* 58, 616, 1986.

55. **De Neve, R.,** Analysis for precious metals combining classical collection procedures with modern instrumentation, *Proc. 10th Int. Precious Met. Inst. Conf.* Rao, U. V., Ed., International Precious Metals Institute, Allentown, PA, 1986, 105-111.

56. **Yang, X.,** Determination of trace platinum and palladium in catalysts by graphite furnace atomic absorption spectroscopy, *Fenxi Ceshi Tongbao,* 5, 59-62, 1986.

57. **Youwei, C.,** Determination of noble metals by graphite furnace atomic absorption spectroscopy. II. Determination of ultramicro amounts of gold, palladium and platinum, *Yankuang Ceshi,* 5, 45, 1986.

58. **Siddik, Z. H., Boxall, F. E., and Harrap, K. R.,** Tissue solubilization in hyamine hydroxide for the flameless atomic absorption spectrophotometric determination of platinum, *Assoc. Int. Cancer Res. Symp. 1985,* 355-360, published 1986.

59. **Branch, C. H. and Hutchinson, D.,** Rapid screening method for the determination of platinum and palladium in geological materials by batch ion-exchange chromatography and graphite furnace atomic absorption spectroscopy, *J. Anal. At. Spectrom.,* 1, 433, 1986.

60. **Parsons, P. J., Morrison, P. F., and LeRoy, A. F.,** Determination of platinum-containing drugs in human plasma by liquid chromatography with reductive electrochemical detection, *J. Chromatogr.,* 385, 323, 1987.

61. **Voth-Beach, L. M., and Shrader, D. E.,** Investigations of a reduced palladium chemical modifier for graphite furnace atomic absorption spectroscopy, *J. Anal. At. Spectrom.,* 2, 45, 1987.

62. **Siddik, Z. H., Boxall, F. E., and Harrap, K. R.,** Flameless atomic absorption spectrophotometric determination of platinum in tissues solubilized in hyamine hydsroxide, *Anal. Biochem.,* 163, 21-26, 1987.

AUTHOR INDEX

SUBJECT INDEX

RARE EARTH ELEMENTS
(DYSPROSIUM, ERBIUM, EUROPIUM, GADOLINIUM, HOLMIUM, LANTHANUM, NEODYMIUM, SAMARIUM, TERBIUM, THULIUM, URANIUM, YTTERBIUM, YTTRIUM)

DYSPROSIUM, Dy   (ATOMIC WEIGHT 162.50)

Instrumental Parameters:

Wavelength	421.2 nm
Slit width	80 μm
Bandpass	0.3 nm
Light source	Hollow Cathode Lamp
Lamp current	8 mA
Purge gas	Argon
Sample size	25 μL
Furnace	Cylindrical cuvette/Pyrolytic graphite

Standard Operating Conditions:

Optimum char temperature	1500°C
Optimum atomization temperature	2700°C
Sensitivity	50pg/1% absorption
Sensitivity check	0.1-2.0 μg/mL
Working range	0.1-5.0 μg/mL
Background correction	Required only to correct light scattering or non-specific absorption from the sample containing high dissolved solids.

General Note:
1. Use commercial standard or a previously analyzed sample as a working standard.
2. Use ultra clean glass and plastic ware, soaked in 1:5 nitric acid solution and rinsed thoroughly with deionized distilled water.
3. All analytical solutions should be prepared in 0.5% (v/v) nitric acid or 1% (v/v) hydrochloric acid.
4. Check for blank values on all reagents including water.
5. Dilution is recommended when sample exhibits greater than 0.5 absorbance units.
6. Use all diluted solutions within 4 h of preparation.
7. Lower atomization temperature will cause reduced sensitivity.

ERBIUM, Er   (ATOMIC WEIGHT 167.26)

Instrumental Parameters:

Wavelength	400.8 nm
Slit width	80 μm
Bandpass	0.3 nm
Light source	Hollow Cathode Lamp
Lamp current	8 mA
Purge gas	Argon
Sample size	25 μL
Furnace	Cylindrical cuvette/Pyrolytic graphite

Standard Operating Conditions:

Optimum char temperature	700°C
Optimum atomization temperature	2700°C
Sensitivity	50 pg/1% absorption
Sensitivity check	0.2-2.0 μg/mL
Working range	0.2- 5.0 μg/mL
Background correction	Required only to correct light scattering or non-specific absorption from the sample-containing high dissolved solids.

General Note:
1. Use commercial standard or a previously analyzed sample as a working standard.
2. Use ultra clean glass and plastic ware, soaked in 1:5 nitric acid solution and rinsed thoroughly with deionized distilled water.
3. All analytical solutions should be prepared in 0.5% (v/v) nitric acid or 1% (v/v) hydrochloric acid.
4. Check for blank values on all reagents including water.
5. Dilution is recommended when sample exhibits greater than 0.5 absorbance units.
6. Use all diluted solutions within 4 h of preparation.
7. Lower atomization temperature will cause reduced sensitivity.

EUROPIUM, Eu   (ATOMIC WEIGHT 162.50)

Instrumental Parameters:

Wavelength	459.4 nm
Slit width	320 μm
Bandpass	1 nm
Light source	Hollow Cathode Lamp
Lamp current	10 mA
Purge gas	Argon
Sample size	25 μL
Furnace	Cylindrical cuvette/Pyrolytic graphite

Standard Operating Conditions:

Optimum char temperature	1300°C
Optimum atomization temperature	2700°C
Sensitivity	20 pg/1% absorption
Sensitivity check	0.06-0.5 μg/mL
Working range	0.02-0.5 μg/mL
Background correction	Required only to correct light scattering or non-specific absorption from the sample containing high dissolved solids.

General Note:
1. Use commercial standard or a previously analyzed sample as a working standard.
2. Use ultra clean glass and plastic ware, soaked in 1:5 nitric acid solution and rinsed thoroughly with deionized distilled water.
3. All analytical solutions should be prepared in 0.5% (v/v) nitric acid or 1% (v/v) hydrochloric acid.
4. Check for blank values on all reagents including water.
5. Dilution is recommended when sample exhibits greater than 0.5 absorbance units.
6. Use of nitrogen as purge gas tends to reduce sensitivity.
7. Lower atomization temperature will cause reduced sensitivity.

GADOLINIUM, Gd   (ATOMIC WEIGHT 157.25)

Instrumental Parameters:

Wavelength	368.4 nm
Slit width	80 µm
Bandpass	0.3 nm
Light source	Hollow Cathode Lamp
Lamp current	10 mA
Purge gas	Argon
Sample size	25 µL
Furnace	Cylindrical cuvette/Pyrolytic graphite

Standard Operating Conditions:

Optimum char temperature	600°C
Optimum atomization temperature	2700°C
Sensitivity	1600 pg/1% absorption
Sensitivity check	2.0-10.0 µg/mL
Working range	2.0-20.0 µg/mL
Background correction	Required only to correct light scattering or non-specific absorption from the sample containing high dissolved solids.

General Note:

1. Use commercial standard or a previously analyzed sample as a working standard.
2. Use ultra clean glass and plastic ware, soaked in 1:5 nitric acid solution and rinsed thoroughly with deionized distilled water.
3. All analytical solutions should be prepared in 0.5% (v/v) nitric acid or 1% (v/v) hydrochloric acid.
4. Check for blank values on all reagents including water.
5. Dilution is recommended when sample exhibits greater than 0.5 absorbance units.
6. Lower atomization temperature will cause reduced sensitivity.

HOLMIUM, Ho   (ATOMIC WEIGHT 164.93)

Instrumental Parameters:

Wavelength	410.4 nm
Slit width	160 μm
Bandpass	0.5 nm
Light source	Hollow Cathode Lamp
Lamp current	12 mA
Purge gas	Argon
Sample size	25 μL
Furnace	Cylindrical cuvette/Pyrolytic graphite

Standard Operating Conditions:

Optimum char temperature	600°C
Optimum atomization temperature	2700°C
Sensitivity	90 pg/1% absorption
Sensitivity check	0.5-2.0 μg/mL
Working range	0.5-3.0 μg/mL
Background correction	Required only to correct light scattering or non-specific absorption from the sample containing high dissolved solids.

General Note:
1. Use commercial standard or a previously analyzed sample as a working standard.
2. Use ultra clean glass and plastic ware, soaked in 1:5 nitric acid solution and rinsed thoroughly with deionized distilled water.
3. All analytical solutions should be prepared in 0.5% (v/v) nitric acid or 1% (v/v) hydrochloric acid.
4. Check for blank values on all reagents including water.
5. Dilution is recommended when sample exhibits greater than 0.5 absorbance units.
6. Lower atomization temperature will cause reduced sensitivity.

LANTHANUM, La   (ATOMIC WEIGHT 138.91)

Instrumental Parameters:

Wavelength	418.7 nm
Slit width	40 μm
Bandpass	0.15 nm
Light source	Hollow Cathode Lamp
Lamp current	10 mA
Purge gas	Argon
Sample size	25 μL
Furnace	Cylindrical cuvette/Pyrolytic graphite

Standard Operating Conditions:

Optimum char temperature	1000°C
Optimum atomization temperature	2700°C
Sensitivity	60 pg/1% absorption
Sensitivity check	0.2-5.0 μg/mL
Working range	2.0-10.0 μg/mL
Background correction	Required only to correct light scattering or non-specific absorption from the sample containing high dissolved solids.

General Note:
1. Use commercial standard or a previously analyzed sample as a working standard.
2. Use ultra clean glass and plastic ware, soaked in 1:5 nitric acid solution and rinsed thoroughly with deionized distilled water.
3. All analytical solutions should be prepared in 0.5% (v/v) nitric acid or 1% (v/v) hydrochloric acid.
4. Check for blank values on all reagents including water.
5. Dilution is recommended when sample exhibits greater than 0.5 absorbance units.
6. Lower atomization temperature will cause reduced sensitivity.

NEODYMIUM, Nd (ATOMIC WEIGHT 162.50)

Instrumental Parameters:

Wavelength	463.4 nm
Slit width	160 μm
Bandpass	0.5 nm
Light source	Hollow Cathode Lamp
Lamp current	8 mA
Purge gas	Argon
Sample size	25 μL
Furnace	Cylindrical cuvette/Pyrolytic graphite

Standard Operating Conditions:

Optimum char temperature	1500°C
Optimum atomization temperature	2700°C
Sensitivity	1800 pg/1% absorption
Sensitivity check	5.0-10.0 μg/mL
Working range	5.0-20.0 μg/mL
Background correction	Required only to correct light scattering or non-specific absorption from the sample containing high dissolved solids.

General Note:

1. Use commercial standard or a previously analyzed sample as a working standard.
2. Use ultra clean glass and plastic ware, soaked in 1:5 nitric acid solution and rinsed thoroughly with deionized distilled water.
3. All analytical solutions should be prepared in 0.5% (v/v) nitric acid or 1% (v/v) hydrochloric acid.
4. Check for blank values on all reagents including water.
5. Dilution is recommended when sample exhibits greater than 0.5 absorbance units.
6. Higher atomization temperature will decrease characteristic sensitivity and graphite tube life.

SAMARIUM, Sm   (ATOMIC WEIGHT 150.35)

Instrumental Parameters:

Wavelength	429.7 nm
Slit width	160 μm
Bandpass	0.5 nm
Light source	Hollow Cathode Lamp
Lamp current	10 mA
Purge gas	Argon
Sample size	25 μL
Furnace	Cylindrical cuvette/Pyrolytic graphite

Standard Operating Conditions:

Optimum char temperature	1400°C
Optimum atomization temperature	2600°C
Sensitivity	240 pg/1% absorption
Sensitivity check	0.5-2.0 μg/mL
Working range	0.5-5.0 μg/mL
Background correction	Required only to correct light scattering or non-specific absorption from the sample containing high dissolved solids.

General Note:

1. Use commercial standard or a previously analyzed sample as a working standard.
2. Use ultra clean glass and plastic ware, soaked in 1:5 nitric acid solution and rinsed thoroughly with deionized distilled water.
3. All analytical solutions should be prepared in 0.5% (v/v) nitric acid or 1% (v/v) hydrochloric acid.
4. Check for blank values on all reagents including water.
5. Dilution is recommended when sample exhibits greater than 0.5 absorbance units.
6. Higher atomization temperature decreases characteristic sensitivity and the tube life.
7. Memory effects due to the formation of stable compounds are possible.

TERBIUM, Tb   (ATOMIC WEIGHT 158.924)

Instrumental Parameters:

Wavelength	214.3 nm
Slit width	160 μm
Bandpass	0.5 nm
Light source	Hollow Cathode Lamp
Lamp current	10 mA
Purge gas	Argon
Sample size	25 μL
Furnace	Cylindrical cuvette/Pyrolytic graphite

Standard Operating Conditions:

Optimum char temperature	1000°C
Optimum atomization temperature	2000°C
Sensitivity	20 pg/1% absorption
Sensitivity check	0.05-2.0 μg/mL
Working range	0.1-1.0 μg/mL
Background correction	Required only to correct light scattering or non-specific absorption from the sample containing high dissolved solids.

General Note:

1. Use commercial standard or a previously analyzed sample as a working standard.
2. Use ultra clean glass and plastic ware, soaked in 1:5 nitric acid solution and rinsed thoroughly with deionized distilled water.
3. All analytical solutions should be prepared in 0.5% (v/v) nitric acid or 1% (v/v) hydrochloric acid.
4. Check for blank values on all reagents including water.
5. Dilution is recommended when sample exhibits greater than 0.5 absorbance units.
6. Use 0.02 mg nickel as nickel nitrate for matrix modification.

THULIUM, Tm   (ATOMIC WEIGHT 168.934)

Instrumental Parameters:

Wavelength	371.8 nm
Slit width	160 μm
Bandpass	0.5 nm
Light source	Hollow Cathode Lamp
Lamp current	10 mA
Purge gas	Argon
Sample size	25 μL
Furnace	Cylindrical cuvette/Pyrolytic graphite

Standard Operating Conditions:

Optimum char temperature	1700°C
Optimum atomization temperature	2700°C
Sensitivity	20 pg/1% absorption
Sensitivity check	0.05-1.0 μg/mL
Working range	0.05-1.0 μg/mL
Background correction	Required only to correct light scattering or non-specific absorption from the sample containing high dissolved solids.

General Note:
1. Use commercial standard or a previously analyzed sample as a working standard.
2. Use ultra clean glass and plastic ware, soaked in 1:5 nitric acid solution and rinsed thoroughly with deionized distilled water.
3. All analytical solutions should be prepared in 0.5% (v/v) nitric acid or 1% (v/v) hydrochloric acid.
4. Check for blank values on all reagents including water.
5. Dilution is recommended when sample exhibits greater than 0.5 absorbance units.
6. Higher atomization temperature will decrease characteristic sensitivity and tube life.

URANIUM, U   (ATOMIC WEIGHT 238.03)

Instrumental Parameters:

Wavelength	358.5 nm
Slit width	80 μm
Bandpass	0.3 nm
Light source	Hollow Cathode Lamp
Lamp current	15 mA
Purge gas	Argon
Sample size	25 μL
Furnace	Pyrolytic graphite

Standard Operating Conditions:

Optimum char temperature	1200°C
Optimum atomization temperature	2700°C
Sensitivity	12000 pg/1% absorption
Sensitivity check	30.0-60.0 μg/mL
Working range	50.0-500.0 μg/mL
Background correction	Required only to correct light scattering or non-specific absorption from the sample containing high dissolved solids.

General Note:
1. Use commercial standard or a previously analyzed sample as a working standard.
2. Use ultra clean glass and plastic ware, soaked in 1:5 nitric acid solution and rinsed thoroughly with deionized distilled water.
3. All analytical solutions should be prepared in 0.5% (v/v) nitric acid or 1% (v/v) hydrochloric acid.
4. Check for blank values on all reagents including water.
5. Dilution is recommended when sample exhibits greater than 0.5 absorbance units.
6. Use of oxidizing acids will suppress sensitivity.
7. Lower atomization temperature will cause reduction in sensitivity.

YTTERBIUM, Yb   (ATOMIC WEIGHT 173.04)

Instrumental Parameters:

Wavelength	398.8 nm
Slit width	80 μm
Bandpass	0.3 nm
Light source	Hollow Cathode Lamp
Lamp current	5 mA
Purge gas	Argon
Sample size	25 μL
Furnace	Cylindrical cuvette/Pyrolytic graphite

Standard Operating Conditions:

Optimum char temperature	1200°C
Optimum atomization temperature	2700°C
Sensitivity	5 pg/1% absorption
Sensitivity check	0.05 μg/mL
Working range	0.05-0.10 μg/mL
Background correction	Required only to correct light scattering or non-specific absorption from the sample containing high dissolved solids.

General Note:
1. Use commercial standard or a previously analyzed sample as a working standard.
2. Use ultra clean glass and plastic ware, soaked in 1:5 nitric acid solution and rinsed thoroughly with deionized distilled water.
3. All analytical solutions should be prepared in 0.5% (v/v) nitric acid.
4. Check for blank values on all reagents including water.
5. Dilution is recommended when sample exhibits greater than 0.5 absorbance units.
6. Lower atomization temperature will cause reduced sensitivity.

YTTRIUM, Y   (ATOMIC WEIGHT 88.905)

Instrumental Parameters:

Wavelength	410.2 nm
Slit width	80 μm
Bandpass	0.3 nm
Light source	Hollow Cathode Lamp
Lamp current	6 mA
Purge gas	Argon
Sample size	25 μL
Furnace	Cylindrical cuvette/Pyrolytic graphite

Standard Operating Conditions:

Optimum char temperature	600°C
Optimum atomization temperature	2700°C
Sensitivity	13000 pg/1% absorption
Sensitivity check	20.0-50.0 μg/mL
Working range	10.0-100.0 μg/mL
Background correction	Required only to correct light scattering or non-specific absorption from the sample containing high dissolved solids.

General Note:
1. Use commercial standard or a previously analyzed sample as a working standard.
2. Use ultra clean glass and plastic ware, soaked in 1:5 nitric acid solution and rinsed thoroughly with deionized distilled water.
3. All analytical solutions should be prepared in 0.5% (v/v) nitric acid or 1% (v/v) hydrochloric acid.
4. Check for blank values on all reagents including water.
5. Dilution is recommended when sample exhibits greater than 0.5 absorbance units.
6. Lower atomization temperature will cause reduced sensitivity.
7. Use 0.02 mg nickel as nickel nitrate for matrix modification.

REFERENCES

1. **Kuga, K. and Tsujii, K.,** Determination of rare earth and alkaline earth elements by graphite furnace atomic absorption spectroscopy using the pyrolytic graphite tube, *Bunseki Kagaku,* 27, 441, 1978.
2. **L'vov, B. V. and Pelieva, L. A.,** Atomic absorption determination of cerium by sample atomization in graphite furnace, *Zh. Anal. Khim.,* 34, 1744, 1979.
3. **Tarui, T. and Tokairin, H.,** Graphite furnace atomic absorption spectroscopy of uranium by a tantalum boat, *Bunseki Kagaku,* 28, 133, 1979.
4. **Berggren, P. O.,** Determination of barium, lanthanum and magnesium in pancreatic islets by electrothermal atomic absorption spectroscopy, *Anal. Chim. Acta,* 119, 161, 1980.
5. **Mazzucotelli, A. and Frache, R.,** Behavior of trace amounts of europium in a silicate matrix in atomic absorption spectrophotometry with electrothermal atomization, *Analyst,* 105, 497, 1980.
6. **Sen Gupta, J. G.,** Determination of yttrium and rare earth elements in rocks by graphite furnace atomic absorption spectroscopy, *Geol. Surv. Can. Pap.,* 80B, 1980.
7. **Horsky, S. J.,** Determination of praseodymium by graphite furnace atomic absorption spectroscopy, *At. Spectrosc.,* 1, 129, 1980.
8. **Wahab, H. S. and Chakrabarti, C. L.,** Studies on the sensitivity of yttrium by electrothermal atomization from metallic and metal carbide surfaces of a heated graphite atomizer in atomic absorption spectroscopy, *Spectrochim. Acta,* 36B, 463, 1981.
9. **Wahab, H. S. and Chakrabarti, C. L.,** Mechanism of yttrium atom formation in electrothermal atomization from metallic and metal-carbide surfaces of a heated graphite atomizer in atomic absorption spectroscopy, *Spectrochim. Acta,* 36B, 475, 1981.
10. **Mazzucotelli, A. and Frache, R.,** Electrothermal atomization for atomic absorption determination of some rare earths in silicate rocks and minerals, *Mikrochim. Acta,* 2, 323, 1981.
11. **L'vov, B. V. Pelieva, L. A., and Novotny, J.,** Electrothermal atomic absorption spectrometric determination of atomization energy of gaseous dysprosium dicarbide, *Zh. Prikl. Spektrosk.,* 35, 403, 1981.
12. **Horsky, S. J. and Fletcher, W. K.,** Evaluation of a combined ion exchange-graphite furnace atomic absorption procedure for determination of rare earth elements in geological samples, *Chem. Geol.,* 32, 335, 1981.
13. **Sen Gupta, J. G.,** Determination of yttrium and rare earth elements in rocks by graphite furnace atomic absorption spectroscopy, *Talanta,* 28, 31, 1981.
14. **Gerardi, M. and Pelliccia, G. A.,** Direct quantitative determination of metallic and rare earth impurities in nuclear solutions containing uranium, thorium and fission products by flameless atomic absorption spectroscopy, *Com. Naz. Energ. Nucl. (Rapp. Tec.),* 81-8, p.23, 1981 (in Italian).
15. **Wu, Z. and Ma, Y.,** Application of pyrolytic graphite coated tube lined with tungsten-tantalum for determination of rare earths (lanthanum, lutetium and holmium) and uranium by graphite furnace atomic absorption spectrometry, *Huanjing Huaxue,* 1, 228, 1982 (in Chinese).
16. **Sneddon, J. and Fuavao, V. A.,** Determination of ytterbium by graphite furnace atomic absorption spectrophotometry, *At. Spectrosc.,* 3, 51-3, 1982.
17. **Fujino, O. and Matsui, M.,** Atomic absorption spectroscopy of ytterbium using carbon tube atomizers. An application to phosphate minerals, *Bunseki Kagaku,* 31, 241-6, 1982.
18. **Fujino, O. and Matsui, M.,** Atomic absorption spectroscopy of dysprosium by a pyrolytic-graphite-coated carbon tube atomizer, *Bunseki Kagaku,* 31, 619, 1982.
19. **Coerdt, W., Crubellati, R., Mainka, E., and Mueller, H. G.,** Determination of gadolinium in nuclear fuels — A comparison of methods, *J. Nucl. Mater.,* 106(1-3), 109-13, 1982.
20. **Sen Gupta, J. G.,** Flame and graphite furnace atomic absorption and optical-emission spectroscopic determination of yttrium and the rare-earth contents of sixteen international reference samples of rocks and coal, *Geostand. Newsl.,* 6(2), 241-8, 1982.
21. **Mo, S.,** Interference in atomic absorption spectrophotometry. VI. Effect of chloride and nitrate of lanthanum and yttrium on the background in flameless graphite furnace atomic absorption spectrophotometry, *Zhongguo Kexueyuan Changchun Ying Yong Huaxue Yanjiuso Jikan,* 19, 58, 1982 (in Chinese).
22. **Daidoji, H. and Tamura, S.,** Graphite furnace atomic absorption spectroscopy with a tantalum boat for the determination of yttrium, samarium and dysprosium in a misch metal, *Bull. Chem. Soc. Jpn.,* 55, 3510, 1982.
23. **Sicinska, P. and Michalewska, M.,** Application of atomic absorption spectrophotometry for the determination of lanthanides in ores and rare earth concentrates, *Fresnius' Z. Anal. Chem.,* 312, 530, 1982.
24. **Zhen, R. and Yang, B.,** Determination of rare earth elements by graphite furnace atomic absorption spectroscopy using a tantalum-coated tube, *Xiyou Jinslu,* 2(2), 178-82, 1983.
25. **Fuavao, V. A. and Sneddon, J.,** Determination of ytterbium by graphite furnace atomic absorption spectrometry with vaporization from a microboat and a platform, *At. Spectrosc.,* 4(5), 179-81, 1983.
26. **Bangia, T. R., Kartha, K. N. R., Varghese, M., Dhawale, B. A., and Joshi, B. D.,** Tracer studies in the recovery of metallic impurities from uranium and plutonium solutions and estimation of cadmium, cobalt and

manganese by electrothermal A.A.S., *Proc. Nucl. Chem. Radiochem. Symp.*, 302-4, 1981, published 1983.

27. **Gerardi, M. and Pelliccia, G. A.**, Direct quantitative determination using flameless atomic absorption spectroscopy of metallic impurities and rare earths in nuclear solutions containing uranium, thorium and fission products, *At. Spectrosc.*, 4(6), 193-8, 1983.

28. **Hamm, U. and Baechmann, K.**, Determination of uranium by graphite furnace atomic absorption spectroscopy, *Fresnius' Z. Anal. Chem.*, 315, 591, 1983.

29. **Wu, Z. and Ma, Y.**, Direct determination of trace neodymium and scandium in soils by tungsten and tantalum coated graphite tube-atomic absorption spectroscopy, *Fenxi Huaxue*, 11, 423, 1983 (in Chinese).

30. **Sen Gupta, J. G.**, Determination of cerium in silicate rocks by electrothermal atomization in a furnace lined with tantalum foil. Application to 19 international geological reference materials, *Talanta*, 31(12), 1053, 1984.

31. **Liang, S., Zhong, Y., and Wang, Z.**, Enrichment of traces of scandium from aqueous solutions by means of flotation, *Z. Anal. Chem.*, 318, 19-21, 1984.

32. **Korde, M., Lee, D. S., and Stallard, M. W.**, Concentration and separation of trace metals (plutonium and technitium) from seawater using a single anion exchange bead, *Anal. Chem.*, 56, 1956, 1984.

33. **Yao, J. and Zeng, J.**, Determination of europium in rare earth oxide mixtures by graphite furnace atomic absorption spectroscopy, *Fenxi Huaxue*, 12, 343, 1984.

34. **Jing, S., Wang, R., Yu, C, Zhang, D., Yan, Y., and Ma, Y.**, Design of model ZM-1 Zeeman effect atomic absorption spectrometer, *Huanjing Kexue*, 5, 41, 1984 (in Chinese).

35. **Fujino, O., Atano, T., Sugiyama, M., and Matsui, M.**, Determination of europium in phosphate minerals by solvent extraction-atomic absorption spectrometry with a pyrolytic-graphite-coated carbon tube atomizer, *Bunseki Kagaku*, 33(11), 604-8, 1984.

36. **Shevchuk, I. A., Alemasova, A. S., and Rokun, A. N.**, Extraction separation and flameless atomic absorption determination of rare earth elements in chloride solutions, *Ukr. Khim. Zh.*, 51(2), 197, 1985, CA 16, 103:31713j, 1985.

37. **Rojas de Olivares, D.**, Standardization of ruthenium solutions, *At. Spectrosc.*, 6, 47, 1985.

38. **Sen Gupta, J. G.**, Determination of rare earths, yttrium and scandium, in silicate rocks and four new geological reference materials by electrothermal atomization from graphite and tantalum surfaces, *Talanta*, 32(1), 1-6, 1985.

39. **Hamm, U. and Baechmann, K.**, Methods of uranium determination using graphite furnace atomic absorption spectroscopy, *Fortschr. Atomspektrom. Spurenanal.*, 1, 115, 1984, CA 12, 102:197152e, 1985.

40. **Itsuki, K., Yagasaki, H., and Fujinuma, H.**, Effect of lanthanum ion on graphitte furnace atomic absorption spectroscopy of indium, *Bunseki Kagaku*, 34(8), T109-T112, 1985.

41. **Haines, J.**, The determination of selected lanthanides by atomic absorption spectroscopy using electrothermal atomization, *Rep. MINTEK*, M2331, 21, 1985.

42. **Modenesi, C. R. and Abrao, A.**, Individual determination of lanthanides in Y and P oxides by GFAAS, *Publ. ACIESP*, 44-1, 74-86, 1984, CA 24, 103:171145a, 1985.

43. **Modenesi, C. R. and Abrao, A.**, Determination of Gd, Sm, Eu and Ds in uranium compounds by GFAAS, *Publ. ACIESP (1984)*, 44-1, 87-94, CA 24, 103:171146b, 1985.

44. **Sneddon, J. and Fuavao, V. A.**, Effect of atomizer surface type in quantitation of molybdenum and ytterbium by electrothermal atomization atomic absorption spectrometry, *Anal. Chim. Acta*, 167, 317, 1985.

45. **Liang, S., Chen, D., and Zhong, Y.**, Electrothermal atomic absorption spectrophotometric determination of ultratrace amounts of ytterbium in water after enrichment with Levextrel, *Huaxue Tongbao*, 8, 33-4, 1986 (in Chinese).

46. **Batistoni, D. A., Erlijman, L. H., and Pazos, A. L.**, Analysis of nuclear grade uranium oxides by electrothermal atomization-atomic absorption spectrometry, *An. Asoc. Quim. Argent.*, 74, 265-75, 1986 (in Spanish).

47. **Yao, J. and Huang, B.**, Tantalum foil-lined graphite furnace atomic absorption spectroscoipy. Effects of small amounts of hydrogen in argon purge gas and the determination of europium in blast furnace slags, *Can. J. Spectrosc.*, 31, 77-80. 1986.

48. **Suzuki, T., Nakagawa, H., and Sawada, K.**, A combined solvent extraction-graphite furnace atomic absorption spectrometry for the determination of scandium in coal fly ash, *Anal. Sci.*, 2(3), 309, 1986.

49. **Ma, Y., Sun, D., and Zhu, M.**, Determination of europium by graphite furnace atomic absorption spectroscopy with Zeeman background correction, *Fenxi Ceshi Tongbao*, 5, 55-9, 1986.

50. **Depu, C., Youhan, Z., and Shuchuan, L.**, Microdetermination of ytterbium with electrothermal atomic absorption spectroscopy, *Guangpuxue Yu Guangpu Fenxi*, 6, 53, 1986 (in Chinese).

51. **Jinyu, Y. and Benli, H.**, Tantalum foil-lined graphite furnace AAS. Effects of small amounts of hydrogen in argon-purge gas and the determination of europium in blast furnace slags, *Can. J. Spectrosc.*, 31, 77, 1986.

52. **Suzuki, T., Nakagawa, H., and Sawada, K.**, A combined solvent extraction-graphite furnace atomic absorption spectroscopy for the determination of scandium in coal fly ash, *Anal. Sci.*, 2, 309, 1986.

53. **Chen, D., Zhong, Y., and Liang, S.**, Microdetermination of ytterbium with electrothermal atomic absorption spectrometry, *Guangpuxue YU Guangpu Fenxi*, 6(1), 53-5, 1986 (in Chinese).

54. **Bettinelli, M., Baroni, U., and Pastorelli, N.**, Determination of scandium in coal fly ash and geologiccal materials by graphite furnace atomic absorption spectroscopy and inductively coupled plasma atomic emission spoectroscopy, *Analyst (London)*, 112, 23, 1987.

AUTHOR INDEX

SUBJECT INDEX

RHENIUM

RHENIUM, Re ATOMIC WEIGHT 186.2

Instrumental Parameters:

Wavelength	346.0 nm
Slit width	160 μm
Bandpass	0.5 nm
Light source	Hollow Cathode Lamp
Lamp current	10 mA
Purge gas	Argon
Sample size	25 μL
Furnace	Cylindrical cuvette/Pyrolytic graphite

Standard Operating Conditions:

Optimum char temperature	1200°C
Optimum atomization temperature	2700°C
Sensitivity	10000 pg/1% absorption
Sensitivity check	5.0-50.0 μg/mL
Working range	10.0-100.0 μg/mL
Background correction	Required only to correct light scattering or non-specific absorption from the sample containing high dissolved solids.

General Note:

1. Use commercial standard or a previously analyzed sample as a working standard.
2. Use ultra clean glass and plastic ware, soaked in 1:5 nitric acid solution and rinsed thoroughly with deionized distilled water.
3. All analytical solutions should be prepared in 0.5% (v/v) nitric acid or 1% (v/v) hydrochloric acid.
4. Check for blank values on all reagents including water.
5. Dilution is recommended when sample exhibits greater than 0.5 absorbance units.
6. Lower atomization temperature will cause reduced sensitivity.

SELENIUM

SELENIUM, Se   (ATOMIC WEIGHT 78.96)

Instrumental Parameters:

Wavelength	196.0 nm
Slit width	640 μm
Bandpass	2 nm
Light source	Hollow Cathode Lamp
Lamp current	12 mA
Purge gas	Argon
Sample size	25 μL
Furnace	Cylindrical cuvette/Pyrolytic graphite

Standard Operating Conditions:

Optimum char temperature	200°C
Optimum atomization temperature	2200°C
Sensitivity	100 pg/1% absorption
Sensitivity check	0.2-0.5 μg/mL
Working range	0.05-1.0 μg/mL
Background correction	Required only to correct light scattering or non-specific absorption from the sample containing high dissolved solids.

General Note:

1. Use commercial standard or a previously analyzed sample as a working standard.
2. Use ultra clean glass and plastic ware, soaked in 1:5 nitric acid solution and rinsed thoroughly with deionized distilled water.
3. All analytical solutions should be prepared in 0.5% (v/v) nitric acid or 1% (v/v) hydrochloric acid.
4. Check for blank values on all reagents including water.
5. Dilution is recommended when sample exhibits greater than 0.5 absorbance units.
6. To avoid hydrolysis, solutions requiring the second dilution (1:25) should be prepared every 4 h.
7. Addition of 10 μg/mL nickel as nickel nitrate to analyte will allow increase in pyrolysis temperature to 1000°C.
8. Use of nitrogen as purge gas will reduce sensitivity.

REFERENCES

1. **Baird, R. B., Pourian, S., and Gabrielian, S. M.,** Determination of trace amounts of selenium in waste waters by carbon rod atomization, *Anal. Chem.,* 44, 1887, 1972.
2. **Baird, R. B. and Gabrielian, S. M.,** A tantalum foil lined graphite tube for the analysis of arsenic and selenium by atomic absorption spectroscopy, *Appl. Spectrosc.,* 28, 213, 1974.
3. **Inhat, M. and Westerby, R. J.,** Application of flameless atomization to the atomic absorption determination of selenium in biological samples, *Anal. Lett.,* 7, 257, 1974.
4. **Vijan, P. N. and Wood, G. R.,** An automated submicrogram determination of selenium in vegetation by quartz tube for atomic absorption spectrophotometry, *Talanta,* 23, 89, 1976.
5. **Inhat, M.,** Selenium in foods: evaluation of atomic absorption spectrometric techniques involving hydrogen selenide generation and carbon furnace atomization, *J. Ass. Offic. Anal. Chem.,* 59, 911, 1976.
6. **Inhat, M.,** Atomic absorption spectrometric determination of selenium with carbon furnace atomization, *Anal. Chim. Acta,* 82, 293, 1976.
7. **Ishizaki, M.,** Determination of selenium in biological materials by flameless atomic absorption spectroscopy using a carbon tube atomizer, *Bunseki Kagaku,* 26, 206, 1977.
8. **Yasuda, K., Taguchi, M., Tamura, S., and Toda, S.,** Determination of selenium in biological samples by solvent extraction-graphite furnace atomic absorption spectroscopy, *Bunseki Kagaku,* 26, 442, 1977.
9. **Aruscavage, P.,** Determination of arsenic, antimony and selenium in coal by atomic absorption spectroscopy with graphite tube atomizer, *J. Res. U.S. Geol. Surv.,* 5, 405, 1977.
10. **Hocquellet, P.,** Application of electrothermal atomization to the determination of arsenic, antimony, selenium and mercury by atomic absorption spectroscopy, *Analusis,* 6, 426, 1978.
11. **Manning, D. C.,** Spectral interferences in graphite furnace atomic absorption spectroscopy. I. The determination of selenium in an iron matrix, *At. Absorp. Newsl.,* 17, 107, 1978.
12. **Ishizaki, M.,** Simple method for determination of selenium in biological materials by flameless atomic absorption spectroscopy using a carbon tube atomizer, *Talanta,* 25, 167, 1978.
13. **Haynes, B. W.,** Arsenic, antimony, selenium and tellurium determinations in high-purity copper by electrothermal atomization, *At. Ab. Newsl.,* 18, 46, 1979.
14. **Saeed, K., Thomassen, Y., and Langmyhr, F. J.,** Direct electrothermal atomic absorption spectrometric determination of selenium in serum, *Anal. Chim. Acta,* 110, 285, 1979.
15. **Elson, C. M. and Macdonald, A. S.,** Determination of selenium in pyrite by an ion exchange-electrothermal atomic absorption spectrometric method, *Anal. Chim. Acta,* 110, 153, 1979.
16. **Vickrey, T. M. and Buren, M. S.,** Factors affecting selenium atomization efficiency in graphite furnace atomic absorption, *Anal. Lett.,* 13, 1465, 1980.
17. **Hocquellet, P.,** Determination of selenium in animal feeds by atomic absorption spectroscopy with electrothermal atomization, *Ann. Fals. Exp. Chim.,* 73, 129, 1980 (in French).
18. **Kamada, T. and Yamamoto, Y.,** Use of transition elements to enhance sensitivity for selenium determination by graphite furnace atomic absorption spectrophotometry combined with a solvent extraction in a uniform magnetic field, *J. Phys. Soc. Jpn.,* 48, 2098, 1980.
19. **Alexander, J., Saeed, K., and Thomassen, Y.,** Thermal stabilization of inorganic and organoselenium compounds for direct electrothermal atomic absorption spectroscopy, *Anal. Chim. Acta,* 120, 377, 1980.
20. **Rail, C. D., Kidd, D. E., and Hadley, W. M.,** Determination of selenium in tissues, serum and blood of wild rodents by graphite furnace atomic absorption spectrophotometry, *Int. J. Environ. Anal. Chem.,* 8, 79, 1980.
21. **Kirkbright, G. F., Shan, H. C., and Snook, R. D.,** Evaluation of some matrix modification procedure for use in the determination of mercury and selenium by atomic absorption spectroscopy with a graphite tube electrothermal atomizer, *At. Spectrosc.,* 1, 85, 1980.
22. **Morita, K., Shimizu, M., Inoue, B., and Ishida, T.,** Graphite furnace atomic absorption spectroscopy of antimony and selenium in whole blood, *Okayama-ken Kankyo Hoken Senta Nempo,* 4, 94, 1980.
23. **Katskov, D. A., Grinshtein, I. L., and Kruglikova, L. P.,** Study of the evaporation of the metals In, Ga, Tl. Ge, Sn, Pb, Sb, Bi, Se and Te from a graphite surface by the atomic absorption method, *Zh. Prikl. Spektrosk.,* 33, 804, 1980.
24. **Ohta, K. and Suzuki, M.,** Determination of selenium in water by electrothermal atomic absorption spectroscopy, *Z. Anal. Chem.,* 302, 177, 1980.
25. **Subramanian, K. S.,** Rapid electrothermal atomic absorption method for arsenic and selenium in geological materials via hydride evolutiuon, *Z. Anal. Chem.,* 305, 382, 1981.
26. **Carelli, G., Cecchetti, G., La Bua, R., Bergamaschi, A., and Iannaccone, A.,** Determination of selenium, cadmium, zinc, cobalt, copper and silver using atomic absorption spectrophotometry in aerosols from the workplace in the color television electronic industry, *Ann. Ist. Super. Sanita,* 17, 505, 1981.
27. **Sanzolone, R. F. and Chao, T. T.,** Determination of submicrogram amounts of selenium in geological materials by atomic absorption spectrophotometry with electrothermal atomization after solvent extraction, *Analyst,* 106, 647, 1981.

28. **Hofsommer, H. J. and Bielig, H. J.,** Experiences in determining selenium in foods, *Atomspektrom. Spurenanal.,* Welz, B., Ed., Springer-Verlag, Weinheim, 1982, 527-39.

29. **Saeed, K. and Thomassen, Y.,** Spectral interferences from phosphate matrixes in the determination of arsenic, antimony, selenium and tellurium by electrothermal atomic absorption spectroscopy, *Anal. Chim. Acta,* 130, 281, 1981.

30. **Szydlowski, F. J. and Vianzon, F. R.,** Further studies on the determination of selenium using graphite furnace atomic absorption spectroscopy, *At. Spectrosc.,* 1, 39, 1980.

31. **Subramanian, K. S. and Meranger, J. C.,** Determination of As(III), As(V), Sb(III), Sb(V), Se(IV) and Se(VI) by extraction with amm.pyrrolidine dithiocarbamate-methyl isobutyl ketone and electrothermal atomic absorption spectroscopy, *Anal. Chim. Acta,* 124, 131, 1981.

32. **Sanzolone, R. F. and Chao, T. T.,** Matrix modification with silver for the electrothermal atomization of arsenic and selenium, *Anal. Chim. Acta,* 128, 225, 1981.

33. **Shan, X. and Ni, Z.,** Matrix modification for the determination of volatile elements of arsenic, selenium, tellurium, silver, antimony and bismuth by graphite furnace atomic absorption spectroscopy, *Huaxue Xuebao,* 39, 575-8, 1981.

34. **Cheam, V. and Asmila, K. I.,** Interlaboratory quality control study No. 26 arsenic and selenium in water, *Rep. Ser. Inland Waters Dir. (Can.),* 68, 8, 1981.

35. **Shan, X., Jin, L., and Ni, Z.,** Determination of selenium in soil digests by graphite furnace atomic absorption spectroscopy after extraction with 1,2-diamino-4-nitrobenzene, *At. Spectrosc.,* 3, 41, 1982.

36. **Subramanian, K. S. and Meranger, J. C.,** Rapid hydride evolution-electrothermal atomization atomic absorption spectrophotometric method for determining arsenic and selenium in human kidney and liver, *Analyst (London),* 107, 157-62, 1982.

37. **Oyamada, N. and Ishizaki, M.,** Determination of trimethyl selenium ion and total selenium in human urine by graphite furnace atomic absorption spectroscopy, *Bunseki Kagaku,* 31, 17-22, 1982.

38. **Tam, G. K. H. and Lacroix, G.,** Dry ashing, hydride generation atomic absorption spectrometric determination of arsenic and selenium in foods, *J. Assoc. Off. Anal. Chem.,* 65, 647, 1982.

39. **Oster, O. and Prelluitz, W.,** A methodological comparison of hydride and carbon furnace atomic absorption spectroscopy for the determination of selenium in serum, *Clin. Chim. Acta,* 124, 277, 1982.

40. **Alfthan, G. and Kumpulainen, J.,** Determination of selenium in small volumes of blood plasma and serum by electrothermal atomic absorption spectroscopy, *Anal. Chim. Acta,* 140, 221-7, 1982.

41. **Julshamn, K., Ringdal, O., Slinning, K. E., and Braekkan, O. R.,** Optimization of the determination of selenium in marine samples by atomic absorption spectroscopy. Comparison of a flameless graphite furnace atomic absorption system with a hydride generation atomic absorption system, *Spectrochim. Acta,* 37B, 473-82, 1982.

42. **Tulley, R. T. and Lehmann, H. P.,** Flameless atomic absorption spectrophotometry of selenium in whole blood, *Clin. Chem. (Winston-Salem, N.C.),* 28, 1448-50, 1982.

43. **Brown, A. A., Ottaway, J. M., and Fell, G. S.,** Determination of selenium in biological material. Comparison of three atomic spectrometric methods, *Anal. Proc. (London),* 19, 321-4, 1982.

44. **Inui, T., Terada, S., Tamura, H., and Ichinose, N.,** Determination of selenium by hydride generation with reducing tube followed by graphite furnace atomic absorption spectroscopy, *Fresnius' Z. Anal. Chem.,* 311, 492-5, 1982.

45. **Dillon, L. J., Hilderbrand, D. C., and Groon, K. S.,** Flameless atomic absorption determination of selenium in human blood, *At. Spectrosc.,* 3, 5-7, 1982.

46. **Gonzalez, C. A. and Tello, A. R.,** Determination of selenium and tellurium in minerals and concentrates by graphite furnace atomic absorption, *Bol. Soc. Chil. Quim.,* 27, 297-9, 1982 (in Spanish).

47. **Chi, X., Lu, Y., and Han, S.,** Determination of trace selenium in surface water by atomic absorption spectrophotometry using a hydride-generation system, *Beijing Shifan Daxue Xuebao Ziran Kexueban,* 2, 31-6, 1982 (in Chinese).

48. **Bye, R. and Lund, W.,** Determination of selenium in pyrite by the atomic absorption-hydride generation technique, *Fresnius' Z. Anal. Chem.,* 313, 211, 1982.

49. **Kaen, E. and Ivanov, G.,** Direct flameless atomic absorption method for determination of trace selenium in air, *Dokl. Bolg. Akad. Nauk,* 35, 1257, 1982 (in French).

50. **Chakraborti, D., Hillman, D. C. J., Irgolic, K. J., and Zingarov, R. A.,** Hitachi Zeeman graphite furnace atomic absorption spectrometer as a selenium-specific detector for ion chromatography. Separation and determination of selenite and selenate, *J. Chromatogr.,* 249, 81, 1982.

51. **Headridge, J. B. and Nicholson, R. A.,** Determination of arsenic, antimony, selenium and tellurium in nickel-base alloys by atomic absorption spectrometry with introduction of solid samples into furnaces, *Analyst (London),* 107, 1200, 1982.

52. **Kellerman, S. P.,** The use of masking agents in the determination by hydride generation and atomic absorption spectrophotometry of arsenic, antimony, selenium, tellurium and bismuth in the presence of noble metals, *Rep. MINTEK,* M39, 14, 1982.

53. **Oyamada, N. and Ishizaki, M.,** Determination of trimethylselenonium ions in human urine by graphite furnace atomic absorption spectroscopy, *Sangyo Igaku,* 24, 320, 1982.

54. **Bye, R. and Holen, B.,** An examination of the hydride generation-atomic absorption spectrometric technique for the determination of selenium in technical sulfuric acid, *Anal. Chim. Acta,* 144, 235, 1982.

55. **Verlinden, M.,** On the acid decomposition of human blood and plasma for the determination of selenium, *Talanta,* 29, 875, 1982.

56. **Saeed, K. and Thomassen, Y.,** Electrothermal atomic absorption spectrometric determination of selenium in blood serum and seminal fluid after protein precipitation with trichloroacetic acid, *Anal. Chim. Acta,* 143, 223, 1982.

57. **Kujirai, O., Kobayashi, T., Ide, K., and Sudo, E.,** Determination of traces of selenium in heat-resisting alloys by graphite furnace atomic absorption spectroscopy after coprecipitation with arsenic, *Talanta,* 30, 9, 1983.

58. **Saeed, K. and Thomassen, Y.,** Flameless atomic absorption determination of selenium in human blood. Comments., *At. Spectrosc.,* 4, 163, 1983.

59. **Hilderbrand, D. C.,** Flameless atomic absorption determination of selenium in human blood. Reply to comments, *At. Spectrosc.,* 4, 164, 1983.

60. **Huguet, C., Chappuis, P., Peynet, J., Thuillier, F., Legrand, A., and Rousselet, F.,** Serum selenium assay by atomic absorption spectrophotometry: its value in the laboratory diagnosis of nonobstructive cardiomyopiathies, *Ann. Biol. Clin. (Paris),* 41, 277, 1983.

61. **Oyamada, N.,** Determination of trimethylselenonium ion in urine by graphite furnace atomic absorption spectrometry, *Ibaraki-ken Eisei Kenkyusho Nempo,* 21, 17-22, 1983.

62. **Welz, W., Melcher, M., and Schelemmer, G.,** Determination of selenium in human blood serum. Comparison of two atomic absorption spectrometric procedures, *Z. Anal. Chem.,* 316, 271, 1983.

63. **Kumpulainen, J., Lehto, J., Koivistoinen, P., and Vuori, E.,** Direct electrothermal atomic absorption spectrometric determination of selenium and chromium in biological fluids, *Proc. 2nd Int. Workshop Trace Elem. Anal. Chem. Med. Biol.,* Braetter, P. and Schramel, P., Eds., Berlin, 1983, 951-67.

64. **Carnrick, G. R., Manning, D. C., and Slavin, W.,** Determination of selenium in biological materials with platform furnace atomic absorption spectroscopy and Zeeman background correction, *Analyst (London),* 108(1292), 1297-312, 1983.

65. **Kumpulainen, J., Raittila, A. M., Lehto, J., and Koivistoinen, P.,** Electrothermal atomic absorption spectrometric determination of selenium in foods and diets, *J. Assoc. Off. Anal. Chem.,* 66, 1129, 1983.

66. **Maher, W. A.,** Determination of selenium in marine organisms using hydride generation and electrothermal atomic absorption spectroscopy, *Anal. Lett.,* 16, 801-10, 1983.

67. **Alexander, J., Kofstad, J., Saeed, K., Thomassen, Y., Overboe, S., and Aaseth, J.,** The application of direct electrothermal atomic absorption spectrophotometric determination of selenium in clinical chemistry, *Proc. 2nd Int. Workshop Trace Elem. Anal. Chem. Med. Biol.,* Braetter P. and Schramel, P., Eds., Berlin, 1983, 729-43.

68. **Norheim, G., Saeed, K., and Thomassen, Y.,** Matrix modification of selenium diamino-naphthalene with organometallic reagents for electrothermal atomic absorption spectrometric determination of selenium in biological matrixes, *At. Spectrosc.,* 4, 99, 1983.

69. **Yu, M. Q., Liu, G. Q., and Jin, Q.,** Determination of trace arsenic, antimony, selenium and tellurium in various oxidation states in water by hydride generation and atomic absorption spectrophotometry after enrichment and separation with thiol cotton, *Talanta,* 30, 265, 1983.

70. **Chung, C. H., Iwamoto, E., Yamamoto, M., and Yamamoto, Y.,** Graphite furnace atomic absorption spectrometry with graphite cloth ribbon for selenium determination, *Anal. Chem.,* 56, 829, 1984.

71. **Welz, B. and Melcher, M.,** Mechanisms of transition metal interferences in hydride generation atomic absorption spectroscopy. I. Influence of cobalt, copper, iron and nickel on selenium determination, *Analyst (London),* 109, 569, 1984.

72. **Imai, N., Terashima, S., and Ando, A.,** Determination of selenium in twenty-eight geological reference materials by atomic absorption spectroscopy, *Geostand. Newsl.,* 8, 39, 1984.

73. **Hatano, S., Nishi, Y., and Usui, T.,** Plasma selenium concentration in healthy Japanese children and adults determined by flameless atomic absorption spectrophotometry, *J. Pediatr. Gastroenterol. Nutr.,* 3(3), 426-31, 1984.

74. **Barrett, P., Barnett, W., and Fernandez, F.,** Advances in atomic absorption graphite furnace analysis, *J. Test Eval.,* 12, 207, 1984.

75. **Imai, N., Terashima, S., and Ando, A.,** Determination of selenium in geological materials by automated hydride generation and electrothermal atomic absorption spectrometry, *Bunseki Kagaku,* 33, 288, 1984.

76. **Robberecht, H. J. and Deelstra, H. A.,** Selenium in human urine. Determination, speciation and concentration levels, *Talanta,* 31, 497, 1984.

77. **Schoenberger, E., Kassovicz, J., and Shenhar, A.,** Micro dry ashing for trace selenium determination in organic matrixes, *Int. J. Environ. Anal. Chem.,* 18, 227, 1984.

78. **Varo, P., Nuurtamo, M., and Koivistoinen, P.,** Selenium content of nonfat dry milk in various countries, *J. Dairy Sci.,* 67, 2071, 1984.

79. **Chiou, K. Y. and Manuel, O. K.,** Determination of tellurium and selenium in atmospheric aerosol samples by graphite furnace atomic absorption spectrometry, *Anal. Chem.,* 56, 2721, 1984.

80. **Ivanov, G.,** Flameless atomic absorption method for total and separate determination of selenium and selenium dioxide in air, *Khim. Ind. (Sofia),* 56, 261, 1984.

81. **Chung, C. H., Iwamoto, E., Yamamoto, M., and Yamomoto, Y.,** Selective determination of arsenic (III, V), antimony (III, V), selenium (IV, V) and tellurium (IV, VI) by extraction and graphite furnace atomic absorption spectrometry, *Spectrochim. Acta,* 39B, 459, 1984.

82. **El-Yazizi, A., Al-Saleh, I., and Al-Mefty, O.,** Concentrations of Ag, Al, Au, Bi, Cd, Cu, Pb, Sb and Se in cerebrospinal fluid of patients with cerebral neoplasma, *Clin. Chem.,* 30, 1358, 1984.

83. **Bombach, H., Luft, B., Weinhold, E., and Mohr, F.,** Determination of arsenic, antimony, bismuth, tin, selenium and tellurium in steels using hydride atomic absorption spectroscopy, *Neue Huette,* 29, 233, 1984 (in German).

84. **Narasaki, H. and Ikeda, M.,** Automated determination of arsenic and selenium by atomic absorption spectroscopy with hydride generation, *Anal. Chem.,* 56, 2059, 1984.

85. **Blackmore, D. J. and Stanier, P.,** Methods for measurement of trace elements in equine blood by electrothermal atomic absorption spectrophotometry, *At. Spectrosc.,* 5, 215, 1984.

86. **Voellkopf, U. and Grobenski, Z.,** Interference in analysis of biological samples using stabilized temperature platform furnace and Zeeman background correction, *At. Spectrosc.,* 5, 115, 1984.

87. **Slavin, W. and Carnrick, G. R.,** Possibility of standardless furnace atomic absorption spectroscopy, *Spectrochim. Acta,* 39B, 271, 1984.

88. **Hudnik, V. and Gomiscek, S.,** Atomic absorption spectrometric determination of arsenic and selenium in mineral waters by electrothermal atomization, *Anal. Chim. Acta,* 157, 135-42, 1984.

89. **Prosbova, M., Vrzgula, L., and Kralicekova, E.,** Determination of selenium in blood serum by flameless atomic absorption photometry, *Chem. Listy,* 78, 861, 1984.

90. **Varo, P., Nuurtamo, M., and Koivistoinen, P.,** Selenium content of nonfat dry milk in various countries, *J. Dairy Sci.,* 67, 2071, 1984.

91. **Imai, N., Terashima, S., and Ando, A.,** Determination of selenium in geological materials by automated hydride generation and electrothermal atomic absorption spectroscopy, *Bunseki Kagaku,* 33, 288, 1984.

92. **Barrett, P., Barnett, W., and Fernandez, F.,** Advances in atomic absorption graphite furnace analysis, *J. Test Eval.,* 12, 207, 1984.

93. **Banslaugh, J., Radzuik, B., Saeed, K., and Thomassen, Y.,** Reduction of effects of structural nonspecific absorption in determination of arsenic and selenium by electrothermal atomic absorption spectrometry, *Anal. Chim. Acta,* 165, 149, 1984.

94. **Pruszkowska, E. and Barrett, P.,** Determination of arsenic, selenium, chromium, cobalt and nickel in geochemical samples using the stabilized temperature platform furnace and Zeeman background correction, *Spectrochim. Acta,* 39B, 485, 1984.

95. **Dulude, G. R. and Sotera, J. J.,** Dealing with interferences in furnace atomizer AAS: selenium in nickel alloys, *Spectrochim. Acta,* 39B, 511, 1984.

96. **Welz, B., Schlemmer, G., and Voellkopf, U.,** Influence of the valency state on the determination of selenium in graphite furnace atomic absorption spectroscopy, *Spectrochim. Acta,* 39B, 501, 1984.

97. **Chung, C. H., Iwamoto, E., Yamamoto, M., Yamamoto, Y., and Ikeda, M.,** Graphite furnace atomic absorption spectrometry with graphite cloth ribbon for selenium determination, *Anal. Chem.,* 56(4), 829-31, 1984.

98. **Ringdal, O., Julshamn, K., Andersen, K. J., and Svendsen, E.,** Determination of selenium in human tissue samples using graphite furnace atomic absorption spectrophotometry based on Zeeman effect background correction, *Proc 3rd Int. Workshop Trace Elem. Anal. Chem. Med. Biol.,* Braetter, P. and Schramel, P., Eds., Berlin, 1984, 189-99.

99. **Ito, N., Funato, Y., and Iwanaga, Y.,** Flameless atomic absorption spectroscopy of selenium in horse serum, *Shimadzu Hyoron,* 40(4), 207-10, 1983, published 1984.

100. **Woo, I. H., Nishiyama, H., Hashimoto, Y., and Lee, Y. K.,** Determination of selenium in coal using graphite furnace atomic absorption spectroscopy after chemical separation, *Bunseki Kagaku,* 34(10), 595-9, 1985.

101. **Edwards, W. C. and Blackburn, T. A.,** Selenium determination by Zeeman atomic absorption spectrophotometry, *Vet. Hum. Toxicol.,* 28, 12, 1985.

102. **Nakata, F., Yasui, Y., Matsuo, H., and Kumamaru, T.,** Determination of selenium (IV, VI) by heated quartz cell AAS with miniaturized suction-flow on-line prereduction/hydride generation system, *Anal. Sci.,* 1(5), 417-21, 1985.

103. **Bye, R. and Lund, W.,** Determination of selenium in biological samples by electrochemical preconcentration and atomic absorption spectrometry, *Z. Anal. Chem.,* 321, 483, 1985.

104. **Ping, L., Lei, W., Matsumoto, K., and Fuwa, K.,** Enhancement effect of palladium addition in the determination of selenium in organic-matrix solutions of electrically heated carbon furnace atomic absorption spectroscopy, *Anal. Sci.,* 1(3), 257-61, 1985.

105. **Sturgeon, R. E., Willie, S. N., and Berman, S. S.,** Preconcentration of selenium and antimony from seawater for determination of GFAAS, *Anal. Chem.,* 57(1), 6-9, 1985.

106. **Branch, C. H. and Hutchinson, D.,** Simultaneous determination of arsenic and selenium in geochemical samples by hydride evolution and atomic absorption spectrometry: success and failure, *Analyst,* 110, 163, 1985.

107. **Horvat, M., Dermelj, M., and Kosta, L.,** Tentative determination of total mercury, methylmercury (Me-Hg) and selenium in the hair of persons from different parts of Yugoslavia, *5th Int. Conf. Heavy Met. Environ.,* Lekkas, T. D., Ed., CEP Consult., Edinburgh, U.K. 1985, 73-5.

108. **Criand, A. and Fouillac, C.,** Use of L'vov platform and molybdenum coating for determination of volatile elements in thermomineral waters by atomic absorption spectrometry, *Anal. Chim. Acta,* 167, 257, 1985.

109. **Chakraborti, D. and Irgolic, K. J.,** Separation and determination of arsenic and selenium compounds by high pressure liquid chromatography with a graphite furnace atomic absorption spectroscopy as the element-specific detector, *5th Int. Conf. Heavy Met. Environ.,* Lekkas, T. D., Ed., CEP Consult., Edinburgh, U.K., 1985, 484.

110. **Shan, X. and Hu, K.,** Matrix modification for determination of selenium in geological samples by graphite furnace atomic absorption spectrometry after preseparation with thiol cotton fiber, *Talanta,* 32, 23, 1985.

111. **Chen, L. and Cai, W.,** Determination of selenium in grain by extraction-graphite atomic absorption spectroscopy, *Fenxi Ceshi Tongbao,* 4, 33-7, 1985.

112. **Droessler, M. S. and Holcombe, J. A.,** Molecular oxygen absorption in the determination of selenium by graphite furnace atomic absorption, *Can. J. Spectrosc.,* 31, 6, 1986.

113. Quantitative determination of selenium, Osaka Hajime, Eur. Pat. Appl. EP 187,717 (Cl.G01N21/74), 11 pp, 16 Jul. 1986, JP Appl. 85/941, 09 Jan. 1985.

114. **Koen, E.,** Flameless atomic absorption spectrometric method for the total and simultaneous determination of selenium and selenium dioxide in air, *Khig. Zdraveopaz,* 28(6), 87-91, 1985, CA 10, 104:154839h, 1986.

115. **Saaranen, M., Suistomaa, U., Kantola, M., Remes, E., and Vanha-Perttula,** Selenium in reproductive organs, seminal fluid and serum of men and bulls, *Hum. Reprod.,* 1(2), 61-4, 1986.

116. **Jowett, P. L. H. and Banton, M. I.,** Iron interference in the measurement of selenium in whole blood, plasma and serum by graphite furnace atomic absorption, *Anal. Lett.,* 19(11-12), 1243-58, 1986.

117. **Voth-Beach, L. M. and Shrader, D. E.,** Graphite furnace atomic absorption spectroscopy: new approaches to matrix modification, *Spectroscopy (Salem, OR),* 1(10), 49, 1986.

118. **Lewis, S. A., Hardison, N. W., and Veillon, C.,** Comparison of isotope dilution mass spectrometry and graphite furnace AAS with Zeeman background correction for the determination of plasma selenium, *Anal. Chem.,* 58(6), 1272-3, 1986.

119. **Edwards, W. C. and Blackburn, T. A.,** Selenium determination of Zeeman atomic absorption spectrophotometry, *Vet. Hum. Toxicol.,* 28(1), 12-13, 1986.

120. **Willie, S. N., Sturgeon, R. E., and Berman, S. S.,** Hydride generation atomic absorption determination of selenium in marine sediments, tissues and seawater with *in situ* concentration in a graphite furnace, *Anal. Chem.,* 58, 1140, 1986.

121. **Inhat, M., Wolynetz, M. S., Thomassen, Y., and Verlinden, M.,** Interlaboratory trial on the determination of total selenium in lyophilized human blood serum, *Pure Appl. Chem.,* 58(7), 1063-76, 1986.

122. **Paschal, D. C. and Kimberly, M. M.,** Automated direct determination of selenium in serum by electrothermal atomic absorption spectroscopy, *At. Spectrosc.,* 7(3), 75-8, 1986.

123. **Neve, J. and Molle, L.,** Direct determination of selenium in human serum by graphite furnace atomic absorption spectroscopy. Improvements due to oxygen ashing in graphite tube and Zeeman effect background correction, *Acta Pharmacol. Toxicol. Suppl.,* 59, 606, 1986.

124. **Saeed, K.,** Direct electrothermal atomic absorption spectrometric determination of selenium in biological fluids. I. Human urine, *Acta Pharmacol. Toxicol. Suppl.,* 59, 593-7, 1986.

125. **Oyamada, N. and Ishizaki, M.,** Fractional determination of dissolved selenium compounds of trimethylselenonium ion, selenium (IV) and selenium (VI) in environmental water samples, *Anal. Sci.,* 2, 365, 1986.

126. **Iwasa, A. and Yonemoto, T.,** Matrix effect on the absorbance of selenium in graphite furnace atomic absorption spectroscopy, *Bunseki Kagaku,* 35(6), 553-5, 1986.

127. **Cedergren, A., Lindberg, I., Lundberg, E., Baxter, D. C., and Frech, W.,** Investigation of reactions involved in graphite furnace atomic absorption procedures. XII. A study of some factors influencing the determination of selenium, *Anal. Chim. Acta,* 180, 373, 1986.

128. **Shand, C. A. and Ure, A. M.,** Graphite furnace atomic absorption spectrometric determination of selenium in plant materials following combustion in a stream of oxygen, *J. Anal. At. Spectrom.,* 2, 143, 1987.

129. **Saeed, K.,** Direct electrothermal atomization atomic absorption spectrometric determination of selenium in whole blood and serum with continuum-source background correction, *At. Spectrom.,* 2, 151, 1987.

130. **Dedina, J., Frech, W., Lindberg, I., Lundberg, E., and Cedergren, A.,** Determination of selenium by graphite furnace atomic absorption spectroscopy. I. Interaction between selenium and carbon, *J. Anal. At. Spectrom.,* 2, 287, 1987.

131. **Eckerlin, R. H., Hoult, D. W., and Carnrick, G. R.,** Selenium determination in animal whole blood using stabilized temperature platform furnace and zeeman background correction, *At. Spectrosc.,* 8, 64, 1987.

132. **Deschuytere, M. A. and Deelstra, H.,** Study of the specific aspects of the analytical determination of the selenium concentration in whole blood of newborns and young children, *Proc. 4th Int. Workshop Trace Elem. Anal. Chem. Med. Biol.,* Braetter, P. and Schramel, P., Eds., Berlin,1987.

133. **Neve, J., Chamart, S., and Molle, L.,** Optimization of a direct procedure for the determination of selenium in plasma and erythrocytes using Zeeman effect atomic absorption spectroscopy, *Proc. 4th Int. Workshop Trace Elem. Anal. Chem. Med. Biol.,* Braetter, P. and Schramel, P., Eds., Berlin, 1987.

134. **Teague-Nishimura, J. E., Tominaga, T., Katsura, T., and Matsumoto, K.,** Direct experimental evidence for *in situ* graphite and palladium selenide formation with improvements on the sensitivity of selenium in graphite furnace atomic absorption spectroscopy, *Anal. Chem.,* 59, 1647, 1987.

AUTHOR INDEX

SUBJECT INDEX

SILICON

SILICON, Si   (ATOMIC WEIGHT 28.086)

Instrumental Parameters:

Wavelength	251.6 nm
Slit width	80 μm
Bandpass	0.3 nm
Light source	Hollow Cathode Lamp
Lamp current	12 mA
Purge gas	Argon
Sample size	25 μL
Furnace	Cylindrical cuvette/Pyrolytic graphite

Standard Operating Conditions:

Optimum char temperature	1400°C
Optimum atomization temperature	2500°C
Sensitivity	200 pg/1% absorption
Sensitivity check	0.2-0.5 μg/mL
Working range	0.05- 5.0 μg/mL
Background correction	Required only to correct light scattering or non-specific absorption from the sample containing high dissolved solids.

General Note:

1. Use commercial standard or a previously analyzed sample as a working standard.
2. Use ultra clean glass and plastic ware, soaked in 1:5 nitric acid solution and rinsed thoroughly with deionized distilled water.
3. All analytical solutions should be prepared in 0.5% (v/v) nitric acid
4. Check for blank values on all reagents including water.
5. Dilution is recommended when sample exhibits greater than 0.5 absorbance units.
6. The presence of hydrofluoric, boric acid and potassium at a 1% or greater level causes severe depression in silicon absorbance.
7. Do not use hydrofluoric acid. If used, then neutralize it with ammonium- or potassium-hydroxide solution. The final potassium concentration should not exceed 1% or greater.
8. To enhance sensitivity to the analyte and allow an increase in pyrolysis temperature to 1000°C, add barium and lanthanum as chloride salts so that the final solution is 100 μg/mL lanthanum and 2 μg/mL barium.
9. Use of nitrogen as purge gas will reduce sensitivity.

REFERENCES

1. **Iida, C. and Yamasaki, K.,** Atomic absorption of spectrometric analysis of silicates by the absorption tube technique, *Anal. Lett.,* 3, 251, 1970.
2. **Ortner, H. M. and Kantuscher, E.,** Metal salt impregnation of graphite tube for improvement of silicon determination by atomic absorption spectrometry, *Talanta,* 22, 581, 1975 (in German).
3. **Lo, D. B. and Christian, G. D.,** Microdetermination of silicon in blood, serum, urine and milk using furnace atomic absorption spectroscopy, *Microchem. J.,* 23, 481, 1978.
4. **Mueller-Vogt, G. and Wendl, W.,** Reaction kinetics in the determination of silicon by graphite furnace atomic absorption spectroscopy, *Anal. Chem.,* 53, 651, 1981.
5. **Lythgoe, D. J.,** Method for improving the determination of silicon by atomic absorption spectroscopy using a tantalum-coated carbon furnace, *Analyst,* 106, 743, 1981.
6. Determination of silicon in gallium arsenide by flameless atomic absorption spectroscopy, Toshiba Corp., Jpn. Kokai Tokkyo Koho JP, 82 26,734 (Cl. GOJN21/31), 3 pp,12 Feb. 1982, Appl. 80/100,525, 24 Jul. 1980.
7. **Powell, L. A. and Tease, R. L.,** Determination of calcium, magnesium, strontium and silicon in brines by graphite furnace atomic absorption spectroscopy, *Anal. Chem.,* 54, 2154, 1982.
8. **Udagawa, T. and Nakashima, T.,** Incorporation of anomalously largle amounts of silicon into chromium-doped semiinsulating gallium arsenide crystals, *Conf. Ser. Inst. Phys.,* 63, 1982.
9. **Berlyno, G. M. and Caruso, C.,** Measurement of silicon in biological fluids in man using flameless furnace atomic absorption spectrophotometry, *Clin. Chim. Acta,* 129, 1983.
10. **Tyson, J. F. and Wan, N. W. S.,** Determination of silicon by an indirect atomic-absorption method using carbon-rod electrothermal atomization, *Talanta,* 30, 117, 1983.
11. **Chen, G. H. and RIsby, T. H.,** Determination of free silica and manganese in airborne particles by flameless atomic absorption spectroscopy, *Anal. Chem.,* 55, 943, 1984.
12. **Xu, X., Li, X., and Cui, X.,** Determination of trace silicon by flameless atomic absorption spectrometry, *Guangpuxue Yu Guangpu Fenxi,* 4, 53, 1984.
13. **McCarthy, J. P., Nunn, E. B., and Kinard, C.,** Atomic absorption spectrophotometric methods for the determination of phosphorus and silicon in steel, *Anal. Chem. Symp. Ser.,* 19, 349, 1984.
14. **Wendl., W. and Mueller-Vogt, G.,** Chemical reactions in the graphite tube for some carbide and oxide forming elements, *Spectrochim. Acta,* 39B, 237, 1984.
15. **Fehse, F.,** The determination of silicon in deionized process water by graphite furnace atomic absorption spectroscopy, *Spectrochim. Acta,* 39B, 597, 1984.
16. **Gorlova, M. N. and Skorskaya, O. L.,** Atomic-absorption determination of silicon in nickel-based alloys, *Fiz. Metody Kontrolya Khim. Sostava Mater.,* 86-92, 1983, CA 12, 102:197097r, 1985.
17. **Lerner, L. A. and Kakhnovich, Z. N.,** Determination of silicon in lysimetric waters by electrothermic atomic absorption spectroscopy, *Pochvovedenie,* 7, 136-9, 1985.
18. **Blinova, E. S., Miskar'yanta, V. G., and Nedler, V. V.,** Atomic absorption determination of silicon (10-3 to 10-4%) in metallic niobium, *Zh. Anal. Khim.,* 40(2), 286, 1985.
19. **Taddia, M.,** Determination of silicon in gallium arsenide by electrothermal atomization atomic absorption spectroscopy using the L'vov platform, *J. Anal. At. Spectrom.,* 1, 437, 1986.
20. **Felby, S.,** Determination of organosilicon oxide polymers in tissue by atomic absorption spectroscopy using an HGA graphite-furnace, *Forensic Sci. Int.,* 32, 61-5, 1986.
21. **Berlyne, G. M., Adler, A. J., Ferran, N., Bennett, S., and Holt, J.,** Silicon metabolism. I. Some aspects of renal silicon handling in normal man, *Nephron,* 43(1), 5-9, 1986.
22. **Taddia, M.,** Determination of silicon in gallium arsenide by electrothermal atomization atomic absorption spectroscopy using the L'vov platform, *J. Anal. At. Spectrom.,* 1, 437, 1986.
23. **Inamoto, I., Turuhara, K., and Uesugi, Y.,** Determination of microgram amounts of silicon in iron oxide or ferrite by fluoride separation-graphite furnace AAS (atomic absorption spectroscopy) and molybdenum blue spectrophotometry, *Bunseki Kagaku,* 35, T67, 1986.
24. **Taddia, M.,** Electrothermal atomic absorption spectrometry of silicon vaporized from different surfaces, *Anal. Chim. Acta,* 182, 231-7, 1986.

AUTHOR INDEX

SUBJECT INDEX

SILVER

SILVER, Ag   (ATOMIC WEIGHT 107.870)

Instrumental Parameters:

Wavelength	328.1 nm
Slit width	320 μm
Bandpass	1nm
Light source	Hollow Cathode Lamp
Lamp current	5 mA
Purge gas	Argon or Nitrogen
Sample size	25 μL
Furnace	Cylindrical cuvette/Pyrolytic graphite

Standard Operating Conditions:

Optimum char temperature	500°C
Optimum atomization temperature	2200°C
Sensitivity	2 pg/1% absorption
Sensitivity check	0.01-0.02 μg/mL
Working range	1.0-5.0 μg/mL
Background correction	Required only to correct light scattering or non-specific absorption from the sample containing high dissolved solids.

General Note:

1. Use commercial standard or a previously analyzed sample as a working standard.
2. Use ultra clean glass and plastic ware, soaked in 1:5 nitric acid solution and rinsed thoroughly with deionized distilled water.
3. All analytical solutions should be at least 0.2-0.5% (v/v) in nitric acid.
4. Do not store solutions in plastic containers over a period of time, since silver will be absorb on to certain plastics.
5. Store solutions in the dark or use on the same day of preparation.
6. Check for blank values on all reagents including water and xylene or kerosene used as diluent.
7. Dilution is recommended when sample exhibits greater than 0.5 absorbance units.
8. Use of nitrogen as purge gas will suppress the sensitivity.
9. For increased sensitivity, step atomization is recommended.
10. Peak area may be used with pressurization to increase sensitivity.
11. For analyzing oil or highly viscous samples, use a maximum 5 μL or 5 μg sample size.

REFERENCES

1. **Chao, T. T. and Ball, J. W.,** Determination of nanogram levels of silver in suspended materials of streams retained by a membrane filter with the "sampling Boat" technique, *Anal. Chim. Acta,* 54, 77, 1971.
2. **Woodriff, R., Culver, B. R., Shrader, D., and Super, A. B.,** Determination of sub-nanogram quantities of silver in snow by furnace atomic absorption spectroscopy, *Anal. Chem.,* 45, 231, 1973.
3. **Terashima, S.,** Determination of microamounts of silver in standard silicates by atomic absorption spectroscopy with a carbon tube atomizer, *Bunseki Kagaku,* 25, 279, 1976.
4. **Bea-Barredo, F., Polo-Polo. C., and Polo-Diaz, C.,** The simultaneous determination of gold, silver and cadmium at ppb levels in silicate rocks by atomic absorption spectroscopy with electrothermal atomization, *Anal. Chim. Acta,* 94, 283, 1977.
5. **Kujirai, O., Kobayashi, T., and Sudo, E.,** Determination of sub-parts-per-million levels of silver in heat-resisting alloys by graphite furnace atomic absorption spectroscopy, *Trans. Jpn. Inst. Met.,* 19, 159, 1978.
6. **Shaeffer, J. D., Mulvey, G., and Skogerboe, R. K.,** Determination of silver in precipitation by furnace atomic absorption spectroscopy, *Anal. Chem.,* 50, 1239, 1978.
7. **Fishkova, N. L. and Vilenkin, V. A.,** Application of an HGA-74 graphite atomizer to atomic absorption determination of gold, silver, platinum and palladium in solutions of complicated composition, *Zh. Anal. Khim.,* 33, 897, 1978 (in Russian).
8. **Aziz-Alrahman, A. M. and Headridge, J. B.,** Determination of silver in irons and steels by atomic absorption spectroscopy with an induction furnace: direct analysis of solid samples, *Talanta,* 25, 413, 1978.
9. **Baeckman, S. and Karlsson, R. W.,** Determination of lead, bismuth, zinc, silver and antimony in steel and nickel-base alloys by atomic absorption spectrophotometry using direct atomization of solid samples in a graphite furnace, *Analyst,* 104, 1017, 1979.
10. **Katskov, D. A. and Grinshtein, I.L.,** Study of the chemical interaction of copper, gold and silver with carbon by an atomic absorption method using an electrothermal atomizer, *Zh. Prikl. Spektrosk.,* 30, 787, 1979.
11. **Sedykh, I. M., Belyaev, Y. I., and Sorokina, E. V.,** Elimination of matrix effects in electrothermal atomic absorption determination of silver, lead, cobalt, nickel and tellurium in samples of complicated composition, *Zh. Anal. Khim.,* 353, 2348, 1980 (in Russian).
12. **Ni, Z., Jin, L., and Wu, D.,** Application of graphite furnace atomic absorption to the determination of trace silver in rain, *Huanjing Kexue,* 1, 48, 1980.
13. **Sterritt, R. M. and Lester, J. N.,** Determination of silver, cobalt, manganese, molybdenum and tin in sewage sludge by a rapid electrothermal atomic absorption spectroscopic method, *Analyst,* 105, 616, 1980.
14. **Carelli, G., Cecchetti, G., La Bua, R., Bergamaschi, A., and Iannaccone, A.,** Determination of selenium, cadmium, zinc, cobalt, copper and silver using atomic absorption spectrophotometry in aerosols from the workplace in the color television electronic industry, *Ann. Ist. Super Sanita,* 17, 505, 1981.
15. **Shan, X. and Ni, Z.,** Matrix modification for the determination of volatile elements of arsenic, selenium, tellurium, silver, antimony and bismuth by graphite furnace atomic absorption spectroscopy, *Huaxue Xuebao,* 39, 575, 1981.
16. **Stryjewska, E. and Rubel, S.,** Determination of silver in industrial wastewaters and wastes. II. Comparison of voltammetric, spectrophotometric and atomic absorption methods, *Chem. Anal. (Warsaw),* 26, 815, 1981.
17. **Sedykh, I. M., Myasoedova, G. V., Fedotova, I. A., Antokol'skaya, I. I., Bol'shakova, L. I., and Savvin, S. B.,** Electrothermal atomic absorption determination of silver in rocks and lunar soil, *Metody Vydeleniya Opred. Blagorodn. Elem.,* Zolotov Y. A., and Petrukhin, O. M., Eds., Akad. Nauk SSSR, Inst. Geokhim. Anal. Khim., Moscow, U.S.S.R., 1981, 96-7.
18. **Katskov, D. A. and Grinshtein, I. L.,** Formation of copper, silver and calcium acetylides in graphite furnace atomic absorption analysis, *Zh. Prikl. Spektrosk.,* 36, 181, 1982 (in Russian).
19. **Li, R.,** Determination of trace silver and cadmium in geochemical reference samples by flameless atomic absorption spectroscopy, *Yankuang Ceshi,* 1, 60, 1982.
20. **Takada, K. and Hirokawa, K.,** Origin of double-peak signals for trace lead, bismuth, silver and zinc in a microamount of steel in atomic absorption spectroscopy with direct electrothermal atomization of a solid sample in a graphite-cup cuvette, *Talanta,* 29, 849, 1982.
21. **Takada, K.,** Determination of small amounts of copper and silver in tin ingot by polarized Zeeman atomic absorption spectroscopy with direct atomization of solid sample in a graphite-cup cuvette, *Bunseki Kagaku,* 32, 197, 1983.
22. **Fazakas, J.,** Determination of silver by graphite furnace atomic absorption spectroscopy with atomization under pressure, *Mikrochim. Acta,* 1, 249, 1983.
23. **Jones, K. C., Peterson, P. J., and Davies, B. E.,** Silver concentrations in Welsh soils and their dispersal from derelict mine sites, *Miner. Environ.,* 5, 122, 1983.
24. **Koshima, H. and Ohnishi, H.,** Adsorption of silver, gold and platinum from aqueous solutions by carbonaceous materials, *Bunseki Kagaku,* 32, E149, 1983.
25. **Bloom, N.,** Determination of silver in marine sediments by Zeeman-corrected graphite furnace atomic absorption spectroscopy, *At. Spectrosc.,* 4(6), 204-7, 1983.

26. **Falk, H., Hoffmann, E., Luedke, C., Ottaway, J. M., and Giri, S. K.,** Furnace atomization with nonthermal excitation-experimental evaluation of detection based on high-resolution Echelle monochromator incorporating automatic background correction, *Analyst,* 108, 1459, 1983.

27. **McHugh, J. B.,** Determination of silver in water by electrothermal atomization, *At. Spectroscopy,* 5, 123, 1984.

28. **Hosking, J. W. and Robert, R. V. D.,** Determination of silver in cyanide solutions by electrothermal atomic absorption spectrophotometry, *S. Afr. J. Chem.,* 37, 129, 1984.

29. **Jones, K. C., Peterson, P. J., and Davies, B. E.,** Extraction of silver from soils and its determination by atomic absorption spectrometry, *Geoderma,* 33, 157, 1984.

30. **El-Azizi, A., Al-Saleh, I., and Al-Mefty, O.,** Concentrations of Ag, Al, Au, Bi, Cd, Cu, Pb, Sb and Se in cerebrospinal fluid of patients with cerebral neoplasma, *Clin. Chem.,* 30, 1358, 1984.

31. **Bloom, N. S. and Crecelius, E. A.,** Determination of silver in seawater by coprecipitation with cobalt pyrrolidine dithiocarbamate and Zeeman graphite furnace atomic absorption spectrometry, *Anal. Chim. Acta,* 156, 139, 1984.

32. **Headridge, J. B. and Riddington, I. M.,** Determination of silver, lead and bismuth in glasses by atomic absorption spectrometry with introduction of solid samples into furnaces, *Analyst,* 109, 113, 1984.

33. **Wennrich, R., Dittrich, K., and Bonitz, U.,** Matrix interference in laser atomic absorption spectrometry, *Spectrochim. Acta,* 39B, 657, 1984.

34. **Sighinolfi, G. P., Gorgoni, C., and Mohamed, A. H.,** Comprehensive analysis of precious metals in some geological standards by flameless atomic absorption spectroscopy, *Geostand. Newsl.,* 8, 25, 1984.

35. **Karmannova, N. G.,** Direct determination of silver in low-mineralized natural waters by atomic-absorption spectroscopy with a graphite furnace, *Zh. Prikl. Spektrosk.,* 40, 904, 1984.

36. **Warren, H. V., Horsky, S. J., and Barakso, J. J.,** Preliminary studies of the biogeochemistry of silver in British Columbia (Canada), *CIM Bull.,* 77, 95, 1984.

37. **Khozhainov, Y. M., Tyurin, O. A., and Delinnikina, N. P.,** Electrothermal atomic absorption determination of silver in cadmium selenide, *Zavod. Lab.,* 51(4), 30, 1985.

38. **Shaojun, L. and Yun, Z.,** Matrix modification for determination of trace silver in geochemical materials by flameless Zeeman atomic absorption spectrometry, *Fenxi Huaxue,* 13, 858, 1985 (in Chinese).

39. **Chorman, F. H., Jr., Spencer, M. J., Lyons, W. B., and Mayewski, P. A.,** A solvent extraction technique for determining concentrations of gold and silver in natural waters, *Chem. Geol.,* 53(1-2), 25-30, 1985.

40. **Jones, K. C., Peterson, P. J., and Davies, B. E.,** Analysis of silver in fresh waters: sample preservation and pre-treatment studies, *Int. J. Environ. Anal. Chem.,* 20(3-4), 2447-54, 1985.

41. **Yukhin, Y. M., Udalova, T. A., and Tsimbalist, V. G.,** Flameless atomic absorption determination of noble metals after liquid-liquid extraction by mixtures of Bis(2 ethylhexyl)dithiophosphate and p-octylaniline, *Zh. Anal. Khim.,* 40, 850, 1985 (in Russian).

42. **Zolotov, Y. A. and Vanifatova, N. G.,** New effective and selective extractants for separation and determination of silver, *Mikrochim. Acta,* 1, 281, 1985.

43. **Yang, G.,** Phase analysis for trace silver in geochemical samples, *Wutan Yu Huatan,* 9(4), 299-302, 1985 (in Chinese).

44. **Kacimi, G., Nguyen, P. L., Fabiani, P., and Truhaut, R.,** Determination of silver by flameless atomic spectrometry: application to biological materials, *C. R. Acvad Sci. Ser. 2,* 302(7), 421-6, 1986.

45. **Lin, S. and Zhou, Y.,** Matrix modification for detrmiation of trace silver in geochemical materials by flameless ZAAS, *Fenxi Huaxue,* 13(11), 858-61, 1985, CA 6, 104:81175, 1986.

46. **Shoeva, O. P., Kucheva, G. P., Kubrakova, I. V., Myasoedova, G. V., Savvin, S. B., and Bannykh, L. N.,** Sorption-atomic absorption determination of gold and silver in natural waters, *Zh. Anal. Khim.,* 41, 2186, 1986.

47. **Andersen, K. J., Wikshaaland, A., Utheim, A., Julshamn, K., and Vik, H.,** Determination of silver in biological samples using graphite furnace atomic absorption spectroscopy based on Zeeman effect background correction and matrix modification, *Clin. Biochem. (Ottawa),* 19, 166, 1986.

AUTHOR INDEX

Al-Mefty, O., 30
Al-Saleh, I., 30
Andersen, K. J., 47
Antokol'skaya, I. I., 17
Aziz-Alrahman, A. M., 8
Baeckman, S., 9
Ball, J. W., 1
Bannykh, L. N., 46
Barakso, J. J., 36
Bea-Barredo, F., 4
Belyaev, Y. I., 11
Bergamaschi, A., 14
Bloom, N., 25
Bloom, N. S., 31
Bol'shakova, L. I., 17
Bonitz, U., 33
Carelli, G., 14
Cecchetti, G., 14
Chao, T. T., 1
Chorman, F. H., Jr., 39
Crecelius, E. A., 31
Culver, B. R., 2
Davies, B. E., 23, 29, 40
Delinnikina, N. P., 37
Dittrich, K., 33
El-Azizi, A., 30
Fabiani, P., 44
Falk, H., 26
Fazakas, J., 22
Fedotova, I. A., 17
Fishkova, N. L., 7
Giri, S. K., 26
Gorgoni, C., 34
Grinshtein, I. L., 10, 18
Headridge, J. B., 8, 32
Hirokawa, K., 20
Hoffmann, E., 26
Horsky, S. J., 36
Hosking, J. W., 28
Iannaccone, A., 14
Jin, L., 12
Jones, K. C., 23, 29, 40
Julshamn, K., 47
Kacimi, G., 44
Karlsson, R. W., 9
Karmannova, N. G., 35
Katskov, D. A., 10, 18
Khozhainov, Y. M., 37
Kobayashi, T., 5
Koshima, H., 24
Kubrakova, I. V., 46
Kucheva, G. P., 46
Kujirai, O., 5

La Bua, R., 14
Lester, J. N., 13
Li, R., 19
Lin, S., 45
Luedke, C., 26
McHugh, J. B., 27
Mohamed, A. H., 34
Mulvey, G., 6
Myasoedova, G. V., 17, 46
Nguyen, P. L., 44
Ni, Z., 12, 15
Ohnishi, H., 24
Ottaway, J. M., 26
Peterson, P. J., 23, 29, 40
Polo-Diaz, C., 4
Polo-Polo. C.,, 4
Riddington, I. M., 32
Robert, R. V. D., 28
Rubel, S., 16
Savvin, S. B., 17, 46
Sedykh, I. M., 11, 17
Shaeffer, J. D., 6
Shan, X., 15
Shaojun, L., 38
Shoeva, O. P., 46
Shrader, D., 2
Sighinolfi, G. P., 34
Skogerboe, R. K., 6
Sorokina, E. V., 11
Sterritt, R. M., 13
Stryjewska, E., 16
Sudo, E., 5
Super, A. B., 2
Takada, K., 20, 21
Terashima, S., 3
Truhaut, R., 44
Tsimbalist, V. G., 41
Tyurin, O. A., 37
Udalova, T. A., 41
Utheim, A., 47
Vanifatova, N. G., 42
Vik, H., 47
Vilenkin, V. A., 7
Warren, H. V., 36
Wennrich, R., 33
Wikshaaland, A., 47
Woodriff, R., 2
Wu, D., 12
Yang, G., 43
Yukhin, Y. M., 41
Yun, Z., 38
Zhou, Y., 45
Zolotov, Y. A., 42

SUBJECT INDEX

STRONTIUM

STRONTIUM, Sr   (ATOMIC WEIGHT 87.62)

Instrumental Parameters:

Wavelength	460.7 nm
Slit width	320 μm
Bandpass	1nm
Light source	Hollow Cathode Lamp
Lamp current	12 mA
Purge gas	Argon or Nitrogen
Sample size	25 μL
Furnace	Cylindrical cuvette/Pyrolytic graphite

Standard Operating Conditions:

Optimum char temperature	1000°C
Optimum atomization temperature	2700°C
Sensitivity	5 pg/1% absorption
Sensitivity check	0.008-0.03 μg/mL
Working range	0.005-0.01μg/mL
Background correction	Required only to correct light scattering or non-specific absorption from the sample containing high dissolved solids.

General Note:
1. Use commercial standard or a previously analyzed sample as a working standard.
2. Use ultra clean glass and plastic ware, soaked in 1:5 nitric acid solution and rinsed thoroughly with deionized distilled water.
3. All analytical solutions should be at least 0.2-0.5% (v/v) in nitric acid.
4. Check for blank values on all reagents including water.
5. Dilution is recommended when sample exhibits greater than 0.5 absorbance units.

REFERENCES

1. **Bek. F., Janouskova, J., and Moldan, B.,** Determination of manganese and strontium in blood serum using the Perkin-Elmer HGA-70 graphite furnace, *At. Absorp. Newsl.,* 13, 47, 1974.
2. **Sedykh, E. M. and Belyaev, Y. I.,** Use of atomic absorption and molecular absorption spectra to study atomization of strontium chlorides in a graphite furnace, *Zh. Anal. Khim.,* 32, 1904, 1977.
3. **Matsusaki, K., Murakami, S., and Yoshino, T.,** Interference effect of chloride on the determination of strontium by atomic absorption spectroscopy with a graphite furnace, *Nippon Kagaku Kaishi,* 7, 1126, 1980.
4. **Katskov, D. A. and Grinshtein, I. L.,** Study of the evaporation of beryllium, magnesium, calcium, strontium, barium and aluminum from a graphite surface by an atomic absorption method, *Zh. Prikl. Spektrosk.,* 33, 1004, 1980.
5. **Lin, S., Gu, J., and Zhang, X.,** Determination of trace amounts of strontium in rocks by flameless atomic absorption spectrophotometry, *Zhongguo Kexue Jishu Daxue Xuebao,* 11, 93, 1981 (in Chinese).
6. **Powell, L. A. and Tease, R. L.,** Determination of calcium, magnesium, strontium and silicon in brines by graphite furnace atomic absorption spectroscopy, *Anal. Chem.,* 54, 2154, 1982.
7. **Rossi, G., Omenetto, N., Pigizzi, G., Vivian, R., Martinez, U., Musty, F., and Crabi, G.,** Analysis of radioactive waste solutions by atomic absorption spectrometry with electrothermal atomization, *At. Spectrosc.,* 4, 113, 1983.

AUTHOR INDEX

SUBJECT INDEX

SULFUR

REFERENCES

1. **Siemer, D. D., Woodriff, R., and Robinson, J.,** Sulfate determination in natural waters by nonresonance line furnace atomic absorption, *Appl. Spectrosc.,* 31, 168, 1977.
2. **Tsunoda, K., Chiba, K., Haraguchi, H., Chakrabarti, C. L., and Fuwa, K.,** Determination of sulfur by molecular absorption spectrometry of metal sulfide bands utilizing an electrothermal graphite furnace, *Can. J. Spectrosc.,* 27, 69, 1982.
3. **Dittrich, K., Vorberg, B., Funk, J., and Beyer, V.,** Determination of some non-metals by using diatomic molecular absorbance in a hot graphite furnace , *Spectrochim. Acta,* 39B, 349, 1984.

AUTHOR INDEX

SUBJECT INDEX

TELLURIUM

TELLURIUM, Te (ATOMIC WEIGHT 127.60)

Instrumental Parameters:

Wavelength	241.3 nm
Slit width	160 µm
Bandpass	0.5 nm
Light source	Hollow Cathode Lamp
Lamp current	10 mA
Purge gas	Argon or Nitrogen
Sample size	25 µL
Furnace	Cylindrical cuvette/Pyrolytic graphite

Standard Operating Conditions:

Optimum char temperature	600°C
Optimum atomization temperature	2500°C
Sensitivity	20 pg/1% absorption
Sensitivity check	0.2 µg/mL
Working range	0.2-2.0 µg/mL
Background correction	Required only to correct light scattering or non-specific absorption from the sample containing high dissolved solids.

General Note:

1. Use commercial standard or a previously analyzed sample as a working standard.
2. Use ultra clean glass and plastic ware, soaked in 1:5 nitric acid solution and rinsed thoroughly with deionized distilled water.
3. All analytical solutions should be at least 0.2-0.5% (v/v) in nitric acid.
4. Addition of 10 µg nickel/mL as nickel nitrate will maximize the charring temperature.
5. Check for blank values on all reagents including water.
6. Dilution is recommended when sample exhibits greater than 0.5 absorbance units.

REFERENCES

1. **Beaty, R. D.,** Atomic absorption determination of nanogram quantities of tellurium using the sampling boat technique, *Anal. Chem.,* 45, 234, 1973.
2. **Beaty, R. D.,** Determination of tellurium in rocks by graphite furnace atomic absorption, *At. Ab. Newsl.,* 13, 38, 1974.
3. **Ohta, K. and Suzuki, M.,** Atomic absorption spectroscopy of tellurium with electrothermal atomization in a molybdenum microtube, *Anal. Chim. Acta,* 110, 49, 1979.
4. **Haynes, B. W.,** Arsenic, antimony, selenium and tellurium in high-purity copper by electrothermal atomization, *At. Ab. Newsl.,* 18, 46, 1979.
5. **Kamada, T., Sugita, N., and Yamamoto, Y.,** Differential determination of tellurium (IV) and tellurium (VI) with sodium diethyldithio carbamate, amm. pyrrolidine dithiocarbamate and dithizone by atomic absorption spectrophotometry with a carbon-tube atomizer, *Talanta,* 26, 337, 1979.
6. **Sedykh, E. M., Belyaev, Y. I., and Sorokina, E. V.,** Elimination of matrix effects in electrothermal atomic absorption determination of silver, lead, chromium, nickel and tellurium in samples of complicated composition, *Zh. Anal. Khim.,* 35, 2348, 1980 (in Russian).
7. **Katskov, D. A., Grinshtein, I.L., and Kruglikova, L. P.,** Study of the evaporation of the metals In, Ga, Tl, Ge, Sn, Pb, Sb, Bi, Se and Te from a graphite surface by the atomic absorption method, *Zh. Prikl. Spektrosk.,* 33, 804, 1980.
8. **Saeed, K. and Thomassen, Y.,** Spectral interferences from phosphate matrixes in the determination of As, Sb, Se and Te by electrothermal atomic absorption spectroscopy, *Anal. Chim. Acta,* 130, 281, 1981.
9. **Xiao-guan, S. and Zheming, N.,** Matrix modification for the differential determination of tellurium (IV) and tellurium (VI) in water samples by graphite furnace atomic absorption spectroscopy, *Acta Sci. Circumst.,* 1, 74, 1981 (in Chinese).
10. **Gonzalez, C. A. and Tallo, A. R.,** Determination of selenium and tellurium in minerals and concentrates by graphite furnace atomic absorption, *Bol. Soc. Chil. Quim.,* 27, 297-9, 1982 (in Spanish).
11. **Headridge, J. B. and Nicholson, R. A.,** Determination of arsenic, antimony, selenium and tellurium in nickel-base alloys by atomic-absorption spectrometry with introduction of solid samples into furnaces, *Analyst (London),* 107, 1200, 1982.
12. **Kujirai, D., Kobayashi, T., Ide, K., and Sudo, E.,** Determination of traces of tellurium in heat-resisting alloys by graphite furnace atomic-absorption spectroscopy after coprecipitation with arsenic, *Talanta,* 29, 27-30, 1982.
13. **Muir, M. K. and Andersen, T. N.,** Determination of selenium and tellurium at ultratrace levels in copper by flameless atomic absorption spectrophotometry, *At. Spectrosc.,* 3, 149, 1982.
14. **Kellerman, S. P.,** The use of masking agents in the determination by hydride generation and atomic-absorption spectrophotometry of arsenic, antimony, selenium, tellurium and bismuth in the presence of noble metals, *Rep. MINTEK,* M39, 14, 1982.
15. **Yu, M. Q., Liu, G. Q., and Jin, Q.,** Determination of trace arsenic, antimony, selenium and tellurium in various oxidation states in water by hydride generation and atomic absorption spectrophotometry after enrichment and separation with thiol cotton, *Talanta,* 30, 265, 1983.
16. **Maher, W. A.,** Determination of tellurium by atomic absorption spectroscopy with electrothermal atomization after preconcentration by hydride generation and trapping, *Analyst (London),* 108, 305, 1983.
17. **Itsuki, K., Ikeda, T., and Kondo, A.,** Atomic absorption spectroscopy of tellurium using graphite furnace with the aid of lanthanum, *Bunseki Kagaku,* 32, 133, 1983 (in Japanese).
18. **Andreae, M. O.,** Determination of inorganic tellurium species in natural waters, *Anal. Chem.,* 56, 2064, 1984.
19. **Maher, W. A.,** Determination of tellurium by electrothermal atomic absorption spectroscopy: isolation of tellurium from potential interferences, *Anal. Lett.,* 17, 979, 1984.
20. **Chiou, K. Y. and Manuel, O. K.,** Determination of tellurium and selenium in atmospheric aerosol samples by graphite furnace atomic absorption spectrometry, *Anal. Chem.,* 56, 2721, 1984.
21. **Bye, R., Engvik, L., and Lund, W.,** Tellurium (IV) as a masking agent for copper in the determination of selenium by hydride generation/atomic absorption spectroscopy, *Fresnius' Z. Anal. Chem.,* 37, 129, 1984.
22. **Chung, C. H., Iwamoto, E., Yamamoto, M., and Yamamoto, Y.,** Selective determination of arsenic (III, V), antimony (III, V), selenium (IV, VI) and tellurium (IV, VI) by extraction and graphite furnace atomic absorption spectrometry, *Spectrochim. Acta,* 39B, 459, 1984.
23. **Bombach, H., Luft, B., Weinhold, E., and Mohr, F.,** Determination of arsenic, antimony, bismuth, tin, selenium and tellurium in steels using hydride atomic absorption spectroscopy, *Neue Huette,* 29, 233, 1984 (in German).
24. **Petit, L. and Petit, G.,** Demonstration of a method for tellurium determination in seawater, *Rev. Int. Oceanogr. Med.,* 79-80, 19-32, 1985 (in French).
25. **Kobayashi, T., Hirose, F., Hasegawa, S., and Okochi, H.,** Determination of trace tellurium and gallium in nickel-base heat-resisting alloys by GFAAS, *Nippon Kinzoku Gakkaishi,* 49(8), 656-62, 1985.

26. **Weng, H. and Qiu, D.,** Determination of tellurium in pyrite by graphite furnace AAS–use of rhodium as a matrix modifier, *Yankung Ceshi,* 5(1), 52-4, 1986.
27. **Muangnoicharoen, S., Chiou, K. Y., and Manuel, O. K.,** Determination of selenium and tellurium in the gas phase using specific columns and atomic absorption spectroscopy, *Anal. Chem.,* 58, 2811, 1986.
28. **Woo, I. H. and Watanabe, K.,** Determination of selenium and tellurium in coal by graphite furnace atomic absorption spectroscopy after coprecipitation with arsenic, *Anal. Sci.,* 3, 49, 1987.

AUTHOR INDEX

SUBJECT INDEX

THALLIUM

THALLIUM, Tl   (ATOMIC WEIGHT 204.37)

Instrumental Parameters:

Wavelength	276.8 nm
Slit width	320 μm
Bandpass	1 nm
Light source	Hollow Cathode Lamp
Lamp current	8 mA
Purge gas	Argon or Nitrogen
Sample size	25 μL
Furnace	Cylindrical cuvette/Pyrolytic graphite

Standard Operating Conditions:

Optimum char temperature	1100°C
Optimum atomization temperature	2200°C
Sensitivity	40 pg/1% absorption
Sensitivity check	0.1 μg/mL
Working range	0.09-0.50 μg/mL
Background correction	Required only to correct light scattering or non-specific absorption from the sample containing high dissolved solids.

General Note:

1. Use commercial standard or a previously analyzed sample as a working standard.
2. Use ultra clean glass and plastic ware, soaked in 1:5 nitric acid solution and rinsed thoroughly with deionized distilled water.
3. All analytical solutions should be at least 0.2-0.5% (v/v) in nitric acid.
4. Check for blank values on all reagents including water.
5. Dilution is recommended when sample exhibits greater than 0.5 absorbance units.
6. Excess acid use will suppress sensitivity.
7. Diluted solutions tend to hydrolyze, therefore, use within 4 h of preparation.

REFERENCES

1. **Kubasik, N. P. and Volosin, M. T.,** Simplified determination of urinary cadmium, lead and thallium with use of carbon rod atomization and atomic absorption spectrophotometry, *Clin. Chem.,* 19, 954, 1973.
2. **Fuller, C. W.,** The effect of acids on the determination of thallium by atomic absorption spectroscopy with a graphite furnace, *Anal. Chim. Acta,* 81, 199, 1976.
3. **Kujirai, O., Kobayashi, T., and Sudo, E.,** Rapid determination of trace quantities of thallium in heat-resisting cobalt and nickel alloys by graphite furnace atomic absorption spectroscopy, *Z. Anal. Chem.,* 29, 398, 1979.
4. **Katskov, D. A., Grinshtein, I. L., and Kruglikova, L. P.,** Study of the evaporation of the metals In, Ga, Tl, Ge, Sn, Pb, Sb, Bi, Se and Te from a graphite surface by the atomic absorption method, *Zh. Prikl. Spektrosk.,* 33, 804, 1980.
5. **Sauer, K. H. and Eckhard, S.,** Traces of thallium in requistic materials, dust and iron metals, *Mikrochim. Acta Suppl.,* 9, 87-98, 1981 (in German).
6. **Keil, R.,** Trace determination of thallium in rocks by flame or flameless atomic absorption spectrophotometry following preconcentration by extraction, *Fresnius' Z. Anal. Chem.,* 309, 181-5, 1981 (in German).
7. **Wronski, R. and Weidhuener, J.,** Thallium intoxication and antidote therapy with reference to chemical detection by atomic absorption spectroscopy, *Aerztl. Lab.,* 27, 316, 1981 (in German).
8. **Elson, C. M. and Albuquerque, C. A. R.,** Determination of thallium in geological materials by extraction and electrothermal atomic absorption spectroscopy, *Anal. Chim. Acta,* 134, 393, 1982.
9. **Voskresenskaya, N. T.,** Thallium occurrence in halide salts, *Geokhimiya,* 3, 450-3, 1982 (in Russian).
10. **Wronski, R. and Weidhuener, J.,** Thallium intoxication and antidote therapy (with regard to chemical detection using atomic absorption spectroscopy), *Atomspektrom. Spurenanal.,* Welz, B., Ed., Springer-Verlag, Weinheim, 1981,111.
11. **Fazakas, J.,** Determination of thallium by graphite furnace atomic-absorption spectroscopy with atomization under pressure, *Anal. Lett.,* 15, 1523, 1982.
12. **Slavin, W., Carnrick, G. R., and Manning, D. C.,** Graphite tube effects on perchloric acid interferences on aluminum and thallium in the stabilized-temperature platform furnace, *Anal. Chim. Acta,* 138, 103-10, 1982.
13. **Elson, C. M. and Albuquerque, C. A. R.,** Determination of thallium in geological materials by extraction and electrothermal atomic absorption spectroscopy, *Anal. Chim. Acta,* 134, 393-6, 1982.
14. **Nessler, F., Glatzel, E., Liebscher, R., and Brueckner, C.,** Thallium – its economic, industrial, medical and environment-toxicological importance for the GDR. First results of analytical studies on finding a method for thallium determination in urine and disintegrated plant materials by flameless AAS-technique, *Mengen Spurenelem.,* Arbeitstag, 93, 1983.
15. **Han, H. and Ni, Z.,** Graphite furnace platform atomic absorption spectrometric determination of thallium in environmental and biological samples, *Fenxi Huaxue,* 11(12), 908, 1983.
16. **Schmidt, W. and Dietl, F.,** Determination of thallium in digested soils using flameless atomic absorption in zirconium-cold graphite tubes, *Fresnius' Z. Anal. Chem.,* 315, 687, 1983 (in German).
17. **Fazakas, J. and Marinescu, D. M.,** Determination of thallium in cadmium and lead by graphite furnace atomic absorption spectrometry with sample vaporization from a platform, *Talanta,* 30(11), 857-60, 1983.
18. **Bessems, G. J. H., Westerhuis, L. W., and Baadenhuijsen, H.,** A comparison of two methods for the determination of thallium in urine. Differential pulse anodic stripping voltammetry and flameless atomic absorption spectrophotometry, *Ann. Clin. Biochem.,* 20, 321, 1983.
19. **Suzuki, T., Suwabe, M., Sawada, K., and Shirai, F.,** Determination of thallium in the human internal organs by solvent extraction using trioctylmethylammonium chloride and graphite furnace atomic absorption spectrometry, *Bunseki Kagaku,* 32(12), 757-60, 1983.
20. **Han, H. and Ni, Z.,** Carrier-free foam-adsorbed graphite furnace atomic absorption spectrophotometric determination of trace thallium in soil, mud, coal fly ash and sediment sample, *Huanjing Huaxue,* 2(2), 44-9, 1983 (in Chinese).
21. **Ikramuddin, M.,** A rapid and precise method for the determination of thallium in geological materials at the one nanogram per gram level, *At. Spectrosc.,* 4, 101, 1983.
22. **Ali, S. L.,** Determination of pesticide residues and other critical impurities such toxic trace metals in medicinal plants. II. Determination of toxic trace metals in drugs, *Pharm. Ind.,* 45, 1294, 1983 (in German).
23. **Botha, P. V. and Fazakas, J.,** Use of nonresonance lines for determination of lead, indium and thallium by electrothermal atomic absorption spectrometry, *Spectrochim. Acta,* 39B, 379, 1984.
24. **Shen, X., Ni, Z., and Zhang, L.,** Application of matrix modification in determination of thallium in waste water by graphite furnace atomic absorption spectrometry, *Talanta,* 31(2), 150-2, 1984.
25. **Chandler, H. A. and Scott, M.,** Determination of low levels of thallium in urine using chelation with sodium diethyldithiocarbamate, extraction into toluene and atomic absorption spectrophotometry with electrothermal atomization, *At. Spectrosc.,* 5, 230, 1984.
26. **Slavin, W., Carnrick, G. R., and Manning, D. C.,** Chloride interferences in graphite furnace atomic absorption spectrometry, *Anal. Chem.,* 56, 163, 1984.

27. **Liem, I., Kaiser, G., Sager, M., and Toelg, G.,** The determination of thallium in rocks and biological materials at ng/g 1 levels by differential-pulse anodic stripping voltammetry and electrothermal atomic absorption spectroscopy, *Anal. Chim. Acta,* 158, 179, 1984.

28. **Matsuno, K., Iwao, S., and Kodama, Y.,** Direct determination of thallium in biological materials by graphite furnace atomic absorption spectrometry, *Bunseki Kagaku,* 33(3), 125-9, 1984.

29. **Chandler, H. A. and Scott, M.,** Determination of low levels of thallium in urine using chelation with sodium diethyldithiocarbamate extraction into toluene and atomic absorption spectrophotometry with electrothermal atomization, *At. Spectrosc.,* 5(6), 230-3, 1984.

30. **Hoenig, M., Scokart, P. O., and Van Hoeyweghen, P.,** Efficiency of L'vov platform and ascorbic acid modifier for reduction of interferences in analysis of plant samples for lead, thallium, antimony, cadmium, nickel and chromium by ETAAS, *Anal. Lett.,* 17A, 1947, 1984.

31. **Hu, K. and Luo, J.,** Graphite-furnace AA spectrometric determination of trace thallium in geological samples after preconcentration with polyurethane foam, *Fenxi Huaxue,* 13(3), 226-9, 1985.

32. **Magyar, B., Wampfler, B., and Wuersch, A.,** Extraction of thallium (III) from iron (III) solutions and analysis of the organic extracts using GFAAS, *Fortschr. Atomspektrom. Spurenanal.,* 1, 659, 1984.

33. **Suzuki, M. and Ohta, K.,** Determination of thallium by electrothermal AAS with a metal atomizer, *Fresnius' Z. Anal. Chem.,* 322(5), 480-5, 1985.

34. **Qin, Z., Wu, Z., Yang, S., and Li, Q.,** Determination of traces of gallium, indium and thallium in geological samples by flameless AAS using vanadium as matrix modifier, *Yanshi Kuangwu Ji Ceshi,* 4(2), 160-3, 1985.

35. **Riley, J. P. and Siddiqui, S. A.,** The determination of thallium in sediments and natural waters, *Anal. Chim. Acta,* 181, 117-123, 1986.

36. **Paschal, D. C. and Bailey, G. G.,** Determination of thallium in urine with Zeeman effect graphite furnace atomic absorption, *J. Anal. Toxicol.,* 10, 252, 1986.

37. **Matsusaki, K. and Yoshino, T.,** Removal of chloride interference in the determination of thallium by AAS with a graphite furnace, *Bunseki Kagaku,* 35, 937, 1986.

38. **Hu, J.,** Determination of trace thallium in rocks and minerals by graphite furnace atomic absorption spectroscopy, *Yankuang Ceshi,* 5, 118, 1986.

39. **Xiaquan, S., Shineng, Y., and Zheming, N.,** Determination of thallium in river sediment, coal, coal fly ash and botanical samples by graphite furnace atomic absorption spectroscopy, *Can. J. Spectrosc.,* 31, 35, 1986.

40. **Berndt, H., Bassner, J., and Messerschmidt, J.,** Comparative atomic absorption spectrometric studies for trace element determination in urine, *Anal. Chim. Acta,* 180, 389-400, 1986 (in German).

41. **Leloux, M. S., Lich, N. P., and Claude, J. R.,** Flame and graphite furnace atomic absorption spectroscopy method for thallium. A review, *At. Spectrosc.,* 8, 71, 1987.

42. **Leloux, M. S., Lich, N. P., and Claude, J. R.,** Determination of thallium in various biological matrices by graphite furnace atomic absorption spectrophotometry using platform technology, *At. Spectrosc.,* 8, 75, 1987.

AUTHOR INDEX

SUBJECT INDEX

TIN

TIN, Sn   (ATOMIC WEIGHT 118.69)

Instrumental Parameters:

Wavelength	235.5 nm
Slit width	160 μm
Bandpass	0.5 nm
Light source	Hollow Cathode Lamp
Lamp current	6 mA
Purge gas	Argon or Nitrogen
Sample size	25 μL
Furnace	Cylindrical cuvette/Pyrolytic graphite

Standard Operating Conditions:

Optimum char temperature	900°C
Optimum atomization temperature	2500°C
Sensitivity	40 pg/1% absorption
Sensitivity check	0.1-1.0 μg/mL
Working range	0.06-0.10 μg/mL
Background correction	Required only to correct light scattering or non-specific absorption from the sample containing high dissolved solids.

General Note:
1. Use commercial standard or a previously analyzed sample as a working standard.
2. Use ultra clean glass and plastic ware, soaked in 1:5 nitric acid solution and rinsed thoroughly with deionized distilled water.
3. All analytical solutions should be prepared in 0.2 % (w/v) potassium dichromate and 1% (v/v) hydrochloric acid or 3% nitric acid (v/v) and 10% (1:1) ammonium hydroxide.
4. Check for blank values on all reagents including water.
5. Dilution is recommended when sample exhibits greater than 0.5 absorbance units.
6. Use of oxidizing acids, such as nitric acid, can suppress sensitivity and enhance tin hydrolysis.

REFERENCES

1. **Meranger, J. C.,** A rapid screening method for the determination of Di-(nOoctyl) tin stabilizers in alcoholic beverages using a heated graphite atomizer, *J. Assoc. Pffic. Anal. Chem.,* 58, 1143, 1975.

2. **Trachman, H. L., Tyberg, A. J., and Branigan, P. D.,** Atomic absorption spectrometric determination of sub-part-per-million quantities of tin in extracts and biological materials with a graphite furnace, *Anal. Chem.,* 49, 1090, 1977.

3. **Varnes, A. W. and Gaylor, V. F.,** Determination of dioctyltin stabilizers in food simulating solvents by atomic absorption spectroscopy with electrothermal atomization, *Anal. Chim. Acta,* 101, 393, 1978.

4. **Tominaga, M. and Umezaki, T.,** Determination of submicrogram amounts of tin by atomic absorption spectroscopy with electrothermal atomization, *Anal. Chim. Acta,* 109, 251, 1979.

5. **Ohta, K. and Suzuki, M.,** Atomic absorption spectroscopy of tin with electrothermal atomization in a molybdenum microtube, *Anal. Chim. Acta,* 107, 245, 1979.

6. **Kojima, S.,** Separation of organotin compounds by using the difference in partition behavior between hexane methanolic buffer solution. I. Determination of butyl tin compounds in textiles by graphite furnace atomic absorption spectrophotometry, *Analyst,* 104, 660, 1979.

7. **Fritzsche, H., Wegscheider, W., Knapp, G., and Ortner, H. M.,** A sensitive atomic absorption spectrometric method for the determination of tin with atomization from impregnated graphite surface, *Talanta,* 26, 219, 1979.

8. **Katskov, D. A., Grinshtein, I. L., and Kruglikova, L. P.,** Study of the evaporation of the metals In, Ga, Tl, Ge, Sn, Pb, Sb, Bi, Se and Te from a graphite surface by the atomic absorption method, *Zh. Prikl. Spektrosk.,* 33, 804, 1980.

9. **Vickrey, T. M., Harrison, G. V., Ramelow, G. J., and Carver, J. C.,** Use of metal carbide-coated graphite cuvettes for the atomic absorption analysis of organo tins, *Anal. Lett.,* 13, 781, 1980.

10. **Itsuki, K. and Iheda, T.,** Atomic absorption spectroscopy of tin using graphite furnace with the aid of lanthanum, *Bunseki Kagaku,* 29, 309, 1980.

11. **Vickrey, T. M., Howell, H. E., Harrison, G. V., and Ramelow, G. J.,** Post column digestion methods for liquid chromatography-graphite furnace atomic absorption speciation of organo lead and organo tin compounds, *Anal. Chem.,* 52, 1743, 1980.

12. **Thiband, Y.,** Analysis of tin by atomic absorption spectrophotometry with electrothermal furnace. Applications to marine organisms, *Rev. Trav. Inst. Peches Marit.,* 44, 349-54, 1980 (in French).

13. **Sterritt, R. M. and Lester, J. N.,** Determination of silver, cobalt, manganese, molybdenum and tin in sewage sludge by a rapid electrothermal atomic absorption spectroscopic method, *Analyst,* 105, 616, 1980.

14. **Glenc, T., Jurczyk, J., and Robosz-Kabza, A.,** Determination of tin and lead in transformer, carbon and alloy steels in the range from 0.0005 to 0.04% by atomic absorption method with electrothermal atomization, *Chem. Anal.,* 25, 515, 1980.

15. **Vickrey, T. M., Harrison, G. V., and Ramelow, G. R.,** Treated graphite surfaces for determination of tin by graphite furnace atomic absorption spectroscopy, *Anal. Chem.,* 53, 1573, 1981.

16. **Torigoe, N. and Matsunaga, K.,** The determination of tin in food by flameless atomic absorption, *Okayama-Ken Kankyo Hoken Senta Nenpo,* 5, 235, 1981 (in Japanese).

17. **Ogihara, K., Chiba, M., and Kikuchi, M.,** Determination of plasma tin by flameless atomic absorption spectroscopy, *Sangyo Igaku,* 23, 420, 1981 (in Japanese).

18. **Brodie, K. G. and Rowland, J. J.,** Trace analysis of arsenic, lead and tin, *Eur. Spectrosc. News,* 36, 41-44, 1981.

19. **Jewett, K. L. and Brinckman, F. E.,** Speciation of trace di- and tri-organotins in water by ion-exchange HPLC-GFAA, *J. Chromatogr. Sci.,* 19, 583-93, 1981.

20. **Kobayashi, T., Ide, K., and Sudo, E.,** Determination of traces of tin in iron and steels by graphite furnace atomic absorption spectroscopy, *Nippon Kinzoku Gakkaishi,* 46, 603-8, 1982 (in Japanese).

21. **Zou, S., Yu, X., and Liang, X.,** Determination of trace tin by flameless atomic absorption spectroscopy with a tungsten-coated carbon tube, *Huanjing Kexue,* 3, 41, 1982 (in Chinese).

22. **Jin, L. and Ni, Z.,** Determination of tin in soil by graphite furnace atomic absorption spectroscopy, *Huanjing Kexue,* 1, 281, 1982.

23. **Lundberg, E., Bergmark, B., and Frech, W.,** Investigations of reactions involved in electrothermal atomic absorption procedures. II. A theoretical and experimental investigation of factors influencing the determination of tin, *Anal. Chim. Acta,* 142, 129, 1982.

24. **Ishii, T.,** Tin in marine algae, *Nippon Suisan Gakkaishi,* 48, 1609, 1982.

25. **Harrison, G. V.,** Graphite furnace atomic absorption analysis of organotin compounds, 247 pp, 1982. Avail. Univ. Microfilms Int., Order No. DA8226092. From *Diss. Abstr. Int. B,* 43, 1835, 1982.

26. **Volynskii, A. B., Sedykh, E. M., Spivakov, B. Y., and Zolotov, Y. A.,** Effect of the composition of aqueous solutions and organic extracts on the atomic absorption determination of tin using a graphite furnace, *Zh. Anal. Khim.,* 38, 435, 1983.

27. **Liu, D., Zhu, W., Huang, T., and Xiong, X.,** Determination of trace triphenyltin chloride from antifouling paints by graphite furnace atomic absorption spectroscopy, *Tuliao Gongye,* 77, 44, 1983.

28. **Pruszkowska, E., Manning, D. C., Carnrick, G. R., and Slavin, W.,** Experimental conditions for the determination of tin with the stabilized temperature platform furnace and zeeman background correction, *At. Spectrosc.,* 4, 87, 1983.

29. **Voronkova, M. A.,Gorlevskaya, N. G., and Voskresenskaya, V. S.,** Atomic absorption determination of tin with pulsed electrothermal atomization, Atomno-absorbtsion, *Metody Analiza Mineral. Syr'ya, M.,* 86-97, 1982 (in Russian). From Ref. *Zh. Metall.,* Abstr. No. 5K16, 1983

30. **Kobayashi, T., Kujirai, O., Hirose, F., and Okochi, H.,** Determination of tin in heat-resisting alloys by graphite furnace atomic absorption spectroscopy, *Nippon Kinzoku Gakkaishi,* 47, 876, 1983 (in Japanese).

31. **Postel, W., Meier, B., and Markert, R.,** Determination of lead, cadmium, aluminum and tin in beer using atomic absorption spectrophotometry, *Monatsschr. Brauwiss,* 36, 300, 1983 (in German).

32. **Du, X. and Shi, Y.,** Determination of trace tin in lead and lead-antimony alloys by silica gel column separation-hydride generation atomic absorption spectroscopy, *Fenxi Huaxue,* 12, 203, 1984.

33. **Liu, Y., Lei, X., Zhang, X., and Luo, C.,** Hydride-generation electrothermal atomic absorption method for the determination of tin and antimony in geological samples, *Fenxi Huaxue,* 12, 218, 1984 (in Chinese).

34. **Yu, Z., Li, X., and Xu, J.,** Determination of trace amounts of tin in rocks, soils and water samples by hydride-graphite furnace atomic absorption spectroscopy, *Yanshi Kuangwu Ji Ceshi,* 3(3), 254, 1984.

35. **Ma, Y. and Liu, G.,** Determination of trace tin in pig iron by hydride-generation-atomic absorption spectroscopy, *Fenxi Huaxue,* 12, 78, 1984 (in Chinese).

36. **Luo, D. and Zhu, M.,** Atomic absorption determination of trace tin in chemical prospecting samples, *Fenxi Huaxue,* 12, 147, 1984.

37. **Zhou, L., Chao, T., and Meier, A. L.,** Determination of total tin in geological materials by electrothermal atomic absorption spectrophotometry using a tungsten-impregnated graphite furnace, *Talanta,* 31(1), 73-6, 1984.

38. **Dong, J. and Cui, D.,** Determination of tin in canned foods by graphite furnace atomic absorption spectroscopy, *Shipin Yu Fajiao Gongye,* 5, 10-23, 1984 (in Chinese).

39. **Jin, L.,** Determination of trace tin in river sediment and coal fly ash by graphite furnace atomic absorption spectroscopy using a mixture of ascorbic acid and iron as matrix modifier, *At. Spectrosc.,* 5, 91, 1984.

40. **Pinel, R., Benabdallah, M. Z., Astruc, A., Potin-Gautier, M., and Astruc, M.,** Automated specific detection for liquid chromatography by electrothermal atomic absorption. Application to organotin compounds, *Analysis,* 12, 344, 1984 (in French).

41. **Cook, L. L., Jacobs, K. S., and Reiter, L. W.,** The distribution in adult and neonatal rat brain following exposure to triethyltin, *Toxicol. Appl. Pharmacol.,* 72(1), 75-81, 1984.

42. **Andrea, M. O. and Byrd, J. T.,** Determination of tin and methyltin species by hydride generation and detection with graphite-furnace atomic absorption or flame emission spectrometry, *Anal. Chim. Acta,* 156, 147-57, 1984.

43. **Fazakas, J.,** Critical study of determination of tin by electrothermal atomic absorption spectrometry using resonance and nonresonance lines, *Talanta,* 31, 573, 1984.

44. **Burns, D. T., Dadgar, D., and Harriott, M.,** Investigations of direct determination of tin in organotin compounds using carbon furnace atomization, *Analyst,* 109, 1099, 1984.

45. **Littlejohn, D., Cook, S., Durie, D., and Ottaway, J. M.,** Investigation of working conditions for graphite probe atomization in electrothermal atomic absorption spectrometry, *Spectrochim. Acta,* 39B, 295, 1984.

46. **Wedl, W. and Mueller-Vogt, G.,** Chemical reactions in the graphite tube for some carbide and oxide forming elements, *Spectrochim. Acta,* 39B, 237, 1984.

47. **Bombach, H., Luft, B., Weinhold, E., and Mohr, F.,** Determination of arsenic, antimony, bismuth, tin, selenium and tellurium in steels using hydride atomic absorption spectroscopy, *Neue Huette,* 29, 233, 1984 (in German).

48. **Mishima, M., Maruyama, T., Koshiyama, M., Murakami, C., Kumagaya, M., Sumiyoshi, M., Asahi, T., Nozawa, T., Tanaka, Y. et. al.,** Determination of trace amount of bis(tri-butyltin) oxide in fish by graphite furnace atomic absorption spectroscopy, *Bunseki Kagaku,* 33, T57, 1984.

49. **Pinel, R., Gandjar, I. G., Benabdallah, M. Z., Astruc, A., and Astruc, M.,** Trace determination of inorganic and organic tin in water by atomic absorption spectroscopy with hydride generation decomposition, *Analysis,* 12, 404, 1984.

50. **Hocquellet, P.,** Use of atomic absorption spectrometry with electrothermal atomization for direct determination of trace elements in oils: cadmium, lead, arsenic and tin, *Rev. Fr. Corps Gras,* 31, 117, 1984 (in French).

51. **Sugawara, H. and Tayama, K.,** Determination of tin and antimony in ander of zinc ore by flameless atomic absorption spectrometry, *Ryusan to Kogyo,* 37(10), 173-6, 1984 (in Japanese).

52. **Hocquellet, P.,** Direct determination of tin at ultratrace levels in edible oils and fats by atomic absorption spectrometry with electrothermal atomization, *At. Spectrosc.,* 6(3), 69-73, 1985.

53. **Pinel, R., Mediec, H., Benabdallah, M. Z., Astruc, A., and Astruc, M.,** A simple method for evaluation of harmful anthropogenic organotin pollution in European aquatic environment, *5th Int. Conf. Heavy Met. Environ.,* Lekkas, T. D., Ed. , CEP Consult, Edinburgh, U.K., 1985, 384

54. **Donard, O. F. X. and Weber, J. H.,** Behavior of methyltin compounds under simulated estuarine conditions, *Environ. Sci. Technol.,* 19, 1104, 1985.

55. **Parks, E. J., Blair, W. R., and Brinckman, F. E.,** GFAAS determination of ultratrace quantities of organotin in seawater by using enhancement methods, *Talanta,* 32, 633, 1985.

56. **Volynsky, A. B., Sedykh, E. M., Spivakov, B. Y., and Havezor, I.,** Factors influencing free oxygen content in electrothermal atomizer, *Anal. Chim. Acta,* 174, 173, 1985.

57. **Xinyuan, D., Chunzlhi, J., and Jinyu, J.,** Atomic absorption spectrophotometric determination of tin in environmental samples with a platform graphite furnace, *Fenxi Huaxue,* 14, 90, 1986.

58. **Randall, L., Donard, O. F. X., and Weber, J. H.,** Speciation of n-butyltin compounds by atomic absorption spectroscopy with an electrothermal quartz furnace after hydride generation, *Anal. Chim. Acta,* 184, 197-203, 1986.

59. **Orren, D. K., Braswell, W. M., and Mushak, P.,** Quantitative analysis of ethyltin compounds in mammalian tissue using HPLC/FAAS, *J. Anal. Toxicol.,* 10(3), 93-7, 1986.

60. **Lundberg, E. and Bergmark, B.,** Determination of total tin in geological materials by electrothermal atomic absorption spectroscopy, *Anal. Chim. Acta,* 188, 111, 1986.

61. **Sturgeon, R. E., McLaren, J. W., Willie, S. N., Beauchemin, D., and Berman, S. S.,** Determination of total tin in Natural Research Council of Canada marine reference materials, *Can. J. Chem.,* 65, 961, 1987.

62. **Chiba, M.,** Determination of tin in biological materials by atomic absorption spectroscopy with a graphite furnace, *J. Anal. Toxicol.,* 11, 125-30, 1987.

AUTHOR INDEX

SUBJECT INDEX

TITANIUM

TITANIUM, Ti   (ATOMIC WEIGHT 204.37)

Instrumental Parameters:

Wavelength	364.3 nm
Slit width	80 μm
Bandpass	0.3 nm
Light source	Hollow Cathode Lamp
Lamp current	8 mA
Purge gas	Argon
Sample size	25 μL
Furnace	Cylindrical cuvette/Pyrolytic graphite

Standard Operating Conditions:

Optimum char temperature	1400°C
Optimum atomization temperature	2700°C
Sensitivity	500 pg/1% absorption
Sensitivity check	0.3-2.0 μg/mL
Working range	0.06-0.30 μg/mL
Background correction	Required only to correct light scattering or non-specific absorption from the sample containing high dissolved solids.

General Note:
1. Use commercial standard or a previously analyzed sample as a working standard.
2. Use ultra clean glass and plastic ware, soaked in 1:5 nitric acid solution and rinsed thoroughly with deionized distilled water.
3. All analytical solutions should be prepared in 5% (v/v) hydrochloric acid and should be used on the day of preparation.
4. Check for blank values on all reagents including water.
5. Dilution is recommended when sample exhibits greater than 0.5 absorbance units.
6. Excess nitric acid use will suppress sensitivity.
7. Use of nitrogen as purge gas will also suppress sensitivity.

VANADIUM

VANADIUM, V   (ATOMIC WEIGHT 50.942)

Instrumental Parameters:

Wavelength	318.5 nm
Slit width	160 μm
Bandpass	0.5 nm
Light source	Hollow Cathode Lamp
Lamp current	10 mA
Purge gas	Argon
Sample size	25 μL
Furnace	Cylindrical cuvette/Pyrolytic graphite

Standard Operating Conditions:

Optimum char temperature	1100°C
Optimum atomization temperature	2500°C
Sensitivity	350 pg/1% absorption
Sensitivity check	0.1-1.0 μg/mL
Working range	0.5-5.0 μg/mL
Background correction	Required only to correct light scattering or non-specific absorption from the sample containing high dissolved solids.

General Note:

1. Use commercial standard or a previously analyzed sample as a working standard.
2. Use ultra clean glass and plastic ware, soaked in 1:5 nitric acid solution and rinsed thoroughly with deionized distilled water.
3. All analytical solutions should be prepared in 0.2% (v/v) hydrochloric acid or nitric acid.
4. Check for blank values on all reagents including water.
5. Dilution is recommended when sample exhibits greater than 0.5 absorbance units.
6. Excess nitric acid (>3%) will suppress sensitivity.
7. Use of nitrogen as purge gas will also suppress sensitivity.

REFERENCES

1. **Cioni, R., Innocenti, F., and Mazzuoli, R.,** The determination of vanadium in silicate rocks with the HGA-70 graphite furnace, *At. Ab. Newsl.,* 11, 102, 1972.
2. **Muzzarelli, R. A. A. and Rocchetti, R.,** The determination of vanadium in seawater by hot graphite atomic absorption spectroscopy on Chitosan after separation from salt, *Anal. Chim. Acta,* 70, 283, 1974.
3. **Schweizer, V. B.,** Determination of Co, Cr, Cu, Mo, Ni and V in carbonate rocks with the HGA-70 graphite furnace, *At. Ab. Newsl.,* 14, 137, 1975.
4. **Krishnan, S. S., Quittkat, S., and Crapper, D. R.,** Atomic absorption analysis for traces of aluminum and vanadium in biological tissue. A critical evaluation of the graphite furnace atomizer, *Can. J. Spectrosc.,* 21, 23, 1976.
5. **Ishizaki, M. and Ueno, S.,** Determination of submicrogram amounts of vanadium in biological materials by extraction with N-cinnamoyl-N-2, 3-xylyl hydroxyl amine and flameless atomic absorption spectroscopy with an atomizer coated with pyrolytic graphite, *Talanta,* 26, 523, 1979.
6. **Lagas, P.,** Determination of beryllium, barium and vanadium and some other elements in water by atomic absorption spectroscopy with electrothermal atomization, *Anal. Chim. Acta,* 100, 139, 1978.
7. **Sutter, E. M. and Leroy, M. J.,** Nature of the interference of nitric acid in the determination of nickel and vanadium by atomic absorption spectroscopy with electrothermal atomization, *Anal. Chim. Acta,* 100, 243, 1978.
8. **Ueno, S. and Ishizaki, M.,** Determination of traces of vanadium in plants and biological samples by N-cinnomoyl-N-(2,3-XYLYL) hydroxyl amine extraction-carbon furnace atomic absorption spectroscopy, *Nippon Kagaku Kaishi,* 2, 217, 1979.
9. **Shimizu, T., Shijo, Y., and Sakai, K.,** Determination of vanadium in airborne particulates and petroleum by graphite furnace atomic absorption spectroscopy with pyrolytic graphite coated tube, *Bunseki Kagaku,* 29, 685, 1980.
10. **Del Monte Tamba, M. G. and Luperi, N.,** Determination of traces of arsenic, antimony, bismuth and vanadium in steel and cast iron by graphite furnace atomic absorption spectroscopy, *Metall. Ital.,* 72, 253, 1980 (in Italian and English).
11. **Kuga, K.,** Determination of vanadium by solvent extraction-graphite furnace atomic absorption spectroscopy using an improved graphite tube, *Bunseki Kagaku,* 29, 90T, 1980.
12. **Studnicki, M.,** Determination of germanium, vanadium and titanium by carbon furnace atomic absorption spectroscopy, *Anal. Chem.,* 52, 1762, 1980.
13. **Hulanicki, A., Karwowska, R., and Stanczak, J.,** Experimental parameters of vanadium determination by atomic absorption spectroscopy with graphite furnace atomization, *Talanta,* 27, 214, 1980.
14. **Ying-qi, G. and Zheming, N.,** Determination of vanadium in water by graphite furnace atomic absorption spectroscopy, *Huan Ching K'o Hsueh,* 2, 241, 1981 (in Chinese).
15. **Shimizu, T., Uchida, Y., Shijo, Y., and Sakai, K.,** Determination of vanadium in river water by combination treatment and graphite furnace atomic absorption spectroscopy, *Bunseki Kagaku,* 30, 113, 1981.
16. **Kawakubo, S., Yamaguchi, S., and Mizuike, A.,** Determination of traces of vanadium in titanium (IV) oxide microsamples by graphite furnace atomic absorption spectroscopy, *Bunseki Kagaku,* 30, 594, 1981.
17. **Shimizu, T. and Sakai, K.,** Determination of vanadium in seawater by graphite furnace atomic absorption spectroscopy with a tube coated with pyrolytic graphite, *Nippon Kagaku Kaishi,* 1, 26, 1981.
18. **Monien, H. and Stangel, R.,** Experience with vanadium determination by atomic absorption spectroscopy in graphite tube furnaces after extraction from seawater, *Fresnius' Z. Anal. Chem.,* 311, 209, 1982 (in German).
19. **Buchet, J. P., Knepper, E., and Lauwerys, R.,** Determination of vanadium in urine by electrothermal atomic absorption spectrometry, *Anal. Chim. Acta,* 136, 243-8,1982.
20. **Barbooti, M. M. and Jasim, F.,** Electrothermal atomic absorption determination of vanadium, *Talanta,* 29, 107-11, 1982.
21. **Stroop, S. D., Helinek, G., and Greene, H. L.,** More sensitive flameless atomic absorption analysis of vanadium in tissue and serum, *Clin. Chem. (Winston-Salem, N.C.),* 28, 79-82, 1982.
22. **Spencer, W. A., Galobardes, J. A., Curtis, M. A., and Rogers, L. B.,** Chromatographic studies of vanadium compounds from Boscan crude oil, *Sep. Sci. Technol.,* 17, 797-819, 1982.
23. **Styris, D. L. and Kaye, J. H.,** Mechanisms of vaporization of vanadium pentoxide from vitreous carbon and tantalum furnace by combined atomic absorption/mass spectrometry, *Anal. Chem.,* 54, 864-9, 1982.
24. **Miyai, Y., Sugasaka, K., and Katoh, S.,** Studies on the extraction of soluble resources in seawater. I. Determination of molybdenum and vanadium in seawater by graphite furnace atomic absorption spectro-photometry with oxine-solvent extraction, *Nippon Kaisui Gakkaishi,* 36, 159, 1982 (in Japanese).
25. **Yoshimura, C. and Matsumura, K.,** Effect of carbon black on flameless atomic absorption spectrophotometry of molybdenum and vanadium, *Nippon Kagaku Kaishi,* 8, 1337, 1982 (in Japanese).
26. **Arnold, D. and Kuennecke, A.,** Determination of 3d elements in high lead-containing glasses, *Silikattechnik,* 33, 312, 1982.
27. **Shijo, Y., Kimura, Y., Shimizu, T., and Sakai, K.,** Determination of vanadium in seawater by graphite furnace atomic absorption spectrometry, *Bunseki Kagaku,* 32, 285E, 1983.

331

28. **Robertiello, A., Petrucci, F., Angelini, L., and Olivieri, R.,** Nickel and vanadium as biodegradation monitors of oil pollutants in aquatic environments, *Water Res.,* 17, 497, 1983.
29. **Fujitani, K. and Nakao, T.,** Vanadium in proteins, *Kenkyu Nenpo-Tpkyo-toritsu Eisei Kenkyusho,* 34, 411, 1983.
30. **Ueno, S.,** Vanadium content in environmental and biological samples, *Ibaraki-ken Eisei Kenkyusho Nempo,* 21, 23-9, 1983.
31. **Shijo, Y., Kimura, Y., Shimizu, T., and Sakai, K.,** Determination of vanadium in seawater by graphite furnace atomic absorption spectroscopy, *Bunseki Kagaku,* 32, E285-E291, 1983.
32. **Carlin, L. M., Colovos, G., Garland, D., Jamin, M. E., and Klenck, M.,** Analytical methods evaluation and validation: arsenic, nickel, tungsten, vanadium, talc and wood dust, Report NIOSH-210-79-0060: Order No. PB83-155325, 223 pp, 1981. Avail. NTIS from Gov. Rep. Announce. Index (U.S.), 83, 2355, 1983.
33. **Ng, K. C. and Caruso, J. A.,** Volatilization of zirconium, vanadium, uranium and chromium using electrothermal carbon cup sample vaporization, *Analyst (London),* 108, 476-80, 1983.
34. **Marecek, J., Braunova, I., and Vachtova, I.,** Determination of molybdenum, vanadium and chromium in brines by atomic absorption spectroscopy, *Chem. Prum.,* 33, 429-35, 1983 (in Czech).
35. **Monsty, F., Omenetto, N., Pietra, R., and Sabbioni, E.,** Atomic absorption spectrometric neutron activation and radioanalytical techniques for determination of trace metals in environmental biochemical and toxicological research, I. Vanadium, *Analyst,* 109, 1451, 1984.
36. **Chen, Z. and Angerer, J.,** Solvent extraction and flameless atomic absorption determination of trace vanadium in urine, *Fenxi Huaxue,* 12, 274, 1984 (in Chinese).
37. **Ueno, S.,** Vanadium content in environmental and biological samples, *Ibaraki-ken Eiseil Kenkyusho Nempo,* 21, 23, 1983.
38. **Wendl, W. and Mueller-Vogt, G.,** Chemical reactions in the graphite tube for some carbide and oxide forming elements, *Spectrochim. Acta,* 39B, 237, 1984.
39. **Littlejohn, D., Duncan, I., Marshall, J., and Ottaway, J. M.,** Analytical evaluation of totally pyrolytic graphite cuvettes for electrothermal atomic absorption spectrometry, *Anal. Chim. Acta,* 157, 291, 1984.
40. **Fish, R. H. and Komlenic, J. J.,** Molecular characterization and profile identifications of vanadyl compounds in heavy crude petroleums by liquid chromatography/graphite furnace atomic absorption spectrometry, *Anal. Chem.,* 56, 510-517, 1984.
41. **Chakrabarti, C. L., Wu, S., Karwowska, R., Chang, S. B., and Bertels, P. C.,** Studies on atomic absorption spectrometry with laboratory-made graphite furnace, *At. Spectrosc.,* 5, 69, 1984.
42. **Fish, R. H., Komlenic, J. J., and Wines, B. K.,** Characterization and comparison of vanadyl and nickel compounds in heavy crude petroleums and asphaltenes by reverse-phase and size-exclusion liquid chromatography/graphite furnace atomic absorption spectrometry, *Anal. Chem.,* 56, 2452-60, 1984.
43. **Abreu, M. and Buenafama, H.,** Matrix effects in the determination of vanadium and nickel in petroleum by FAAS, *Collect. Colloq. Semin. (Inst. Fr. Pet.),* 40, 179-84, 1984.
44. **Jing, S., Wang, R., Yu, C., Zhang, D., Yan, Y., and Ma, Y.,** Design of model ZM-1 Zeeman effect atomic absorption spectrometer, *Huanjing Kexue,* 5, 41, 1984 (in Chinese).
45. **Pyy, L., Hakala, E., and Lajunen, L. H. J.,** Screening for vanadium in urine and blood serum by electrothermal atomic absorption spectrometry and d.c. plasma atomic emission spectroscopy, *Anal. Chim. Acta,* 158, 292-303, 1984.
46. **Manning, D. C. and Slavin, W.,** Factors influencing atomization of vanadium in graphite furnace AAS, *Spectrochim. Acta,* 40B(3), 461-73, 1985.
47. **Buratti, M., Pellegrino, O., Caravelli, G., Calzaferri, G., Bettinelli, M., Colombi, A., and Maroni, M.,** Sensitive determination of urinary vanadium by solvent extraction and atomic absorption spectroscopy, *Clin. Chim. Acta,* 150, 53, 1985.
48. **Wendl, W. and Mueller-Vogt, G.,** Chemical reactions of chromium and vanadium in graphite furnace atomic absorption spectrometry, *Spectrochim. Acta,* 40B(3), 527-31, 1985.
49. **Tominaga, M., Bansho, K., and Umezaki, Y.,** Electrothermal atomic absorption spectrometric determination of lead, manganese, vanadium and molybdenum in seawater with ascorbic acid to reduce matrix effects, *Anal. Chim. Acta,* 169, 171, 1985.
50. **Zamilova, L. M., Biktimirova, T. G., and Sokolova, V. I.,** Atomic absorption determination of vanadium and nickel in gas-oil fractions of crude oils, *Neftekhimiya,* 25, 159, 1985 (in Russian).
51. **Bermejo-Barrera, D., Bermejo-Martinez, F., and Cocho de Juan, J. A.,** Determination of vanadium in urine by electrothermal atomization atomic absorption spectroscopy, *J. Anal. At. Spectrom.,* 2, 163, 1987.

AUTHOR INDEX

SUBJECT INDEX

ZINC

ZINC, Zn   (ATOMIC WEIGHT 65.37)

Instrumental Parameters:

Wavelength	213.9 nm
Slit width	320 µm
Bandpass	1 nm
Light source	Hollow Cathode Lamp
Lamp current	5 mA
Purge gas	Argon
Sample size	25 µL
Furnace	Cylindrical cuvette/Pyrolytic graphite

Standard Operating Conditions:

Optimum char temperature	400°C
Optimum atomization temperature	2200°C
Sensitivity	2 pg/1% absorption
Sensitivity check	0.005 µg/mL
Working range	0.005-0.020 µg/mL
Background correction	Required only to correct light scattering or non-specific absorption from the sample containing high dissolved solids.

General Note:
1. Use commercial standard or a previously analyzed sample as a working standard.
2. Use ultra clean glass and plastic ware, soaked in 1:5 nitric acid solution and rinsed thoroughly with deionized distilled water.
3. All analytical solutions should be prepared in 0.5% (v/v) nitric or phosphoric acid.
4. Check for blank values on all reagents including water.
5. Dilution is recommended when sample exhibits greater than 0.5 absorbance units.
6. Use step atomization for increased sensitivity.

REFERENCES

1. **Jensen, F. O., Dolezal, J., and Langmyhr, F. J.,** Atomic absorption spectrometric determination of cadmium, lead and zinc in salts or salt solutions by hanging mercury drop electrodeposition and atomization in a graphite furnace, *Anal. Chim. Acta,* 72, 245, 1974.

2. **Vondenhoff, T.,** Determination of lead, cadmium, copper and zinc in plant and animal material by atomic absorption in a flame and in a graphite tube after sample decomposition by the Schoniger technique, *M.HBl. GDCh. Fachgr. Lebensmittelchem. Gerichtl. Chem.,* 29, 341, 1975 (in German).

3. **Evenson, M. A. and Anderson, C. T., Jr.,** Ultramicro analysis for copper, cadmium and zinc in human liver tissue by use of atomic absorption spectrophotometry and the heated graphite tube atomizer, *Clin. Chem.,* 21, 537, 1975.

4. **Del-Castiho, P. and Herber, R. F.,** The rapid determination of cadmium, lead, copper and zinc in whole blood by atomic absorption spectroscopy with electrothermal atomization. Improvements in precision with a peak-shape monitoring device, *Anal. Chim. Acta,* 94, 269, 1977.

5. **Campbell, W. C. and Ottaway, J. M.,** Direct determination of cadmium and zinc in seawater by carbon furnace atomic absorption spectrometry, *Analyst,* 102, 495, 1977.

6. **Mazzucotelli, A., Galli, M., Benassi, E., Loeb, C., Ottonello, G. A., and Tanganelli, P.,** Atomic absorption determination of zinc in cerebrospinal fluid by ion exchange chromatography and electrothermal atomization, *Analyst,* 103, 863, 1978.

7. **Sturgeon, R. E., Berman, S. S., Desaulniers, A., and Russell, D. S.,** Determination of iron, manganese and zinc in seawater by graphite furnace atomic absorption spectroscopy, *Anal. Chem.,* 51, 2364, 1979.

8. **Baeckman, S. and Karlsson, R. W.,** Determination of lead, zinc, silver and antimony in steel and nickel base alloys by atomic absorption spectrophotometry using direct atomization of solid sample in a graphite furnace, *Analyst,* 104, 1017, 1979.

9. **Sturgeon, R. E., Berman, S. S., Desaulniers, A., and Russell, D. S.,** Determination of iron, manganese and zinc by graphite furnace atomic absorption spectroscopy. Reply to comments, *Anal. Chem.,* 52, 1767, 1980.

10. **Segar, D. A. and Cantillo, A. Y.,** Determination of iron, manganese and zinc in sea water by graphite furnace atomic absorption spectroscopy. Comments, *Anal. Chem.,* 52, 1766, 1980.

11. **Chakrabarti, C. L., Wan, C. C., and Li, W. C.,** Atomic absorption spectrometric determination of cadmium, lead, zinc, copper, cobalt and iron in oyster tissue by direct atomization from the solid state using the graphite furnace platform technique, *Spectrochim. Acta,* 35B, 547, 1980.

12. **Yang, F. and Ni, Z.,** Determination of zinc in seawater by graphite furnace atomic absorption using citric acid as a matrix modifier, *Huanjing Kexue,* 2, 423-7, 1981 (in Chinese).

13. **Carelli, G., Cecchetti, G., La Bua, R., Bergamaschi, A., and Iannaccone, A.,** Determination of selenium, cadmium, zinc, cobalt, copper and silver using atomic absorption spectrophotometry in aerosols from the work place in the color television electronic industry, *Ann. Ist. Super Sanita,* 17, 505, 1981.

14. **Sekiya, T., Tanimura, H., and Hkiasa, Y.,** Simplified determination of copper, zinc and manganese in plasma and bile by flameless atomic absorption spectroscopy, *Archiv. Jpn. Chir.,* 50, 729-39, 1981.

15. **Arafat, N. M. and Glooschenko, W. A.,** Method for the simultaneous determination of arsenic, aluminum, iron, zinc, chromium and copper in plant tissue without the use of perchloric acid, *Analyst (London),* 106, 1174-8, 1981.

16. **Bouzanne, M.,** Determination of traces of inorganic pollutants (copper, lead, cadmium, zinc) in waters, by differential pulse anodic stripping voltammetry at a rotating mercury film electrode, *Analysis,* 9, 461-7, 1981 (in French).

17. **Shaw, J. C., Bury, A. J., Barber, A., Mann, L., and Taylor, A.,** Micromethod for the analysis of zinc in plasma or serum by atomic absorption spectrophotometry using graphite furnace, *Clin. Chim. Acta,* 118, 229, 1982.

18. **Kumamaru, T., Riordan, J. F., and Vallee, B. L.,** Determination of picogram quantities of zinc in zinc metalloproteins by atomic absorption spectroscopy using a graphite furnace atomizer, *Anal. Biochem.,* 126, 214, 1982.

19. **Ebdon, L., Ellis, A. T., and Ward, R. W.,** Aspects of chloride interference in zinc determination by atomic absorption spectroscopy with electrothermal atomization, *Talanta,* 29, 297-302, 1982.

20. **Busheiva, I. S. and Headridge, J. B.,** Determination of cadmium, indium and zinc in nickel-base alloys by atomic absorption spectroscopy with introduction of solid samples into furnaces, *Anal. Chim. Acta,* 142, 197, 1982.

21. **Bourcier, D. R., Sharma, R. P., Bracken, W. M., and Taylor, M. J.,** Cadmium-copper interaction: effect of copper pretreatment and cadmium-copper chronic exposure on the distribution and accumulation of cadmium, copper, zinc and iron in mice, *Trace Substr. Environ. Health,* 16, 273, 1982.

22. **Whitehouse, R. C., Prasad, A. S., Rabbani, P. I., and Cossack, Z. T.,** Zinc in plasma, neutrophils, lymphocytes and erythrocytes as determined by flameless atomic absorption *spectrophotometry, Clin. Chem. (Winston-Salem, N.C.),* 28, 475-80, 1982.

23. **Shaw, J. C. L., Bury, A. J., Barber, A., Mann, L., and Taylor, A.,** A micromethod for the analysis of zinc in plasma or serum by atomic absorption spectrophotometry using graphite furnace, *Clin. Chim. Acta,* 118, 229, 1982.

24. **Prasad, A. S. and Cossack, Z. T.,** Neutrophil zinc: an indicator of zinc status in man, *Trans. Assoc. Am. Physicians,* 95th, 165, 1982.

25. **Gardiner, P. E., Roesick, E., Roesick, U., Braetter, P., and Kynast, G.,** Application of gel filtration, immunonephelometry and electrothermal atomic absorption spectroscopy to the study of the distribution of copper, iron and zinc bound constituents in human amniotic fluid, *Clin. Chim. Acta,* 120, 103, 1982.

26. **Lieser, K. H., Sondermeyer, S., and Kliemchen, A.,** Precision and accuracy of analytical results in determining the elements cadmium, chromium, copper, iron, manganese and zinc by flameless atomic absorption spectroscopy, *Fresnius' Z. Anal. Chem.,* 312, 517, 1982 (in German).

27. **Foote, J. W. and Delves, H. T.,** Determination of zinc in small volumes of serum using atomic absorption spectrophotometry with electrothermal atomization, *Analyst (London),* 107, 1229, 1982.

28. **Prasad, A. S.,** Recent developments in the diagnosis of zinc deficiency in man, *Curr. Top. Nutr. Dis.,* 7, 141, 1982.

29. **Matsuura, S.,** Serum zinc determination by a micromethod using a graphite furnace atomizer, *Kurame Med. J.,* 29, 35-47, 1982.

30. **Takada, K. and Hirokawa, K.,** Origin of double-peak signals for trace lead, bismuth, silver and zinc in a microamount of steel in atomic absorption spectroscopy with direct electrothermal atomization of a solid sample in a graphite cup cuvette, *Talanta,* 29, 849,1982.

31. **Takada, K. and Shoji, T.,** Trace analysis for zinc and bismuth in high purity tin by polarized Zeeman atomic absorption spectroscopy with direct atomization of solid samples, *Fresnius'Z. Anal. Chem.,* 315, 34-7, 1983.

32. **Wawschinek, O.,** Micromethod using flameless atomic absorption for determining zinc in serum, *Mikrochim. Acta,* 3, 77-81, 1983 (in German).

33. **Uthurriague, C., Jouzier, E., and Crockett, R.,** Comparative study on the evaluation of zinc levels in man using flame or graphite furnace atomic absorption spectrophotometry, *Bull. Soc. Pharm. Bordeaux,* 122(1-2), 3-9, 1983 (in French).

34. **Foote, J. W. and Delves, H. T.,** Distribution of zinc amongst human serum proteins determined by affinity chromatography and atomic absorption spectrophotometry, *Analyst (London),* 108, 492-504, 1983.

35. **Whitehouse, R. C., Prasad, A. S., and Cossack, Z. T.,** Determination of ultrafiltrable zinc in plasma by flameless atomic absorption spectrophotometry, *Clin. Chem.,* 29(11), 1974-7, 1983.

36. **Castledine, C. G. and Robbins, J. C.,** Practical field-portable atomic absorption analyzer, *J. Geochem. Explor.,* 19, 689, 1983.

37. **Foote, J. W. and Delves, H. T.,** Measurement of zinc in serum proteins, *Proc. 2nd Int. Workshop, Trace Elem. Anal. Chem. Med. Biol.,* Braetter, P. and Schramel, P., Eds., Berlin, 1983, 877-84.

38. **Kawano, Y.,** Determination of plasma zinc, magnesium and copper by atomic absorption spectrometry with a flame or flameless system, *Kyushu Yakugakkai Kaiho,* 37, 43-52, 1983.

39. **Hinks, L. J., Colmsee, M., and Delves, H. T.,** Measurement of zinc and copper on leukocytes, *Proc. 2nd Int. Workshop.Trace Elem. Anal. Chem. Med. Biol.,* Braetter, P. and Schramel, P., Eds., Berlin, 1983, 885-92.

40. **Hayashi, Y., Yabuta, Y., Tanaka, T., and Nose, T.,** Determination of copper, iron and zinc in cutaneous leg lymph of rabbits by atomic absorption spectrophotometry, *Bunseki Kagaku,* 32, 212, 1983.

41. **Brune, D., Gjerdet, N., and Paulsen, G.,** Gastrointestinal and *in vitro* release of copper, cadmium, indium, mercury and zinc from conventional and copper-rich amalgams, *Scand. J. Dent. Res.,* 91, 66, 1983.

42. **Favier, A. and Ruffieux, D.,** Physiological variations of serum levels of copper, zinc, iron and manganese, *Biomed. Pharmacother.,* 37, 462, 1983.

43. **Patel, B. M., Goyal, N., Purohit, P., Dhobale, A. R., and Joshi, B. D.,** Direct determination of magnesium, manganese, nickel and zinc in uranium by electrothermal atomic absorption spectroscopy, *Fresnius' Z. Anal. Chem.,* 315, 42-6, 1983.

44. **Stevens, B. J.,** Determination of aluminum, copper and zinc in human hair, *At. Spectrosc.,* 4(5), 176-8, 1983.

45. **Subramanian, K. S. and Meranger, J. C.,** Blood levels of cadmium, copper, lead and zinc in children in a British Columbia community, *Sci. Total Environ.,* 30, 231-44, 1983.

46. **Mazzucotelli, A., Minoia, C., and Frache, R.,** Electrothermal atomic absorption spectrophotometry of aluminum, lead and zinc in CSF samples, *Proc. 2nd Int. Workshop Trace Elem. Anal. Chem. Med. Biol.,* Braetter, P. and Schramel, P., Eds., Berlin, 1983, 975-80.

47. **Postel, W., Meier, B., and Markert, R.,** Lead, cadmium, aluminum, tin, zinc, iron and copper in bottled and canned beer, *Moatsschr. Brauwiss,* 36, 360-7, 1983 (in German).

48. **Bahrenyi-Toosi, M. H., Dawson, J. B., Chilvers, D. C., and Ellis, D. J.,** An improved electrothermal atomic absorption technique for the determination of copper and zinc in plasma protein fractions, *Proc. 2nd Int. Workshop Trace Elem. Anal. Chem. Med. Biol.,* Braetter, P. and Schramel, P., Eds., Berlin, 1982, 811-18

49. **Gardiner, P. E., Gessner, H., Braetter, P., Stoeppler, M., and Nuernberg, H. W.,** The distribution of zinc in human erythrocytes, *J. Clin. Chem. Biochem.,* 22(2), 159-63, 1984.

50. **Lewis, S. A., O'Haver, T. C., and Harnly, J. M.,** Simultaneous multielement analysis of microliter quantities of serum for copper, iron and zinc by graphite furnace atomic absorption spectroscopy, *Anal. Chem.,* 56, 1651, 1984.

51. **Batz, L., Ganz, S., Hermann, G., Scharmann, A., and Wirz, P.,** Measurement of stable isotope distribution using Zeeman atomic absorption spectroscopy, *Spectrochim. Acta,* 39B, 993, 1984.
52. **Alcock, N. W.,** Flame and flameless atomic absorption spectrophotometry: application to the measurement of tissue zinc, *Neurol. Neurobiol.,* 11A, 305, 1984.
53. **Foote, J. W. and Delves, H. T.,** Distribution of zinc human serum globulins determined by gel filtration-affinity chromatography and atomic absorption spectrophotometry, *Analyst (London),* 109, 709, 1984.
54. **Boniforti, R., Ferraroli, R., Frigieri, P., Hettai, D., and Queirazza, G.,** Intercomparison of five methods for determination of trace metals in seawater, *Anal. Chim. Acta,* 162, 33, 1984.
55. **Ding, Y., Yao, M., Che, K., and Yang, W.,** Content of zinc, calcium, magnesium, copper and manganese in the hair of Huangshi (China) inhabitants, *Yingyang Xuebao,* 6(3), 298, 1984 (in Chinese).
56. **Backstrom, K., Danielsson, L. G., and Nord, L.,** Sample work-up for graphite furnace atomic absorption spectrometry using continuous flow extraction, *Analyst,* 109, 323, 1984.
57. **Bahrenyi-Toosi, M. H., Dawson, J. B., Ellis, D. J., and Duffield, R. J.,** Examination of instrumental systems for reducing cycle time in atomic absorption spectroscopy with electrothermal atomization, *Analyst,* 109, 1607, 1984.
58. **L'vov, B. V. and Yatsenko, L. F.,** Carbothermal reduction of zinc, cadmium, lead and bismuth oxides in graphite furnaces for atomic absorption analysis in the presence of organic substances, *J. Phys. C.,* 17(35), 6415-34, 1984.
59. **L'vov, B. V. and Yatsenko, L. F.,** Carbothermal reduction of zinc, cadmium, lead and bismuth oxides in graphite furnaces for atomic absorption analysis in presence of organic substances, *Zh. Anal. Khim.,* 39, 1773, 1984 (in Russian).
60. **Chung, C.,** Atomization mechanism with arrhenius plots taking the dissipation function into account in graphite furnace atomic absorption spectrometry, *Anal. Chem.,* 56, 2714, 1984.
61. **Carrondo, M. J. T., Reboredo, F., Ganho, R. M. B., and Olieveira, J. F. S.,** Analysis of sediments for heavy metals by rapid electrothermal atomic absorption procedure, *Talanta,* 31, 561, 1984.
62. **Danielsson, L. G.,and Westerlund, S.,** Short term variations in trace metal concentrations in Baltic, *Mar. Chem.,* 15, 273, 1984.
63. **Sperling, K. R.,** Tube-in-tube technique in electrothermal atomic absorption spectrometry. III. Atomization behavior of Zn, Cd, Pb, Cu and Ni, *Spectrochim. Acta,* 39B, 371, 1984.
64. **Matousek, J. P.,** Removal of sample vapor in electrothermal atomization studied at constant-temperature conditions, *Spectrochim. Acta,* 39B, 205, 1984.
65. **Lewis, S. A., O'Haver, T. C., and Harnly, J. M.,** Simultaneous multielement analysis of microliter quantities of serum for copper, iron and zinc by graphite furnace atomic absorption spectrometry, *Anal. Chem.,* 56, 1651, 1984.
66. **Blackmore, D. J. and Stanier, P.,** Methods for measurement of trace elements in equine blood by electrothermal atomic absorption spectrophotometry, *At. Spectrosc.,* 5, 215, 1984.
67. Release of metals (lead, cadmium, copper, zinc, nickel and chromium) from kitchen blenders, Statens Levnedsmiddelinstitut REPORT 1984. PUB-88, Order No. PB84-193416, 26 pp (in Danish). Avail. NTIS From Gov. Rep. Announce. Index (U.S.), 84, 67, 1984.
68. **Kitagishi, K. and Obata, H.,** Extra situ quantitative histochemical determination of zinc and manganese in plant tissues by flameless atomic absorption spectrophotometry, *Mie Daigaku Kankyo Kagaku Kenkyu Kiyo,* 10, 171-9, 1985.
69. **Arnaud, J., Favier, A., and Alary, J.,** Evaluation of a flameless atomic absorption method for determination of zinc in human milk, *Proc. 5th Int. Symp. Trace Elem. Man Anim.,* 645-7, 1984, published in 1985.
70. **Wennrich, R. and Feustel, A.,** Determination of cadmium and zinc in human prostatic tissues by flameless atomic absorption spectroscopy, *Z. Med. Laboratoriumsdiagn.,* 26(7), 365-9, 1985.
71. **Kluessendorf, B., Rosopulo, A., and Kreuzer, W.,** Study of the distribution and rapid determination of lead, cadmium and zinc in livers of slaughtered pigs by Zeeman AAS of solid samples, *Fresnius' Z. Anal. Chem.,* 322(7), 721-7, 1985.
72. **Yinying, W., Cuiping, M., Yunxia, Y., and Genzao, F.,** Determination of zinc at ppb levels in high purity gallium with flameless atomic absorption spectroscopy, *Fenxi Huaxue,* 14 (4), 304-6,1986 (in Chinese).
73. **Chen, J., Zhoung, X., and Gao, Y.,** Preconcentration and determination of trace zinc in high-purity gallium by selective dissolution-graphite furnace atomic absorption spectroscopy, *Xiyou Jinshu,* 4(4), 63-8, 1985 (in Chinese), CA 16, 105:34718s, 1986.
74. **Maroof, F. B. A., Hadi, D. A., Khan, A. H., and Chowdhury, A. H.,** Cadmium and zinc concentrations in drinking water supplies of Daka City, Bangladesh, *Sci. Total. Environ.,* 53, 233, 1986.
75. **El-Azizi, A., Al-Saleh, I., and Al-Mefty, O.,** Concentrations of zinc, iron, molybdenum, arsenic and lithium in cerebrospinal fluid of patients with brain tumors, *Clin. Chem.,* 32, 2187, 1986.
76. **Gomez Coedo, A. and Dorado Lopez, M. T.,** Determination of tin, chromium and zinc. Electrothermal atomic absorption spectroscopy, *Rev. Metal. (Madrid),* 22, 90, 1986.
77. **Feustel, A., Wennrich, R., and Dittrich, M.,** Studies of cadmium, zinc and copper levels in human kidney tumors and normal kidney, *Urol. Res.,* 14(2), 105-8, 1986.

AUTHOR INDEX

SUBJECT INDEX

CLINICAL ANALYSIS

REFERENCES

1. **Greenfield, S. and Smith, P. B.,** The determination of trace metals in microliter samples by plasma torch excitation with special references to oil, organic compounds and blood samples, *Anal. Chim. Acta*, 59, 341, 1972.
2. **Amos, M. D., Bennett, P. A., Brodie, K. G., and Lung, P. W.,** Carbon rod atomizer in atomic absorption and fluorescence spectroscopy and its clinical applications, *Anal. Chem.*, 43, 211, 1971.
3. **Gross, S. B. and Parkinson, E. S.,** Analyses of metals in human tissues using base (TMAH) digests and graphite furnace atomic absorption spectrophotometrry, *At. Ab. Newsl.*, 13, 107, 1974.
4. **Maeda, T., Nakagawa, M., Kawakatsu, M., and Tanimoto, Y.,** Determination of metallic elements in serum and urine by flame and graphite furnace atomic absorption spectrophotometry, *Shimadzu Hyoron*, 37, 1980 (in Japanese).
5. **Halls, D. J.,** Applications of graphite furnace atomic absorption spectroscopy in clinical analysis, *Anal. Proc. (London)*, 18, 344, 1981.
6. **Sutton, D. C., Rosa, W. C., and Legotte, P. A.,** Analytical measurements of selected metals in samples from a human metabolic study, *Trace Subst. Environ. Health*, 15, 270-8, 1981.
7. **Takahashi, Y., Shiozawa, Y., Kawai, T., Shimizu, K., Kato, H., and Hasegawa, J.,** Elements dissolved from three types of amalgams in synthetic saliva, *Aichi Gakuin Daigaku Shigakkaishi*, 19, 107-18, 1981 (in Japanese).
8. **Jin, K., Matsuda, K., and Chiba, Y.,** Concentrations of some metals in whole blood and plasma of normal adult subjects, *Hokkaidoritsu Eisei Kenkyushoho*, 31, 16-22, 1981 (in Japanese).
9. **Riley, C. M., Sternson, L. A., and Repta, A. J.,** Assessment of cisplatin reactivity with peptides and proteins using reverse-phase high-performance liquid chromatography and flameless atomic absorption spectroscopy, *Anal. Biochem.*, 124, 167-79, 1982.
10. **Aziz, A., Broeckaert, J. A. C., and Leis, F.,** Analysis of microamounts of biological samples by evaporation in a graphite furnace and inductively coupled plasma atomic emission spectroscopy, *Spectrochim. Acta Part B*, 37B, 369-79, 1982.
11. **Game, G. I., Balabanoff, K. L., Valdebenito, S. M. R., and Vivaldi, Q. L.,** Technique for the preparation of mixed human saliva samples for the determination of copper, zinc and manganese by flameless atomic absorption spectroscopy, *Bol. Soc. Chil. Quim.*, 27, 340, 1982.
12. **Brodie, K. and Routh, M. W.,** Trace metal analysis of biological samples with the GTA-95 graphite tube atomizer, *Varian Instrum. Appl.*, 16, 18-20, 1982.
13. **Gardiner, P. E., Roesick, E., Roesick, U., Braetter, P., and Kynast, G.,** The application of gel filtration, immunonephelometry and electrothermal atomic absorption spectroscopy to the study of the distribution of copper-, iron- and zinc-bound constituents in human amniotic fluid, *Clin. Chim. Acta*, 120, 103-17, 1982.
14. **Smith, P., Stubley, D., and Blackmore, D. J.,** Measurement of superoxide dismutase, diamine oxidase and ceruloplasmin oxidase in the blood of thoroughbreds, *Res. Vet. Sci.*, 35(2), 160-4, 1983.
15. **Hecquet, B., Adenis, L., and Demaille, A.,** *In vitro* interactions of TNO6 with human plasma, *Cancer Chemother. Pharmacol.*, 11(3), 177-81, 1983.
16. **Hudnik, V., Marolt-Gomiscek, M., Zargi, R., and Gomiscek, S.,** Some aspects of metal determination in liver diseases, *Proc. 2nd Int. Workshop Trace Elem. Anal. Chem. Med. Biol.*, 1982, 389-400, published 1983.
17. **Buratti, M., Calzaferri, G., Caravelli, G., and Colombi, A.,** Aluminum assay in biological fluids, *Med. Lav.*, 74, 70, 1983.
18. **Clavel, J. P., Chalvignac, B., and Thuillier, A.,** Selenium determination in biological fluids by electrothermal atomic absorption spectrophotometry, *J. Pharm. Clin.*, 2, 55-64, 2983 (in French).
19. **Dungs, K. and Neidhart, B.,** Analysis of urine samples by electrothermal atomization atomic absorption spectroscopy: a comparison of natural and control material, *Analyst (London)*, 109, 877, 1984.
20. **Bengtsson, G. and Gunnarsson, T.,** A micromethod for the determination of metal ions in biological tissues for furnace atomic absorption spectrophotometry, *Microchem. J.*, 29, 282, 1984.
21. **Voelkopf, U. and Grobenski, Z.,** Interference in the analysis of biological samples using the stabilized temperature platform furnace and Zeeman background correction, *At. Spectrosc.*, 5, 115, 1984.
22. **Alcock, N. W.,** Practical aspects of graphite furnace analysis of biological specimens, *At. Spectrosc.*, 5, 78, 1984.
23. **Kurfuerst, U., Grobecker, H. H., and Stoeppler, M.,** Homogeneity studies in biological reference and control materials with solid sampling and direct Zeeman-Atomic Absorption Spectroscopy, *Proc. 3rd Int. Workshop Trace Elem. Anal. Chem. Med. Biol.*, Braetter, P. and Schramel, P., Eds., Berlin, 1984, 591.
24. **Yang, X.,** Digestions of biological samples and their element analysis, *Shengwu Huaxue Yu Shengwu Wuli Jinzhan*, 59, 56-9, 1984 (in Chinese).
25. **Delves, H. T. and Shuttler, L. L.,** Interferences in the measurement of platinum in body tissues and fluids using electrothermal atomization and atomic absorption spectrophotometry, *Assoc. Int. Cancer Res. Symp.*, 4, 329, 1985.

26. **Narayanan, S., Kassay, S., Liu, F. C., Pearson, K. H., Shermaier, A. J., and Tytko, S. A.,** Assessment of levels of contamination from evacuated blood collection tubes, *Proc. 5th Int. Symp.Trace Elem. Man Anim.,* 1984, 657-9, published 1985.

27. **Levin, D. Z.,** An *in vitro* microperfusion study of distal tubule bicarbonate reabsorption in normal and ammonium chloride rats, *J. Clin. Invest.,* 75(2), 588-95, 1985.

28. **Aksnes, A., Galbrandsen, K. E., and Julshamn, K.,** Contents and biological availability of selenium in oxidized and protected fish meals from mackerel, *Fiskeridir. Skr., Ser. Ernaer.,* 2(4), 117, 1983, CA 11, 102:165526p, 1985.

29. **Lindh, U., Juntti-Berggren, L., Berggren, P. O., and Hellman, B.,** Proton microprobe analysis of pancreatic beta-cells, *Biomed. Biochem. Acta,* 44(1), 55-61, 1985.

30. **Casey, C. E., Hambridge, K. M., and Neville, M. C.,** Studies in human lactation: Zn, Cu, Mn and Cr in human milk in the first month of lactation, *Am. J. Clin. Nutr.,* 41(6), 1193-200, 1985.

31. **Norval, E.,** The graphite furnace in atomic absorption spectroscopy and some applications in clinical analysis, *S. Afr. J. Sci.,* 81(4), 169-71, 1985.

32. **Weiss, K. C. and Linder, M. C.,** Copper tranport in rats involving a new plasma protein, *Am. J. Physiol.,* 249, E77-E88, 1985.

33. **Lewis, S. A., O'Haver, T. C., and Harnly, J. M.,** Determination of metals at the microgram-per-liter level in blood serum by simultaneous multielement atomic absorption spectroscopy with graphite furnace atomization, *Anal. Chem.,* 57(1), 2-5, 1985.

34. **Ping, L., Lei, W., Matsumoto, K., and Fuwa, K.,** A study of the inter-elemental effect in the carbon-furnace AAS of selenium and mercury in biological samples by the palladium-addition method, *Bull. Chem. Soc. Japan,* 58(11), 3259-63, 1985.

35. **Hesseltine, G. R., Wolff, R. K., Hanson, R. L., McClellan, R. O., and Mauderly, J. L.,** Comparison of lung burdens of inhaled particles of rats exposed during the day or night, *J. Toxicol. Environ. Health,* 16(2), 323-9, 1985.

36. **Shiraishi, K., Kawamura, H., and Tanaka, G.,** Daily intake of elements as estimated from analysis of total diet samples in relation to refernce Japanese man, *Radiat. Res.,* 27(1), 121-9, 1986.

37. **Voelkopf, U. and Grobenski, Z.,** Use of direct solid sample feed in graphite-tube furnace AAS for screening analysis of marine animals, *Fortschr. Atomspektrom. Spurenanal.,* 2, 465, 1986.

38. **Harnly, J. M.,** Simultaneous multielement AAS with carbon furnace atomization for the analysis of biological samples, *Fresnius' Z. Anal. Chem.,* 323(7), 759-61, 1986.

39. **Yin, X., Schlemmer, G., and Welz, B.,** Cadmium determination in biological materials using graphite furnace atomic absorption spectroscopy with palladium nitrate-ammonium nitrate modifier, *Anal. Chem.,* 59, 1462, 1987.

AUTHOR INDEX

SUBJECT INDEX

HEAVY METALS

REFERENCES

1. **Sperling, K. R.,** Determination of heavy metals in seawater and in marine organisms by flameless atomic absorption spectrophotometry. V. Methane pyrolysis treatment of the graphite tube atomizer, *Z. Anal. Chem.,* 283, 30, 1977.

2. **Sperling, K. R. and Bahr, B.,** Determination of heavy metals in seawater and in marine organisms by flameless atomic absorption spectrophotometry. XI. Quality criteria for graphite tubes — a warning, *Z. Anal. Chem.,* 299, 206, 1977.

3. **Nichols, J. A. and Woodriff, R.,** Coprecipitation of heavy metals directly in graphite crucibles for furnace atomic absorption spectroscopy, *J. Assoc. Off. Anal. Chem.,* 63, 500, 1980.

4. **Mojo, L., Martella, S., and Martino, G.,** Seasonal concentrations of heavy metals (mercury, cadmium, lead) in some marine organisms in the Central Mediterranian Sea, *Mem. Biol. Mar. Oceanogr.,* 10, 27-39, 1980 (in Italian).

5. **Hoshino, Y., Utsunomiya, T., Mise, N., and Sakabe, K.,** Preconcentration of heavy metals ions onto metals with high melting points for graphite furnace atomic absorption spectroscopy, *Nippon Kagaku Kaishi,* 1, 19, 1981.

6. **Reboredo, F., Carrondo, M. J. T., Ganho, R. M. B., and Oliviera, J. F. S.,** Use of a rapid flameless atomic absorption method for the determination of the metallic content of sediments in the Tejo Estuary, Portugal, *3rd Int. Conf. Heavy Met. Environ,* CEP Consult. Ltd., Edinburgh, U.K., 1981, 587-90.

7. **Foerster, M. and Lieser, K. H.,** Determination of traces of heavy metals in inorganic salts and organic solvents by energy-dispersive X-ray fluorescence analysis or flameless atomic absorption spectroscopy after enrichment on a cellulose exchanger, *Fresnius' Z. Anal. Chem.,* 309, 355-8, 1981 (in German).

8. **Shishido, M. and Yamaka, M.,** Determination of heavy metals in highly concentrated solutions by flameless atomic absorption spectrophotometry, *Kanagawa-ken Taiki Osen Chosa Kenkyu Hokoku,* 23, 120, 1981 (in Japanese).

9. **Marlier-Geets, O., Heck, J. P., Barideau, L., and Rocher, M.,** Method for determination of heavy metals in sewage sludge, their distribution according to their origin and their concentration variation with time, *Comm. Eur. Communities (Rep) EUR,* 7076, 284-90, 1980 (in French).

10. **Sakata, M. and Shimoda, O.,** Atomic absorption determination of heavy metals in sediment by a digestion method with a teflon-lined bomb, *Bunseki Kagaku,* 31, T81-T86, 1982 (in Japanese).

11. **Kurfuerst, U.,** Direct determination of heavy metals (lead, cadmium, nickel, chromium, mercury) in blood and urine using Zeeman atomic absorption, *Fresnius' Z. Anal. Chem.,* 313, 97, 1982.

12. **Sztraka, A.,** Determination of heavy metals in the activated sludge of a laboratory scale model wastewater treatment plant, *Atomspektrom. Spurenanal. Vortr. Kolloq.,* Welz, B., Ed., Springer-Verlag, Weinheim, 1982, 335-40 (in German).

13. **Ma, Y., Sun, D., and Zhu, M.,** Determination of heavy metals in paints by graphite furnace atomic absorption and atomic emission spectrographic methods, *Huanjing Huaxue,* 6, 474, 1982.

14. **Sperling, K. R.,** Determination of heavy metals in seawater and in marine organisms by flameless atomic absorption spectrophotometry. XVII. On the usefulness of signal integration — a warning, *Fresnius' Z. Anal. Chem.,* 314, 417, 1983.

15. **Sperling, K. R. and Bahr, B.,** Determination of heavy metals in seawater and in marine organisms by flameless atomic absorption spectrophotometry. XVIII. Measures for improved independency from graphite tube quality variations, *Fresnius' Z. Anal. Chem.,* 314, 760, 1983.

16. **Schonberger, E., Kassovicz, J., and Shenhar, A.,** Preconcentration of trace heavy metals in Dead Sea for determination by GFAAS, *4th Int. Conf. Heavy Met. Environ.,* 1, 218, 1983.

17. **Haraldsson, C. and Magnusson, B.,** Heavy metals in rainwater. Collection, storage and analysis of samples, *4th Int. Conf. Heavy Met. Environ.,* 1, 82, 1983.

18. **Tsushida, T.,** Content of heavy metals in tea. Application of flame and flameless atomic absorption spectrophotometry to tea, *Chagyo Shikenjo Kenkyu Hokoku,* 19, 59, 1983.

19. **Nuurose-Ervasto, L., Pitkaniemi, M., and Uusi-Rauva, E.,** Determination of certain heavy metals in edible fats using flameless AAS. A comparison of three prehandling methods, *Meijeritiet. Aikak.,* 41(2), 34-9, 1983.

20. **Drapeau, A. J., Laurence, R. A., Harbec, P. S., Saint-Germain, G., and Lambert, N. G.,** Bioaccumulation of heavy metals by some microorganisms, *Sci. Tech. Eau,* 16(4), 359-63, 1983 (in French).

21. **Aurand, K., Drews, M., and Seifert, B.,** A passive sampler for the determination of the heavy metal burden of indoor environments, *Environ. Technol. Lett.,* 4(10), 433, 1983.

22. **Sakai, Y.,** Determination of trace heavy metals in seawater by flameless atomic absorption spectrometry. I, *Kogai to Taisaku,* 19(11), 1051-61, 1983.

23. **Heinrich, R. and Angerer, J.,** Determination of heavy metals in urine-advantages of a cleanup procedure, *Proc. 2nd Int. Workshop Trace Elem. Anal. Chem. Med. Biol.,* Braetter, P. and Schramel, P., Eds., Berlin, 1983, 675-83.

24. **Kurfuerst, U.,** Heavy metal determination in solids using direct Zeeman atomic absorption spectroscopy. III. Importance of the graphite boat technique for the analysis of solids, *Fresnius' Z. Anal. Chem.,* 376, 1-7, 1983 (in German).

25. **Hiraide, M., Mizutani, J., and Mizuike, A.,** Rapid separation by flotation of suspended solids in fresh waters for the determination of absorbed heavy metals, *Anal. Chim. Acta,* 151, 329-37, 1983.

26. **Sakai, Y.,** Analysis of trace heavy metals in seawaters by flameless atomic absorption spectroscopy. II, *Kogai to Taisaku,* 20(4), 345, 1984.

27. **Batifol, F. M. and Boutron, C. F.,** Atmospheric heavy metals in high altitude surface snows from Mont Blanc, French Alps, *Atmos. Environ.,* 18, 2507, 1984.

28. **Carrondo, M. J. T., Reboredo, F., Ganho, R. M. B., and Oliviera, J. F. S.,** Analysis of sediments for heavy metals by a rapid electrothermal atomic absorption procedure, *Talanta,* 31, 561, 1984.

29. **De Leeuw, S., Robbiani, R., and Buechi, W.,** Screening for heavy metals and fat soluble pesticides in milk by sample and accurate methods, *Spec. Publ. R. Soc. Chem.,* 49, 333, 1984.

30. **Carl, M.,** Determination of heavy metal contaminants in milk and milk products, *Spec. Publ. R. Soc. Chem.,* 49, 323, 1984.

31. **Bengtsson, M. and Johansson, G.,** Preconcentration and matrix isolation of heavy metals through a two-stage solvent extraction in a flow system, *Anal. Chim. Acta,* 158, 147, 1984.

32. **Cha, K. W., Back, S. O., and Jin, D. S.,** Study on the determination of heavy metals in human body by atomic absorption spectroscopy, *Kich'o Kwahak Yonguso Nonmumjip,* 5, 75-80, 1984.

33. **Kato, T., Aozima, K., Kubota, H., Tanaka, Y., Terashima, H., and Kasuya, M.,** Simple pretreatment for determination of heavy metals in urine using chelate resins, *Hokuriku Koshu Eisei Gakkaishi,* 11(2), 42, 1984 (in Japanese).

34. **Rosopulo, A., Grobecker, K. H., and Kurfuerst, U.,** Studies of the analysis for heavy metals in solids by direct Zeeman atomic absorption spectroscopy. IV. Methodology of the direct solid analysis of biological materials, *Fresnius' Z. Anal. Chem.,* 319(5), 540-6, 1984.

35. **Hoffmann, E. W. and Poll, K. G.,** Heavy metal determination in uncontaminated clay sediments of the Rhine Valley near Dinslaken, *Z. Wasser Abwasser Forsch.,* 18(1), 31-4, 1985.

36. **Allegrini, M., Gallorini, M., Lanzola, E., and Oruni, E.,** Toxic and essential element intakes of formula fed infants, *5th Int. Conf. Heavy Met. Environ.,* Lekkas, T. D., Ed., CEP Consult., Edinburgh, U.K., 1985, 76.

37. **Sperling, K. R.,** Determination of heavy metals in seawater and in marine organisms by flameless AAS. XIX. Determination of cadmium traces in environmental samples, *Fortschr. Atomspektrom. Spurenanal.,* 1, 385-401, 1984, CA 16, 103:31704g, 1985.

38. **Spankova, M. and Smirnova, L.,** Determination of microgram quantities of heavy metals in fruit, vegetables and other products, *Acta Fac. Rerum Nat. Univ. Comenianae, Form. Prot. Nat.,* 9, 233-43, 1984, CA 22, 103:12849n, 1985.

39. **Legret, M., Divet, L., and Demare, D.,** Reduction of interferences in the determination of trace heavy metals in river sediments and sewage sludges by electrothermal AAS, *Anal. Chim. Acta,* 175, 203-10, 1985.

40. **Janssen, A., Brueckner, B., Grobecker, K. H., and Kurfuerst, U.,** Determination of heavy metals (lead, uranium, copper and nickel) in polyethylene by direct Zeeman AAS, *Fresnius' Z. Anal. Chem.,* 322(7), 713-6, 1985 (in German).

41. **Rosopulo, A.,** Determination of heavy metals directly in the solid sample and after chemical digestion — a comparison of the methods, *Fresnius' Z. Anal. Chem.,* 322(7), 669-72, 1985.

42. **Stoeppler, M., Kurfuerst, U., and Grobecker, K. H.,** Investigations on the analysis for heavy metals in solids by direct Zeeman AAS. V. The homogeneity factor as a criterion for pulverized solid samples, *Fresnius' Z. Anal. Chem.,* 322(7), 687-91, 1985.

43. **Keller, H. G. and Ebertz, U.,** Determination of heavy metals in lime products for environmental protection, *Fortschr. Atomspektrom. Spurenanal.,* 2, 649-56, 1986.

44. **Nakamura, S. and Kubota, M.,** Determination of some heavy metals in biological samples by chemical interference-free electrothermal AAS with a fast response system, *Bunseki Kagaku,* 35, 982, 1986.

45. **Zauke, G. P., Jacobi, H., Gieseke, U., Suengerlaub, G., Baeumer, H. P., and Butte, W.,** Sequential multielement determination of heavy metals in brackish-water organisms, *Fortschr. Atomspektrom. Spurenanal.,* 2, 543, 1986 (in German).

46. **Janssen, E.,** Experience with the sequential determination of heavy metals by the Zeeman graphite-tube furnace technique taking into account the STPF (stabilized temperature platform furnace) concept, *Fortschr. Stomspektrom. Spurenanal,* 2, 429, 1986.

AUTHOR INDEX

SUBJECT INDEX

MISCELLANEOUS ANALYSIS

REFERENCES

1. **Brodie, K. G. and Matousek, J. P.,** Application of the carbon rod atomizer to atomic absorption spectroscopy of petroleum products, *Anal. Chem.,* 43, 1557, 1971.
2. **Newton, M. P. and Davis, D. G.,** An indirect determination of some ions by flameless atomic absorption using a wire atomizer, *Anal. Lett.,* 6, 923, 1973.
3. **Nemets, A. M., Nikolaev, G. I., Flisyuk, V. G., and Bodrov, N. V.,** Distribution of the concentration of metal vapors in an open cuvette with a graphite heater during vaporization in an inert medium, *Z. Angew. Spektrosk.,* 21, 795, 1974 (in Russian).
4. **Katskov, D. A., Kruglikova, L. P., L'vov, B. V., and Polzik, L. K.,** Use of a graphite furnace with a circular cavity for the atomic absorption analysis of ultrapure materials, *Zh. Prikl. Spektrosk.,* 20, 739, 1974 (in Russian).
5. **Ishibashi, W., Sato, M., and Hashimoto, K.,** Atomic absorption spectroscopy of impurities in high purity tantalum with a heated graphite atomizer, *Bunseki Kagaku,* 23, 597, 1974.
6. **Gomiscek, S., Lengar, Z., Cernetic, J., and Hudnik, V.,** Behavior of metal tetramethylenedithiocarbamates in the graphite tube furnace for atomic absorption spectroscopy, *Anal. Chim. Acta,* 73, 97, 1974.
7. **Ediger, R. D., Peterson, G. E., and Kerber, J. D.,** Application of the graphite furnace to saline water analysis, *At. Ab. Newsl.,* 13, 61, 1974.
8. **Yamasaki, S. and Kishita, A.,** Determination of sub-microgram amounts of elements in soil solution by flameless atomic absorption spectrophotometry with a heated graphite atomizer, *Soil Sci. Plant Nutr.,* 21, 63, 1975.
9. **Crisp, P. T., Eckert, J. M., Gibson, N. A., Kirkbright, G. F., and West, T. S.,** The determination of anionic detergents at ppb levels by graphite furnace atomic absorption spectroscopy, *Anal. Chim. Acta,* 87, 97, 1976.
10. **L'vov, B. V. and Pelieva, L. A.,** Atomic absorption determination of carbide-forming elements in a graphite furnace lined with a tantalum foil, *Zavod. Lab.,* 44, 173, 1978.
11. **Parks, E. J., Brinckman, F. E., and Blair, W. R.,** Application of a graphite furnace atomic absorption detector automatically coupled to a high-performance liquid chromatograph for speciation of metal-containing macromolecules, *J. Chromatogr.,* 185, 563, 1979.
12. **Carrondo, M. J., Perry, R., and Lester, J. N.,** Influence of conditioning agents on the determination of metallic content of sewage sludge by atomic absorption spectrophotometry with electrothermal atomization, *Analyst,* 104, 937, 1979.
13. **Noller, B. N. and Bloom, H.,** Application of graphite furnace atomic absorption spectroscopy to the determination of metals in air particulates, *Clean Air (Melbourne),* 14, 9, 1980.
14. **Katskov, D. A., Burtseva, I. G., Grinshtein, I. L., and Kruglikova, L. P.,** Widening of range of element content to be determined by electrothermal atomic absorption analysis using automatic evaporator temperature control, *Zh. Anal. Khim.,* 35, 2289, 1980 (in Russian).
15. **Goda, A., Moriyama, K., and Hariyama, S.,** Steel by atomic absorption spectroscopy with a graphite tube atomizer, *Trans. Iron Steel Inst. Jpn.,* 20, B405, 1980.
16. **Lieu, V. T. and Woo, D. H.,** Application of flame and graphite furnace atomic absorption spectroscopy to the determination of metals in oil-field injection water and its suspended solids, *At. Spectrosc.,* 1, 149, 1980.
17. **Calapaj, R., Ciraolo, L., Berdar, A., and Cavaliere, A.,** Metal content of some bathial fish species preserved in museum, collected from the Strait of Messina (Italy), *Atti Soc. Peloritana Sci. Fis. Mat. Nat.,* 26(1-2), 37-42, 1980 (in Italian).
18. **Rcheulishvili, A. N.,** Graphite furnace atomic absorption spectroscopy with separation of the determined element by volatilization, *Zh. Anal. Khim.,* 36, 1889-94, 1981 (in Russian).
19. **Garbett, K., Goodfellow, G. I., and Marshall, G. B.,** Application of atomic absorption spectroscopy in the analysis of metallic sodium. I. Volatilization characteristics of sodium salts during electrothermal atomization, *Anal. Chim. Acta,* 126, 135, 1981.
20. **Kane, J. S. and Smith, H.,** Analysis of Egyptian Geological Survey and Mining Department samples by rapid rock and atomic absorption procedures, *Geol. Surv. Open File Rep. (U.S.),* 81-99, 59, 1981.
21. **Foland, K. A. and Wagner, M. E.,** A simple method of preparing glasses from rock powders, *Am. Mineral.,* 66, 1086, 1981.
22. **Borg, H., Edin, A., Holm, K., and Skoeld, E.,** Determination of metals in fish livers by flameless atomic absorption spectroscopy, *Water Res.,* 15, 12981-5, 1981.
23. **Hinderberger, E. J., Kaiser, M. L., and Koirtyohann, S. R.,** Furnace atomic absorption analysis of biological samples using the L'vov platform and matrix modification, *At. Spectrosc.,* 2, 1, 1981.
24. **Benoit, O. and Lamathe, J.,** Comparison of methods for determining anionic detergents in water by flameless atomic absorption spectrometry, *Bull. Liasion Lab. Points Chausseees,* 115, 25-32, 1981 (in French).
25. **Oddo, N. E.,** Modern atomic spectroscopy for the analysis of metal toxicity, *Not.-Soc. Ital. Biochbim. Clin.,* 9 (Suppl.), 633-48, 1981 (in Italian).

26. **Schulze, H. D.,** Fully automated determination of metal traces directly in flowing sample streams, *Int. Lab.,* 11, 92, 1981 (in German).

27. **Hocquaux, H.,** Flameless atomic absorption spectrophotometry application to metallurgical analysis, *Met. Corros. Ind.,* 56, 216, 1981 (in French).

28. **Kmiec, G. E.,** The washout of gaseous pollutants (sulfur dioxide, nitrogen oxides) in the vicinity of coal fired power plants by cooling towers and the effects upon the quantitative balance in the cooling system, *Proc. Int. Clean Air Conf.,* 7, 799-807, 1981.

29. **Brovko, I. A.,** Diphenylcarbazine as an extraction reagent in atomic absorption, *Deposited Doc.,* 1981, VINITI 2842-81, 66pp, Avail. VINITI (in Russian).

30. **Kaegler, S. H.,** Statue of the standardization work using AAS and NFAAS in mineral oil analysis, *Atomspektrom. Spurenanal., Vortr. Kolloq.,* Welz, B., Ed., Springer-Verlag, Weinheim, 1982, 493.

31. **Torsi, G., Palmisano, F., Desimoni, E., and Rinaldi, R.,** A simple and effective field sampler for metal determination in air by means of the electrostatic accumulation furnace for electrothermal atomic spectrometry (EAFEAS), *Ann. Chim. (Rome),* 72, 365, 1982.

32. **Alder, J. F. and Batoreu, M. C.,** Ion exchange resin beads as solid standards for electrothermal atomic absorption spectrometric determination of metals in hair, *Anal. Chim. Acta,* 135, 229-34, 1982.

33. **Matsusaki, K.,** Mechanism and removal of halide interferences in the determination of metals by atomic absorption spectroscopy with electrothermal atomization, *Anal. Chim. Acta,* 141, 233-40, 1982.

34. **Tominaga, M. and Umezaki, Y.,** Comparison of ascorbic acid and related compounds as interference suppressors in electrothermal atomic absorption spectroscopy, *Anal. Chim. Acta,* 139, 279, 1982.

35. **Torsi, G. and Desimoni, E.,** Electrostatic accumulation furnace for electrothermal atomic spectroscopy: a new apparatus for the determination of metals in the atmosphere, *Comm. Eur. Communities (Rep),* EUR 7624, 47-53, 1982.

36. **Lendermann, B. and Hundeshagen, D.,** Usefulness of multielement standards for calibration in AAS water analysis, *Fresnius' Z. Anal. Chem.,* 310, 415-22, 1982 (in German).

37. **Nikolaev, G. I. and Nemets, A. M.,** Atomic-absorption spectroscopy in the study of the vaporization of metals, *Atomno-Absorbtsionnaya Spektroskopiya v Issledovanii Ispareniya Metallov,* 151 pp, 1982 (in Russian).

38. **Folio, M. R., Hennigan, C., and Errera, J.,** A comparison of five toxic metals among rural and urban children, *Environ. Pollut. Ser. A,* 29, 261, 1982.

39. **Van Stekelenburg, G. J., Valk, C., and De Boer, G. J.,** Determination of deferoxamine chelated iron, *Clin. Chem.,* 28, 23328, 1982.

40. **Gockley, G. B. and Skriba, M. C.,** The monitoring of cationic species in a nuclear power plant using on-line atomic absorption spectroscopy, *Autom. Stream Anal. Stream Anal. Process Control,* 1, 69, 1982.

41. **Shrader, D., Voth, C., and Covick, L.,** The determination of toxic metals in waters and wastes by furnace atomic absorption, *Varian Instrum. Appl.,* 16, 1218, 1982.

42. **Balaes, G. E. E. and Robert R. V. D.,** Determination by atomic absorption spectrophotometry of impurities in manganese dioxide, Report 1981, NIM-20094; Order No. DE82903814, 15 pp, 1982. Avail NTIS From energy Res. Abstr. 415-17, 7, 1982.

43. **Bye, R.,** On the storage of the sodium borohydride solution used in the hydride-generation atomic absorption technique, *Talanta,* 29, 797, 1982.

44. **Sugiyama, K., Sugogo, Y., Shigeta, M., Yamazaki, Y., and Torii, K.,** Behavior of organometallic compounds in red tide, *Yamaguchi-ken Kogai Sento Nenpo,* 8, 147-50, 1982 (in Japanese).

45. **Sawada, K., Inomoto, S., Gobara, B., and Suzuki, T.,** Extraction and determination of anionic surfactants with copper (II)-ethylenediamine derivative complexes, *Talanta,* 30, 155-9, 1983.

46. **Nash, A. M., Mounts, T. L., and Kwolek, W. F.,** Determination of ultratrace metals in hydrogenated vegetable oils and fats, *J. Am. Oil Chem. Soc.,* 60, 811, 1983.

47. **Ahluwalia, B. and Duffus, C. M.,** A simple and rapid method for determination of metal ions in developing barley seeds, *J. Inst. Brew.,* 89, 24-7, 1983.

48. **Watling, R. J. and Watling, H. R.,** Metals and estuarine sediments: Studies using secondary geochemical metal interrelationships and differential thermal release analysis, *4th Int. Conf. Heavy Met. Environ.,* 2, 964, 1983.

49. **Grobenski, Z., Voellkopf, U., and Erler, W.,** The stabilized temperature platform furnace with and without Zeeman-effect background correction for trace analysis of environmental samples, *4th Int. Conf. Heavy Met. Environ.,* 1, 229, 1983.

50. **Park, M. K., Kim, B. K., Park, J. H., and Kim, H. J.,** A study of mineral elements in mineral waters (II), *Soul Taehakkyo Yakhak Nonmunjip,* 8, 34-6, 1983.

51. **Flanagan, M.,** Analysis of metals in coal pile runoff, *Proc. Coal Test. Conf.,* 3, 151, 1983.

52. **Caravajal, G. S., Mahan, K. I., Goforth, D., and Leyden, D. E.,** Evaluation of methods based on acid extraction and atomic absorption spectroscopy for multielement determinations in river sediments, *Anal. Chim. Acta,* 147, 133, 1983.

53. **Karimiann-Teherani, D., Kiss, I., Altmann, H., Wallisch, G., and Kapeller, K.,** Accumulation and distribution of elements in plants (paprika), *Acta Aliment.,* 12(4), 301-18, 1983.

54. **Kijewski, H. and Bock, G.,** Possibilities and limits of a shooting distance determination from smoke element concentration in the contusion ring for shooting distance up to 400 m, *Beitr. Gerichtl. Med.,* 41, 383-9, 1983 (in German).

55. **Munoz, F. E., Calvo, A., and Leon, L. E.,** Direct analysis of solid samples by flameless atomic absorption spectrometry. Application to some siliceous materials, *Anal. Lett.,* 16, 835, 1983.

56. Application of flameless atomic absorption spectroscopy in iron metallurgical analysis, Centro Sperimentale Metallurgico S. P. A. (1-00129 Rome, Italy), *Comm. Eur. Communities (Rep.),* EUR 7886, 77, 1983.

57. **Pelieva, L. A. and Bukhantsova, V. G.,** Atomic-absorption determination of carbide-forming elements using a graphite furnace and a tantalum boat, *Zavod. Lab.,* 49, 35-7, 1983.

58. **Ward, N., Bryce-Smith, D., Minski, M., Zaaijman, J. T., and Pim, B.,** Multielement neutron activation analysis of amniotic fluid in relation to varying gestation membrane ruptures, *Proc. 2nd Int. Conf. Trace Elem. Anal. Chem. Med. Biol.,* Braetter, P. and Schramel, P., Eds., Berlin, 1983, 483-98.

59. **Yudelevich, I. G. and Papina, T. S.,** Atomic absorption methods for determination of doping impurities in germanium single crystals and films, *Izv. Sib. Otd. Akad. Nauk SSSR Ser. Khim. Nauk,* 3, 74-9, 1983.

60. **Hendrickson, L. L. and Corey, R. B.,** A chelating-resin method for characterizing soluble metal complexes, *Soil Sci. Soc. Am. J.,* 47, 467, 1983.

61. **Hocquaux, H.,** Atomic absorption spectrometry for analyses of alloyed steels, *Colloq. Metall.,* 25, 207-16, 1982, published 1983 (in French).

62. **Adachi, A. and Kobayashi, T.,** Atomic absorption spectrophotometric determination of polyoxyethylene type nonionic surfactants in water, *Eisei Kagaku,* 29, 123-9, 1983

63. **Rossi, G., Omenetto, N., Pigozzi, G., Vivian, R., Mattiuz, U., Mousty, F., and Crabi, G.,** Analysis of radioactive waste solutions by atomic absorption spectroscopy with electrothermal atomization, *At. Spectrosc.,* 4, 113-117, 1983.

64. **Carlin, L. M., Colovos, G., Garland, D., Jamin, M. E., and Klenck, M.,** Analytical methods evaluation and validation: arsenic, nickel, tungsten, vanadium, talc wood dust, Report NIOSH-210-79-0060, Order No. PB83-155325, 223pp, 1981. Avail. NTIS from Gov. Rep. Announce. Index (U.S.), 83, 2835, 1983.

65. **Karmanova, N. G.,** Use of rock and ore standards for atomic absorption and emission determination of elements with atomizers of graphite capsule type, *Metody Spectr. Anal. Miner. Syr'ya (Mater. Vses. 2nd Konf. Nov. Metodam Spektr. Anal. Ikh Primen.),* Lontsikh, S. V., Ed., Nauka, Sib. Otd.: Novosibirsk, U.S.S.R., 1984.

66. **Kurosawa, S., Homma, Y., Yamawaki, M., and Tanaka, T.,** Determination of impurities in gallium arsenide by using multielement-doped GaAs standards, *Denshi Tsushin Gakkai Ronbunshi,* C. J.-67-C(12), 977-84, 1984 (in Japanese).

67. **Parks, E. J., Manders, W. F., Johannessen, R. B., and Brinckman, F. E.,** Characterization of organometallic polymers by chromatographic methods and nuclear magnetic resonance. II, Report-NBSIR-83-2802, 1984. Order No. PB84-183599, 56pp. Avail. NTIS. From Gov. Rep. Announce. Index (U.S.), 84, 89, 1984.

68. **Hoenig, M., Scokart, P. O., and Van Hoeyweghen, P.,** Efficiency of L'vov platform and ascorbic acid modifier for reduction of interferences in the analysis of plant samples for lead, thallium, antimony, cadmium, nickel and chromium by ET-AAS, *Anal. Lett.,* 17(A17), 1947-62, 1984.

69. **Dornemann, A. and Kleist, H.,** Determination of microtraces of various foreign metals in titanium oxide pigments, *Forbe Lack,* 90(9), 750, 1984.

70. **Turner, M. A., Hendrickson, L. L., and Corey, R. B.,** Use of chelating resins in metal adsorption studies, *Soil Sci. Soc. Am. J.,* 48, 763, 1984.

71. **Harnly, J. M., Miller-Ihli, N. J., and O'Haver, T. C.,** Simultaneous multielement atomic absorption spectrometrry with graphite furnace atomization, *Spectrochim. Acta,* 39B, 305, 1984.

72. **Judelevich, J. G., Beisel, N. F., Papina, T. S., and Dittrich, K.,** Layer-by-layer and film analysis of semiconductor materials by atomic absorption with electrothermal atomization, *Spectrochim. Acta,* 39B, 467, 1984.

73. **Smith, R.,** A laboratory manual for the determination of metals in water and wastewater by atomic absorption spectrophotometry, Report 1983, CSIR-K-63, 35pp. Avail. INIS. From INIS Atomindex 15, 1984. Abstr. No. 15:030051.

74. **Murnane, M. M.,** Analysis of brine by atomic absorption with the graphite furnace using direct sample injection, *Chem. N. Z.,* 48, 39, 1984.

75. **Dittrich, K., Vorberg, B., Funk, J., and Beyer, V.,** Determination of some nonmetals by using diatomic molecular absorbance in a hot graphite furnace, *Spectrochim. Acta,* 39B, 349, 1984.

76. **Blinova, E. S., Guzeev, I. D., and Smirnova, I. V.,** Analysis of alloys by atomic absorption with electrothermal atomization, *Nauch. Tr. N.-i, i Proekt. In-t Redkomet. Prom. Sti Giredmet,* 119, 76-9, 1983. From Ref. *Zh. Khim.,* Abstr. No. 16G167, 1984.

77. **Charykov, A. K., Osipov, N. N., and Panichev, N. A.,** Determination of metals in water, USSR SU:057,858 (Cl. G01N31/22), 30 Nov. 1983. Appl. 3,478,101, 14 Jun. 1982. From Otkrytiya, Izobreet., Prom. Obraztsy, Tovarnys Znaki, 44, 166, 1983. CA 7, 100:91059y, 1984.

78. **Barrais, P., Largeaud, P., and Lamathe, J.,** Determination of anionic detergent in waters by Furnace Atomic Absorption Spectroscopy, *Bull. Liasion Lab. Ponts Chaussees,* 133, 91-6, 1984 (in French).

79. **Zhao, Y.,** Determination of impurities in gold alloys by graphite furnace atomic absorption spectrometrry, *Guangpuxue Yu Guangpu Fenxi*, 4(5), 48-52, 1984 (in Chinese).

80. **Drwiega, I., Jedrzejewska, H., Barczkkowicz, A., and Malusecka, M.,** Investigations on lowering of the detection limits of some impurities in high purity copper, *Pr. Inst. Met. Niezelaz*, 12(1-2), 201, 1983, CA 15, 103:16118a, 1985.

81. **Zieba, J.,** Possibility of application of FAAS in forensic analysis, *Z. Zagadnien Kryminal*, 16-17, 76-83, 1983, CA 15, 103:1583z, 1985.

82. **Liese, T.,** Analysis of elements for the determination of soil plant transfer factors, *Kern Forschungszent. Karlsruhe (Ber.)*, KfK 3830, 47, 1985, CA 16, 103:21650, 1985.

83. **Oreshkin, V. N., Tatsii, Y. G., and Belyaev, Y. I.,** Improvement of the direct atomic absorption determination of pollutant metals in environmental samples, *Fiz. Metody Kontrolya Khim. Sostava Mater.*, 22-7, 1983, CA 12, 102:197091j, 1985.

84. **Lightowlers, E. C. and Davies, G.,** Spectroscopy of excitons bound to isoelectronic defect complexes in silicon, *Solid State Commun.*, 53(12), 1055, 1985.

85. **Schindler, E.,** Determination of Ag, Al, As, Cd, Cr, Cu, Fe, Mn, Pb, Se and Zn in drinking water. Procedure for the study of drinking water by graphite tube AAS (Atomic Absorption Spectroscopy). I, *Dtsch. Lebensm. Rundsch.*, 81(1), 1-11, 1985.

86. **Vavilin, L. N., Filimonov, V. V., Bessonov, G. D., Kostennich, A. V., Suvernev, V. E., Turkin, Y. I., Gunchenko, A. I., and Panichev, N. A.,** Aerosol geochemical method of prospecting for ore deposits, *Geokhim. Metody Poiskov Sev. Raionakh Sib.*, 153, 1984. **Polikarpochkin, V. V., Ed.,** Nauka Sib. Otd. Novosibirsk, U.S.S.R., CA 12, 102:188248p, 1985.

87. **Tsunoda, K., Haraguchi, H., and Fuwa, K.,** Halide interferences in an electrothermal graphite furnace atomic absorption spectroscopy with group III B elements as studied by atomic absorption signal profiles, *Spectrochim. Acta*, 40B, 1651, 1985.

88. **Littlejohn, D., Stephen, S. C., and Ottaway, J. M.,** Slurry sample introduction procedures for the analysis of foods by electrothermal atomization atomic absorption spectrometry, *Anal. Proc. (London)*, 22(12), 376-8, 1985.

89. **Burba, P. and Willmer, P. G.,** Analytical multielement preconcentration on metal hydroxide-coated cellulose, *Fresnius' Z. Anal. Chem.*, 321(2), 109-18, 1985.

90. **Suzuki, M., Ohta, K., and Isobe, K.,** Mechanism of interference elimination by thiourea in electrothermal AAS, *Anal. Chim. Acta*, 173, 321-5, 1985.

91. **De Kersabiec, A. M., Blanc, G., and Pinta, M.,** Water analysis by Zeeman AAS, *Fresnius' Z. Anal. Chem.*, 322(7), 7331-5, 1985.

92. **Wibetoe, G. and Langmyhr, F. J.,** Spectral interferences and background over-compensation in inverse Zeeman corrected AAS. II. The effect of cobalt, manganese and nickel on 30 elements and 53 element lines, *Anal. Chim. Acta*, 176, 33-40, 1985.

93. **Easer, P.,** Direct analysis of inorganic solids: experiences of a lab in cement industries, *Fresnius' Z. Anal. Chem.*, 322(7), 677-80, 1985.

94. **Frech, W., Lundberg, E., and Cedergren, A.,** Investigation of interference effects in the presence of sodium and copper chlorides using platform graphite furnace AAS, *Can. J. Spectrosc.*, 30(5), 123-9, 1985.

95. **Carroll, J., Miller-Ihli, N. J., Harnly, J. M., Littlejohn, D., Ottaway, J. M., and O'Haver, T. C.,** Simultaneous multielement analysis by continuum source atomic absorption spectrometry with graphite probe electrothermal atomization, *Analyst*, 110, 1153, 1985.

96. **Yudelevich, I. G., Katskov, D. A., Papina, T. S., Kopeikin, V. A., and Vasil'eva, L. A.,** Determination of volatile elements in hydrobromic acid solutions by atomic absorption spectrometrry with electrothermal atomization, *Izv. Sib. Otd. Akad. Nauk SSSR Ser. Khim. Nauk*, 3, 104, 1985 (in Russian).

97. **Chiba, I.,** Clean analysis of toxic metals — determination of toxic metals by zirconium hydroxide coprecipitation. I, *Fukushima-ken Eisei Kogai Kenkyusho Nenpo*, 2, 105-11, 1984, published 1985.

98. **Tsunoda, K., Haraguchi, H., and Fuwa, K.,** Halide interferences in an electrothermal graphite furnace atomic absorption with Group III A elements as studied by atomic and molecular absorption signal profiles, *Spectrochim. Acta*, 40B (10-12), 1651-61, 1985.

99. Analysis of impurities in molybdenum, Nippon Telegraph & Telephone Public Corporation, Jpn. Kokai Tokkyo Koho JP 60 86,466 (85 86,466) (Cl. G01N33/20), 16 May 1985. Appl. 83/193,386, 18 Oct. 1983, 4pp.

100. **Hirota, S. and Seguchi, K.,** Adsorption of metallic ions on wool, *Mukogawa Joshi Daigaku Kiyo, Hifuku-hen*, 32, 69-76, 1984, published in Mukogawa Joshi Daigaku Kiyo, Shokumotsuhen, 32. CA 24, 103:161733b, 1985.

101. **Okamoto, K. and Fuwa, K.,** Mussel tissue powder, a certified reference material, *Analyst (London)*, 110(7), 785-9, 1985.

AUTHOR INDEX

SUBJECT INDEX

TRACE ELEMENTS

REFERENCES

1. **Welz, B. and Wiedeking, E.,** Determination of trace elements in serum and urine with flameless atomization, *Z. Anal. Chem.,* 252, 171, 1970.
2. **Morrison, G. H. and Talmi, Y.,** Microanalysis of solids by atomic absorption and emission spectroscopy using a radiofrequency induction furnace, *Anal. Chem.,* 42, 8009, 1970.
3. **Segar, D. A.,** The determination of trace metals in saline waters and biological tissues using the heated graphite atomizer, AIAA Paper No. 71-1051, 1971.
4. **Belyaev, U. I., Pehlintsev, A. M., Zvereva, N. F., and Kostin, B. I.,** Atomic absorption determination of traces of elements in rocks using the pulsed thermal atomization of solid samples. Determination of cadmium, *Zh. Anal. Khim.,* 26, 492, 1971.
5. **Segar, D. A. and Gonzalez, J. G.,** Evaluation of atomic absorption with a heated graphite atomizer for the direct determination of trace transition metals in sea water, *Anal. Chim. Acta,* 58, 7, 1972.
6. **Goncharova, N. N., Shipitsyn, S. A., and Shpeizer, G. M.,** Direct atomic absorption determination of some trace elements in natural waters using a furnace flame atomizer, *Gidrokhim. Mater.,* 59, 158, 1973.
7. **Medina, R.,** Determination of some trace elements in pure tin by atomic absorption with a graphite furnace, *Z. Anal. Chem.,* 271, 346, 1974 (in German).
8. **Ashy, M. A., Headridge, J. B., and Sowerbutts, A.,** Determination of trace and minor elements in alloys by atomic absorption spectroscopy using an induction heated graphite well furnace as an atom source, *Talanta,* 21, 649, 1974.
9. **Lundgren, G. and Johansson, G.,** A temperature controlled graphite tube furnace for the determination of trace metals in solid biological tissue, *Talanta,* 21, 257, 1974.
10. **Dudas, M. J.,** Effect of drying parameters on the heated graphite furnace determination of trace metals in MIBK-APDC solutions, *At. Ab. Newsl.,* 13, 67, 1974.
11. **Runnels, J. H., Merryfield, R., and Fisher, H. B.,** Analysis of petroleum for trace metals. Method for improving detection limits for some elements with the graphite furnace atomizer, *Anal. Chem.,* 47, 1258, 1975.
12. **Pardhan, S. I. and Ottaway, J. M.,** Determination of trace elements in soaps and phosphate materials by carbon furnace atomic absorption spectroscopy, *Proc. Anal. Div. Chem. Soc.,* 12, 291, 1975.
13. **Van der Piepen, H. and Radzuik, B.,** A quartz "T" tube furnace-atomic-absorption spectroscopy system for metal speciation studies, *Can. J. Spectrosc.,* 21, 46, 1976.
14. **Koirtyohann, S. R., Wallace, G., and Hinderberger, E.,** Multielement analysis of *Drosophila* for environmental monitoring purposes using carbon furnace atomic absorption, *Can J. Spectrosc.,* 21, 61, 1976.
15. **Sperling, K. R.,** Determination of heavy metals in seawater and marine organisms by flameless atomic absorption spectrophotometry. I. Attempts to obtain improved reproducibility by a control of gas movements in the graphite furnace, *At. Ab. Newsl.,* 15, 1, 1976.
16. **Noller, B. N. and Bloom, H.,** Sampling of metal air particulates for analysis by furnace atomic absorption spectroscopy, *Anal. Chem.,* 49, 346, 1977.
17. **Holak, W.,** Temperature programmable furnace for ashing of foods in trace metal analysis, *J. Assoc. Off. Anal. Chem.,* 60, 239, 1977.
18. **Hudnik, V., Gomiscek, S., and Gorenc, B.,** The determination of trace metals in mineral waters. I. Atomic absorption spectrometric determination of cadmium, cobalt, chromium, copper, nickel and lead by electrothermal atomization after concentration by coprecipitation, *Anal. Chim. Acta,* 98, 39, 1978.
19. **Danielsson, L. G., Magnusson, B., and Westerlund, S.,** An improved metal extraction procedure for the determination of trace metals in sea water by atomic absorption spectroscopy with electrothermal atomization, *Anal. Chim. Acta,* 98, 47, 1978.
20. **Golimowski, J., Valenta, P., Stoeppler, M., and Nuernberg, H. W.,** Toxic trace metals in food. II. A comparative study of the levels of toxic trace metals in wine by differential pulse anodic stripping voltammetry and electrothermal atomic absorption spectroscopy, *Z. Lebensm. Unters. Forsch.,* 168, 439, 1979.
21. **Forrester, J. E., Lehecka, V., Johnston, J. R., and Oh, W. L.,** Direct determination of trace quantities of antimony, arsenic, bismuth, cadmium, lead, selenium, silver, tellurium and thallium in high-purity nickel by electrothermal atomic absorption spectroscopy, *At. Ab. Newsl.,* 18, 73, 1979.
22. **Headridge, J. B.,** Determination of trace elements in metals by atomic absorption spectroscopy with introduction of solid samples into furnaces: an appraisal, *Spectrochim. Acta,* 35B, 785, 1980.
23. **Sturgeon, R. E., Berman, S. S., Desaulniers, A., and Russell, D. S.,** Preconcentration of trace metals from seawater for determination by graphite furnace atomic absorption spectroscopy, *Talanta,* 27, 85, 1980.
24. **Hoenig, M. and Dupire, S.,** Effect of complex matrixes in trace element determination by electrothermal atomic absorption spectroscopy. I. The background, *Analysis,* 8, 16, 1980.
25. **Chakrabarti, C. L., Wan, C. C., and Li, W. C.,** Direct determination of traces of Cu, Zn, Pb, Co, Fe and Cd in bovine liver by graphite furnace atomic absorption spectroscopy using the solid sampling and the platform techniques, *Spectrochim. Acta,* 35B, 93, 1980.

26. **Uratani, F., Yoshinaka, T., and Miyagi, M.,** The determination of trace amounts of elements by atomic absorption spectrochemical analysis. IV. Graphite furnace atomizer equipped with temperature feedback controller, *Osaka-Furitsu Kogyo Gijutsu Kenkyusho Hokoku,* 77, 22-7, 1980 (in Japanese).

27. **Sekiya, T., Tanimura, H., and Hikasa, Y.,** Study on the dosage of trace elements in total parenteral nutrition, *Jutsugo Taisha Kenkyu Kaishi,* 14, 114, 1980.

28. **Slovak, Z. and Docekal, B.,** Determination of trace metals in aluminum oxide by electrothermal atomic absorption spectroscopy with direct injection of aqueous suspension, *Anal. Chim. Acta,* 129, 263, 1981.

29. **Khrapai, V. P., Provodenko, L. B., Kharitonova, V. A., and Babayants, T. A.,** Determination of trace impurities in high-purity gold and silver by electrothermal atomic absorption spectroscopy, *Zavod. Lab.,* 47, 33, 1981.

30. **Shinji, H.,** A preliminary study for determination of trace elements in primary teeth. Chromatographic separation of cadmium, zinc, lead and copper on a cation-exchange resin with a mixed eluent of hydrochloric acid and acetone and determination of trace elements by flameless atomic absorpotion spectrophotometry, *Kanagawa Shigaku,* 16, 1-14, 1981 (in Japanese).

31. **Satsmadjis, J. and Voutsinou-Taliadouri, F.,** Determination of trace metals at concentrations above the linear calibration range by electrothermal atomic absorption spectrometry, *Anal. Chim. Acta,* 131, 83-90, 1981.

32. **Koster, P. B., Raats, P., Hibbert, D., Phillipson, R. T., Schiweck, H., and Steinle, G.,** Collaborative study on the determination of trace elements in dried sugar beet pulp and molasses. III. Lead, *Zuckerindustrie (Berlin),* 106, 895, 1981 (in German).

33. **Sturgeon, R. E., Desaulniers, J. A., Berman, S. S., and Russell, D. S.,** Determination of trace metals in estuarine sediments by graphite furnace atomic absorption spectroscopy, *Anal. Chim. Acta,* 134, 283-91, 1982.

34. **Danielsson, L. G., Magnusson, B., and Zhang, K.,** Matrix interference in the determination of trace metals by graphite furnace AAS after Chelex-100 preconcentration, *At. Spectrosc.,* 3, 39, 1982.

35. **Voellkopf, U., Grobenski, Z., and Welz, B.,** Ways of interference-free determination of trace elements in wastewaters using graphite furnace AAS, *GIT Fachz. Lab.,* 26, 444, 1982 (in German).

36. **Erzinger, J. and Puchelt, H.,** Methods for the determination of trace elements in geological materials, *Erzmetall,* 35, 173-9, 1982 (in German).

37. **Hoenig, M. and Wollast, R.,** The possibilities and limitations of electrothermal atomization in atomic absorption spectroscopy for the direct determination of trace metals in seawater, *Spectrochim. Acta,* 37B, 399-415, 1982.

38. **Raptis, S. E., Kaiser, G., and Toelg, G.,** The decomposition of oils and fats in a stream of oxygen for the determination of trace elements, *Anal. Chim. Acta,* 138, 93-101, 1982.

39. **Takada, K. and Hirokawa, K.,** Atomization and determination of traces of copper, manganese, silver and lead in microamounts of steel. Atomic-absorption spectroscopy using direct atomization of solid sample in a graphite-cup cuvette, *Fresnius' Z. Anal. Chem.,* 312, 109-13, 1982.

40. **Mahalingam, T. R., Geetha, R., Thiruvengadasamy, A., and Mathews, C. K.,** Analysis of trace metals in sodium by flameless atomic absorption spectrophotometrry, *Proc. Conf. Mater. Behav. Phys. Chem. Liq. Met. Syst.,* Borgstedt, H. U., Ed., Plenum Press, N.Y., 1982, 329-34.

41. **Sturgeon, R. E., Berman, S. S., and Willie, S. N.,** Concentration of trace metals from seawater by complexation with 8-hydroxyquinoline and adsorption on C-18 bonded silica gel, *Talanta,* 29, 167-71, 1982.

42. **Nath, R.,** Preparation of biological samples and use of standard reference material in trace analysis, *Bull. Postgrad. Inst. Med. Educ. Res. Chandigarh,* 16, 134, 1982.

43. **Menz, D. and Conradi, G.,** Determination of trace elements as a geochemical method for classifying crude oils, *Atomspektrom. Spurenanal., Vortr., Kolloq.,* Welz, B., Ed., Verlag, Springer-Weinheim, 1982, 489-510.

44. **Raptis, S. E., Wrgscheider, W., and Knapp, G.,** Comparison of methods for determining trace elements in biological and organic samples by using wet chemical decomposition and graphite-furnace atomic absorption spectroscopy, *Atomspektrom. Spurenanal., Vortr. Kolloq.,* Welz, B., Ed., Verlag, Springer-Weinheim, 1982, 511-22

45. Determination of trace elements in coal and coal products, National Energy Research Development and Demonstration Council, Report 1982, NERDDP-EG-81-20. Avail. Natl. Energy Res. Dev. Demonstr. Council., Canberra, Aust. From Energy Res. Abstr., 9, 1984. Abstr. No. 4324.

46. **Yates, J.,** Determination of trace metals (cadmium, copper, lead, nickel and zinc) in filtered saline water samples, *Tech. Rep. TR-Water Res. Cent. (Medmenham, England)* TR181, 63, 1982.

47. **Wennrich, R. and Dittrich, K.,** Simultaneous determination of traces in solid samples with laser-AAS, *Spectrochim. Acta,* 37B, 913, 1982.

48. **Mahalingam, T. R., Geetha, R., Thiruvengadasamy, A., and Mathews, C. K.,** Determination of trace metals in sodium by electrothermal atomic absorption spectroscopy, *Anal. Chim. Acta,* 142, 189, 1982.

49. **Akguen, E. and Pindur, U.,** Trace analysis of lead, cadmium and thallium in active ingredients and pharmaceutical preparations by atomic absorption spectroscopy, I, *Pharm. Ind.,* 44, *930, 1982.*

50. **Kingston, H. M., Greenberg, R. R., Beary, E. S., Hardas, B. R., and Moody, J. R.,** The characterization of the Chesapeake Bay: a systematic analysis of toxic trace elements, Report 1982, EPA-600/3-82-085; Order No. PB82-26565, 210pp, 1982. Avail. NTIS. From Gov. Rep. Announce. Index (U.S.), 82, 5450, 1982.

361

51. **Danielsson, L. G., Magnusson, B., Westurlund, S., and Zhang, K.,** Trace metal determinations in estuarine waters by electrothermal atomic absorption spectroscopy after extraction of dithiocarbamate complexes into freon, *Anal. Chim. Acta,* 144, 183, 1982.
52. **Boutron, C.,** Atmospheric trace metals in the snow layers deposited at the South Pole from 1928 to 1977, *Atmos. Environ.,* 16, 2451, 1982.
53. **Nomoto, S.,** Sample preparation for analysis of trace metals in biological materials for electrothermal atomic absorption spectrophotometry and its detection limits, *Proc. 11th Int. Congr. Clin. Chem.,* Kaiser, E., Gabl. F., and Mueller, M. M., Eds., Berlin, 1982, 1097.
54. **Hocquaux, H.,** Determination of some trace elements or elements present in low amounts in steels, ferrous metallurgical products, and industrial waters by flameless atomic absorption spectrophotometry, *Comm. Eur. Communities (Rep.) EUR 1982,* EUR7885, 71 1982 (in French).
55. **Pszonnicki, L., Veglia, A., and Suschny, O.,** Report on intercomparison W-3/1 of the determination of trace elements in water, Report 1982 IAEA-RL-94, 62pp. Avail. INIS. From INIS Atomindex 13(18), 1982. Abstr. No. 694862.
56. **Ward, N. I. and Minski, M. J.,** Comparison of trace elements in whole blood and scalp hair of multiple sclerosis patients and normal individuals, *Trace Subst. Environ. Health,* 16, 252, 1982.
57. **Danielsson, L. G., Magnusson, B., and Zhang, K.,** Matrix interference in the determination of trace metals by graphite furnace AAS after chelex-100 preconcentration, *At. Spectrosc.,* 3, 39-40, 1982.
58. **Burton, J. D., Maher, W. A., and Statham, P. J.,** Some recent measurements of trace metals in Atlantic ocean waters, *NATO Conf. Ser.,* 4, 9, 1983.
59. **Willie, S. N., Sturgeon, R. E., and Berman, S. S.,** Comparison of 8-quinolinol-bonded polymer support for the preconcentration of trace metals from seawater, *Anal. Chim. Acta,* 149, 59, 1983.
60. **Parks, E. J., Johannesen, R. B., and Brinckman, F. E.,** Characterization of organometallic coplymers and coplymerization by size-exclusion chromatography coupled with trace metal- and mass-sensitive detectors, *J. Chromatogr.,* 255, 439, 1983.
61. **Manning, D. C. and Slavin, W.,** The determination of trace elements in natural waters using the stabilized temperature platform furnace, *Appl. Spectrosc.,* 37, 1-11, 1983.
62. **Patel, B. M., Page, A. G., Bangia, T. R., Sastry, M. D., and Joshi, B. D.,** Determination of trace metals in uranium by electrothermal atomic absorption spectroscopy, *1st Int. Symp. Trace Anal. Technol. Dev. Spec. Contrib. Pap.,* Sanker, D. M., Ed., John Wiley & Sons, N.Y., 1983, 290-5.
63. **Mahalingam, T. R., Geetha, R., Thiruvengadasamy, A., and Mathews, C. K.,** Determination of trace metals in sodium by flameless atomic absorption spectrophotometry, *1st Int. Symp. Trace Anal. Technol. Dev. Spec. Contrib. Pap.,* Sanker, D. M., Ed., John Wiley & Sons, N.Y., 1983, 285-9.
64. **Schroen, W., Bombach, G., and Benge, P.,** Fast method for determination of trace elements in geological samples by graphite furnace AAS, *Spectrochim. Acta,* 38B, 1269, 1983 (in German).
65. **Brodie, K., Doidge, P., McKenzie, T., and Routh, M.,** Trace metal analysis by graphite furnace atomic absorption, *Varian Instrum. Appl.,* 17, 24-6, 1983.
66. **Ali, S. L.,** Determination of pesticide residues and other critical impurities- such as toxic trace metals-in medicinal plants. II. Determination of toxic trace metals in drugs, *Pharm. Ind.,* 45(12), 1294, 1983.
67. **Voellkopf, U., Grobenski, Z., and Welz, B.,** The stabilized temperature platform furnace with Zeeman effect background correction for trace analysis in wastewater, *At. Spectrosc.,* 4(5), 165-70, 1983.
68. **Dungs, K., Lippmann, C., and Neidhart, B.,** Speciation analysis by graphite furnace AAS, *Proc. 2nd Int. Workshop Trace Elem. Anal. Chem. Med. Biol.,* Braetter, P. and Schramel, P., Eds., Berlin, 1983.
69. **Hatano, S.,** Studies on trace element metabolism in childhood. II. Trace elements (copper, zinc, manganese and selenium), in blood relation to dietary intake during infancy, *Hiroshima Daigaku Igaku Zasshi,* 31(5), 825-33, 1983.
70. **Atsuya, I.,** Direct determination of trace elements in powdered biological samples by atomic absorption spectrometry with a graphite miniature cup, *Kogai to Taisaku,* 19(10), 939-45, 1983.
71. **Bykov, I. V., Skvortsov, A. B., Tatsii, Y. G., and Chekalin, N. V.,** Metal trace analysis by flame/graphite furnace OG spectroscopy, *J. Phys. Colloq.,* C7, 345-52, 1983.
72. **Mohr, F., Luft, B., and Bombach, H.,** Determination of trace concentrations in highly purified phosphorus using a hydride system and the AAS-1N atomic absorption spectrophotometer, *Jena Rev.,* 28(3), 120-2, 1983.
73. **Page, A. G., Godbole, S. V., Kulkarni, M. J., Porwal, N. K., Shelar, S. S., and Joshi, B. D.,** Trace metal assay of uranium oxide (U_3O_8) powder by electrothermal AAS, *Talanta,* 30(10), 783-6, 1983.
74. **Bettinelli, M.,** Determination of trace metals in siliceous standard reference materials by electrothermal atomic absorption spectroscopy after lithium tetraborate fusion, *Anal. Chim. Acta,* 148, 193, 1983.
75. **Shroen, W., Bombach, G., and Beuge, P.,** Rapid flameless atomic absorption spectroscopic study of trace elements in geological samples, *Spectrochim. Acta,* 38B(10), 1269-76, 1983.
76. **Sarx, B. and Baechmann, K.,** Speciation of trace elements in biological matrices, *Proc. 2nd Int. Workshop Trace Elem. Anal. Chem. Med. Biol.,* Braetter, P. and Schramel, P., Eds., Berlin, 1983, 713-19.
77. **Headridge, J. B.,** Determination of ultratrace elements in nickel alloys using atomic absorption spectrometric methods and solid samples, *Anal. Proc. (London),* 20, 207, 1983.

78. **Okuda, S., Hoshiba, E., Terata, K., and Sakai, S.,** Chemotaxonomic studies on dermapterous and orthopteroid insects. VIII. Some trace elements of Dermapterous and Orthpteroid species (2), Daito Bunka Daigaku Kiyo, *Shakai Shizen Kagaku,* 21, 85, 1983 (in Japanese).

79. **Fuchs, C., Armstrong, V. W., Hein, H., and Kraft, B.,** Problems in measuring and determining trace elements in body fluids, *Nieren-Hochdruckkrankh,* 12, 164, 1983.

80. **Toshiba Corpn.,** Determination of trace elements in gallium arsenide, Jpn. Kokai Tokkyo Koho JP58 00,740 [83 00,740] (Cl. G01N21/31), 5pp, 5 Jan. 1983. Appl. 81/98,356, 26 Jun. 1981.

81. **Jervis, R. E., Landsberger, S., Aufreiter, S., Van Loon, J. C., Lecomte, R., and Monaro, S.,** Trace elements in wet atmospheric deposition: application and comparison of PIXE, INAA and graphite-furnace AAS techniques, *Int. J. Environ. Anal. Chem.,* 15, 89-106, 1983.

82. **Kunic, M. and Fable, T.,** Determination of trace metals in petroleum distillates by electrothermal atomic absorption, *Goriva Maziva,* 22, 81-8, 1983 (in Serbo-Croatian).

83. **Rains, T. C., Rush, T. A., and Butler, T. A.,** Determination of selected trace elements in human liver samples by atomic absorption spectroscopy, *NBS Spec. Publ. (U.S.),* 656, 39-45, 1983.

84. **Kanipayor, R., Naranjit, D. A., Radzuik, B. A., Van Loon, J. C., and Thomassen, Y.,** Direct analysis of solids for trace elements by combined electrothermal furnace/quartz T-tube/flame atomic absorption spectrometrry, *Anal. Chim. Acta,* 166, 39, 1984.

85. **Blackmore, D. J. and Stanier, P.,** Methods for the measurement of trace elements in equine blood by electrothermal atomic absorption spectrophotometry, *At. Spectrosc.,* 5, 215, 1984.

86. **Marumo, F., Tsukamoto, Y., Iwanami, S., Kishimoto, T., and Yamagami, S.,** Trace element concentrations in hair, fingernails and plasma of patients with chronic renal failure on hemodialysis and hemofiltration, *Nephron,* 38, 267, 1984.

87. **Grobenski, Z., Lehmann, R., Radzuik, B., and Voellkopf, U.,** Determination of trace metals in seawater using Zeeman graphite furnace AAS, *At. Spectrosc.,* 5, 87, 1984.

88. **Niwa, M., Kajimoto, M., Tmai, T., Hirasawa, K., and Inoue, M.,** Analysis of trace metals in human teeth. III. Cadmium concentration in human teeth by flameless atomic absorption spectroscopy, *Shigaku,* 71, 1083, 1984.

89. **Hudnik, V., Marolt-Gomiscek, M., and Gomiscek, S.,** The determination of trace metals in human fluids and tissues. I. Estimation of 'normal values' for copper, zinc, cadmium and manganese in blood serum and liver tissue, *Anal. Chim. Acta,* 157, 143-50, 1984.

90. **Karmanova, N. G. and Vratkovskaya, S. V.,** Atomic absorption determination of trace elements in organic matter of rocks using electrothermal atomization, *Zh. Anal. Khim.,* 39(2), 349, 1984.

91. **Goncharova, N. N., Shpeizer, G. M., Ivanova, E. I., and Potanina, L. V.,** Atomic absorption determination of trace metals in natural and industrial waters, *Khim. Tekhnol. Vody,* 6, 152, 1984.

92. **Shimazaki, A., Hiratsuka, H., Matsushita, Y., and Yoshii, S.,** Chemical analysis of ultratrace impurities in silicon dioxide films, *Ext. Abstr. Conf. Solid State Devices Mater.,* 16, 281-4, 1984.

93. **Chappuis, P. and Rousselet, F.,** Determination of trace elements by flameless atomic absorption spectro-photometry: principles, significance, problems and applications, *Tech. Biol.,* 10(4), 158, 1984 (in French).

94. **Mishima, M.,** Improved accuracy and precision in the determination of trace metals by GFAAS, *Koshu Eiseiin Kenkyu Hokoku,* 33(1), 16-23, 1984 (in Japanese).

95. **Voellkopf, U., Grobenski, Z., Schlemmer, G., and Welz, B.,** Determination of trace elements in biological samples using a GF in thermal equilibrium, *Fortschr. Atomspektrom. Spurenanal.,* 1, 375, 1984.

96. **Knutti, R.,** Matrix modification for the direct determination of ultra-trace elements in blood and urine by GFAAS, *Fortschr. Atomspektrom. Spurenanal.,* 1, 327, 1984 (in German).

97. **El Hossadi, A., Alian, A., Ali, S. S., Farooq, R., Hamid, A., and Majed, T. A.,** Studies on pollution of water and air dust of Benghazi using various analytical methods. I. The analysis of air dust samples from Benghazi for the trace elements using neutron activation analysis and flameless atomic absorption, *J. Radioanal. Nucl. Chem.,* 81(2), 359-68, 1984.

98. **Trefry, J. H. and Metz, S.,** Selective leaching of trace metals from sediments as a function of pH, *Anal. Chem.,* 56(4), 745-9, 1984.

99. **Hadula, E.,** Analytical problems connected with the content of trace elements in fodder plants, *Zesz. Probl. Postepow Nauk Roln.,* 276, 241-5, 1983, CA 6, 100:66703w, 1984.

100. **Hayashi, Y., Yabuta, Y., Shimizu, Y., and Shimao, S.,** Behavior of trace metals in leg lymph of rabbits: copper and zinc contents, *Igaku to Seibutsugaku,* 105(6), 405-8, 1982. CA 4, 100:325064u, 1984.

101. **Danielsson, L. G. and Westerlund, S.,** Short-term variations in trace metal concentrations in the Baltic, *Mar. Chem.,* 15(3), 273, 1984.

102. **Schlemmer, G., Voellkopf, U., and Welz, B.,** Trace element determination in marine biological tissue using graphite furnace AAS solid sampling versus sample decomposition, *5th Int. Conf. Heavy Met. Environ.,* 2, 487, 1985.

103. **Lam, B. and Zinger, M.,** Use flameless atomic absorption for trace metals with no sample work-up, *Res. Dev.,* 27(2), 150-3, 1985.

104. **Wan, C. C., Chiang, S., and Corsini, A.,** Two-column method for preconcentration of trace metals in natural waters on acrylate resin, *Anal. Chem.,* 57(3), 719-23, 1985.

105. **Maddalone, R. F., Moyer, P. W., and Scott, J. W.,** Aqueous discharges from steam electric power plants: trace metal sampling and analysis reference guide. Final report, Report EPRI-CS-3739, 1984. Order No. T1855992001133, 251pp. Avail. RRC, POB 50490, Palo Alto, CA. From Energy Res. Abstr., 10(4), Abstr. No. 7404, 1985.

106. **McWeeny, D. J., Crews, H. M., Massey, R. C., and Burrell, J. A.,** Effects of digestive enzymes on the solubility of trace elements in food, Proc. 5th Int. Symp.Trace Elem. Man Anim., 628-30, 1984, published 1985.

107. **Gunshin, H., Yoshikawa, M., Doudou, T., and Kato, N.,** Trace elements in human milk, cow's milk and infant formula, *Agric. Biol. Chem.,* 49(1), 21-6, 1985.

108. **Navas, M. J., Jiminez-Trillo, J. L., and Asnero, A. G.,** Determination of trace elements in milk by using atomic absorption spectrophotometry with electrothermal atomization, *Aliinentaria (Madrid),* 22, 103, 1985.

109. **Dittrich, K., Mothes, W., Yudelevich, I. G., and Papina, T. S.,** Investigation of trace analysis of $A_{III}B_V$ semiconductor microsamples by atomic spectroscopy. VII. Investigation of trace and thin-layer analysis of doping elements (Ag, Au, Bi, Cd, Sn, Tl) in InAs by atomic absorption with electrothermal atomization, *Talanta,* 32, 195, 1985.

110. **Heininger, P., Duennbier, V., and Henrion, G.,** Trace metal determination in hydrofluoric acid after sub-boiling distillation, *Z. Chem.,* 25(1), 33, 1985.

111. **Stoeppler, M.,** Trace metal analysis for the German environmental specimen bank, *Symp. Biol. Ref. Mater.,* Wolf, W. R., Ed., John Wiley & Sons, New York, 1985, 281-97.

112. **Dittrich, K., Hanisch, B., and Staerk, H. J.,** Trace analysis by nonthermal excitation methods, *ATOMKI Kozl.,* 27, 361, 1985 (in German).

113. **Benfenati, L.,** Use of the analytical technique of atomic absorption in the determination of trace elements present in cosmetic products and raw materials, *Prod. Chim. Aerosol Sel.,* 26, 25, 1985 (in Italian).

114. **Saito, N. and Kumatani, S.,** Analyses of trace metal compositions in blood. II. Contaminant levels in analytical containers in the determination of cadmium, lead and manganese, *Iwateken Eisei Kenkyusho Nenpo,* 26, 9-14, 1983 (in Japanese), CA 11, 102:163221m, 1985.

115. **Matsumoto, K.,** Development of new techniques in x-ray fluorescence and atomic absorption spectrometries for trace elements in biological materials, *Mem. Sch. Sci. Eng. Waseda Univ.,* 49, 1-19, 1985.

116. **Graf, H. J. and Reynolds, W. L.,** Techniques for ultratrace metals analyses in dopant materials, *Solid State Technol.,* 28(3), 141-7, 1985.

117. **Granat, L. and Ross, H. B.,** Methodology for the collection and analysis of trace metals in atmospheric precipitation, Report 1984, CM-67, 41pp. Avail. NTIS from Sci. Tech. Aerosp. Rep. 23(9), Abstr. No. N85-18491, 1985.

118. **Sztraka, A.,** Application of the L'vov platform to trace analysis in wastewater and sludge, Fortschr. Atomspektrom. Spurenanal., 1, 511-21, 1984, CA 17, 103:42222z, 1985.

119. **Borg, H.,** Background levels of trace metals in Swedish fresh waters, Report 1984, SNV-PM-1817; Order No. DE85750373, 57pp. Avail. NTIS. From Energy Res. Abstr., 10(8), Abstr. No. 14491, 1985.

120. **Kapeller, R., Purtscheller, F., and Schnell, E.,** Characterization of filter collected dust samples, *Mikrochim. Acta,* 1(1-2), 115-34, 1985.

121. **Korsrud, G. O., Meldrum, J. B., Salisbury, C. D., Houlahan, B. J., Saschenbrecker, P. W., and Tittiger, F.,** Trace element levels in liver and kidney from cattle, swine and poultry slaughtered in Canada, *Can. J. Comp. Med.,* 49(2), 159-63, 1985.

122. **Grobecker, K. H. and Kluessendorf, B.,** Routine determination of trace metals with Zeeman Atomic Absorption Spectroscopy (ZAAS), *Labor Praxis,* 9(11), 1306-8, 1985.

123. **Grobecker, K. H. and Kluessendorf, B.,** Trace heavy metals in marine food stuffs of various origin. Determination of cadmium, lead and mercury in fresh and dried material by direct solid sampling analysis, *Fresnius' Z. Anal. Chem.,* 322(7), 673-6, 1985.

124. **Grobecker, K. H. and Muntau, H.,** Direct Zeeman Atomic Absorption Spectroscopic determination of trace element in suspended matter collected on organic filter materials, *Fresnius' Z. Anal. Chem.,* 322(7), 728-30, 1985.

125. **Desaulniers, J. A. H., Sturgeon, R. E., and Berman, S. S.,** Atomic absorption determination of trace metals in marine sediments and biological tissues using stabilized temperature platform furnace, *At. Spectrosc.,* 6, 125, 1985.

126. **Takada, S., Demura, R., and Yamamoto, I.,** Rapid determination of trace metals in foods by microwave oven digestion method. I. Determination of arsenic, *Eisei Kagaku,* 31, 37, 1985.

127. **Auerswald, D. C. and Cutler, F. M.,** Characterization of steam cycle chemistry at Southern California Edison steam plants, Report EPRI-CS=4005, 1985.

128. **Del Monte Tamba, M. G.,** Determination of trace elements in materials of interest to steel making by flameless AAS directly on compact specimens, *Comm. Eur. Communities [Rep.] EUR 1985,* 9893, 218, 1985 (in French).

129. **Ososkov, V. K., Plintus, A. M., Kornelli, M. E., and Zakhariya, A. N.,** Application of finely-dispersed ion exchangers for flotational preconcentration of trace elements, *Ukr. Khim. Zh.,* 51(12), 1298-301, 1985 (in Russian).

130. **Benfenati, L.,** Use of the analytical technique of atomic absorption in the determination of trace elements present in cosmetic products and raw materials, *Prod. Chim. Aerosol Sel.,* 26(1-2), 25-8, 1985.

131. **Sokolova, I. S. and Georgievskii, V. V.,** Flameless atomic absorption determination of traces of nonionic synthetic surfactants in seawater, *Aktual Probl. Okeanol.,* M. 116-18, 1984, From Ref. *Zh. Khim.* Abstr. No. 121191, 1985.

132. **Hinds, M. W., Jackson, K. W., and Newman, A. P.,** Electrothermal atomization atomic absorption spectroscopy with the direct introduction of slurries. Determination of trace metals in soil, *Analyst (London),* 110(8), 947-50, 1985.

133. **Itai, K., Tsunoda, H., and Ikeda, M.,** Effect of matrix modifier and furnace material on the determination of traces of fluoride by electrothermal molecular absorption spectrometry of aluminum monofluoride, *Anal. Chim. Acta,* 171, 293-301, 1985.

134. **Ottaway, J. M., Carroll, J., Cook, S., Littlejohn, D., Marshall, J., and Stephen, S. C.,** Some novel approaches to trace element analysis, *Anal. Proc. (London),* 22(7), 192-4, 1985.

135. **Mingorance, M. D. and Lachica, M.,** Direct determination of some trace elements in milk by EAAS, *Anal. Lett.,* 18(A12), 1519-31, 1985.

136. **Jaffar, M., Ashraf, M., and Tariq, M.,** A basic medium reduction-based method for multimetal trace analysis of natural waters, *At. Spectrosc.,* 7(4), 96, 1986.

137. **Welz, B., Schlemmer, G., and Voelkopf, U.,** Trace element determination in biological materials using stabilized temperature platform furnace AAS and Zeeman effect background correction, *Acta Pharmacol. Toxicol. Suppl.,* 59, 589, 1986.

138. **Berger, H., Meyberg, F., and Dannecker, W.,** Effects of some typical environmentally relevant matrixes on trace element analysis by graphite-tube furnace AAS, *Fortschr. Atomspektrom. Spurenanal.,* 2, 607, 1986.

139. **Seong Lee, B., Yoshimura, E., Tanaka, Y., Saitoh, J., Yamazaki, S., and Toda, S.,** Combination of wet charring and dry ashing suitable for the determination of trace metals in oily foodstuffs by graphite furnace AAS, *Bunseki Kagaku,* 35, T120-T123, 1986 (in Japanese).

140. **Hiraide, M., Tschoepel, P., and Toelg, G.,** Separation of trace elements from high-purity metals by simultaneous electrolytic dissolution and electrodeposition of the matrix on a mercury cathode. I. Determination of aluminum in iron by electrothermal atomic absorption spectroscopy, *Anal. Chim. Acta,* 186, 261-6, 1986.

141. **Kobayashi, T., Ide, K., and Okochi, H.,** Determination of trace elements in steels by graphite furnace atomic absorption spectroscopy with solution method using very small samples of mg order, *Nippon Kinzoku Gakkaishi,* 50, 921-6, 1986 (in Japanese).

142. **Nagohrney, S. J., Heit, M., and Bogen, D. C.,** Electrothermal atomic absorption spectroscopic analysis of trace metals in Adirondack Mountain Lakes, *Spectroscopy (Springfield, OR),* 1, 41-5, 1986.

143. **Bettinelli, M., Pastorelli, N., and Baroni, U.,** Determination of trace metals in sediment standard reference materials by graphite-furnace atomic absorption spectroscopy with a stabilized temperature platform, *Anal. Chim. Acta,* 185, 109-17, 1986.

144. **Magnusson, I., Axner, O., Lindgren, I., and Rubinsztein-Dunlop, H.,** Laser-enhances ionization detection of trace elements in a graphite furnace, *Appl. Spectrosc.,* 40, 968-71, 1986.

145. **Tamura, H., Inui, T., and Fudagawa, N.,** Determination of trace elements in coals and fly ashes by metal furnace AAS, *Shizuoka-ken Kogyo Gijutsu Senta Kenkyo Hokoku,* 30, 73-9, 1986 (in Japanese), CA 23, 105:155862t, 1986.

146. Critical evaluation of analytical methods for determination of trace elements in various matrixes. Determination of manganese in biological materials, *Int. Union Pure Appl. Chem.,* 58(9), 1307-16, 1986.

147. **Hoenig, M. and Van Hoeyweghen, P.,** Alternative to solid sampling for trace metal determination by platform electrothermal AAS: direct dispensing of powdered samples suspended in liquid medium, *Anal. Chem.,* 58, 2614, 1986.

148. **Hung, J. J.,** Chemical speciation of dissolved trace metals in soils and sludge-amended soils, *Chung-kuo Nung Yeh Hua Hsueh Hui Chih,* 23(1-2), 73-84, 1986.

149. **Stein, K. and Umland, F.,** Trace analysis of lead, cadmium and manganese in honey and sugar, *Anal. Chem.,* 323(2), 176, 1986.

150. **Awadallah, R. M., Sherif, M. K., Amrallah, A. H., and Grass, F.,** Determination of trace elements of some Egyptian crops by instrumental neutron activation, ICP-AES and flameless AAS analysis, *J. Radioanal. Nucl. Chem.,* 98(2), 235-46, 1986.

151. **Bettinelli, M., Pastorelli, N., and Broni, U.,** STPF determination of trace metals in fly ash samples, *At. Spectrosc.,* 7(2), 45-8, 1986.

152. **Grobecker, K. H. and Kluessendorf, B.,** Routine determination of trace metals with ZAAS (Zeeman Atomic Absorption spectroscopy), *LP Spec: Chromatogr. Spektrosk.,* 214-17, 1986 (in German).

153. **Subramanian, K. S.,** Determination of trace metals in human blood by graphite furnace atomic absorption spectrometrry, *Prog. Anal. Spectrosc.,* 9(2), 237-334, 1986.

154. **Ciurea, I. C., Humbert, B. E., and Lipka, Y. F.,** Trace analysis of mineral elements in raw materials and finished products in the chocolate industry. I. Study of the mineralization of high fat content samples with HPA apparatus for the determination of lead and cadmium ultratrace content by electrothermal atomic absorption spectroscopy, *Mitt. Geb. Lebensmittelunters. Hyg.,* 77, 509, 1986.

155. **Fishman, M. J., Perryman, G. R., Schroder, L. J., and Matthews, E. W.,** Determination of trace metals in low ionic strength waters using Zeeman and deuterium background correction for graphite furnace atomic spectroscopy, *J. Assoc. Off. Anal. Chem.,* 69(4), 704-8, 1986.

156. **Nakamura, Y. and Kobayashi, Y.,** Determination of trace amounts of impurities in gallium arsenide crystals by flameless atomic absorption spectrometrry, *Bunseki Kagaku,* 35(5), 446-50, 1986 (in Japanese).

157. **Nakamura, Y., Kobayashi, Y., and Abe, K.,** Determination of trace impurities in gallium by tungsten furnace AAS (atomic absorption spectroscopy), *Bunseki Kagaku,* 35, 685, 1986.

158. **Muzgin, V. N., Adriashev, V. B., Pupyshev, A. A., and Atnashev, Y. B.,** Atomic absorption trace analysis with a coil tungsten atomizer modified with carbon, *Zh. Anal. Khim.,* 41, 1798, 1986.

159. **Lin, J., Johnson, K. W., and Huang, Z.,** Determination of trace metals in seawater by using graphite-furnace atomic absorption spectroscopy after extraction with APDC/DDDC into Freon TF, *Huanjing Kexue,* 7, 75-9, 1986.

160. **Fuwa, K. and Tsunoda, K.,** Recent advances on the study of trace elements and their spectrochemical analysis, *Yokugaku Zasshi,* 106, 951, 1986.

161. **Brodie, K. G., Routh, M. W., and Pohl, B.,** Trace element analysis with graphite tube atomic absorption spectroscopy. Lead in whole blood, manganese and aluminum in serum and chromium in urine, *Labor Med.,* 9, 218-229, 1986 (in German).

162. **Brueggerhoff, S. and Jackwerth, E.,** Multielement trace analysis of optical glass, *Fresnius' Z. Anal. Chem.,* 326, 528, 1987.

AUTHOR INDEX

SUBJECT INDEX

WEAR METALS

REFERENCES

1. **Reeves, R. D., Molnar, C. J., Glenn, M. T., Ahlstrom, J. R., and Winefordner, J. D.,** Determination of wear metals in engine oils by atomic absorption spectroscopy with a graphite rod atomizer, *Anal. Chem.,* 44, 2205, 1972.
2. **Sotera, J. J., Corum, T. L., and Kahn, H. L.,** The determination of wear metals and metal contaminants in organic materials, IL Inc., Rep. No. 16, Dec. 1979.
3. **Suhardono, E. and Djunaeni, A.,** Spectrophotometric determination of several kinds of contaminants in used lubricating oils, *Lembaran Publ. Lemigas,* 16, 26, 1982.
4. **Holding, S. T. and Rowson, J. J.,** The determination of additive elements in unusual lubricating oils by flameless atomic absorption spectroscopy, *Proc. Inst. Pet. London,* 81, 227-31, 1982.
5. **Rovid, K., Graf Harsanyi, E., Polos, L., Fodor, P., and Pungor, E.,** Determination of metallic components in lubricating oils by atomic absorption spectroscopy, *Magy. Kem. Foly,* 88, 39-42, 1982 (in Hungarian).
6. **Lima, E.,** Evaluation of a continuous current plasma for determination of elements in crude oil and its products, *Rev. Tec. INTEVEP,* 3(2), 125-34, 1983.
7. **Miwa, T., Maiya, T., and Mizuike, A.,** Low temperature plasma ashing of petroleum pitch, *Bunseki Kagaku,* 32, 393, 1983.
8. **Komlenic, J. J., Vermeulen, T., and Fish, R. H.,** Molecular characterization and finger printing of vanadyl porphyrin and nonporphyrin compounds in heavy crude petroleums by HPLC-GFAA analysis, REPORT 1982; LBL-15399; Order No. DE83006964, 86pp. Avail. NTIS. From Energy Res. Abstr., 8, Abstr. No. 17249, 1983.
9. **Abu-Elgheit, M. A.,** Flameless atomic-absorption spectroscopic determination of wear metals in spent motor oils, *Indian J. Technol.,* 21, 128, 1983.
10. **Brown, R. M., Pickford, C. J., and Davison, W. L.,** Speciation of metals in soils, *Int. J. Environ. Anal. Chem.,* 18, 135, 1984.
11. **Hocquellet, P.,** Use of atomic absorption spectroscopy with electrothermal atomization for the direct determination of trace elements in oils: cadmium, lead, arsenic and tin, *Rev. Fr. Corps Gras,* 31, 117, 1984 (in French).
12. **Fish, R. H. and Komlenic, J. J.,** Molecular characterization and profile identifications of vanadyl compounds in heavy crude petroleums by liquid chromatography/graphite furnace atomic absorption spectrometry, *Anal. Chem.,* 56(3), 510-17, 1984.
13. **Saba, C. S., Rhine, W. E., and Eisentraut, K. J.,** Determination of wear metals in aircraft lubricating oils by atomic absorption spectroscopy using a graphite furnace atomizer, *Appl. Spectrosc.,* 39(4), 689-93, 1985.
14. **Schulz, H., Langenbeck, K., and Wagner, G.,** Monitoring of gear wear by lubricant studies, *Tribol. Schmierungstech,* 33(2), 113-19, 1986.
15. **Seong Lee, B., Yoshimura, E., Tanaka, Y., Saitoh, J., Yamazaki, S., and Toda, S.,** Combination of wet charring and dry ashing suitable for the determination of trace metals in oily foodstuffs by graphite furnace atomic absorption spectroscopy, *Bunseki Kagaku,* 35, T120, 1986 (in Japanese).

AUTHOR INDEX

SUBJECT INDEX

APPENDICES

APPENDIX 1
GLOSSARY OF TERMS AND DEFINITIONS

Absorbance, A
Logarithm to base 10 of the reciprocal of the transmittance

$$A = \log_{10}(1/T)$$

Absorbance Scale Expansion
The ratio of the magnitude of the scale reading to the actual absorbance

Absorption Cells
A simple tube system in which the flame from a total consumption burner is directed in at one end and the consumption producers emerge from the other

Absorption Coefficient (kv)
The absorption coefficient at a discrete frequency, v, is defined by

$$I_v = I_v{}^o e{-}^{kvL}$$

where $I_v{}^o$ and I_v are the initial and final intensities of radiation of frequency v passed through an absorption cell of length L

Absorptivity (a)
Absorbance divided by the product of concentration (c, in g/L) times the path length (b in cm)

$$a = A/bc$$

Note: Absorptivity is a constant to normalize the units and is experimentally determined

Accuracy
The measure of the agreement between a measured value and the value expected as "true"

Angstrom (Å)
Unit of length equal to 1/6438.4694 of the wavelength of the red line of cadmium (for practical purposes, it is considered equal to 10^{-8} cm)

Atomic Absorption Spectroscopy (AAS or AA)
An analytical method for the determination of elements, based on the absorption of radiation of free atoms

Atomic Vapor
A vapor that contains free atoms of the analysis element

Atomization
The process that converts the analysis element, or its compounds, to an atomic vapor

Atomizer
The device used to produce and stabilize or maintain a population of free atoms

Background Correction
An interference due to light scattering by particles in the flame, and molecular absorption of light from the lamp by molecules in the flame

Bandpass
The width of the observed peak, in wavelength units, between the two points where the transmitted intensity is equal to half that at the maximum

Beer's Law
Absorptivity of a substance is constant with respect to changes in concentration

$$I = I_0 10^{-abc}$$

where, I = transmitted radiation power; I_0 = incident radiant power; a = absorptivity; b = path length; and c = concentration of the absorbing species in the analyzing beam

Blank
That which contains no analyte element

Boat Sampling
Boat sampling is based on the principle of evaporating the sample completely and quickly in the flame or furnace in order to record a high narrow absorption peak

Bohr's Equation
Transition between two quantized state corresponds to the absorption or emission of energy in the form of electromagnetic radiation, the frequency v of which is determined by

$$\Delta E = E_1 - E_2 = h v$$

where E_1 and E_2 are the energies in the initial and final states, respectively, and h is Planck's constant

Boosted Output Lamp
A lamp in which a secondary discharge is used to increase the emission of characteristics absorbed radiation (this increase is normally both absolute and relative to other radiation from the lamp)

Calibration
The relationship between the absorption indicated by the instrument and the concentration of the element that produces it is established in the calibration curve

Carrier Gas
The gas used to convey the sample mist to the atomizer

Characteristic absorbed radiation
Radiation that is specifically absorbed by free atoms of the analysis element

Chemical interference
If the sample being analyzed contains a thermally stable compound with the analyte that is not totally decomposed by the energy of the flame or furnace, a chemical interference exists and as a result the number of atoms in the flame capable of absorbing light is reduced

Concentration
A continuum source emits light over a broad spectrum of wavelengths instead of at specific lines, e.g., deuterium arc lamp in UV or tungsten-iodide lamp for visible wavelengths

Curvature Corrector
A device or factor that alters the relationship between the signal and the readout as a function of signal, such that there is a linear response between the readout and the desired physical measurement, e.g., concentration

Depressions
An interference that causes a decreased instrument response

Detection Limit
The detection limit is defined as the concentration of an element that will produce an absorbance equal to twice the SD of a series of measurements of a solution, the concentration of which is distinctly detectable above the base line

$$\text{Detection limit} = \frac{\text{standard concentration} \times 2 \text{ standard deviations}}{\text{mean}}$$

Deviation from Beer's Law
The deviation from Beer's law is largely due to the result of unabsorbed and unabsorbable light (stray light, non-homogeneities of temperature and space in the absorption cell, line broadening, and absorption at nearby lines) reaching the detector

Diffracting Grating
Light is deflected by a pattern of fine slits at diffrent angles according to its wavelength

Direct Injection
A sample is directly nebulized into the flame or furnace

Double-Beam Spectrometer
The light from the source lamp is divided into a sample beam, which is focused through the sample cell, and a reference beam, which is directed around the sample cell, and the readout represents the ratio of the sample and the reference beams

Double-Channel Spectrometers
These allow simultaneous measurements at two wavelengths either in absorption or emission and require two separate optical systems — primary source, monochromator, and detector

Electrical Damping
Electrical damping reduces the noise shown by a meter or on a recorder trace. The degree of damping is usually 0.2 to 2 sec with a meter and 1 to 4 sec with a recorder

Electrodeless Discharge Tube (EDT or EDL)
A tube containing the element to be determined in a readily vaporized form and constructed so as to enable a discharge to be induced in the vapor (this discharge can be used as a source of characteristic radiation)

Emission Interference

At high analyte concentrations, the atomic absorption analysis for highly emissive elements sometimes exhibits poor analytical precision, if the emission signal falls within the spectral bandpass being used

Enhancement

An interference that causes an increased response

Fluctuational Concentration Limit/Fluctuational Sensitivity

See "detection limit"

Grating

See "diffraction grating"

Hollow Cathode Lamp (HCL)

A discharge lamp with a hollow cathode, usually cylindrical, used in atomic spectroscopy to provide characteristic radiation

Interference

A general term for an effect that modifies the instrumental response to a particular concentration of the analysis element

Ionization Buffer

A spectroscopic buffer used to minimize or stabilize the ionization of free atoms of the analysis element

Ionization Interference

Interference occurs when the flame or furnace temperature has enough energy to cause the removal of an electron from the atom, creating an ion. As these electronic rearrangements deplete the number of ground state atoms, absorption is reduced

Limit of Detection

The minimum concentration or amount of an element that can be detected with 95% certainty, assuming a normal distribution of errors

Long Tube Device

A device in which an atomizing flame is directed into a tube lying along the optical axis of an atomic absorption spectrometer

Matrix Effect

An interference caused by differences between the sample and a standard containing only the analysis element and, where appropriate, a solvent

Matrix Matching

To match the standards and samples with respect to the materials that are present in excess of 1% in the total solution

Mean (X)

A statistical term for the central tendency of a group or series of data (the mean is defined as the sum of all of the individual test results divided by the number of data)

$$X = (X1)/N$$

Mean Error
An accuracy reference that is defined as the difference between the true or accepted value and mean average value resulting from a series of determinations

Nanometer
Units of length equal to 10^{-9} m, almost, but not exactly, equal to 10 Å

Nonabsorbing Line
A line from the conventional hollow cathode lamp that lies closer to the resonance line, but which is not absorbed by the sample atom

Nonspecific Absorption
This effect is due to the presence of dried and semidried salt particles in the flame/furnace that scatter or absorb the incident radiation from the source

Precision
The measure of agreement among test results as measured in terms of the SD or relative SD

Radiation Generator
Equipment for producing characteristic absorbed radiation (the equipment normally) consists of a lamp and power supply unit

Radiation Scattering
An interference effect caused by scattering of radiation from drops or particles associated with the atomic vapor

Range
A statistical term signifying the difference between the highest and lowest values in a series of test results

Relative Error
The mean error expressed as a percentage of the true result

Repeatability
A statistical term to measure the index of precision of a single aspect of a complete analytical scheme (normally, repeatability is the SD of a series of test results without varying the operator or apparatus)

Resonance Radiation
Characteristic absorbed radiation that corresponds to the transfer of an electron from the ground state level to a higher energy level in the atom

Sampling Unit
The part of an atomic absorption spectrometer that accepts the sample solution and prepares it for atomization

Scale Expansion
In atomic absorption instruments, scale expansion simply involves an electrical expansion of the presented signal by a chosen factor, e.g., 2, 5, 10, or 20. Scale expansion is valuable when the noise level is 0.5% absorption (0.002 absorbance units) or less and facilitates the reading of small scale deflexions

Scatter
Scatter is the result of the presence of small solid particles in the resonance beam. These solid particles may be caused by the inability of the flame/furnace to vaporize a high dissolved solid content of the sample solution or may be due to the formation of particles

Sensitivity
Sensitivity is defined as that concentration in solution of the analysis element which will produce a change, compared to pure solvent, of 0.00044 absorbance units (i.e., 1% absorption) in the optical transmission of the atomic vapor at the wavelength of the radiation used

$$\text{Sensitivity} = \frac{\text{concentration of standard} \times 0.0044}{\text{measured absorbance}}$$

Sensitivity Check
The sensitivity check (in mg/L) value is the concentration of an element that will produce a signal of approximately 0.2 absorbance units under optimum conditions at the wavelength listed

Spectral Bandwidth (Δv)
The wavelength or frequency interval of radiation leaving the exit slit of a monochromator between limits set at a radiant power level half way between the continuous background and the peak of an emission line or an absorption band of negligible intrinsic width

Spectral Interference
Spctral interference occurs when an absorbing wavelength of an element present in the sample, but not being determined, falls within the bandwidth of the absorption line of the element of interest

Standard Deviation
A statistical term indicating the distribution of test results in a Gaussian distribution of a series of results (it is defined as the square root of the variance)

Stray Radiation Energy
All radiant energy that reaches and is sensed by the detector at wavelengths that do not correspond to the spectral energy under consideration

Transmittance (T)
The ratio of the radiant power transmitted by a sample to the radiant power incident on the sample

Wavelength
The distance along the line of propagation, between two points that are in phase on adjacent waves

APPENDIX 2
ABSTRACTS AND REVIEWS

1. **Menzies, A. C.,** Trends in automatic spectrochemical analysis, *Acta Suppl.,* 106, 1957.
2. **Malmstadt, H. V.,** Atomic absorption spectrochemical analysis, in *Encyclopedia of Spectroscopy,* Clark, G. L., Ed., Reinhold, New York, 1960.
3. **David, D. J.,** The application of atomic absorption to chemical analysis (a review), *Analyst (London),* 85, 779, 1960.
4. **Menzies, A. C.,** A study of atomic absorption spectroscopy, *Anal. Chem.,* 32, 898, 1960.
5. **Robinson, J. W.,** Atomic absorption spectroscopy, *Anal. Chem.,* 32, 17a, 1960.
6. **Robinson, J. W.,** Recent advances in atomic absorption spectroscopy, *Anal. Chem.,* 33, 1067, 1961.
7. **Robinson, J. W.,** Flame photometry and atomic absorption spectroscopy, in *Analytical Chemistry,* Crouthamel, C. E., Ed., Pergamon Press, Philadelphia, 1961.
8. **Walsh, A.,** Application of atomic absorption spectra to chemical analysis, in *Advances in Spectroscopy,* Vol. 2, Thompson, H. W., Ed., Pergamon Press, Philadelphia, 1961.
9. **Butler, L. R. P.,** Atomic absorption spectroscopy, *S. Afr. Ind. Chem.,* 15, 162, 1961.
10. **David, D. J.,** Emission and atomic absorption spectrochemical methods, in *Modern Methods of Plant Analysis,* Vol. 5, Peach and Tracey, Eds., Springer-Verlag, Berlin, 1962.
11. **David, D. J.,** Atomic absorption spectrochemical analysis with particular reference to plant analysis, *Rev. Univ. Ind. Santander,* 4, 207, 1962 (in Spanish).
12. **Franswa, C. E. M.,** Atomic absorption spectrophotometry, *Chem. Weekbl.,* 58, 177, 1962.
13. **Johansson, A.,** Atomic absorption spectrophotometry, *Svensk. Kem. Tidskr.,* 74, 415, 1962.
14. **L'vov, B. V.,** Theory and method of atomic absorption analysis, *Zavod. Lab.,* 28, 931, 1962 (in Russian).
15. **Milazzo, G.,** Atomic absorption, *Chim. Ind. (Milan),* 44, 493, 1962 (in Italian).
16. **Allan, J. E.,** A review of recent work in atomic absorption spectroscopy, *Spectrochim. Acta,* 18, 605, 1962.
17. **Kahn, H. L. and Slavin, W.,** Atomic absorption analysis, *Int. Sci. Technol.,* 1962.
18. **Robinson, J. W.,** Atomic absorption spoectroscopy, *Ind. Chem.,* 38, 226, 1962.
19. **Tabeling, R. W. and Devany, J. J.,** Factors influencing sensitivity in atomic absorption, in *Developments in Applied Spectroscopy,* Vol. 1, Ashby, W. D., Ed., Plenum Press, New York, 1962.
20. **Willis, J. B.,** Atomic absorption spectroscopy, in *Proc. 10th Coll. Spectroscopium Internationale,* Lippincott, E. R. and Margoshes, M., Eds., Spartan Books, Rochelle Park, N.J., 1963.
21. **Willis, J. B.,** Analysis of biological materials by atomic absorption spectroscopy, in *Methods of Biochemical Analysis,* Vol. 2, Glick, D., Ed., Interscience, New York, 1963.
22. **Gilbert, P. T.,** Atomic absorption spectroscopy: a review of recent developments, in Proc. 6th Conf. Anal. Chem. Nuclear Reactor Technology, AEC TID-7655, Gatlinburg, TN, 1963.
23. **David, D. J.,** Recent developments in atomic absorption analysis, *Spectrochim. Acta,* 20, 1185, 1964.
24. **Slavin, W.,** Quantitative metal analysis by atomic absorption spectrophotometry, *Chim. Ind.,* 46, 60, 1964 (review in Italian).
25. **Slavin, W.,** Atomic absorption instrumentation and technique. A review, in *Analysis Instrumentation,* Fowler, L., Roe, D. K., and Harmon, R. G., Eds., Plenum Press, New York, 1964; At. Ab. Newsl., No. 24, Sept. 1964.
26. **Lockyer, R.,** Atomic absorption spectroscopy, in Advances in *Analytical Chemistry and Instrumentation,* Reilley, C. N., Ed., Interscience, New York, 1964.

27. **Robinson, J. W.,** The future of atomic absorption spectroscopy, in *Developments in Applied Spectroscopy,* Vol. 4, Plenum Press, New York, 1965.

28. **Ulrich, W. F. and Shifrin, N.,** Atomic absorption spectrophotometry, *Analyzer,* 6, 10, 1965.

29. **Slavin, W.,** The application of atomic absorption spectroscopy to analytical, biochemistry and toxicology, *Occup. Health Rev.,* 17, 9, 1965 (in French and English).

30. **Slavin, W.,** The application of atomic absorption spectroscopy to geochemical prospecting and mining, *At. Ab. Newsl.,* 4, 243, 1965.

31. **Zettner, A.,** Principles and applications of atomic absorption spectroscopy, in *Advances in Clinical Chemistry,* Vol. 7, Sobotka, H., Ed., Academic Press, New York, 1965.

32. **Herrmann, R.,** Principles and applications of atomic absorption spectroscopy in flames, *Z. Klin. Chem.,* 6, 178, 1965 (in German).

33. **Rubeska, I. and Velicka, I.,** Atomic absorption spectrophotometry, *Chem. Listy,* 59, 769, 1965.

34. **Schleser, F. H.,** Atomic absorption spectrophotometry, *Z. Instrumentenkd.,* 73, 25, 1965 (in German).

35. **Adams, P. B.,** Flame and atomic absorption spectrometry, in *Standard Methods of Chemical Analysis, Instrumental Analysis,* Vol. 3B, 6th ed., Welcher, F. J., Ed., D. Van Nostrand, Princeton, N.J., 1966.

36. **Slavin, W.,** Recent developments in analytical atomic absorption spectroscopy, *At. Ab. Newsl.,* 5, 42, 1966.

37. **Walsh, A.,** Some recent advances in atomic absorption spectroscopy, *J. N. Z. Inst. Chem.,* 30, 7, 1966.

38. **Walsh, A. and Willis, J. B.,** Atomic absorption spectrometry, in *Standard Methods of Instrumental Methods,* Vol. 3, Part A, Welcher, F. J., Ed., D. Van Nostrand, Princeton, N.J., 1966.

39. **Suzuki, M. and Takeuchi, T.,** Recent advances in atomic absorption spectroscopic analysis, *Bunseki Kagaku,* 15, 1003, 1966 (in Japanese).

40. **West, T. S.,** Inorganic trace analysis, *Chem. Ind. (London),* 25, 1005, 1966.

41. **Weberling, R. P. and Cosgrove, J. F.,** Atomic absorption spectroscopy, in *Trace Analysis,* Morrison, G. H., Ed., Interscience, New York, 1966.

42. **Koirtyohann, S. R.,** Recent developments in atomic absorption and flame emission spectroscopy, *At. Absorp. Newsl.,* 6, 77, 1967.

43. **Slavin, W.,** Atomic absorption: recent developments and applications, *Chim. Ind.,* 49, 60, 1967 (in Italian).

44. **Allaire, R. F., Brachett, F. P., and Shafer, J. T.,** Analytical benefits of atomic absorption spectrophotometric methods, *J. Soc. Motion Pict. Television Eng.,* Oct. 1967.

45. **Walsh, A.,** Atomic absorption spectroscopy, *Aust. Phys.,* 4, 185, 1967.

46. **Willis, J. B.,** Recent advances in atomic absorption spectroscopy, *Rev. Pure Appl. Chem.,* 17, 111, 1967.

47. **Panday, V. K. and Ganguly, A. K.,** A short review on the determination of elements by atomic absorption spectrophotometry, *Trans. Bose Res. Inst. Calcutta,* 30, 131, 1867.

48. **Kahn, H. L.,** Principles and practice of atomic absorption, in *Trace Inorganics in Water,* Baker, R. A., Ed., Advances in Chemistry Series 73, American Chemical Society, Washington, D. C., 1968.

49. **Lewis, L. L.,** Atomic absorption spectrometry — applications and problems, *Anal. Chem.,* 40, 28a, 1968.

50. **Scholes, P. H.,** The application of atomic absorption spectrophotometry to the analysis of iron and steel, *Analyst (London),* 93, 197, 1968.

51. **Walsh, A.,** Atomic absorption spectroscopy: a foreword, *Appl. Opt.,* 7, 1259, 1968.

52. **Reynolds, R. J.,** Atomic absorption spectroscopy, its principles, applications and future developments, *Lab. Equip. Dig.,* 6, 79, 1968.

53. **Beamish, F. E., Lewis, C. L., and Van Loon, J. C.,** A critical review of atomic absorption spectrochemical and X-ray fluorescence methods for the determination of noble metals. II, *Talanta,* 16, 1, 1969.

54. **Slavin, W. and Slavin, S.,** Recent trends in analytical atomic absorption spectroscopy, *Appl. Spectrosc.,* 23, 421, 1969.

55. **Walsh, A.,** Physical aspects of atomic absorption, *ASTM STP,* 443, 3, 1969.

56. **Britske, M. E.,** Atomic absorption spectrophotometric analysis, *Zavodsk. Lab.,* 35, 1329, 1969.

57. **West, T. S.,** Flame emission and atomic absorption spectrometry, *Chem. Ind.,* 21, 387, 1970.

58. **L'vov, B. V.,** Progress in atomic absorption spectroscopy employing flame and graphite cuvette techniques, *Pure Applied Chem.,* 23, 11, 1970.

59. **Pinta, M.,** New prospects in atomic absorption, *Meth. Physe. Analyse,* 6, 368, 1970 (in French).

60. **Mallett, R. C.,** Review of techniques for the determination of gold and silver by atomic absorption spectroscopy, *Miner. Sci. Eng.,* 2, 28, 1970.

61. **Winefordner, J. D., Sviboda, V., and Cline, L. J.,** A critical comparison of atomic emission, atomic absorption and atomic fluorescence flame spectrometry, *CRC Crit. Rev. Anal. Chem.,* 12, 233, 1970.

62. **Platt, P.,** *Annual Reviews in Analytical Chemistry,* SAC, London, 1971.

63. **Ando, A.,** Flame and atomic absorption spectrochemical analysis, *Bunseki Kagaku,* 20, 112, 1971.

64. **L'vov, B. V.,** Modern state and main problems of atomic absorption analysis, *Zh. Anal. Khim.,* 26, 590, 1971.

65. **Yudelevich, I. G., Shelpakova, I. R., Brusentsev, F. A., and Zayakina, S. B.,** Dispersion analysis of chemical spectrographic flame photometric and atomic absorption method for determining trace impurities, *Zh. Anal. Khim.,* 26, 2075, 1971.

66. **Kirkbright, G. F.,** Application of non-flame atom cells in atomic absorption and atomic fluorescence spectroscopy. A review, *Analyst (London),* 96, 609, 1971.

67. **Slavin, S.,** An atomic absorption bibliography for 1971, *At. Absorpt. Newsl.,* 11, 7, 1972.

68. **Levine, S. L.,** Atomic absorption spectrophotometry, *Chem. Technol.,* 110, 1972.

69. **Amos, M. D.,** Nonflame atomization in AAS. A current review, *Am. Lab.,* 4, 57, 1972.

70. **DeWaele, M. and Droeven, G.,** Atomic absorption spectrometry, *Ind. Chim. Belge,* 37, 322, 1972.

71. **Slavin, S.,** An atomic absorption bibliography for January-June 1972, *At. Absorpt. Newsl.,* 11, 74, 1972.

72. **Belcher, R.,** New methods for the determination of elements in trace amounts, *Z. Anal. Chem.,* 263, 257, 1973.

73. **Slavin, S.,** An atomic absorption bibliography for July-December 1972, *At. Ab. Newsl.,* 12, 9, 1963.

74. **Buttgereit, G.,** Modified atomic absorption spectroscopic methods for trace metal analysis, *Z. Anal. Chem.,* 267, 81, 1973.

75. **Barnard, A. J., Jr. and Dudley, R. W.,** Tracing the elements, *Ind. Res.,* March 1973.

76. **Stone, R. G. and Warren, J.,** Atomic absorption spectrometry: a critical review of modern instrumentation and techniques, *Lab. Equip. Dig.,* 12, 49, 1973.

77. **Horncastle, D. C. J.,** Atomic absorption spectrophotometry, *Med. Sci. Law,* 13, 3, 1973.

78. **Sen Gupta, J. G.,** A review of the methods for the determination of the platinum group metals, silver and gold by atomic absorption spectroscopy, *Miner. Sci. Eng.,* 5, 207, 1973.

79. **Reif, I., Fassel, V. A., and Kniseley, R. N.,** Spectroscopy flame temperature measurements and their physical significance. I. Theoretical concepts — a critical review, *Spectrochim. Acta,* 287b, 105, 1973.

80. **Slavin, S.,** An atomic absorption bibliography for January-June 1973, *At. Absorpt. Newsl.,* 12, 77, 1973.

81. **Lagesson, V.,** Trace metal determination by atomic absorption spectrometry, *Kem. Tidskr.,* 85, 66, 1973.

82. **Slavin, S.,** An atomic absorption bibliography for July-December 1973, *At. Absorpt. Newsl.,* 13, 11, 1973.

83. **Thomerson, D. R. and Price, K. C.,** Recent develoments in atomic absorption spectrometry, *Am. Lab.,* 6, 53, 1974.

84. **Lisk, D. J.,** Recent developments in the analysis of toxic elements, *Science,* 184, 1137, 1974.

85. **Slavin, S.,** An atomic absorption bibliography for January-June 1974, *At. Absorpt. Newsl.,* 13, 84, 1974.

85. **Thomerson, D. R. and Price, W. J.,** Atomic absorption behavior of some of the rare earth elements, *Anal. Chim. Acta,* 72, 188, 1974.

86. **Walsh, A.,** Atomic absorption spectroscopy — stagnant or pregnant, *Anal. Chem.,* 46, 698a, 1974.

87. **Woodriff, R.,** Atomization fluorescence and atomic absorption spectrochemical analysis. Review., *Appl. Spectrosc.,* 28, 413, 1974.

88. **West, T. S.,** Atomic fluorescence and atomic absorption spectrometry for chemical analysis, *Analyst (London),* 99, 886, 1974.

89. **Hurlbut, J. A.,** History, Uses, Occurrences, Analytical Chemistry and Biochemistry of Beryllium. A review. Rep. 2152, U. S. Atomic Energy Commission, 1974.

90. **Price, W. J.,** Atomic absorption an essential tool in modern metallurgy, *Metals Mater.,* 8, 485, 1974.

91. **Woodriff, R.,** Atomization chambers for atomic absorption spectrochemical analysis. Review, *Appl. Spectrosc.,* 28, 413, 1974.

92. **Kharlamov, I. P. and Eremina, G. V.,** Use of an atomic absorption method for analyzing steels and alloys. Review of research published in 1966-1971, *Zavod. Lab.,* 40, 385, 1974.

93. **Reinhold, J. G.,** Review: trace elements — a selective survey, *Clin. Chem. (Winston-Salem, N.C.),* 21, 476, 1974.

94. **Slavin, S. and Lawrence, D. M.,** An atomic absorption bibliography for July - December 1974, *At. Absorp. Newsl.,* 14, 1, 1975.

95. **Sunshine, I.,** Analytical toxicology, *Anal. Chem.,* 47, 212A, 1975.

96. **Ure, A. M.,** The determination of mercury by non-flame atomic absorption and fluorescence spectrometry — a review, *Anal. Chim. Acta,* 76, 1, 1975.

97. **Slavin, S. and Lawrence, D. M.,** An atomic absorption bibliography for January - June 1975, *At. Absorp. Newsl.,* 14, 81, 1975.

98. **Winefordner, J. D., Fitzgerald, J. J., and Omenetto, N.,** Review of multielement atomic spectroscopic methods, *Appl. Spectrosc.,* 29, 369, 1975.

99. **Walker, G. W.,** Forensic science atomic absorption spectroscopy, *Chem. Br.,* 11, 440, 1975.

100. **Wondzinski, W.,** Atomic absorption and atomic fluorescence spectrophotometry, *GIT Fachz. Lab.,* 19, 671, 1975 (in German).

101. **Hieftje, G. M., Copeland, T. R., and De-Olivaren, D. R.,** Flame emission, atomic absorption and atomic fluorescence spectrometry, *Anal. Chem.,* 48, 142r, 1976.

102. **L'vov, B. V.,** Trace characterization of powders by atomic absorption spectrometry. The state of the art, *Talanta,* 23, 109, 1976.

103. **Goldstein, S. A. and Walters, J. P.,** A review of considerations of high-fidelity imaging of laboratory spectroscopic sources. I, *Spectrochim. Acta.,* 31b, 201, 1976.

104. **Goldstein, S. A. and Walters, J. P.,** A review of considerations of high-fidelity imaging of laboratory spectroscopic sources. II, *Spectrochim. Acta,* 31b, 201, 1976.

105. **Slavin, S. and Lawrence, D. M.,** An atomic absorption bibliography for January - June 1976, *At. Absorp. Newsl.,* 15, 77, 1976.

106. **Kharlamov, I. P., Eremina, G. V., and Neimark, V. Y.,** Atomic absorption determination of harmful impurities of certain non-ferrous metals. A review, *Zavod. Lab.,* 42, 1320, 1976.

107. **O'Laughlin, J. W., Hemphill, D. D., and Pierce, J. O.,** *Analytical Methodology for Cadmium in Biological Matter - A critical review,* International Lead Zinc Research Organization, New York, 1976.

108. **Schuller, P. L. and Egan, H.,** Cadmium, Lead, Mercury and Methyl Mercury Compounds. Review of Methods of Trace Analysis and Sampling with Special Reference to Food, Rep. No. 92-5-1-00094-M-84, Food and Agriculture Organization, Rome, 1976.

109. **Slavin, S. and Lawrence, D. M.,** An atomic absorption bibliography for January - June 1977, *At. Absorpt. Newsl.,* 16, 89, 1977.

110. **Walsh, A.,** Atomic absorption spectroscopy and its applications, old and new, *Pure Appl. Chem.,* 49, 1621, 1977.

111. **Olwin, J. H.,** Metals in the life of man, *J. Anal. Toxicol.,* 1, 245, 1977.

112. **Kalman, S. M.,** The pathphysiology of lead poisoning: a review and a case report, *J. Anal. Toxicol.,* 1, 277, 1977.

113. **Fernandez, F. J.,** Metal speciation using atomic absorption as a chromatography detector. A review, *At. Absorpt. Newsl.,* 16, 33, 1977.

114. **Slavin, S. and Lawrence, D. M.,** An atomic absorption bibliography for July - December 1976, *At. Absorpt. Newsl.,* 16, 4, 1977.

115. **Aidarov, T. K.,** Status of developments and some trends in developing atomic absorption spectrometer instrumentation (review), *Zh. Prikl. Spektrosk.,* 26, 779, 1977.

116. **Jenkins, R.,** Recent developments in wavelength and energy dispersive spectrometry, *Pure Appl. Chem.,* 49, 1583, 1977.

117. **Langmyhr, F. J.,** Direct atomic absorption spectrometric analysis of geological materials. A review, *Talanta,* 24, 277, 1977.

118. **Peterson, G. E.,** The application of atomic absorption spectrophotometry to the analysis of non-ferrous alloys, *At. Absorpt. Newsl.,* 16, 133, 1977.

119. **Harrison, R. M. and Laxen, D. P. H.,** Comparative study of methods for analysis of total lead in soils, *Water Air Soil Pollut.,* 8, 2387, 1977.

120. **Kavanova, M. A.,** Atomic absorption spectrophotometry basis for standardization of methods for determination of elements in natural waters, *Probl. Sovrem. Anal. Khim.,* 2, 23, 1977 (in Russian).

121. **Slavin, S. and Lawrence, D. M.,** An atomic absorption bibliography for July - December 1977, *At. Absorpt. Newsl.,* 17, 7, 1978.

122. **Walsh, A.,** Atomic spectroscopy — What next?, *At. Absorpt. Newsl.,* 17, 97, 1978.

123. **Eller, P. M. and Haartz, J. C.,** Sampling and analytical methods for antimony and its compounds — a review, *J. Am. Ind. Hyg. Assoc.,* 39, 790, 1978.

124. **Slavin, S. and Lawrence, D. M.,** An atomic absorption bibliography for January - June 1978, *At. Absorpt. Newsl.,* 17, 73, 1978.

125. **Weiss, E. B. and Rosenthal, I. M.,** Clinical use of laboratory determinations for metals, anti-convulsive agents and cardiovascular drugs, *J. Anal. Toxicol.,* 2, 166, 1978.

126. **Pinta, M.,** Present tendencies in atomic absorption spectroscopy, *Analysis,* 6, 227, 1978 (in French).

127. **Jones, E. A. and Dixon, K.,** Review of the literature on the separation and determination of rare earth elements, Natl. Inst. Metall. Repub. S. Afr. Rep., No. 1943, 1978.

128. **Slavin, S. and Lawrence, D. M.,** An atomic absorption bibliography for July - December 1978, *At. Absorpt. Newsl.,* 18, 18, 1979.

129. **Willis, J. B.,** Review analytical atomic spectroscopy at the CSIRO Division of Chemical Physics, *Anal. Chim. Acta,* 106, 175, 1979.
130. **Wilson, D. L.,** Separation and concentration techniques for atomic absorption: a guide to the literature, *At. Absorpt. Newsl.,* 18, 13, 1979.
131. **Zingaro, R. Z.,** How certain trace elements behave, *Environ. Sci. Technol.,* 13, 282, 1979.
132. **Gladney, E. S., Perrin, D. R., Owens, J. W., and Knab, D.,** Elemental concentrations in the United States Geological Survey's geochemical exploration reference samples — a review, *Anal. Chem.,* 51, 1557, 1979.
133. **Lawrence, D. M.,** An atomic absorption bibliography for January - June 1979, *At. Absorpt. Newsl.,* 18, 77, 1979.
134. **Razumov, V. A. and Zvyagintsev, A. M.,** Nonselective weakening of light in atomic absorption and atomic fluorescence analysis. Review, *Zh. Prikl. Spektrosk.,* 31, 381, 1979.
135. **El-Shaarawy, M. I.,** Selected features of atomic absorption spectrophotometry and their pertinence to oil analysis, *Arabian J. Sci. Eng.,* 4, 75, 1979.
136. **Fan, P. Y.,** Atomic absorption spectroscopy and its application in medical examination, *Chung-Hua I Hsueh Chien Yen Tsa Chih,* 2, 175, 1979.
137. **Kinnunen, J.,** Determination of precious metals in mining and metallurgical products, A review, *Kem. Kemi,* 6, 617, 1979.
138. **Price, W. J.,** New techniques of atomic absorption in clinical and biochemical analysis, *Sci. Ind.,* 12, 22, 1979.
139. **Wang, P. L., Wu, T. C., Yau, W. C., Liu, C. L., and Ma, K. C.,** Thirty years development of atomic absorption analysis in China, *Fen Hsi Hua Hsueh,* 7, 378, 1979.
140. **Marcus, L. H.,** Atomic absorption spectroscopy (June 1970 - June 1980), Report 1980 NERACUSGNT 0456: Order No. PB80-859614, 1980; available from NTIS, Springfield, Va; Gov. Rep. Announce. Index (U.S.), 80, 5637, 1980.
141. **Johansson, A.,** Atomic absorption spectromery, *Kem. Tidskr.,* 92, 26, 1980.
142. **Johansson, A.,** More study on chemical literature. II. Atomic absorption spectrometry, *Kem. Tidskr.,* 92, 48, 1980.
143. **Lawrence, D. M.,** An atomic absorption bibliography for July - December 1979, *At. Spectrosc.,* 1, 8, 1980.
144. **Owens, J. W., Gladney, E. S., and Purtymun, W. D.,** Modification of trace element concentrations in natural waters by various field sampling techniques, *Anal. Lett.,* 13a, 253, 1980.
145. **Ali, S. L.,** Atomic absorption spectrophotometry — state of the art after 25 years, *Pharm. Ztg.,* 125, 450, 1980.
146. **De Galan, L. and Van Dalen, J. P. J.,** Atomic absorption spectrometry, *Pharm. Weekbled.,* 115, 689, 1980.
147. **Gilbert, R. K. and Platt, R.,** Measurement of calcium and potassium in clinical laboratories in the United States, 1971-1978, *Am. J. Clin. Pathol.,* 74, 508, 1980.
148. **Hughes, H.,** Analysis — survey. I. Oxide materials, *Iron Steel Int.,* 53, 13, 1980.
149. **Lawrence, D. M.,** An atomic absorption bibliography for January - June 1980, *At. Spectrosc.,* 1, 94, 1980.
150. **L'vov, B. V.,** Twenty five years of analytical atomic absorption spectrometry, *Zh. Anal. Khim.,* 35, 1575, 1980.
151. **Walsh, A.,** The birth of modern atomic absorption spectroscopy, *Chimia,* 34, 427, 1980.
152. **Godden, R. G. and Thomerson, D. R.,** Generation of covalent hydrides in atomic absorption spectroscopy. A review, *Analyst (London),* 105, 1137, 1980.
153. **Grobenski, Z. and Schultz, H.,** Modern atomic spectroscopy techniques for trace metal determination in foods, *GIT Fachz. Lab.,* 24, 1156, 1980 (in German).
154. **Koirtyohann, S. R.,** A history of atomic absorption spectroscopy, *Spectrochim. Acta,* 35b, 663, 1980.

155. **Magyar, B. and Aeschbach, F.,** Why not ICP as atom reservoir for AAS?, *Spectrochim. Acta,* 35b, 663, 1980.

156. **Tonini, C.,** Atomic absorption spectroscopy in determining metals in textiles, *Tincoria,* 77, 358, 1980 (in Italian).

157. **Walsh, A.,** Atomic absorption spectroscopy. Some personal recollections and speculations, *Spectrochim. Acta,* 35b, 639, 1980.

158. **Willis, J. B.,** Some memories of the early days of atomic absorption spectroscopy, *Spectrochim. Acta,* 35b, 653, 1980.

159. **Yoneda, S.,** Ppm and ppb. Detection limitations of elements, *Kagaku to Seibutsu,* 18, 312, 1980 (in Japanese).

160. **Blake, C. J.,** Sample preparation methods for the analysis of metals in food by atomic absorption spectrometry. A literature review, *Sci. Tech. Surv. Br. Food Manufact. Ind. Res. Assoc.,* 122, 57, 1980.

161. **Young, E. F.,** A review of the spectrophotometer, *Opt. Spectra,* 14, 44, 1980.

162. **Kirkbright, G. F.,** Current status and future needs in atomic absorption instrumentation, *Anal. Chem.,* 52, 736a, 1980.

163. **Marcus, L. H.,** Atomic absorption spectroscopy, Rep. NERACUSGNT 0 56: Order No. PB 80-859614, 1980: Available from the NTIS, Springfield, Va,; Govt. Rep. Announce. Index (U.S.), 80(26), 5637, 1980.

164. **Berndt, H. and Messerschmidt, J.,** Loop atomic absorption spectrometry. A brief review, *Aerztl. Lab.,* 28, 133, 1981.

165. **Cresser, M. S. and Sharp, B. L.,** Annual reports on analytical atomic spectroscopy. *Reviewing,* 10, 1, 1981.

166. **Lawrence, D. M.,** An atomic absorption bibliography for July - December 1980, *At. Spectrosc.,* 2, 22, 1981.

167. **Lawrence, D. M.,** An atomic absorption bibliography for January - June 1980, *At. Spectrosc.,* 2, 101, 1981.

168. **Verlinden, M., Deelstra, H., and Adriaenssens, E.,** Determination of selenium by atomic absorption spectrometry. A review, *Talanta,* 28, 637, 1981.

169. **Snook, R. D.,** A critical appraisal of the hydride generation method, *Anal. Proc. (London),* 18, 342, 1981.

170. **Miller, H. C., James, R. H., Dickson, W. R., Neptune, M. D., and Carter, M. H.,** Evaluation of methodology for survey analysis of solid wastes, *ASTM STP,* 760, 240, 1981.

171. **Van Loon, J. C.,** Review of methods for elemental speciation using atomic spectrometry detectors for chromatography, *Can. J. Spectrosc.,* 26, 22a, 1981.

172. **Varma, A.,** A guide to atomic absorption spectrophotometric analysis of alkali metals, Instrumentation Labs. Inc., Waltham, Mass., 1982.

173. **Varma, A.,** A reference guide to antimony using atomic absorption, Instrumentation Labs. Inc., Waltham, Mass., 1982.

174. **Varma, A.,** A literature guide for atomic absorption analysis of boron, aluminum, gallium, indium and thallium, Instrumentation Labs. Inc., Waltham, Mass., 1982.

175. **Varma, A.,** A reference guide for calcium using atomic absorption, Instrumentation Labs. Inc., Waltham, Mass., 1982.

176. **Horlick, G.,** Atomic absorption, atomic fluorescence and flame photometry, *Anal. Chem.,* 54, 276r, 1982.

177. **Komarek, J. and Sommer, L.,** Organic complexing agents in atomic absorption spectrometry. A review, *Talanta,* 29, 159, 1982.

178. **Lawrence, D. M.,** An atomic spectroscopy bibliography for July - December 1981, *At. Spectrosc.,* 3, 13, 1982.

179. **Eaton, A., Oelker, G., and Leong, L.,** Comparison of AAS and ICP for analysis of natural waters, *At. Spectrosc.,* 3, 152, 1982.

180. **Falk, H.,** Limiting factors for intensity and line profile of radiation sources for atomic absorption spectrometry, *Prog. Anal. At. Spectrosc.,* 5, 205, 1982.

181. **Kirkbright, G. F.,** Some recent studies in optical emission and absorption spectroscopy for trace analysis, *Pure Appl. Chem.,* 54, 769, 1982.

182. **Lawrence, D. M.,** An atomic spectroscopy bibliography for January - June 1982, *At. Spectrosc.,* 3, 93, 1982.

183. **Schaller, K. H.,** Methods of quantitative determination of mercury in human biological materials, *Staub Reinhalt. Luft,* 42, 142, 1982.

184. **Werbicki, J. J., Jr.,** How much gold?, *Prod. Finish. (Cincinnati),* 46, 42, 1982.

185. **Yudelevich, I. G., Startseva, E. A., and Gordaev, G. A.,** Atomic absorption determination of platinum group metals, *Zavod. Lab.,* 48, 23, 1982.

186. **Horlick, G.,** Atomic absorption, atomic fluorescence and flame spectrometry, *Anal. Chem.,* 54, 276r, 1982.

187. **Baranov, S. V., Baranova, I. V., and Ivanov, N. P.,** Spectral lamps for atomic absorption spectrometry. Review, *Zh. Prikl. Spektrosk.,* 36, 357, 1982.

188. **Koirtyohann, S. R. and Kaiser, M. L.,** Furnace atomic absorption — a method approaching maturity, *Anal. Chem.,* 54, 1515a, 1982.

189. **Lawrence, D. M.,** An atomic spectroscopy bibliography for July - December 1982, *At. Spectrosc.,* 4, 10, 1983.

190. **Katskov, D. A.,** Current concepts on the mechanism of thermal atomization of substances in atomic absorption analysis (review), *Zh. Prikl. Spektrosk.,* 38, 181, 1983.

191. **Headridge, J. B.,** Determination of ultra-trace elements in nickel alloys using atomic spectrometric methods and solid samples, *Anal. Proc. (London),* 20, 207, 1983.

192. **Lawrence, D. M.,** An atomic spectroscopy bibliography for July - December 1982, *At. Spectrosc.,* 4, 10, 1983.

193. **Fijalkowski, J.,** Worldwide development trends in analytical atomic spectrometry in recent years, *Mater. Konwersatorium Spektrom. At. Emisyjnej, Absorpc. Spektrom. Mas.,* 120, 39-49, 1983.

194. **Barbooti, M. M. and Jasim, F.,** A review of nonflame atomization for atomic absorption spectrochemical analysis. I. Atomizers, *J. Iraqi Chem. Soc.,* 8, 1-17, 1983.

195. **Petrakiev, A.,** Critical analysis of the methods and apparatus for Zeeman atomic absorption spectral analysis (AASA), *Iotov, Ts. God. Sofii, Univ., "Kliment Okhridski" Fiz. Fak.,* 70-71, 1979, published in 1983.

196. **Mori, K.,** Atomic absorption spectroscopy. Pretreatment, *Kagaku to Kogyo (Osaka),* 57(12), 486-91, 1983.

197. Annual reports on analytical atomic spectroscopy, Vol. 12: Reviewing 1982, Cresser, M. S. and Ebdon, L., Eds., Royal Society of Chemistry, London, U.K., 404 pp, 1983.

198. **Voelkopf, U. and Schulze, H.,** Graphite tube furnace atomic absorption spectrometry with Zeeman effect background corrections. I. Review of Zeeman systems, *Labor Praxis,* 7, 410, 1983.

199. **Bulska, E. and Kaczmarczyk, K.,** New trends in flameless atomic absorption spectrometry, Mater. Konwersatorium Spektrom. Mas. Emisyjnej, *Absorpc. Spektrom. Mas,* 120, 118-32, 1983 (in Polish).

200. **Herber, R. F. M.,** Some instrumental improvements in electrothermal atomization — atomic absorption spectroscopy. Application in biomedical research, *Spectrochim. Acta,* 38B, 783-9, 1983.

201. **Lawson, S. R., Dewalt, F. G., and Woodriff, R.,** Influence of furnace design on operation, sensitivity and matrix interferences in electrothermal atomic absorption spectroscopy, *Prog. Anal. At. Spectrosc.,* 6, 1-48, 1983.

202. **Voelkopf, U. and Schulz, H.,** Graphite tube furnace atomic absorption spectroscopy with Zeeman effect background correction. II, *Labor Praxis,* 7, 544, 1983 (in German).

203. **Chappuis, P. and Rousselet, F.,** Determination of trace elements by flameless atomic absorption spectrophotometry: principles, significance, problems and applications, *Tech. Biol.,* 10(4), 158, 1984 (in French).

204. **Price, W. J.,** Atomic absorption spectroscopy — some recent perspectives, *Kem. Ind.,* 33, 239, 1984.

205. **Lawrence, D. M.,** An atomic spectroscopy bibliography for July - December 1983, *At. Spectroscopy,* 5(1), 10-31, 1984.

206. **Lawrence, D. M.,** An atomic spectroscopy bibliography for January - June 1984, *At. Spectrosc.,* 5, 156, 1984.

207. **Katskov, D. A.,** Study of atomization processes as a way to use atomic absorption spectroscopy in physiochemical analysis, *Nov. Metody Spektr. Anal.* 48-52, 1983 (in Russian), Lontsikh, S. V., Ed., *Izd. Nauka, Sib. Ord.; Novosibirsk, USSR* CA 16 101:32412f, 1984.

208. **Suzuki, M. and Ohta, K.,** Furnace atomic absorption spectroscopy, *Bunseki Kagaku,* 6, 412, 1984.

209. **Cai, D.,** Atomic absorption spectroscopic analysis. I, *Huaxue Shijie,* 25, 234, 1984.

210. **Dungs, K. and Neidhart, B.,** Graphite furnace chemistry — ways to metal species analysis, *Fortschr. Atomspektrom. Spurenanal,* 1, 411-19, 1984.

211. **Schulze, H.,** Atomic absorption spectrometry is now routine even for complex samples, *Labor Praxis,* 8(10), 1008, 1984.

212. **Norval, E.,** The graphite furnace in AAS and some applications in clinical analysis, *S. Afr. J. Sci.,* 81(4), 169-71, 1985.

213. **Maddalone, R. F., Moyer, P. W., and Scott, J. W.,** Aqueous discharges from steam electric power plants trace metal sampling and analysis reference guide. Final report, Report EPRI-CS-3739, 1984: Order No. T185920133, 251 pp. Avail. RRC, POB 50490, Palo Alto, CA. From *Energy Res. Abstr.,* 10(4), abstr. no. 7404, 1985.

214. **Navas, M. J., Jiminez-Trillo, J. L., and Asuero, A. G.,** Determination of trace elements in milk by using atomic absorption spectrophotometry with electrothermal atomization, *Alimentaria (Madrid),* 22, 103, 1985.

215. **Chang, S. B. and Chakrabarti, C. L.,** Factors affecting atomization in graphite furnace AAS, *Prog. Anal. At. Spectrosc.,* 8(2), 83-91, 1985.

216. **De Loos-Vollebregt, M. T. C.,** Zeeman atomic absorption spectrometry, *Prog. Anal. At. Spectrosc.,* 8(1), 47-81, 1985.

217. **Zieba, J.,** Possibility of application of FAAS in forensic analysis, *Z. Zagadnien Kryminal,* 16-17, 76-83, 1983, CA 15, 103:1583z, 1985.

218. **Ottaway, J. M., Carroll. J., Cook, S., Littlejohn, D., Marshall, J., and Stephen, S. C.,** Some novel approaches to trace element analysis, *Anal. Proc. (London),* 22(7), 192-4, 1985.

219. **Fuwa, K. and Tsunoda, K.,** Recent advances on the study of trace elements and their spectrochemical analysis, *Yokugaku Zasshi,* 106, 951, 1986.

220. **Schlemmer, G. and Welz, B.,** Basic atomic spectrometric analysis, *Laboratoriumsmedizin,* 10(5), 160-5, 1986 (in German).

221. **Sturgeon, R.,** Graphite furnace AAS: fact and fiction, *Fresnius' Z. Anal. Chem.,* 324(8), 807-18, 1986.

222. **Liu, J.,** Probe atomization technique in graphite furnace AAS, *Fenxi Ceshi Tongbao,* 4(2), 1-6, 1985 (in Chinese), CA 23, 105: 163916d, 1986.

223. **He, H.,** Application of Zeeman AAS, *Fenxi Ceshi Tangbao,* 4(1), 23-7, 1985, CA 23, 105: 163914b, 1986.

224. **De Galan, L.,** New directions in optical atomic spectrometry, *Anal. Chem.*, 58(6), 697A-698A, 1986.

225. **Caroli, S.,** Hollow cathode lamps as excitation sources for analytical atomic spectrometry, *Fresnius' Z. Anal. Chem.*, 324(5), 442-7, 1986.

226. **Koshy, V. J. and Garg, V. N.,** Atomic absorption spectroscopy and trace elements analysis, *J. Sci. Ind. Res.*, 45(6), 294-314, 1986.

227. **Agness, D.,** Recent advances in graphite furnace analysis, *Aher Publ. R. Soc. Chem.*, 61, 223, 1986.

228. **Atomic Spectroscopy,** bibliography published twice a year, Perkin Elmer Corp., Norwalk, CT.

229. **Chemical Abstracts,** Am. Chem. Soc., published bimonthly. CA Selects on Atomic Spectroscopy, Trace Element analysis.

230. **Analytical Chemistry,** annual reviews, American Chemistry Society.

231. **Analytical Abstracts,** Society for Analytical Chemistry, London, monthly publication.

232. **Masek, P. R., Sutherland, I., and Grivell, S.,** *Atomic Absorption and Flame Emission Spectroscopy Abstracts,* Sci. & Tech. Agency, London. Bimonthly publications.

233. Annual reports on **Analytical Atomic Spectroscopy,** S. A. C., London (From 1972 forward).

APPENDIX 3
REFERENCE BOOKS

1. **Smith, E. A.,** *The Sampling and Assay of the Precious Metals,* 2nd ed., Griffin, London, 1947.
2. **Sandell, E. B.,** *Colorimetric Determination of Traces of Metals,* 3rd ed., Wiley Interscience, New York, 1959.
3. **Gibson, J. H., Grossman, W. E. L., and Cooks, W. D.,** The use of continuous source in atomic absorption spectroscopy, in *Anal. Chem. Proc. Feigl Anniversary Symposium,* Elsevier, Amsterdam, 1962.
4. **David, D. J.,** Emission and atomic absorption spectrochemical methods, in *Modern Methods of Plant Analysis,* Vol. 5, Peach and Tracey, Eds., Springer-Verlag, Berlin, 1962.
5. **Roth, D. J.,** *Developments in Applied Spectroscopy,* Vol. 4, Plenum Press, New York, 1965.
6. **Walsh, A.,** Some recent studies in atomic absorption spectroscopy, in *12th Coll. Spectro. Intern. Exeter. 1965,* Hilger & Watts, London, 1965.
7. **Zettner, A.,** Principles and applications of atomic absorption spectroscopy, in *Advances in Clinical Chemistry,* Vol. 7, Sobotka, H., Ed., Academic Press, New York, 1965.
8. **Beamish, F. E.,** *The Analytical Chemistry of the Noble Metals,* Pergamon Press, Oxford, 1966.
9. **Elwell, W. T. and Gidley, J. A. F.,** *Atomic Absorption Spectrophotometry,* 1st ed., 1961, 2nd ed., Pergamon Press, Oxford, 1966.
10. **Movrodineau, R.,** Atomic absorption spectrophotometry, in *Encyclopedia of Industrial Chemical Analysis,* Snell, F. D. and Hilton, C. L., Ed., Interscience, New York, 1966.
11. **Robinson, J. W.,** *Atomic Absorption Spectroscopy,* Edward Arnold, London, 1966.
12. **Rousselet, F.,** *Spectrophotometric par Absorption Atomique Appliquee a Biologie,* Sedes, Paris, 1966.
13. **Weberling, R. P. and Cosgrove, J. F.,** *Trace Analysis,* Morrison, G. H., Ed., Interscience, New York, 1966.
14. **Angino, E. E. and Billings, G. K.,** *Atomic Absorption Spectrometry in Geology,* Elsevier, Amsterdam, 1967.
15. **Ramirez-Munoz, J.,** *Atomic Absorption Spectroscopy,* Elsevier, New York, 1968.
16. **Slavin, W.,** *Atomic Absorption Spectroscopy,* Vol., 25, Elving, P. J. and Kolthoff, I. M., Eds., Interscience, New York, 1968.
17. **Dean, J. A. and Rains, T. C.,** *Flame Emission and Atomic Absorption Spectrometry, Vol. 1, Theory,* Marcell Dekker, New York, 1969.
18. **Price, W. J.,** *Chapters on Atomic Absorption and Fluorescence in Spectroscopy,* Browning, D. R., Ed., McGraw-Hill, London, 1969.
19. **Rubeska, I. and Moldan, B.,** *Atomic Absorption Spectrophotometry,* SNTL, Prague, 1967, Engl. ed., Hiffe Books, London, 1969.
20. **Christian, G. D. and Feldman, J. J.,** *Atomic Absorption Spectroscopy: Applications in Agriculture, Biology and Medicine,* Wiley Interscience, New York, 1970.
21. **Eardley, R. P. and Mountford, A. H.,** Application of atomic absorption spectroscopy in the analysis of ceramic materials, in *Automation in Production, Sampling and Testing of Silicate Materials, Symposium Proc.,* Society of Chemical Industry Publications, London, 1970.
22. **L'vov, B. V.,** *Atomic Absorption Spectrochemical Analysis,* Transl. from the Russian by Divon, J. H., Adam Hilger, London, 1970.
23. **Reynolds, R. J., Aldoua, K., and Thompson, K. C.,** *Atomic Absorption Spectroscopy,* Griffin, London, 1970.

24. **Beamish, F. E. and Van Loon, J. C.,** *Recent Advances in the Analytical Chemistry of Noble Metals,* Pergamon Press, Oxford, 1972.

25. **Roelandt, I. and Guillaume, M.,** Comparative study on atomic absorption spectrometry and neutron activation: determination of rubidium in geochemistry, in *Proc. 3rd Int. Congr. on Atomic Absorption and Atomic Flame Spectrometry,* Adam Hilger, London, 1972.

26. **Ediger, R. D. and Coleman, R. L.,** An evaluation of anodic stripping voltammetry and nonflame atomic absorption as routine analytical tools, in *Trace Substances in Environmental Health,* Vol. 6, Hemphill, D. D., Ed., University of Missouri, Columbia, MO, 1973.

27. **Green, H. C.,** *Atomic Absorption Spectroscopy in Metallurgical Research,* Publ. No. A. I. D. D. G97, Department of Scientific and Industrial Research, Auckland, New Zealand, 1973.

28. **Kirkbright, G. F. and Sargent, M.,** *Atomic Absorption and Fluorescence Spectroscopy,* Academic Press, New York, 1974.

29. **Price, W. J.,** *Analytical Atomic Absorption Spectrometry,* 2nd ed., Heyden & Son, London, 1974.

30. **Willis, J. B.,** Atomic absorption, atomic fluorescence and flame emission spectroscopy, in *Handbook of Spectroscopy,* Robinson, J. W., Ed., CRC Press, Boca Raton, FL., 1974, 799.

31. **Welz, B.,** *Atomic Absorption Spectroscopy,* Verlag Chemie, Berlin, 1976.

32. **Beamish, F. E. and Van Loon, J. C.,** *Analysis of Noble Metals,* Academic Press, New York, 1977.

33. **Reeves, R. D. and Brooks, R. R.,** *Trace Element Analysis of Geological Materials,* Vol. 51, Elving, P. J. and Winefordner, J. D., Eds., John Wiley & Sons, New York, 1978 , 160.

34. **Berman, E.,** *Toxic Metals and Their Analysis,* Heydon & Son, London, 1980.

35. **Bratter, P. and Schramel, P.,** *Trace Elements Analytical Chemistry in Medicine & Biology,* Walter de Gruyter, Berlin, 1980.

36. **Van Loon, J. C.,** *Analytical Atomic Absorption Spectroscopy,* Academic Press, New York, 1980.

37. **Klein, A. A.,** Analysis of low-alloy steel using a sequential atomic absorption spectrophotometer equipped with an autosample, in *New Analytical Techniques for Trace Constituents of Metallic and Metal-bearing Ores,* Javier-Son, A., Ed., American Society for Testing and Materials, Philadelphia, PA, 1981, 29.

38. **Rains, T. C.,** Determination of aluminum, barium, calcium, lead, magnesium and silver in ferrous alloys by atomic absorption spectrometry, in *New Analytical Techniques for Trace Constituents of Metallic and Metal-bearing Ores,* Javier-Son, A., Ed., American Society for Testing and Materials, Philadelphia, PA, 1981.

39. **Magyar, B.,** Guidelines to planning of atomic absorption analysis, in *Studies in Analytical Chemistry,* Vol. 4, Elsevier, Amsterdam, 1982, 274.

40. **Methods for Analytical Atomic Spectroscopy,** ASTM publications, Philadelphia, PA., 1982.

41. **Vanasse, G. A.,** *Spectrometric Techniques,* Vol.3, Academic Press, New York, 1983, 334.

42. **Slavin, W.,** *Graphite Furnace AAS, A Source Book,* The Perkin-Elmer Corpn., Norwalk, CT 06856, 1984.

43. **Saad, S. M. H.,** *Organic Analysis using Atomic Absorption Spectrometry* (Ellis Horwood Series in Analytical Chemistry), John Wiley & Sons, New York, 1984, 384.

44. **Sergio, C.,** *Improved Hollow Cathode Lamps for Atomic Spectroscopy,* John Wiley & Sons, New York, 1984, 232.

45. **Varma, Asha,** *Handbook of Atomic Absorption Analysis,* Vol. 1, CRC Press, Boca Raton, FL, 1984.
46. **Varma, Asha,** *Handbook of Atomic Absorption Analysis,* Vol. II, CRC Press, Boca Raton, FL, 1984.
47. **Welz, B.,** *Atomic Absorption Spectrometry,* 2nd ed., Verlagsgellschaft mbH, Weinheim, FRG, 1985, 506.
48. **Caroli, S., Ed.,** *Improved Hollow Cathode Lamps for Atomic Spectroscopy,* John Wiley & Sons, New York, 1985.
49. **Berman, S., Ed.,** *Instrumentation in Analytical Chemistry, 1982-86,* American Chemical Society, ISBN O-8412-0969-3, 1986, 326.
50. **Beyer, H. J. and Kleinpoppen, H., Eds.,** *Progress in Atomic Spectroscopy,* Plenum Press, New York, 1987.
51. **Anderson, R.,** *Sample Pretreatment and Separation,* John Wiley & Sons, New York, 1987, 632.

APPENDIX 4
HOLLOW CATHODE LAMPS (SINGLE ELEMENT)

Element	Gas fill	Exit window	Wavelengths (nm)
Aluminum	Neon	Pyrex	309.2, 396.2
Antimony	Neon	Quartz	217.6, 231.1
Arsenic	Argon	Quartz	193.7, 197.2
Barium	Neon	Pyrex	350.1, 553.6
Beryllium	Neon	Quartz	234.8
Bismuth	Neon	Quartz	223.1, 306.8
Boron	Argon	Quartz	249.7
Cadmium	Neon	Quartz	228.8, 326.1
Calcium	Neon	Pyrex	422.7
Cerium	Neon	Quartz	520.0, 569.7
Chromium	Neon	Quartz/Pyrex	357.9, 425.4
Cobalt	Neon	Quartz	240.7, 345.4
Copper	Neon	Pyrex	324.7, 327.4
Dysprosium	Neon	Quartz	404.6, 418.7
Erbium	Neon	Quartz	386.3, 400.8
Europium	Neon	Quartz	459.4, 462.7
Gadolinium	Neon	Quartz	368.3, 400.8
Gallium	Neon	Quartz	287.4, 417.2
Germanium	Neon	Quartz	259.2, 265.1
Gold	Neon	Quartz	242.8, 267.6
Hafnium	Neon	Quartz	286.6, 307.2
Holmium	Neon	Quartz	405.4, 410.4
Indium	Neon	Quartz	304.0, 410.1
Iridium	Neon	Quartz	264.0, 285.0
Iron	Neon	Quartz	248.3, 372.0
Lanthanum	Neon	Quartz	392.8, 550.1
Lead	Neon	Quartz	217.0, 283.3
Lithium-natural	Neon	Pyrex	670.8
Lithium-6	Neon	Pyrex	670.8
Lithium-7	Neon	Pyrex	670.8
Lutetium	Neon	Quartz	335.9, 337.6
Magnesium	Neon	Quartz	202.5, 285.2
Manganese	Neon	Quartz	279.5, 280.1
Mercury	Argon	Quartz	253.7
Molybdenum	Neon	Quartz	313.3, 317.0
Neodymium	Neon	Quartz/Pyrex	463.4, 492.5
Nickel	Neon	Quartz	232.0, 341.5
Niobium	Neon	Quartz/Pyrex	405.9, 408.0
Osmium	Argon	Quartz	290.9, 301.8
Palladium	Neon	Quartz	244.8, 247.6
Phosphorus	Neon	Quartz	213.6, 214.9
Platinum	Neon	Quartz	265.9, 299.8
Potassium	Neon	Pyrex	404.4, 766.5
Praseodymium	Neon	Quartz	495.1, 513.3
Rhenium	Neon	Quartz/Pyrex	346.0, 346.5
Rhodium	Neon	Quartz/Pyrex	343.5, 350.7
Rubidium	Neon	Pyrex	420.1, 780.0

Ruthenium	Neon	Quartz/Pyrex	349.9, 392.5
Samarium	Neon	Quartz/Pyrex	429.6, 476.0
Scandium	Neon	Quartz/Pyrex	390.7, 391.2
Selenium	Neon	Quartz	196.0, 204.0
Silicon	Neon	Quartz	251.6, 288.1
Silver	Argon	Quartz/Pyrex	328.1, 338.3
Sodium	Neon	Pyrex	330.2, 589.0
Strontium	Neon	Pyrex	460.2
Tantalum	Argon	Quartz	271.4, 277.5
Tellurium	Neon	Quartz	214.3, 238.6
Terbium	Neon	Quartz/Pyrex	431.9, 432. 6
Thallium	Neon	Quartz	258.0, 276.7
Thorium	Neon	Quartz	
Thulium	Neon	Quartz	371.7, 409.4
Tin	Neon	Quartz	2246, 286.3
Titanium	Neon	Quartz/Pyrex	364.3, 399.8
Tungsten	Neon	Quartz	255.1, 400.9
Uranium	Neon	Quartz/Pyrex	348.9, 351.5
Vanadium	Neon	Quartz	318.4, 385.5
Ytterbium	Argon	Quartz/Pyrex	346.4, 398.8
Yttrium	Neon	Quartz	407.7, 410.2
Zinc	Neon	Quartz	213.9, 307.6

APPENDIX 5
HOLLOW CATHODE LAMPS (MULTI-ELEMENTS)

Aluminum-Calcium-Copper-Iron-Magnesium-Silicon-Zinc
Aluminum-Calcium-Iron-Magnesium
Aluminum-Calcium-Lithium-Magnesium
Aluminum-Calcium-Iron-Titanium
Aluminum-Calcium-Magnesium
Aluminum-Iron-Magnesium
Antimony-Arsenic-Bismuth
Arsenic-Nickel
Arsenic-Selenium-Tellurium**
Barium-Calcium-Strontium
Barium-Calcium-Strontium-Magnesium
Cadmium-Lead-Silver-Zinc
Calcium-Aluminum-Magnesium
Calcium-Iron-Aluminum-Magnesium
Calcium-Magnesium
Calcium-Magnesium-Aluminum-Lithium
Calcium-Magnesium-Zinc
Calcium-Zinc
Chromium-Cobalt-Copper-Iron-Manganese-Nickel
Chromium-Cobalt-Copper-Manganese-Nickel
Chromium-Copper
Chromium-Copper-Iron-Nickel-Silver
Chromium-Copper-Nickel-Silver
Chromium-Iron-Manganese-Nickel
Cobalt-Chromium-Copper-Iron-Nickel-Manganese
Cobalt-Copper-Iron-Manganese-Molybdenum
Copper-Cadmium-Lead-Zinc
Copper-Cobalt
Copper-Gallium
Copper-Iron
Copper-Iron-Lead-Nickel-Zinc
Copper-Iron-Manganese-Zinc
Copper-Iron-Nickel
Copper-Lead-Tin-Zinc
Copper-Manganese
Copper-Zinc-Molybdenum
Copper-Zinc-Molybdenum-Cobalt
Gold-Copper-Iron-Nickel
Gold-Nickel
Gold-Silver
Iron-Copper-Manganese
Manganese-Zinc
Molybdenum-Copper-Iron
*Sodium-Potassium**
Zinc-Silver-Lead-Cadmium

Note: (1) The gas fill in all the hollow cathode lamps is neon. (2) The exit window is quartz in all HCL. (3) The HCL* is filled with argon. (4) The HCL** has a Pyrex window.

APPENDIX 6
LIST OF ELECTRODELESS DISCHARGE LAMPS

Aluminum
Antimony
Arsenic
Bismuth
Cadmium
Cesium
Germanium
Lead
Mercury

Phosphorus
Potassium
Rubidium
Selenium
Tellurium
Thallium
Tin
Titanium
Zinc

APPENDIX 7
MANUFACTURERS OF HOLLOW CATHODE LAMPS

1. **Acton Research Corporation**
    P. O. Box 215
    525 Main St., Acton, MA. 01720
    Phone (617) 263-3584
    Telex 94-0787

2. **Advanced Radiation Corp.**
    2210 Walsh Av., Santa Clara, CA. 95050
    Phone (408) 727-9200
    FAX (408) 727-9255
    TWX 910 338-7441

3. **Anspec Co. Inc.**
    P. O. Box 7730
    Ann Arbor, MI. 48107
    Phone (313) 665-9666; (800) 521-1720

4. **Applitek N. V.**
    K. Piquelaan 84
    3-9800 Deinze, Belgium
    Phone 32-31-86-34-02
    Telex 12987
    FAX 32-31-86-72-37

5. **Atomic Spectral Lamps Pty. Ltd.**
    23-31 Islington Street
    Melbourne, Victoria, Australia

6. **Baxter Healthcare Corp.**
    Scientific Product Division
    1430 Wankegan Rd
    McGaw Park, IL. 60085
    Phone (312) 689-8410

7. **Beckman Instruments Inc.**
    2500 Harbor Blvd.
    Fullerton, CA. 92634
    Phone (714) 871-4848/ (714 773-7645

8. **Buck Scientific Inc.**
    58 Fort Point Street
    East Norwalk, CT. 06855-1097
    Phone (203) 853-9444; (800) 562-5566
    Telex 643589

9. **Cathodeon Ltd.**
    Nuffield Road
    Cambridge CB4 ITF, England, U.K.
    Phone 44(0223) 460100
    Telex 81685

10. **E G & G Electro-optics Div.**
    35 Congress Street
    Salem, MA. 01970
    Phone (617) 745-3200
    Telex 681-7405 EGGSA UW

11. **Fisher Scientific Co.**
    30 Water Street
    West Haven, CT. 06516
    Phone (203) 934-5271/ (412) 562-8300

12. **Fisher Scientific Co.**
    Fisher International Div.
    101 Thomson Rd.
    # 21-04 Gold Hill Sq.
    1130 Singapore
    Phone 2509766
    Telex FSC SPR RS 26479
    FAX (065) 253-2286

13. **FIVRE**
    Via Panciatichi 70
    Firenze-Castello, Italy

14. **Gallenkamp**
    Belton Road West
    Loughborough, LE 11 OTR, England, U.K.
    Phone 0509-237371
    Telex 34391
    FAX 0509-231893

15. **Hamamatsu Corp.**
    360 Foothill
    P.O. Box 6910
    Bridgewater, N.J. 08807
    Phone (201) 231-0960
    Telex 833403
    FAX (201) 231-1539

16. **Hamatsu Photonics K. K.**
    1126 Ichino-Cho
    Hamamatsu 435, Japan
    Phone (201) 231-0960
    Telex 833403
    FAX (201) 231-1639

17. **Hamatsu Systems Inc.**
    P. O. Box 648
    Waltham, MA. 02154
    Phone (617) 890-3440
    Telex 923461

18. **HANOVIA**
A subsidiary of Conrad Inc.
100 Chestnut Street
Newark, N.J. 07105
Phone (201) 589-4300
Telex 13-8531
FAX (201) 589-4430

19. **Hitachi Ltd.**
1-4 Marunouchi
Chiyoba-ku, Tokyo 160, Japan

20. **International Crystal Labs**
11 Erie Street
Garfield, N.J. 07206
Phone (201) 478-8944
Telex 833231

21. **Dr. Kern und Sprenger GmbH**
Post Box 751
Florenz Sartorious Strasse 5
Gottingen, Fed. Rep. of Germany

22. **McPherson**
Div. of S. I. Corp.
530 Main Street
Acton, MA. 01720
Phone (617) 263-7733; (800) 255-1055
Telex 928435

23. **Perkin-Elmer & Co. GmbH**
130 Denseewerk
P.O. Box 1120
D-7770, Uberlingen, West Germany
Phone 07551-810
Telex 0733902
FAX 011-44-7551-1612

24. **Quartzlampen GmbH**
Hoehensonnen Strasse Hanau/Main,
Fed. Rep. of Germany

25. **Rank Precision Industries Inc.**
98 St. Pancras Way
London, England, U.K.

26. **Starna Ltd.**
33 Station Road
Chadwell Health, Ramford
Essex RM6 4BL, England, U.K.
Phone (01) 599-5115
FAX (01) 599-5415
Telex 895 11 54 STARNAG

27. **Starna Cells Inc.**
P.O. Box 1919
Atascadero, CA. 93423
Phone (805) 461-8855
FAX (805) 461-1575
Telex 704611 STARNA CELLS

28. **Starna Pty. Ltd.**
P. O. Box 113
Thornleigh, NSW 2120, Australia
Phone (02) 875-3544
FAX (02) 875-3268
Telex 24619 STARNA AA

29. **S & J Juniper & Co.**
7 Potter Street
Harlow, Essex CM 17 9AD,
England, U.K.
Phone 0279 22456
Telex 817907

30. **Sherba Analytical Lab Product Inc.**
P. O. Box 880
New Port Richey, FL. 34656-0880
Phone (813) 849-9456
    (800) 228-5085

31. **Spectra Instrumentation**
11850 Industrial Court
Auburn, CA. 95603
Phone (916) 878-8506

32. **Thermo Jarrell Ash**
8E Forge Parkway
Franklin, MA. 02038
Phone (617) 520-188-
Telex 92-3443

33. **Varian Associate Inc.**
220 Humboldt Court
Sunnyvaly, CA. 94089
Phone (408) 752-2163
    (800) 231-5772
TWX 910-373-1731

34. **Varian Associates**
28 Manor Road
Walton-on-Thames
KT 12 2QF, Great Britain, U.K.
Phone 09 32 24 3741
Telex 928070
FAX 0932-228769

35. **Westinghouse Electric Corp.**
Industrial & Government Tube Division
Westinghouse Circle
Horseheads, N.Y. 14845
Phone (607) 796-3211
TWX 510252 1588

APPENDIX 8
MANUFACTURERS OF ATOMIC ABSORPTION
SPECTROPHOTOMETERS AND ACCESSORIES

1. **Allied Analytical**
   590 Lincoln Street
   Waltham, MA. 02254
   Phone (617) 890-4300

2. **Analabs/Foxboro Analytical**
   80 Republic Drive
   North Haven, CT. 06473
   Phone (203) 288-8463
   Telex 96-3518, TWX 710-4650-267

3. **Analect Instruments**
   17819 Gillette Avenue
   Irvine, CA. 92714
   Phone (714) 660-9269
   Telex 646/803

4. **Analyte Corporation**
   611 Southeast L Street
   Grants Pass, OR. 97526
   Telex 1 811 6045

5. **Anaspec Ltd.**
   P.O. Box 25
   Newbury, Berks., RG14 5LL
   England, U.K.
   Phone (0635) 44329
   Telex 849266

6. **Applied Research Labs Inc.**
   P. O. Box 129
   9545 Wentworth Street
   Sunland, CA. 91040
   Phone (818) 352-6011
        (800) 551-8741
   Telex 4720130
   FAX (818) 352-8241

7. **BCM Inc. of Tennessee**
   135 W. Adams Suite 306
   Kirkwood, MO. 63122
   Phone (314) 965-6757
        (800) 325-1235

8. **Buck Scientific**
   58 Fort Point Street
   East Norwalk, CT. 06855
   Phone (203) 853-9444/9441
        (800) 562-5566
   Telex 643589 BUCK SCIENLK

9. **Buck Scientific Ltd.**
   Unit # 6, Upper Wingbury Courtyard
   Wingrave, Aylesburg
   Bucks, England, U.K.
   Phone (0296) 681892

10. **Burkard Instruments AG**
    Buckhauserstrasse 26, CH-8048
    Zurich, Switzerland
    Phone (01) 491-5000
    Telex 56025 buin ch

11. **CEA Instruments Inc.**
    Box 303
    Emerson, N.J. 07630
    Phone (201) 967-5660
    Telex 642128

12. **Cecil Instruments Ltd.**
    Milton Industrial Estate
    Milton, Cambridge, CB4 4A2
    England, U.K.
    Phone (0223) 66821

13. **CHEMetrics, Inc.**
    Rt. 28
    Calverton, VA. 22016

14. **Custom Sample Systems**
    7534 Watson Road
    St. Louis, MO 63119
    Phone (314) 962-4555

15. **Elico Pvt. Ltd.**
    Instruments B-17
    Sanathanagar Industrial Estate
    Hyderabad, 500 018, Andhra
    Pradesh, India
    Phone (0842) 26 0285
    Telex 0155-714 PH HD IN

16. **Fisher Scientific Co.**
    711 Forbes Avenue
    Pittsburgh, PA. 15219
    Phone (412) 562-8300

17. **FIATRON Systems Inc.**
    Office of Research Labs
    6651 N. Sidney Plaza
    Milwaukee, WI 53209
    Phone (414) 351-6650
    Telex 26-774FIAtron

18. **Foss Electric (Ireland) Ltd.**
    Sandyford Industrial Estate
    Foxrock, Dublin. 18. Ireland
    Phone 953301
    Telex 24316

19. **Hamamatsu Corp.**
    360 Foothill Road
    P.O. Box 6910
    Bridgewater, N.J. 08807
    Phone (2012) 231-0960
    Telex 833403
    FAX 9201 231-1539

20. **Hamatsu Photonics K. K.**
    1126 Ichino-Cho Hamamatsu 435
    Japan
    Telex 4225-185
    FAX 0534-56-7889

21. **Hamamatsu Systems Inc.**
    P. O. Box 648
    Waltham, MA 02154
    Phone (617) 890-3440
    Telex 923461

22. **Hi-Tech Scientific Instruments**
    Analytical Instruments Department
    460 East Middlefield Road
    Mountain View, CA. 94043
    Phone (415) 969-1100
    Telex 171429

23. **Hitachi Instruments, Inc.**
    15 Miry Brook Road
    Danbury, CT. 06810
    Phone (203) 748-9001
          (800) 548-9001
    FAX (203) 748-4669

24. **Hitachi Ltd.**
    Tokyo, Japan

25. **Instruments S A Inc.**
    8 Olsen Avenue
    Edison, N.J. 08820
    Phone (201) 494-8660
    Telex 844 516
    FAX (201) 494-8796

26. **Instrumentation Laboratory (UK) Ltd.**
    Kelvin Close
    Birchwood Science Park
    Warrington, Cheshire, England, U.K.
    Phone (0925) 81 0141
    Telex 627713
    FAX (0925) 826708

27. **JASCO Inc.**
    314 Commerce Drive
    Easton, MD. 21601
    Phone (301) 822-1220

28. **Jobin-Yvon**
    Longjumeau, France

29. **Kontron AG, Anal. Division**
    Bernerstrasse-Sued 169,8048
    Zurich, Switzerland
    Phone (01) 435 41 11
    Telex 822191 kon

30. **LKB Produkter, Inc.**
    BOX 305
    S-161 26 Bromma, Sweden
    Phone (08)98 00 40
    Telex 330896 LP I

31. **LP Italiana SpA**
    via Carlo Riale
    1514, 20157, Milan, Italy
    Phone (02) 376-4646/3641
    Telex 330896 LP I

32. **Lamotte Chemical Products Co.**
    Box 329
    Chestertown, MD. 21620
    Phone (301) 778-3100
          (800) 344-3100

33. **Monitek Inc.**
    1495 Zephyr Avenue
    Haywood, CA. 94544
    Phone (415) 471-8300

34. **Optical S. P. A.**
    Milan, Italy

35. **Oxford Instruments Ltd.**
    Osney Mead
    Oxford, OXZ ONX, England, U.K.
    Phone (203) 762-1000
    Telex 00096-5854
    TWX 710-468-3213

36. **PCP Inc.**
    2155 Indian Road
    West Palm Beach, FL. 33409
    Phone (305) 683-0507

37. **Perkin-Elmer Corporation**
    761 Main Avanue
    Norwalk, CT. 06859-0012
    Phone (203) 762-1000
          (800) 762-4000
    Telex 00096-5954
    TWX 710-468-3213

38. **Perkin-Elmer Portable Wear Metal Analyzer (PWMA)**
    Applied Science Division
    2771, North Garey Avenue
    Pomona, CA 91767
    Phone (714) 593-3581
    TWX 910-581-3826

39. **Bodenseewerk Perkin-Elmer & Co., GmbH**
Postfach 1120, 7770 Ueberlingen
Federal Republic of Germany
Phone (755) 810

40. **Perkin-Elmer Ltd.**
Post Office Lane
Beaconsfield, Buckinghamshire
 HP 9 1QA, England
Phone (049) 46 6161

41. **Perkin-Elmer International Inc.**
P.O. Box 5329
Amman, Jordan
Phone 672323

42. **Perkin-Elmer Far East Pte. Ltd.**
70 Bendemeer Rd
02-02, Hiap Huat House
Singapore 1233
Phone 2972411

43. **Perkin-Elmer East Asia Ltd.**
Instrument Sales & Service Div.
Suite 8, 5th Floor, Tower A
Hung Hom, Commercial Center
37-39 Ma Tau Wai Road
Hunghom, Kowloon
Phone 3-622-368/9

44. **Perkin-Elmer Japan Co. Ltd.**
Yokohama Nishiguchi
K. N. Building
2-8-4, Kuita-Saiwai
Ni shi-ku, Yokohama 220
Phone (045) 314-8010

45. **Perkin-Elmer AG**
Postfach 1438, 8700
Kuesnacht. Switzerland
Phone (01) 913 31 11
Telex 53970
FAX (01) 910 60 34

46. **Perkin-Elmer de Mexico S.A.**
P.O. Box 19-333
Mexico, C.P. 03900 D.F.
Phone (525) 6-51-70-77

47. **Philips Analytical, Div. of Phillips**
Bldg. HKF, NL-5600
MD Eindhoven, The Netherlands
Phone 040/785213
Telex 36591

48. **Philips Industries, S. A.**
Spectrometry Department
Blvd. de l'Europe 131
B-1301 Wavre, Belgium
Phone (010) 41 65 11
Telex 59058

49. **P.S. Analytical**
Orpington, Kent, U.K.

50. **Perkin-Elmer Instruments**
Miles Street, Mulgrave
3170, P.O. Box 216
Glen, Waverly 3150, Australia

51. **Philips Mexicana S. A. de C. V.**
Div. Cientifico Industrial
Durango 167, 10 Piso
Aptdo. Postal 24-238
06700 Mexico D. F.
Phone 525-15-40; 554-14-22
Telex 71227

52. **Preiser Scientific Inc.**
900 MacCorkle Avenue, S. W.
Charleston, W. Va. 25322
Phone (304) 344-4031
TWX 710-938-1634

53. **Pye Unicam Ltd.**
A Sc. & Ind. Co. of Phillips
York Street
Cambridge, CBI 2PX, England, U.K.
Phone 44 223 358866
Telex 817331

54. **Questron Corp.**
P.O. Box 2387
Princeton, N.J. 08540
Phone (609) 275-0779; 587-6869
Telex 9103503424

55. **Savant Instruments, Inc.**
110-103 Bi-County Blvd.
Farmingdale, N.Y. 11735
Phone (516) 249-4600
     (800) 634-8886
Telex and TWX 6973199 SAVANT UW
FAX (516) 249-4639

56. **SCINTREX Ltd.**
222 Sniderscroft Road
Concord, ON. L4K 135, Canada
Phone (416) 669-2280
Telex 06-964570

57. **Shandon Southern Ltd.**
Camberley, England, U.K.

58. **Shimadzu Corporation**
Shinjoku Mitsui Bldg., 40th Floor
No.1-1, Nishi-Shinjuku 2-Chome
Shinjuku-ku, Tokyo 163, Japan
Phone (03) 346-5641
Telex 0232-3291 SHMDT J

59. **Shimadzu (Europa) GmbH**
Albert-Hahn-Strasse, 6 10, D-4100
Duisburg 29 (Grossenbaun)
West Germany
Phone (0203) 7687-0
Telex 855220
FAX (0203) 766625

60. **Shimadzu Scientific Instruments, Inc.**
7102 Riverwood Road
Columbia, MD. 21046
Phone (301) 381-1227
Telex 87-959
FAX (301) 381-1222
Chicago, IL. (312) 860-7094
Atlanta, GA. (404) 441-0018
San Francisco, CA. (415) 685-4545

61. **Spectra Instrumentation**
11850 Industrial Ct.
Auburn, CA. 95603-0000
Phone (916) 823-1638

62. **Spectrex Corporation**
3594 Haven Avenue
Redwood City, CA. 94063
Phone (415) 365-6567
Telex 750589

63. **Spex Industries, Inc.**
3880 Park Avenue
Edison, N.J. 08820
Phone (201) 549-7144; (800) 438-7739
TWX 910-373-1731

64. **Starna Cells Inc.**
BOX 1919
Atascadero, CA. 93423
Phone (805) 466-8855; (800) 228-4482
Telex 704611; FAX (805) 461-1575

65. **Syconex Corp.**
433 W. Allen Avenue
San Dimos, CA. 91773
Phone (714) 592-5684

66. **Tennelc Inc.**
601 Oakridge Turnpike
Oakridge, TN.37830
Phone (615) 483-8405
TWX 810-572-1018

67. **Thermo Jarrell Ash Corp.**
P. O. Box 9101
Franklin, MA. 02038-9907
Phone (617) 520-1880
Telex 92-3443

68. **U.S. Analytical Instruments**
1511 Industrial Road
San Carlos, CA. 94070
(415) 595-8200; (800) 437-9701

69. **V G Instruments**
6 North Kwun Hang Tsuen
Sai Kung N. T.
Phone 3-281-4094

70. **Varian AG**
Steinhauser strasse
CH-6300 Zug, Switzerland
Phone (042) 44 88 44
Telex 868 841

71. **Varian Canada, Inc.**
45 River Drive
Georgetown, ON. L7G 2J4, Canada
Phone (416) 457-4130
Telex 069-7502

72. **Varian Analytical Corp.**
3rd Matsuda Bldg., 2-2-6 Ohkubu
Shinjuku, Tokyo 160, Japan
Phone 03-204-1211
Telex J26471

73. **Varian Associates Ltd.**
28 Manor Road
Walton-on-Thames, Surrey
KT 12 2QF
England, U.K.
Phone (0932)24 37 41
Telex 928070
FAX 0932-228769

74. **Varian GmbH**
Alsfelderstrasse 6, D-6100
Darmstadt, FRG
Phone (06151) 7031
Telex 419429

75. **Varian Industria a Comercio Ltd.**
Avenida Drive
Cardoso de Melo 1644
CEP 04548 Sao Paulo, Brazil
Phone (905) 545-4077

76. **Varian Instruments Div.**
     220 Humbolt Ct.
     Sunnyvale, CA. 94089-9922
     Phone (408) 734-5370; (800) 231-5772
     TWX 910-373-1731

77. **Varian S. A.**
     Francisco Petrarca 326
     Mexico D. F., Mexico 5
     Phone (905) 545-4077
     Telex 177-4410

78. **Varian Pty Ltd.**
     679 Springvale Rd
     Mulgrave, Victoria, Australia 3170
     Phone (03) 560 71 33

APPENDIX 9
CHEMICAL SUPPLIERS (STANDARD SOLUTIONS, COMPLEXING AGENTS, ORGANOMETALLICS, AND CERTIFIED STANDARDS)

1. **Aldrich Chemical Co.**
   2371 North 30th Street
   Milwaukee, WI, 53210
2. **Aldrich Chemical Co. Inc.**
   940 West St. Paul Avenue
   Milwaukee, WI, 53233
   Phone (414) 273-3850; (800) 558-9160
3. **Alfa Inorganics**
   8 Congress Street
   Beverly, MA. 01915
4. **Alfa Products Thiokol/Ventron Division**
   P.O. Box 299
   152 Andover Street
   Danvers, MA. 01923
   Phone (617) 777-1970
5. **Alpha Analytical Laboratories**
   Division of Alpha Metals Inc.
   Jersey City, N.J. 07604
6. **Analytical Standards Lab**
   Box 21
   434 02 Kiungsbacka 2, Sweden
7. **Analytical Standards Ltd.**
   Fjallagatan 18
   41317 Goteborg, Sweden
8. **Anderman & Co. Ltd.**
   Battlebudge House
   87-95 Tooley Street
   London, SE1, England, U.K.
9. **Anderson Laboratories Inc.**
   5901 Fitzhugh Avenue
   Fort Worth, TX. 76119
   Phone (817) 457-4474
10. **Angstrom Inc.**
    P. O. Box 248
    Belleville, MI. 48111
    Phone (313) 697-8058
11. **Apache Chemicals Inc.**
    Grant Street, P. O. Box 126
    Seward, IL. 61077
    Phone (815) 247-8491
    TWX 910-642-0601
12. **Baird Corporation**
    125 Middlesex Turnpike
    Bedford, MA. 01730
    Phone (617) 276-6000

13. **J. T. Baker Chemical Co.**
    222 Red School Lane
    Phillipsburg, N.J. 08865
    Phone (201) 859-2151
    Telex 847480 (Domestic; 831644 (Export)
14. **Baker Chemikalien**
    6080 Gross Gerau
    Postfach 1661
    Federal Republic of Germany
    Phone (06152) 710371
    Telex 0 419 1113
15. J. T. Baker Chemicals BV
    Rijster Borgher Weg 20
    P. O. Box 1
    7400 AA Deventer, The Netherlands
    Phone (05700) 11341
    Telex 49072
16. **J. T. Baker S. A. de C. V.**
    Apartado Postal No. 75595
    Col. Lindvista. Deleg. Gustav A.
    Madero 07300, Mexico D. F.
    Phone 569-1100
    Telex 1772336
17. **J. T. Baker Chemical Co.**
    Suite 1102, World Trade Center
    Telok Blangah Road
    Singapore 0409
    Phone 2735285
    Telex RS 39323 BACHEM
18. **Barnes Engineering Co.**
    30 Commerce Road
    Stamford, CT. 06902
    Phone (203) 348-5381
19. **B D H Chemicals Ltd.**
    Poole, Dorset BH12 4NN, England, U.K.
20. **B D H**
    Via Greda
    142 Milano, Utaly
21. **Bie and Berntsen**
    Sandbaekvej 7
    2610 Roedovre, Denmark
22. **Bio-Rad Laboratories**
    2200 Wright Avenue
    Richmond, CA. 94804
    Phone (415) 234-4130
    Telex 337-732

23. **Bio-Rad Laboratories Ltd.**
Caxton Way
Holywell Industrial Estate
Watford, Herts. WDI 8RP, England,
U.K.
Phone (0923) Watford 40322
Telex 8813192

24. **Bracco**
Via Folli 50
Milano, Italy 20134

25. **Brammer Standard Co., Inc.**
5607 Fountainbridge Lane
Houston, TX. 77069
Phone (713) 440-9396
Telex 775-376

26. **Buck Scientific, Inc.**
58 Fort Point Street
East Norwalk, CT. 06855
Phone (203) 853-9444
Telex 643589 BUCK SCIENLK

27. **Bundesanstalt fur Material
Prufung**
Unter der Eichen 85
1000 Berlin 45, Federal Republic of
Germany

28. **Burt & Harvey Ltd.**
Poole, Dorset BH12 4NN, England,
U.K.

29. **Carlo Erba Farmattalia**
Analytical Division
Via C. Imbonati
24 20159 Milano, Italy
Telex 330314

30. **Chemplex Indstries Inc.**
160 Marbledale Road
Tuckahoe, N.Y. 10707
Phone (914) 337-4200

31. **Conostan, Conoco, Inc.**
P. O. Box 1267
Ponka City, OK. 74601

32. **Curtis Mathis Scientific, Inc.**
9999 Stuebner-Airline
Houston, TX. 77001
Phone (713) 820-1661

33. **Deutsch Vertretung**
Dipl-Met. G. Winopal
Universal-Forschungsbedarf
Echternfeld 25, Postfach 40
3000 Hannover 51, Fed. Rep. of
Germany

34. **Durham Raw Materials Ltd.**
1-4 Great Tower Street
London, EC3 R4AB, England, U.K.

35. **E G & G Gamma Scientific**
3777 Ruffin Road
San Dieago, CA 92123
Phone (619) 279-8034

36. **Eastman Organic Chemicals**
Eastman Kodak Co.
343 State Street
Rochester, N. Y. 14650
Phone (716) 724-4000

37. **Extrel Corporation**
240 Alpha Drive
Box 11512
Pittsburgh, PA. 15238
Phone (412) 782-3884

38. **F & J Scientific**
79 Far Horizon Drive
Monroe, CT. 06468

39. **Firma Merck AG**
Postfach 4119
6100 Darmstadt 2, Fed. Rep. of
Germany

40. **Firma Riedel-de-Haen AG**
Wunstorfer Strasse
316 Seelze-Hannover, Fed. Rep. of
Germany

41. **Fisons Scientific Equipment**
Bishop Meadow Road
Loughborough, Leics., LE11 ORG
England, U.K.

42. **FIAtron Process Systems**
510 S. Worthington
Oconomowoc, WI. 53066
Phone (414) 567-3810

43. **Fluka AG**
Chemische Fabrik
CH-9470 Buchs, Switzerland
Phone 085 6 0275
Telex 855282, 855283

44. **Fluka Feinchemikalien GmbH**
Postfach 1346
7910 New-Ulm, Fed. Rep. of
Germany

45. **Foxboro Co.**
330 Neponset Avenue
Foxboro, MA. 02035
Phone (617) 543-8750; (800) 343-
0933

46. **GAF Chemicals Corp.**
    1361 Alps Road
    Wayne, N.J. 07470
    Phone (201) 628-3000
47. **GBC Scientific Equipment Pty. Ltd.**
    22 Brooklyn Avenue
    Dandenone, Victoria 3175, Australia
    Phone 61-3-793-1448
    Telex AA37123
    FAX 61-3-794-9008
48. **Hach Co.**
    Box 389
    Loveland, CO. 80539
    Phone (303) 669-3050; (800) 525-5940
49. **Hopkins & Williams Ltd.**
    P. O. Box 1
    Romford, Essex, RMI 1HA,
    England, U.K.
50. **Inorganic Ventures Inc.**
    Toms River, N.J. 08573
51. **Johnson Matthey/AESAR**
    P. O. Box 1087
    Seabrook, N.H. 03874
    Phone (800) 343-1990
52. **Johnson Matthey Chemicals Ltd.**
    74 Hatton Garden
    London, EC1P 1AE, England, U.K.
53. **A. Johnson and Co.**
    Box 57
    201 20 Malmo, Sweden
54. **K & K Laboratories**
    121 Express Street
    Plainville, N.Y. 11803
55. **Kebo-grave**
    Domnarvag 4
    16391 Spaanga, Sweden
56. **Koch Light Laboratories Ltd.**
    37 Hollands Road
    Haverhill, Suffolk, England, U.K.
    Phone Haverhill 702436
    Telex 81504
57. **Labassco ab**
    Almedahlsv, Id.
    Box 5205
    A02 24 Gothenberg, Sweden

58. **Mallinckrodt. Inc**
    Science Products Division
    675 McDonnell Boulevard
    P. O. Box 5840
    St. Louis, MO.
    Phone (314) 895-2340
59. **May & Baker Ltd.**
    Dagenham
    Essex, RM10 7XS, England, U.K.
60. **M B H Analytical Ltd.**
    Station House
    Potters Bar, Herts., EN6 IAL,
    England, U.K.
61. **Merck**
    65 Rue Victoire
    Paris, France
62. **Merck Chemicals Pty. Ltd.**
    Wrench Road
    Isando, Rep. of South Africa
63. **E. Merck**
    Frankfurter Strasse 250
    Postfach 4119
    D-6100 Darmstadt, Fed. Rep. of
    Germany
    Phone 0615/721
    Telex 04 -19-325 emd d. cable
    emerck darmstadt
64. **Molybdenum Corp. of America**
    280 Park Avenue
    New York, N.Y. 10017
65. **National Bureau of Standards**
    Office of Standard Reference
    Materials
    Room B 311, Chemistry Bldg.
    Gaithersberg, MD. 20878-9950
    Phone (301) 975-OSRM (6776)
66. **National Spectrographic Labs. Inc.**
    19500 South Miles Road
    Cleveland, OH. 44128
67. **pH-tamm Laboratoriet ab**
    Liljeborgsv 12 752 36
    Uppsala, Sweden
68. **Prolabo**
    12 Rue Pelee
    Paris, France
69. **Regine Brooks RBS**
    Pariser Strasse 5
    5300 Bonn 1. Fed. Rep. of Germany

70. **Research Organic/Inorganic Chemical Corp.**
11686 Sheldon Street
Sun Valley, CA 91352

71. **Ricca Chemical Co.**
P. O. Box 13090
Arlington, TX 76013
Phone (817) 461-5601

72. **Riedel de Haen**
171 Av Jean Jaures
Aubervilliers, France

73. **Riedel de Haen AG**
Wunstorter Strasse 40
D 3016 Seelze 1, Fed. Rep. of
Germany

74. **Spectrum Chemical Mfg. Corp.**
14422 S. San Pedro Street
Gardena, CA. 90248-9985
Phone (800) 772-8786

75. **Spex Industries Inc.**
3880 Park Av.
Edison, N.J. 08820
Phone (201) 549-7144; (800) LAB-SPEX
Telex 178341

76. **Struers K/S**
Volhojs Alle 176
DK-2610 Rodovre
Copenhagen, Denmark
Phone (45) 1 70 80 90
Telex 19625

77. **U. S. Geological Survey National Center**
Reston, VA 22092

78. **Ventron GmbH**
Postfach 6540
7500 Karlsruhe 1, Fed. Rep. of
Germany
Phone 0721-557061
Telex 7826579

APPENDIX 10
STANDARD STOCK SOLUTIONS FOR FAAS ANALYSIS

Aluminum

To prepare 1000 mg/L solution, dissolve 1.000 g of high purity aluminum wire in 50 mL of concentrated hydrochloric acid with gentle heating, adding a small drop of mercury as a catalyst. Filter the solution to remove the catalyst, and dilute to 1 L with deionized distilled water.

Antimony

To prepare 1000 mg/L solution, either dissolve 1.000 g of metallic antimony in 100 mL of hydrochloric acid + 2 mL nitric acid and dilute to 1 L or dissolve 2.743 g of potassium antimonyl tartrate hemohydrate $(K(SbO)C_4H_4O_6 \cdot 1/2H_2O$ in deionized water and dilute to 1 L with deionized distilled water.

CAUTION: This element is very toxic and should be handled with care.

Arsenic

For 1000 mg/L solution, dissolve 1.3203 g of arsenious oxide (As_2O_3) in 25mL of 20% (w/v) potassium hydroxide solution. Neutralize it with 20% (v/v) sulphuric acid to a phenolphthalein end point and dilute to 1 L with 1% (v/v) sulfuric acid.

CAUTION: This element is toxic, take precautions in handling it.

Barium

To prepare 1000 mg/L, dissolve 1.437 g of barium carbonate, $BaCO_3$ in 20 mL of 1+1 hydrochloric acid. Dilute to 1 L with 1% (v/v) hydrochloric acid.

Beryllium

To prepare 1000 mg/L, dissolve 1.000 g of metallic beryllium in 20 mL of 1+1 hydrochloric acid. Dilute to 1 L with 1% (v/v) hydrochloric acid.

CAUTION: Take extra precaution in handling this element due to high toxicity.

Bismuth

To prepare 1000 mg/L, dissolve 1.0000 g of bismuth metal in a minimum volume of 1+1 nitric acid and dilute to 1 L with 2% (v/v) nitric acid.

Boron

To prepare 500 mg/L, dissolve 28.60 g of boric acid (H_3BO_3) in deionized water and dilute to 1 L.

Cadmium

To prepare 1000 mg/L, dissolve 1.000 g of cadmium metal in 50 mL of 1:1 hydrochloric acid, dilute to 1 L with 1% (v/v) hydrochloric acid or dissolve 1.142 g of cadmium oxide (CdO) in a minimum volume of 1+1 nitric acid and dilute to 1 liter with 1% (v/v) nitric acid.

CAUTION: Care should be taken to handle this element due to toxicity.

Calcium

To prepare 1000 mg/L, dissolve 2.497 g of calcium carbonate in 50 mL of water , add dropwise 10 mL of hydrochloric acid to effect complete solution of the calcium carbonate. Dilute to 1 L with deionized distilled water.

Cesium

To prepare 1000 mg/L, dissolve 1.267 g of cesium chloride in deionized water and dilute to 1 L with deionized distilled water.

Chromium

To prepare 1000 mg/L, dissolve 3.735 g of potassium chromate (K_2CrO_4) in deionized water or 1.000 g of metallic chromium in 50 mL of 1:1 hydrochloric acid with gentle heating and dilute to 1 L with deionized water.

Cobalt

To prepare 1000 mg/L, dissolve 1.000 g of cobalt metal in a minimum volume of 1+1 hydrochloric acid and dilute to 1 L with 1% (v/v) hydrochloric acid.

Copper

To prepare 1000 mg/L, dissolve 1.000 g of high purity copper metal in 50 mL of 1+1 nitric acid and dilute to 1 L with 1% (v/v) nitric acid.

Dysprosium

To prepare 1000 mg/L, dissolve 1.148 g of dysprosium oxide in 20 mL of hydrochloric acid and dilute to 1 L with 1% (v/v) hydrochloric acid.

Erbium

To prepare 1000 mg/L, dissolve 1.143 g of erbium oxide in a minimum volume of hydrochloric acid and dilute to 1 L with 1% (v/v) hydrochloric acid.

Europium

To prepare 1000 mg/L, dissolve 1.158 g of europium oxide in 20 mL of hydrochloric acid and dilute with 1% (v/v) hydrochloric acid solution.

Gadolinium

To prepare 1000 mg/L, dissolve 1.153 g of gadolinium oxide in 20 mL of hydrochloric acid and dilute to 1 L with 1% (v/v) hydrochloric acid.

Gallium

To prepare 1000 mg/L, dissolve 1.000 g of metallic gallium in 20 mL of volume of aqua regia with heating and dilute to 1 L with 1% (v/v) hydrochloric acid.

Germanium

To prepare 1000 mg/L, dissolve 0.1000 g of germanium metal in a Teflon® beaker, add 5 mL concentrated hydrofluoric acid and add concentrated nitric acid dropwise until dissolution is just complete. Dilute to 100 mL with deionized water.

Gold

To prepare 1000 mg/L, dissolve 0.1000 g of gold metal in a minimum volume of aqua regia (freshly prepared mixture of 3 parts hydrochloric acid + 1 part nitric acid). Place it on a hot plate and heat to dryness. Dissolve the residue in 5 mL hydrochloric acid, cool, and dilute to 100 mL of deionized distilled water. Use 10% (v/v) hydrochloric acid for further dilutions.

Holmium

To prepare 1000 mg/L, dissolve 1.146 g of holmium oxide in 20 mL of hydrochloric acid and dilute to 1 L with deionized distilled water.

Indium

To prepare 1000 mg/L, dissolve 1.000 g of indium metal in a minimum volume of 1+1 hydrochloric acid. Add a few drops of nitric acid with gentle heating. Cool and dilute to 1 L with 1% (v/v) hydrochloric acid.

Iridium

To prepare 1000 mg/L, dissolve 2.294 g of ammonium hexachloriiridate ($(NH_4)_2IrCl_6$) in 10 mL of 1% (v/v) hydrochloric acid and dilute to 1 L with the same acidic solution.

Iron

To prepare 1000 mg/L, dissolve 1.000 g of high purity iron wire in 50 mL of 1+1 nitric acid and dilute to 1 L with deionized distilled water.

Lanthanum

To prepare 10,000 mg/L, wet 11.730 g of lanthanum oxide with deionized water. Add cautiously and slowly, 250 mL of concentrated hydrochloric acid to dissolve the lanthanum oxide and dilute to 1 L with deionized water.

Lead

To prepare 1000 mg/L, dissolve 1.000 g of metallic lead or 1.598 g of lead nitrate in 50 mL of 1% (v/v) nitric acid and dilute to 1 L with the same acid solution.

CAUTION: Take extra care in handling this toxic element.

Lithium

To prepare 1000 mg/L, dissolve 5.324 g of lithium carbonate in a 20 mL of 1+1 hydrochloric acid and dilute to 1 L with deionized distilled water.

Magnesium

To prepare 1000 mg/L, cautiously dissolve 1.000 g of magnesium ribbon in 20 mL of 1+1 hydrochloric acid and dilute to 1 L with 1% (v/v) hydrochloric acid.

Manganese

To prepare 1000 mg/L, dissolve 1.0000 g of manganese metal in a 50 mL of 1+1 nitric acid and dilute to 1 L with 1% (v/v) hydrochloric acid.

Molybdenum

To prepare 1000 mg/L, dissolve 1.000 g of metallic molybdenum in 50 mL hot concentrated nitric acid and dilute to 1 L or dissolve 1.840 g of ammonium heptamolybdate tetrahydrate ($(NH_4)_6Mo_7O_{24}.4H_2O$) in 1 L of 1% (v/v) ammonium hydroxide or dissolve 1.500 g molybdenum trioxide in a minimum volume of (1+1) hydrochloric acid and dilute to 1 L.

Neodymium

To prepare 10,000 mg/L, dissolve 1.167 g of neodymium oxide in 20 mL of hydrochloric acid and dilute to 100 mL with 1% (v/v) hydrochloric acid.

Nickel

To prepare 1000 mg/L, dissolve 1.000 g of nickel metal in 50 mL of 1+1 nitric acid and dilute to 1 L with 1% (v/v) nitric acid.

Osmium

To prepare 1000 mg/L, dissolve 1.340 g of osmium tetraoxide in 100 mL of deionized water. Add a sodium hydroxide pellet for complex dissolution and dilute to 1 L.

CAUTION: Osmium tetraoxide is highly toxic, take extra precautions when handling it.

Palladium

To prepare 1000 mg/L, dissolve 0.2672 g of ammonium chloro paladite ($(NH_4)_2PdCl_4$) in water and dilute to 1 mL water, or dissolve 0.1000 g of palladium wire in a minimum volume of aqua regia (freshly prepared 3 parts hydrochloric acid and 1 part nitric acid). Heat the sample gently to dryness, add 5 mL of concentrated hydrochloric acid and 25 mL water. Dilute to 100 mL with deionized distilled water.

Phosphorus

To prepare 50,000 mg/L, dissolve 21.32 g of dibasic ammonium phosphate in deionized water and dilute to 100 mL with deionized water or dissolve 21.969 g KH_2PO_4 in deionized distilled water and dilute to 100 mL.

Plantinum

To prepare 1000 mg/L, dissolve 2.275 g of ammonium chloroplatinate $((NH_4)_2PtCl_6)$ in deionized distilled water, or dissolve 0.1000 g of platinum metal in a 20 mL of aqua regia (3 parts hydrochloric acid+1 part nitric acid) and evaporate to dryness on hot plate. Add 5 mL of hydrochloric acid and 0.1 g of sodium chloride and evaporate to near dryness, cool and dissolve the residue in 20 mL of 1+1 hydrochloric acid and dilute to 100 mL with deionized distilled water.

Potassium

To prepare 1000 mg/L, dissolve 1.907 g of potassium chloride in deionized distilled water and dilute to 1 L.

Praseodymium

To prepare 10,000 mg/L solution, dissolve 1.170 g of praseodymium oxide in 20 mL of hydrochloric acid and dilute to 1 L with 1% (v/v) hydrochloric acid.

Rhenium

To prepare 1000 mg/L, dissolve 1.554 g of potassium perrhenate in 250 mL of deionized water and dilute to 1 L with 1% (v/v) sulfuric acid.

Rhodium

To prepare 1000 mg/L, dissolve 3.86 g of ammonium hexachlororhodate in 50 mL of 10% (v/v) hydrochloric acid or dissolve 0.412 g of ammonium hexachlororhodate trihydrate in 20 mL of 50% (v/v) hydrochloric acid and warm until dissolved. Dilute to 1 L.

Rubidium

To prepare 1000 mg/L, dissolve 1.415 g of rubidium chloride in deionized water and dilute to 1 L.

Ruthenium

To prepare 1000 mg/L, dissolve 0.2052 g of ruthenium chloride in 50 mL of 20% (v/v) hydrochloric acid and dilute to 100 mL with the same acidic solution.

Samarium

To prepare 10,000 mg/L, dissolve 1.159 g of samarium oxide in 20 mL of hydrochloric acid and dilute to 100 mL with 10% (v/v) hydrochloric acid.

Scandium

To prepare 1000 mg/L, dissolve 1.534 g of scandium oxide in 20 mL of 1:1 hydrochloric acid and dilute to 1 L with 1% (v/v) hydrochloric acid.

Selenium

To prepare 1000 mg/L, dissolve 1.000 g of selenium metal in 20 mL of1:1 nitric acid and evaporate to dryness. Add 2 mL of water and heat to dryness, repeat the process 2 to 3 times. Dissolve the residue in 10% v/v) hydrochloric acid

CAUTION: This metal is highly toxic, take extra precautions.

Silicon

To prepare 1000 mg/L, dissolve 10.112 g of sodium metasilicate ($Na_2SiO_3.9H_2O$) in 50 mL water + 10 mL concentrated hydrochloric acid or fuse 0.2139 g of silicon dioxide with 2 g of sodium carbonate in a platinum crucible. Dissolve the melt in deionized water and dilute to 100 mL volume.

Silver

To prepare 1000 mg/L, dissolve 1.000 g of pure silver metal in 20 mL of 1:1 nitric acid or dissolve 0.787 g of silver nitrate in 50 mL of deionized water for 500 mg/L of silver solution. Dilute to 1 L with 5% (v/v) nitric acid.

Strontium

To prepare 1000 mg/L, dissolve 2.415 g of strontium nitrate in 20 mL of 1:1 nitric acid or in 100 mL water and add 10 mL of concentrated hydrochloric acid. Dilute to 1 L.

Terbium

To prepare 1000 mg/L, dissolve 1.176 g of terbium oxide in a 20 mL of concentrated hydrochloric acid and dilute to 1 L with 1% (v/v) hydrochloric acid.

Tellurium

To prepare 1000 mg/L dissolve 1.000 g of tellurium metal in 50 mL of1:1:1 mixture of hydrochloric, nitric, and water. Add 20 mL of hydrochloric acid to redissolve the precipitate, and heat to expel nitrogen oxides. Cool and dilute to 1 L with 1% (v/v) hydrochloric acid.

CAUTION: Take extra care in handling this toxic material.

Thalium

To prepare 1000 mg/L, dissolve 1.303 g of thallium nitrate in 20 mL of nitric acid and dilute to 1 L with deionized water.

CAUTION: Take extra care in handling this toxic material.

Uranium

To prepare 10,000 mg/L, dissolve 2.110 g of uranyl nitrate hexahydrate ($UO_2(NO_3)_2.6H_2O$) in deionized water and dilute to 100 mL.

Vanadium

To prepare 1000 mg/L, dissolve 1.000 g vanadium metal in 50 mL of nitric acid and dilute to 1 L with 1% (v/v) nitric acid.

Ytterbium

To prepare 1000 mg/L, dissolve 1.139 g of ytterbium oxide in hot 20 mL of hydrochloric acid and dilute to 1 L with 1% (v/v) hydrochloric acid.

Yttrium

To prepare 1000 mg/L, dissolve 1.270 g of yttrium oxide in 20 mL of hydrochloric acid and dilute to 1 L with 1% (v/v) hydrochloric acid.

Zinc

To prepare 1000 mg/L, dissolve 1.000 g of zinc metal in a minimum volume of (1 + 1) hydrochloric acid and dilute to 1 L.

INDEX

T

U

V

W

Y

Z

Printed and bound by CPI Group (UK) Ltd, Croydon, CR0 4YY

22/10/2024

01777638-0008